Harnessing Biodiversity Tourism for Regenerative Conservation Management

Parinya Nakpathom
International College, Burapha University, Thailand

Kannapat Kankaew
International College, Burapha University, Thailand

Krittipat Pitchayadejanant
International College, Burapha University, Thailand

Minenkova Vera Vladimirovna
Kuban State University, Russia

Evgeniia Rotai
Kuban State University, Russia

IGI Global
Scientific Publishing
Publishing Tomorrow's Research Today

Vice President of Editorial Melissa Wagner
Managing Editor of Acquisitions Mikaela Felty
Managing Editor of Book Development Jocelynn Hessler
Production Manager Mike Brehm
Cover Design Phillip Shickler

Published in the United States of America by
IGI Global Scientific Publishing
701 East Chocolate Avenue
Hershey, PA, 17033, USA
Tel: 717-533-8845
Fax: 717-533-8661
Website: https://www.igi-global.com E-mail: cust@igi-global.com

Library of Congress Cataloging-in-Publication Data

Library of Congress Cataloging-in-Publication Data

Names: Nakpathom, Parinya, 1983- editor.
Title: Harnessing biodiversity tourism for regenerative conservation management / edited by Parinya Nakpathom, Kannapat Kankaew, Krittipat Pitchayadejanant, Vera Minenkova, Evgeniia Rotai.
Description: Hershey, PA : IGI Global Scientific Publishing, 2025. | Includes bibliographical references and index. | Summary: "This book aims to address the new normal tourism paradigm, behavior of visitors, and new generations trends toward protected areas"-- Provided by publisher.
Identifiers: LCCN 2024054494 (print) | LCCN 2024054495 (ebook) | ISBN 9798369382325 (hardcover) | ISBN 9798369382332 (paperback) | ISBN 9798369382349 (ebook)
Subjects: LCSH: Sustainable tourism. | Sustainable tourism--Case studies. | Ecotourism. | Ecotourism--Case studies. | Biodiversity conservation. | Biodiversity conservation--Case studies.
Classification: LCC G156.5.S87 H37 2025 (print) | LCC G156.5.S87 (ebook) | DDC 333.95/16--dc23/eng/20250111
LC record available at https://lccn.loc.gov/2024054494
LC ebook record available at https://lccn.loc.gov/2024054495

British Cataloguing in Publication Data
A Cataloguing in Publication record for this book is available from the British Library.

Table of Contents

Detailed Table of Contents

Protected areas are the most important element of the territorial tourism and recreational system. The development, creation and improvement of ecotourism routes are carried out in many protected areas. The dynamics of tourist visits, being one of the indicators of the effectiveness of the functioning of protected areas, demonstrates a rapidly positive character throughout the world. It is important that the development of ecotourism in protected areas has both a positive (mainly in the economic aspect) and a negative (ecological aspect) impact on natural complexes. The purpose of the chapter is to analyze the approaches and principles of ecotourism development in specially protected natural areas, and to recommend basic content positions of a typical Ecotourism Development Strategy in Specially Protected Natural Areas.

Based on a theoretical framework about interaction between applied regional geography and proximity tourism, the proposed article reveals geographical superlatives in the Friulian Alps (Italy). As a contribution to applied regional geography, the valorization of this in proximity tourism (marketing) is discussed. The lowest lying glacier, the most colorful ethnolinguistic mosaic, the greatest rainfall volume or the highest depopulation numbers give the Montagna friulana a special geographical position within the Alps. This extraordinary diversity, however, is not expressed in any way in the regional development: Rather, the north of Friuli is one of the structurally weakest areas of the Alpine region. In the sense of applied geography, this contribution should be understood as an advisory function and is intended to inspire professional marketing experts to offer suggestions that geographical particularities can be applied specifically in proximity tourism marketing and therefore to regional development of peripheral areas.

Virtual reality and environmental conservation in tourist sites can provide an innovative solution for sustainable tourism. VR enable to observe natural wonders in person without even being there and causing environmental harm. Whether this is through the simulation of real-world landscapes to educate "visitors" about conservation or in conveying how important it is that it preserves what natural habitats are out there leading to a wider sense of awareness and education on their ecological footprints. The service could also be used as a way to combat over-tourism at main tourist destinations, save biodiversity and cultural heritage. Also, enabling the use of VR in eco-tourism can encourage responsible actions which enable users to dive into ecosystems remotely such as endangered species and so on thereby instilling a very relevant connection with conservation causes. This chapter explores the design of an innovative approach to reduce environmental degradation and develop novel tourist experiences through hitching biodiversity focused sustainable tourism combined with VR technology.

The purpose of this investigation is to clarify the ways in which partnership of public and private, development of tourism, and sustainability of tourism are interrelated within the tourist sector. The investigation makes use of quantitative techniques to investigate the dynamics by using regression analysis and exploratory factor analysis. The results show that successful partnership of public and private have a considerable affirmative influence on tourist development and sustainability. Also, measures focused at promoting tourism can result in better sustainable practices since development of tourism acts as a critical mediator among partnership of public and private and sustainability outcomes. The investigation's findings highlight how crucial stakeholder involvement and cooperative governance are to the development of sustainable tourism practices. Policymakers and sector stakeholders can benefit greatly from these insights, which imply that enhancing partnership of public and private and giving sustainable development of tourism top priority.

The purpose of the chapter is to define modern approaches and methods of sustainable development of mountain tourist destinations, and also to share Russian experience of sustainable tourism development on the example of post-Olympic Sochi (Russia). The chapter is presented in two important parts (theoretical and practical). The first one is dedicated to theory and methods: approaches and systems for sustainable tourism development in mountain tourist destinations in the world. The chapter outlines systems that define the criteria for sustainable tourism development, indicators of sustainable development for tourism destinations and also issues of particular concern to mountain environments. The second part explores successful Russian practices: experience of sustainable tourism development in post-Olympic Sochi (Russia). The study is devoted to the analysis of sustainable tourism development practices at the resorts of Krasnaya Polyana and Rosa Khutor.

This research seeks to draw up the state of the art technological innovations that facilitate and aid the practice of regenerative tourism. The review aims at determining the major technologies, its usages, and on how these technologies can assist in the regeneration of the nature,wildlife and the social aspects of tourism-destinations The approach used in the study is a systematic literature review with a focus towards the types of technological innovations that can facilitate tourism for the benefit of wildlife. Histories relevant to regeneration and technology strategies were sought in the Scopus and Web of Science databases using the words 'regenerative tourism' and 'technological innovations'. Examples include studies by Frías-Jamilena et al (2022) who regarding sustainable tourism, sought to enhance tourist behavior through the Gamification of environmental education.

The study aims to address an important contemporary issue related to the impact of tourism on nature. With the growth of tourist activities and the load on recreational areas, the sustainable use of natural resources is necessary; otherwise, it can lead to serious problems related to ecology and public health. The study presents an assessment method through chemical analysis of water and soil to evaluate how tourism and recreational activities affect water quality. Interpretation of the results of the experiment allowed us to draw conclusions about the impact of tourist and recreational activities on hydrological resources. If emissions are not controlled, then poisoning of the local population can be observed in the near future. These results can be used to develop management strategies and measures to reduce the environmental impact of tourism activities.

As global biodiversity declines and tourism-related pressures on ecosystems increase, the airline industry has a unique opportunity to support regenerative conservation through biodiversity tourism. This study

The work considers reducing the manifestation of an environmental hazard factor of the energy type through measures that reduce the likelihood of environmental risks. It is this type of factor that is the most destructive, since the basis of energy is natural resources. Consuming a huge amount of primary resources in the form of solid, liquid and gaseous fuels, the energy sector, along with the production of electrical energy and heat, emits a large amount of waste into the environment in the form of gaseous and solid combustion products. In addition, there is the problem of exhaustion of resources for traditional energy. In this regard, much attention is paid to energy saving and energy efficiency based on renewable energy sources. Solar energy is one of the most developed branches of renewable energy. Therefore, the work examines various heat supply schemes using solar systems, which will not only reduce dependence on traditional fossil fuels, but also reduce greenhouse gas emissions, that is, reduce environmental risks.

Mud volcanism is a rather rare, complex, and in many ways still mysterious geological phenomenon. Currently, the Taman Peninsula is the region with the greatest development of mud volcanoes in the Russian Federation. The Taman Peninsula belongs to one of the youngest land areas. Back in the first century AD, this area was an archipelago of several islands. However, due to the high dynamism of the Kuban River delta, the high intensity of sediment coming from the river (up to several million tons per year) and the rise of tectonic plates, the islands joined together. Currently, the Taman Peninsula is the region with the greatest development of mud volcanoes in the Russian Federation. All known forms of mud volcanic manifestations (continental and marine, buried and open, extinct and active) are found here. Many of them are highly active and are located in close proximity to populated areas, so the mud volcanoes of the Taman Peninsula are currently considered as objects of sightseeing display.

This study investigates the synergistic effects of climate change and tourism on biodiversity loss, focusing on the complex interactions between these two factors and their collective impact on various ecosystems. The research examines the role of tourism in exacerbating environmental degradation and the additional pressures placed on biodiversity by changing climate conditions. By analyzing data from different regions and ecosystems, the study identifies critical areas where the intersection of tourism and

Preface

The global tourism landscape has undergone a profound transformation following the COVID-19 pandemic. As travel restrictions eased, the tourism sector entered a "new normal," characterized by a shift in travel preferences toward smaller groups, open and remote destinations, and experiences that prioritize cleanliness, sustainability, and safety. This evolution in tourism behavior aligns with the growing popularity of ecotourism and rural tourism, where travelers seek unspoiled environments, fresh air, and immersive outdoor activities such as hiking, cycling, and nature exploration. Indeed, the future of tourism is intrinsically linked to spatial awareness, health security, environmental consciousness, and the integration of technology into travel experiences.

In response to these emerging trends, destinations worldwide have adopted their tourism models to emphasize sustainability and regeneration. The Biosphere Guide 2022-2023 highlights how post-pandemic tourism focuses on discovering new locations while respecting local cultures, natural resources, and environmental sustainability. Furthermore, newer generations of travelers prioritize meaningful experiences over mere sightseeing and demonstrate a stronger commitment to environmental stewardship compared to previous generations.

This paradigm shift in tourism naturally leads to a greater emphasis on protected areas, which serve as vital hubs for biodiversity conservation and ecological balance. However, despite their significance, protected areas, particularly in developing nations, often suffer from inadequate planning, management, and development. These challenges stem from the intricate interplay of governmental and non-governmental stakeholders, as well as the complexities of balancing conservation efforts with tourism activities. As protected areas encompass over 15.4% of the global terrestrial landscape, they play a crucial role in preserving biodiversity and maintaining ecosystem functionality (Namsrai et al., 2019). Nonetheless, research on effective tourism management within these areas remains limited.

This edited volume, *Harnessing Biodiversity Tourism for Regenerative Conservation Management*, seeks to address the contemporary challenges and opportunities associated with tourism in protected areas. It delves into the "new normal" tourism paradigm, visitor behavior trends, and the evolving preferences of modern travelers toward sustainability-driven tourism. The book explores key themes, including innovative planning and management strategies, theoretical frameworks, and practical tools for enhancing protected area tourism while ensuring ecological integrity.

Chapter 1

This chapter analyzes the approaches and principles of ecotourism development in specially protected natural areas. It discusses both the positive economic impact and the negative ecological consequences of ecotourism. The chapter also presents recommendations for a standard Ecotourism Development Strategy to balance conservation and tourism.

Chapter 2

This chapter explores the role of applied regional geography in proximity tourism, using the Friulian Alps (Italy) as a case study. It examines unique geographical features and their potential for regional tourism development, offering insights into how geographical particularities can be integrated into marketing strategies to boost local tourism.

Chapter 3

Using geoinformation systems (GIS), this chapter evaluates the recreational potential of mountain forests in Russia's Mostovsky district. It highlights the need for effective management strategies to optimize the use of these forests for tourism while ensuring sustainability.

Chapter 4

This chapter investigates the barriers to tourism growth in KwaZulu-Natal. Using qualitative research methods, it identifies key obstacles affecting the sector and suggests strategic recommendations to enhance the region's tourism industry.

Chapter 5

This chapter examines the significant role of women and children in environmental conservation, particularly in coastal communities. It highlights their contributions to waste management, recycling, and community-driven sustainability initiatives, while also addressing challenges such as resource constraints and institutional support.

Chapter 6

Focusing on the Krasnodar Territory near Lake Abrau, this chapter discusses geophysical studies for groundwater exploration. It highlights the growing demand for clean water in tourism regions and underscores the need for sustainable water resource management.

Chapter 7

This chapter explores how virtual reality (VR) can contribute to sustainable tourism. By simulating real-world landscapes, VR offers an innovative approach to environmental education, reducing the ecological impact of tourism while enhancing visitor experiences.

Chapter 8

Through quantitative analysis, this chapter evaluates the role of public-private partnerships in tourism sustainability. It demonstrates how collaboration between stakeholders can lead to enhanced tourism development and improved environmental practices.

Chapter 9

This chapter presents theoretical approaches and practical applications of sustainable tourism in mountain destinations, using post-Olympic Sochi as a case study. It outlines criteria for sustainable tourism and highlights successful initiatives in Krasnaya Polyana and Rosa Khutor.

Chapter 10

This chapter examines emerging technologies that support regenerative tourism. Through a systematic literature review, it identifies key technological advancements that aid in environmental conservation and sustainable tourism practices.

Chapter 11

Using chemical analysis, this chapter assesses how tourism and recreational activities affect water quality. It provides insights into managing tourism's environmental footprint to prevent long-term ecological degradation.

Chapter 12

This chapter explores how airlines can contribute to regenerative conservation through biodiversity-focused tourism. It discusses sustainable travel practices, carbon offset programs, and partnerships with conservation organizations to minimize environmental harm.

Chapter 13

Examining Kartarpur as a model for religious tourism, this chapter highlights its role in fostering interfaith harmony and public diplomacy. It suggests ways to enhance cultural exchange programs and responsible tourism practices to promote peace and biodiversity conservation.

Chapter 14

The final chapter presents research on endemic plant species and their conservation strategies in tourist regions. It highlights the ecological importance of preserving plant diversity and integrating conservation efforts into tourism development.

Chapter 15

This chapter explores the critical role of renewable energy in mitigating environmental hazards associated with traditional energy production. Given that the energy sector is a major consumer of natural resources and a significant contributor to environmental pollution, the chapter examines strategies for reducing environmental risks

Chapter 16

Mud volcanism is a rare and complex geological occurrence, and the Taman Peninsula in Russia is home to some of the most developed mud volcanoes in the region. This chapter provides an in-depth examination of the geological evolution of the peninsula, tracing its transformation from an ancient archipelago to its present state due to sediment accumulation and tectonic activity.

Chapter 17

This chapter investigates the interconnected impacts of climate change and tourism on biodiversity loss, emphasizing the compounded pressures these factors place on ecosystems. Through an analysis of regional case studies and empirical data, the study identifies areas where the effects of tourism-driven environmental degradation are most pronounced in the context of shifting climate conditions.

Aimed at a global audience, this book is particularly valuable for higher education institutions offering undergraduate and graduate courses in tourism, hospitality, and environmental management. Additionally, it serves as a crucial resource for policymakers, scholars, tourism practitioners, and conservationists seeking to align tourism development with biodiversity conservation goals.

By bringing together diverse perspectives and expertise, this volume aspires to foster a deeper understanding of how tourism can serve as a catalyst for regenerative conservation management. Through innovative approaches and informed strategies, we hope to contribute to a future where tourism and conservation not only coexist but thrive in synergy.

Parinya Nakpathom
International College, Burapha University, Thailand

Kannapat Kankaew
International College, Burapha University, Thailand

Krittipat Pitchayadejanant
International College, Burapha University, Thailand

Minenkova Vera Vladimirovna
Kuban State University, Russia

Evgeniia Rotai
Kuban State University, Russia

Chapter 1
Ecotourism Development Strategy in Specially Protected Natural Areas

Vera Minenkova
https://orcid.org/0000-0003-4184-2522
Kuban State University, Russia

Anatoly Filobok
https://orcid.org/0000-0002-7745-9472
Kuban State University, Russia

Anna Mamonova
https://orcid.org/0000-0001-8222-4718
Kuban State University, Russia

Maya Kaitamba
Abkhazian State University, Abkhazia

Evgeniia Rotai
Kuban State University, Russia

ABSTRACT

Protected areas are the most important element of the territorial tourism and recreational system. The development, creation and improvement of ecotourism routes are carried out in many protected areas. The dynamics of tourist visits, being one of the indicators of the effectiveness of the functioning of protected areas, demonstrates a rapidly positive character throughout the world. It is important that the development of ecotourism in protected areas has both a positive (mainly in the economic aspect) and a negative (ecological aspect) impact on natural complexes. The purpose of the chapter is to analyze the approaches and principles of ecotourism development in specially protected natural areas, and to recommend basic content positions of a typical Ecotourism Development Strategy in Specially Protected Natural Areas.

DOI: 10.4018/979-8-3693-8232-5.ch001

INTRODUCTION

Human activity transforms landscapes. Today, one third of all land is degraded or degrading, which harms biodiversity and jeopardises essential ecosystem services such as carbon storage. Protected areas offer a solution through effective management and government that can save nature and cultural resources, protect human health and well-being, provide sustainable livelihoods and support sustainable development. Protected area is defined as geographical space, recognized, dedicated, and managed, through legal or other effective means, to achieve the long-term conservation of nature with associated ecosystem services and cultural values (definition of International Union for Conservation of Nature).

Protected areas are important for many environmental, cultural, social and health benefit reasons such as:

- conserving nature and providing refuges for diverse lifeforms including threatened species;
- providing ecosystem services for human populations such as fresh air, clean water, pollination, flood control and productive soils and oceans;
- carbon sequestration in areas such as old growth forests and mangroves and playing a functional role in mitigating climate change;
- enabling active connection with nature and culture and nature-based recreational activities that increase physical activity, foster psychological wellbeing, reduce anxiety and stress and improve self-esteem;
- providing economic benefits to local communities through employment across multiple sectors including on parks as rangers and contractors, jobs associated with hospitality and tourism, education, leisure, fitness, outdoor recreation, wellbeing, conservation and research. (Ecotourism Plan for Queensland's Protected Areas 2023-2028).

Protected natural areas perform important scientific and socio-cultural functions, being ecological standards of natural communities, testing grounds for studying natural ecosystems and the processes occurring in them. Protected areas that preserve ecosystems are natural reserves for the restoration of lost natural complexes or their components. Protected areas contribute to the education and environmental training of the population, and have significant recreational potential. All over the world, visiting objects of the so-called wild nature is one of the most popular and mass types of human activity in free time. Recently, it has been called ecotourism and it is developing very intensively (annual growth is estimated at 20-30%) (Ivanov, A.N., Chizhova, V.P., 2024).

Because of their ecological value, protected areas contain many of the world's greatest ecotourism attractions. These attractions may consist of one or a combination of rare or endemic species of flora or fauna, abundant wildlife, high indices of species diversity, unusual or spectacular geomorphological formations, or unique historic or contemporary cultural manifestations in a natural context.

Despite its relatively low impact on the natural environment, ecotourism can have a significant impact on it, especially under high recreational loads. Protected area managers face the challenge of controlling and limiting the impacts of unfettered ecotourism while at the same time have to decide where and how to plan adequately for the development of ecotourism as a compatible economic development option.

By integrating ecotourism development into a systematic approach to conservation one can ensure that ecotourism is only initiated when it is the most effective strategy to achieve tangible, lasting results at scale. These distinct but intimately interrelated aspects of ecotourism – conservation management and business development – must be fully understood by ecotourism planners and protected area managers before moving ahead with plans to implement ecotourism activities. Conservationists have typically approached ecotourism with a limited understanding of business issues and an incomplete understanding of the management mechanisms that are available and necessary to ensure the sustainability of tourism in protected areas. Business focuses on economic benefits, leaving environmental conservation in the background. Now it`s obvious that in order for ecotourism to be successful, conservationists need a greater understanding of business considerations; likewise, developers need a greater awareness of the management mechanisms that are necessary to ensure the sustainability of the activity. (Drumm, A. & Moore, A., 2005)

Combining both perspectives is essential for a successful ecotourism program which should form the basis of the Ecotourism Development Strategy in Specially Protected Natural Areas.

LITERATURE REVIEW AND TOPIC BACKGROUND

A significant modern scientific base in the study of approaches to the analysis of ecotourism activities, including in specially protected natural areas, was formed by scientists from different schools of eco-tourism around the world, which were presented in the works of Andy Drumm & Alan Moore (Ecotourism Development. A Manual for Conservation Planners and Managers, 2005), Bharat Gotame (Economic Impacts of Ecotourism in Protected Areas, 2012), Tom Sengalama (Environmental and social impacts of ecotourism around protected areas, 2012), Henok Bekele (Ecotourism Development in Protected Areas, 2013), Roy Ballantyne & Jan Packer (International Handbook on Ecotourism, 2013), Stephen Momanyi (Ecotourism & Socio-Economic Development in Wildlife Protected Areas, 2015) and others. In 2024 IGI Global Scientific Publisher published the book "Supporting Environmental Stability through Ecotourism". The book serves as a comprehensive guide for both academics and professionals. It presents solutions and insights that are essential for establishing a sustainable and responsible tourism industry, ensuring the preservation of our environment and the well-being of local communities. Books "Ecotourism and Community Intervention: Emerging Research and Opportunities" (IGI Global, 2019), "Destination Branding and Bias in Ecotourism" (IGI Global, 2024) are also of scientific interest.

Many publications related to ecotourism development issues are initially devoted to sustainable development and sustainable tourism: "Sustainable Tourism: A Global Perspective" (Rob Harris, Tony Griffin & Peter Williams, 2002), "Ecotourism and Sustainable Development" (Martha Honey, 2008), "Ecotourism and Sustainable tourism. New Perspectives and Studies" (Jaime A. Seba, 2012), "Tourism and Sustainability" (Martin Mowforth & Ian Munt, 2016), "Special Interest Trends for Sustainable Tourism" (IGI Global, 2024) and many others.

Russian scientists Khrabovchenko V.V., Chizhova V.P., Drozdov A.V., Kuskov A.S., Sergeeva T.K., Nekrasova M.L., Ivlieva O.V. and others, studied the problems of development of ecological tourism.

Paying tribute to the great contribution of scientists to solving problems related to the ecological tourism development, it should be mentioned that in the methodological and practical results they have obtained, a unified approach to the concepts of "ecological tourism" and "resources of ecological tourism" has not yet been developed, and models of rational use of resources of ecological tourism have not been considered. On the other hand, this is explained by the lack of universal solutions since each natural territory is unique and unrepeatable.

What concerns the ecotourism development strategies, it can be noted that such a document has already been adopted in many countries. It may be called in different ways, have different content, and relate to territories of different hierarchical levels. For example – National Strategic Plan for Ecotourism Development in the Dominican Republic, National Ecotourism Strategy & Action Plan (Philippines), National Ecotourism Strategy for Malawi, National Strategy for Ecotourism in India, Ecotourism development strategy for cross-border areas (Latvia and Lithuania), Development of Ecotourism in the Košice Region (Slovakia), Ecotourism Plan for Queensland's Protected Areas (Australia), National Strategy for Sustainable Tourism Development (Albania) etc. If the issues of the development of ecotourism are not regulated by documents, then methodological recommendations for this type of activity are developed (for example, Ecotourism Handbook for Ireland, ets.). Often, issues of ecotourism development are included in documents on biodiversity conservation or tourism development in general.

In the Russian Federation, the issues of tourism development in specially protected natural areas are enshrined in the following documents:

- Federal Law of the Russian Federation On Specially protected natural areas (No. 33-ФЗ, March 14, 1995);
- Model regulations on the organization and implementation of tourism, including tourism safety in specially protected natural areas of regional and local significance (approved by the Decree of the Government of the Russian Federation of December 21, 2023 No. 2230);
- Calculation rules for the maximum permissible recreational capacity of specially protected natural areas of federal significance (approved by the Decree of the Government of the Russian Federation of October 31, 2023 No. 1811);
- Model calculation rules for the maximum permissible recreational capacity of specially protected natural areas of regional and local significance (approved by the Order of the Government of the Russian Federation of October 31, 2023 No. 1809).

The Ecotourism Development Strategy in Specially Protected Natural Areas of Federal Significance is still being developed. The approaches proposed in this project formed the basis for the main thesis of this publication.

RESEARCH METHODS

The study is devoted to the analysis of strategic approaches and principles for the development of ecotourism in specially protected natural areas. Conceptually, the basis of these approaches and principles is proposed to consider the points of the concept of sustainable development, which will allow maintaining a balance between tourist consumption and the preservation of fragile natural complexes.

During the study, documents on strategic development of ecotourism in different countries were studied, and common provisions were identified. At the same time, the Russian experience was taken as the basis for the proposed directions of the Ecotourism Development Strategy.

The study used general scientific methods: dialectical method of knowledge, system approach, analysis of the existing base of information sources, including the regulatory framework on the essence of the topic and related interdisciplinary problems, synthesis, comparisons, generalizations, descriptive analysis.

DISCUSSION

Specially protected natural areas and ecotourism

The main goal of creating protected areas is to preserve natural complexes, ensure their natural development and monitor these processes. Protected lands and their natural resources are completely or partially, permanently or temporarily excluded from economic use.

Specially protected natural areas are land areas, water surface and air space above them where natural complexes and objects, flora and fauna, natural ecological systems are located and have special environmental, scientific, cultural, aesthetic, recreational and health significance, which are withdrawn by decisions of state authorities in whole or in part from economic use and for which a special protection regime has been established (Federal Law of the Russian Federation On Specially protected natural areas).

According to federal legislation of the Russian Federation, there are several categories of protected natural areas: state nature reserves (including biosphere reserves), national parks, nature parks, natural monuments, dendrological parks and botanical gardens. Each category has rules for zoning the territory, including the allocation of recreational zones (they are in state nature reserves, national parks, nature parks, dendrological parks and botanical gardens).

The strategy of isolated functioning of protected areas is currently losing its relevance. And specially protected natural areas are becoming one of the most important elements of territorial tourism and recreation systems. Tourism in protected areas is one of the effective tools of environmental education. The dynamics of tourist arrivals is one of the indicators of the effectiveness of the functioning of protected areas, and demonstrates a rapidly positive character throughout the world.

In Russia, tourism in specially protected natural areas is defined as temporary trips (travels) of citizens of the Russian Federation, foreign citizens and stateless persons from their permanent place of residence to specially protected natural areas for the purpose of visiting unique natural complexes and objects and for other purposes. Ecotourism is one of the most promising types of tourism and in the modern world is inseparably linked with specially protected natural areas.

Ecotourism is an activity aimed at organizing travel in various forms of nature tourism, focused on studying nature and cultural attractions, caring for nature and assisting in its conservation (National standard of the Russian Federation "Tourist services. Ecological Tourism. General terms. No. 56642-2021). Educational tourism in protected areas is a type of ecotourism carried out for the purpose of studying natural and cultural-historical attractions located on the territory of protected areas, as well as with activities to preserve and study natural and cultural heritage.

The specific features of ecotourism include:

– focus on nature (natural heritage in the form of natural landscapes and their components is the main in ecological travel, nature is the main value);
– environmental safety (ecological travel should not harm the natural environment);
– educational component, environmental education (tourists not only observe natural phenomena, but also acquire modern scientific knowledge about them);
– cultural aspect (the relationship between the natural features of travel areas and the culture of their population);
– socialization (ensuring the attractiveness of ecological travel for the local population, taking into account its economic interests); (Dyozhkin V.V. et al., 2006)

– ecotourists personally participate in actions that preserve or restore wildlife resources;
– income from ecotourism is directed to financial support for environmental protection.

Ecotourism typically involves engaging with the biotic elements of the natural world. Minimal environmental impact on the places visited is the goal of ecotourism. In addition to promoting reverence for the environment, ecotourism aims to generate socio-economic advantages for the local inhabitants. For this reason, ecotourism requires not only the assessment of environmental and cultural aspects but also the encouragement of recycling, energy conservation, water conservation, and the development of local economies. Proponents of social and environmental responsibility are drawn to ecotourism for these reasons.

Long ago experts came to the conclusion about the possibility and necessity of developing ecological tourism in specially protected natural areas. Communication with untouched nature, the opportunity to feel and study the surrounding world often turn visitors into active assistants, and sometimes even investors of nature conservation organizations. In addition to direct income (entrance fees, various fees for tourist services, funds from the sale of souvenirs, guidebooks, etc.), ecological tourism brings in additional funds, which, if well managed, can be directed to the maintenance of the protected area, salaries for employees, repair of railways and roads, provision of an adequate level of tourist service, etc. Tourism in special protected areas produces benefits and costs. These effects interact often in complex ways. It is the responsibility of the protected area planner to maximize benefits while minimizing costs.

Protected areas are established primarily to preserve some type of biophysical process or condition such as a wildlife population, habitat, natural landscape, or cultural heritage such as a community's cultural tradition. Tourists visit these protected areas to understand and appreciate the values for which the area was established and to gain personal benefits.

Tourism planning and development aims to take advantage of the interest shown by tourists so as to: enhance economic opportunities, protect the natural and cultural heritage, and advance the quality of life of all concerned. These goals are expanded upon in Table 1.

Table 1. Potential benefits of tourism in protected areas

Direction of impact	Benefits
Enhancing economic opportunity	– Increases jobs for local residents – Increases income – Stimulates new tourism enterprises, and stimulates and diversifies the local economy – Encourages local manufacture of goods – Obtains new markets and foreign exchange – Improves living standards – Generates local tax revenues – Enables employees to learn new skills – Increases funding for protected areas and local communities – Provides a multiplier effect from tourism development – Improves the investment climate and attracts investors – Stimulates regional and territorial development
Protecting natural and cultural heritage	– Protects ecological processes and watersheds – Conserves biodiversity (including genes, species and ecosystems) – Protects, conserves and values cultural and built heritage resources – Creates economic value and protects resources which otherwise have no perceived value to residents, or represent a cost rather than a benefit – Transmits conservation values, through education and interpretation – Helps to communicate and interpret the values of natural and built heritage and of cultural inheritance to visitors and residents of visited areas, thus building a new generation of responsible consumers – Supports research and development of good environmental practices and management systems to influence the operation of travel and tourism businesses, as well as visitor behaviour at destinations – Improves local facilities, transportation and communications – Helps develop self-financing mechanisms for protected area operations – Creates a sense of pride among local people for their culture and the desire to preserve its identity – Provides global intercultural exchange
Enhancing quality of life	– Promotes aesthetic, spiritual, and other values related to well-being – Supports environmental education for visitors and locals – Establishes attractive environments for destinations, for residents as much as visitors, which may support other compatible new activities, from fishing to service – or product-based industries – Improves intercultural understanding – Encourages the development of culture, crafts and the arts – Increases the education level of local people – Encourages people to learn the languages and cultures of foreign tourists – Encourages local people to value their local culture and environments

Source: Eagles P.F. J. et al. (2022), added by the authors.

The main way to increase the efficiency of ecotourism is to extract greater economic benefit from each visitor. To do this, it is necessary to define a list of events that will create incentives for greater consumer activity of tourists, while ensuring a balance of economic and environmental interests. The United Nations Environment Programme and the International Council carry out the systematization of such events for the Conservation of Nature.

The UN Environment Programme's recommendations for ecotourism are presented in the publication "Sustainable Tourism in Protected Areas: Guidelines for Management and Planning". It suggests the following measures to increase economic benefits (Eagles P.F. J. et al. (2022):

– increase the number of visitors (increasing visitation is risky unless the financial benefits from the visitors exceed their costs. It may increase other impacts, some negatively);
– increase the length of stay (increased length of stay provides more opportunity to sell local products and services);

– attract richer market niches (different marketing tactics may bring in consumers with strong abilities to spend);
– increase purchases per visitor (offering more locally-made goods for sale, available directly and indirectly to the visitor, helps increase visitor expenditure and local incomes);
– provide lodging (the costs of overnight accommodation are relatively large and are paid for locally. Local lodging also increases expenditures on meals, and local goods and services);
– provide guides or other services (since much tourist activity in protected areas is information intensive, there are usually good opportunities for guide services);
– host events (artwork, crafts and festivals based on local culture can increase local economic impact);
– purchase local food and drink (when visitors, park staff and tourism employees consume locally grown food and drink, they provide important income to local farmers).

The proposals in the collection "Building and Operating Diversity-Friendly Hotels" (UICN) are built in a similar logic. It provides solutions for the development of hotel complexes that allow for a combination of sustainable development of the territory and an increase in the value of the product created in the protected area:

– in hotel restaurants give preference to local producers, especially those who use advanced sustainable development methods;
– in guest rooms and public areas, give preference to decoration made of natural materials, especially wood, use medicinal and aromatic plants for spa products, ornamental plants for public areas;
– in souvenir shops, avoid souvenirs made from endangered or protected species of plants, animals species;
– in hotel grounds and gardens, use indigenous plants for landscaping, choosing species that can be maintained with low or no use of chemicals, and minimize light and noise;
– in tourist destinations, inform visitors about responsible behavior and the need to preserve biodiversity.

The document "Guidelines and Criteria for National Geoparks seeking UNESCO assistance to join the Global Geoparks Network" qualifies protected areas as geoparks based on a number of requirements:

– value: the area must contain a geological heritage site of international significance;
– management: the area must be managed by a registered organization and developed according to a comprehensive management plan;
– promotion: the geopark must implement an information campaign through various channels and media resources;
– partnership network: the geopark must develop connections not only within the local community, but also work as part of a regional and global network of geoparks.

The key value of the decision-making process for geoparks is the organization of horizontal interactions between various participants who act as founders of the management organization. Such an approach changes the logic of the management of protected areas: they are understood not as departmental institutions, but as a tool for coordinating the interests of a wide range of stakeholders. (International experience in the development of ecological tourism in specially protected natural areas, 2022)

Recommendations for organizing transformational activities are contained in the book «Sustainable Tourism for Development» (World Tourism Organization). It outlines strategies by which the tourism sector can be a means of stimulating economic development while minimizing negative social, cultural and environmental impacts.

The authors propose the intervention method as a key tool. Unlike the complex transformation method, which involves developing the entire territory through the implementation of large-scale projects, it is based on the development and implementation of point changes and the collection of feedback on the effects of their implementation.

The ecotourism intervention method is based on five basic pillars:

- tourism policy and governance;
- economic performance, investment and competitiveness;
- employment, decent work and human capital;
- poverty reduction and social inclusion;
- sustainability of the natural and cultural environment.

Point interventions are implemented according to the following algorithm:

- assess country situation;
- identify areas for intervention;
- select interventions;
- formulate actions;
- implement actions;
- evaluate actions taken.

Following the evaluation, the intervention cycle is restarted. (Sustainable Tourism for Development, 2013)

Protected area managers should aim to develop tourism development policies which support long-term economic development and encourage repeat visits. They should try to maximize local employment, social and cultural benefits through high visitor spending and low local leakage.

At the same time, it is necessary to note the critics who are against the tourism development in specially protected natural areas. Their main counterargument is that the main goal of creating such areas is to preserve biological, landscape, historical and cultural diversity, as well as unique, reference and typical natural objects and complexes, improve the state of the environment, and provide environmental education. Specially protected natural areas are strictly protected reserves. At the same time, tourism is considered as a type of economic activity that in any case has an impact on the environment, even if minimal. Therefore, the intervention of tourists in the life of specially protected natural areas is unacceptable.

General patterns of the process of recreational impact on nature are shown in Fig. 1, which shows the complexity and multifactorial nature of the problem of preserving natural nature when it is visited by a large number of people.

Figure 1. Scheme of the impact of recreational activities on the natural environment

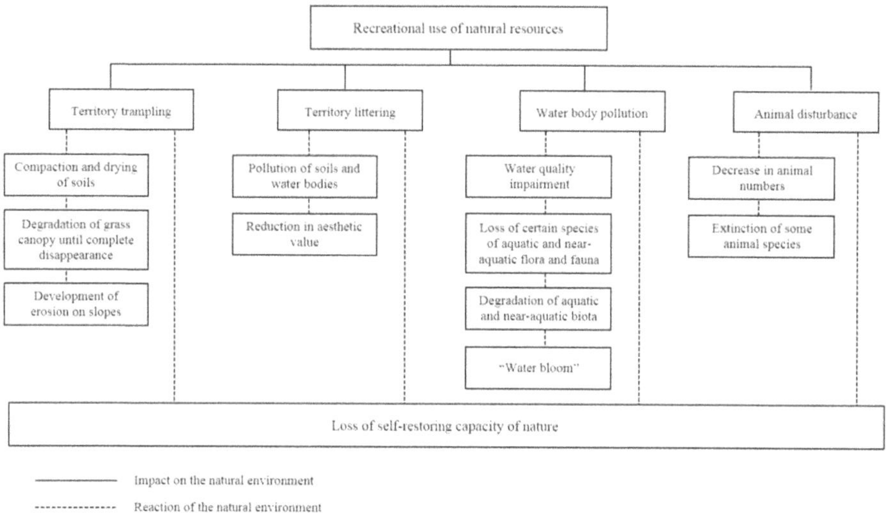

Source: Ivanov, A.N., Chizhova, V.P., 2024.

Absolute denial of the opinion about the negative factors of tourism impact on natural complexes of protected areas is obviously incorrect. It means that this opinion should be taken into account when making a decision on the possibility of organizing tourism in the territory of protected areas. An important task of this activity is the desire to minimize the impact on the environment, which should be reflected in the content of the Ecotourism Development Strategy in Specially Protected Natural Areas.

Main positions of the Ecotourism Development Strategy in Specially Protected Natural Areas

The Ecotourism Development Strategy in Specially Protected Natural Areas should determine the main directions for the development of tourism in the system of specially protected natural areas, as well as measures to improve the efficiency of state policies in this area and the preservation of natural and cultural heritage in the course of the development of ecological tourism, should be aimed at creating legal, organizational and economic conditions for the development of this type of tourism activity, promoting the development of the material base of tourism, reflecting the tasks and priorities for the development of these types of tourism.

The main criteria and features of the organization and implementation of tourism in specially protected natural areas in Russia are:

a. preservation of unique and typical natural complexes and objects, flora and fauna, natural ecological systems, biodiversity;
b. minimizing the negative impact on the environment during tourism;

c. following the established maximum permissible recreational capacity of a specially protected natural area when implementing tourism (the maximum number of visitors who can arrive to a protected area or its parts as a tourist per unit of time without degradation of natural complexes and objects, flora and fauna, natural ecological systems), the regime of its special protection;
d. implementation of tourism in specially equipped places and routes;
e. preservation of cultural heritage sites (historical and cultural monuments). (Federal Law of the Russian Federation On Specially protected natural areas)

The Ecotourism Development Strategy in Specially Protected Natural Areas should determine:

– goals, objectives, principles and main directions of tourism and recreational activities;
– procedure of environmental education activities in protected areas;
– regulations of organizing and implementing tourism, including ensuring tourism safety in protected areas;
– principles and procedure for forming tourism infrastructure in protected areas;
– principles and approaches to determining the maximum permissible recreational capacity of protected areas;
– directions of socio-cultural programming of protected areas;
– institutional principles and mechanisms for organizing ecotourism in protected areas.

The Strategy establish the basis for understanding the long-term objectives facing the state system of specially protected natural areas in terms of the development of ecotourism. It is a tool for forming work plans for executive authorities and participants in the tourism business at all levels, as well as for orienting citizens' entrepreneurial initiative in the area under consideration. In accordance with the positions of the Strategy, it is necessary to develop an action plan for its implementation.

The Goal, Objectives, Principles And Strategic Guidelines For The Development Of Ecological Tourism In Specially Protected Natural Areas

The ecotourism development goal in specially protected natural areas is to develop in broad population an understanding of the modern role of specially protected natural areas in preserving biological and landscape diversity as the basis of the biosphere, natural and associated with nature cultural heritage, as well as their place in the socio-economic development of regions. This should ensure effective public support for the system of specially protected natural areas as objects of national and world heritage.

Achieving this goal requires solving the following tasks:

– optimization of the regulatory and methodological framework for ecotourism in specially protected natural areas;
– mandatory strategic and territorial planning for the development of ecotourism for each s protected area;
– scientific and information support to develop ecotourism;
– control and monitoring system in the field of ecotourism development;
– creation and modernization of modern ecotourism infrastructure;

– development and implementation of a marketing strategy, promotion of the relevant tourism product in the domestic and international tourism markets;
– strengthening of the social role of ecotourism, including through the development of children's and youth tourism, as well as tourism for people with disabilities;
– development and improvement of a system for control and monitoring the consequences of tourism in specially protected natural areas;
– promoting the socio-economic development of the territory and ensuring an increase in the quality of life of the local population living in specially protected natural areas and in close proximity to them, by involving the local population in the development of tourism;
– volunteerism as an effective tool for popularizing ecotourism and supporting the system of protected areas by civil society;
– international cooperation and exchange of experience in the field of ecotourism;
– appropriate professional training and advanced training for specialists working in the field of ecotourism in specially protected natural areas.

The principles of development of ecotourism in protected areas:

– ensuring the priority of preserving biological and landscape diversity and cultural heritage, minimizing anthropogenic impact on natural and historical-cultural complexes and objects;
– popularizing the role of specially protected natural areas in preserving natural and closely related to nature cultural heritage, including the traditional culture of the local population;
– educational focus of ecotourism;
– preservation of the socio-cultural environment;
– promoting sustainable development of areas where ecotourism is carried out;
– partnership with regional and local governments, educational, scientific and cultural institutions, travel agents, tour operators and other structures interested in the development of ecotourism;
– involving the local population, including indigenous peoples and their communities, in activities related to the development of ecotourism.

MAIN DIRECTIONS OF DEVELOPMENT OF ECOLOGICAL TOURISM IN SPECIALLY PROTECTED NATURAL AREAS

Legal framework and methodological support for the development of ecotourism

All regulatory legal acts in the sphere of development of specially protected natural areas and in the sphere of development of tourism and ecotourism should be related to and reflect issues on the specifics of tourist activities in protected areas. They should contain economic mechanisms for development of the ecotourism system, including taking into account international experience, mechanisms of public-private partnership, regulation of relations between federal institutions and tour operators.

Methodological recommendations for development of ecotourism in protected areas should be developed and approved. They should be devoted to issues of minimizing the impact of tourism on natural ecosystems, assessment of permissible recreational load on natural complexes, on the functional organization and socio-cultural programming of protected areas, creation and arrangement of excursion trails and viewpoint platforms, organization and holding of children's ecological camps and expeditions, work of tour guides, provision of paid services in ecotourism, etc.

Planning in ecotourism

Each specially protected natural area must have its own strategy for the development of ecological tourism. It is developed on the basis of an analysis of the potential for the ecotourism development and taking into account the need to minimize anthropogenic impact on natural complexes, as well as scientifically substantiated size, spatial distribution and seasonal use of areas of the territory used for tourism. To determine the key objects to display, visitors must be allocated unique landscapes, objects of historical and cultural significance, attractive objects of the animal and plant world.

It is important to identify the target groups of visitors to the territory, their needs and demand, as well as the degree of priority of each of the group. Taking this data into account, the tourism product of the protected area is formed and promoted, which should be based on the use of local cultural traditions and the involvement of the local population. Packages of excursion programs, a network of routes for various categories of visitors are developed, giving priority to demonstrating wildlife.

Planning the development of ecotourism infrastructure in the protected area requires a lot of work.

All levels and divisions of the organizational system of the protected area should be involved in the planning work, their roles and degrees of involvement in the development of ecological tourism should be determined in accordance with their competence.

Ecotourism information and analytical support

The analysis of the ecotourism potential for the development is necessary for each specially protected natural area. The Ecotourism Development Strategy should include the recommendations in the following areas:

- unification of the list of materials recommended for posting in the section "Development of Ecotourism" on the official websites of protected areas in the Internet;
- content and design of information booklets and guides to protected areas;
- creation of electronic guides (including audio guides) for each protected area;
- preparation of a single basic package of information and advertising materials on the possibilities of ecotourism in protected areas;
- placement of advertising and information materials about protected areas in airports, railway stations and other public places;
- expansion of the practice of using social networks to post information on about protected areas and the possibilities of ecotourism.

On a national scale, it is important to create a federal website with information about the possibilities of ecotourism in protected areas, natural and historical-cultural attractions, tourist routes and tours, and services for the tourism sector. Taking measures to improve the image of specially protected natural areas that have international status (UNESCO World Heritage, UNESCO biosphere reserves, Ramsar lands, etc.) is also an important issue. In general, it contributes to the formation of a positive image of a country as a territory favorable for the development of ecotourism.

Preservation of natural and cultural heritage, control and monitoring of activities in the field of ecotourism

This objective is associated with the definition of areas for a protected area where the development of ecotourism is recommended, as well as possible restrictions for these areas related to seasonal and other natural phenomena, and the purposes of visiting the territory. After this, studies should be conducted to assess the recreational potential and recreational capacity of areas of protected areas where tourism development is recommended.

In order to increase the efficiency of preserving natural and cultural heritage sites in protected areas, it is necessary to monitor their condition. For each protected area, proposals are prepared to minimize the impact on natural complexes and cultural and historical sites during the development of ecotourism, and complex recreational monitoring systems are developed and implemented on all excursion and tourist routes and sites.

Currently, it is relevant to create a problem-oriented geographic information system (GIS) for planning the development of ecotourism in protected areas, forecasting and determining ways to minimize negative impacts on natural complexes.

It is also important to involve specially protected natural areas in the system of voluntary certification of elements of tourism and recreational activities, to involve specialists and experts from specialized organizations (environmental and scientific state and public organizations) in the implementation of research work related to the development of ecotourism and the assessment of its impact on natural ecosystems.

In any protected area, the development of ecoltourism should be carried out taking into account its size, landscape and environmental specifics, determined by individual features of these specially protected natural areas.

Development of environmental education activities

Specially protected natural areas occupy key positions in the field of environmental education, the formation of a competent and responsible attitude to the environment, play an important role in matters of growing a sense of patriotism, in raising the level of environmental culture of the population and forming in citizens an understanding of the problems of preserving biological and landscape diversity, the role of protected areas in solving these problems, as well as their place in the socio-economic development of the region, provides public support for protected areas as objects of national heritage. Despite the fact that environmental education is officially recognized as one of the primary areas of improving the activities of educational systems in many countries, the greatest damage to the environment is caused by the low level of environmental education of the population.

Effective organization of environmental education activities is based on the implementation of the following positions:

- – promoting education and awareness raising on biodiversity issues;
- – reflecting in communication, training and educational programs the importance of plant and animal diversity, as well as the need to preserve them;

– ensuring close interaction between the system of regional protected areas and educational authorities at different levels;
– forming a scientific methodological base for environmental education work, using national and foreign experience;
– ensuring the development and implementation of a program for forming a positive public opinion regarding protected areas, including the popularization of knowledge about protected areas, through holding thematic exhibitions with an environmental education focus, creating photo and video products, appearing in the media, etc.;
– developing a base for effective environmental education work (creating and equipping visitor centers with modern, including interactive, expositions, information centers and a network of environmental, educational trails and tourist routes);
– holding competitions for children and young people to implement joint scientific and environmental projects, and for journalists – competitions for the best publications, programs and films about protected areas;
– introducing the principle of open information about the state of natural monuments, creating new protected areas for the purpose of effective protection of areas with preserved natural landscapes and the biotic component of the ecosystem;
– supporting continuous environmental education, promoting the introduction of issues related to environmental protection and nature conservation into educational programs at all levels of education;
– developing a public environmental movement.

Development and implementation of rules for the organization and implementation of ecotourism in specially protected natural areas

The organization of ecotourism in protected areas is a set of measures taken by tourism organizers aimed at ensuring the implementation of a tourist product within the boundaries of protected areas, taking into account the established regime of special protection of such territory.

The organization and implementation of tourism in protected areas is carried out on the basis of the following criteria and features:

– preservation of unique and typical natural complexes and objects, flora and fauna, natural ecological systems, biodiversity;
– minimization of negative impact on the environment during tourism;
– compliance with the established maximum permissible recreational capacity of protected areas during tourism, its special protection regime;
– implementation of tourism in specially equipped places and routes;
– preservation of cultural heritage sites. (Model regulations on the organization and implementation of tourism, 2023)

The visiting rules are a set of requirements for being in a protected area, applicable to visitors, based on the established regime of its special protection. The rules must include:

a. prohibitions and restrictions for tourists when visiting the protected area, including prohibitions on tourists being outside specially designated areas;

b. information on the general operating mode of the area;
c. the procedure for using tourists' personal equipment when visiting the area;
d. rules for waste disposal on the area;
e. requirements imposed to ensure fire safety on the area;
f. rules for the proper tourists behavior, their appearance and condition;
g. established restrictions on economic and other activities of citizens in the area;
h. other issues related to informing visitors to the protected area about the established regime of its special protection.

The main types of services provided within the framework of tourism in protected areas are:

– excursions, educational and leisure events (fairs, festivals, game competitions, exhibitions, etc.);
– visiting nature museums, visit-centers and nature corners by tourists;
– providing equipped and controlled ecoroutes, ecotrails, viewpoints platforms, camping grounds;
– services of tour guides;
– accommodation services;
– tent camp sites, overnight stays in hoetels, tourist bases, mountain shelters guest houses and other similar accommodation facilities;
– conducting film, video and photography during visits and study of objects of the natural reserve fund, natural and cultural heritage;
– sanitary cleaning and improvement of recreation areas and accommodation of tourists;
– production and sale of food products, organization of public catering facilities for tourists, cooking, sale of ecologically clean local food products;
– services for tourists transportation by various types of transport;
– production and sale of handicrafts and souvenirs to tourists;
– information, educational and consulting services on environmental education;
– rental of tourist equipment, sports equipment;
– organization of collection of permitted mushrooms, berries, medicinal plants.

Organization of bivouacs, toilets, fires, cooking areas should be carried out in designated and marked places in compliance with sanitary and fire safety requirements.

Tourist transport used in ecotours should mainly meet environmental requirements (priority is given to the use of bicycles, motorless boats, paragliders and other environmentally friendly types of transport).

The rules should reflect issues of ensuring the safety of tourism in specially protected natural areas. Ecotrails and ecotours should pass mainly through ecologically favorable natural and cultural landscapes. Ecotrails must provide the possibility of providing first medical aid or organizing timely evacuation for the purpose of providing medical aid, the availability of accessible communication with rescue services and the readiness of rescue services to conduct timely evacuation from all sections of the ecotrail, and possible evacuation routes.

As an additional measure to ensure safety, it is possible to develop a number of alternative routes, the visitation of which can be regulated, reducing excessive loads on particularly popular routes, temporarily closing them and offering tourists other routes of their choice.

The start of an ecotrail should preferably be located near the entrance center of a protected area or in such a way that equipped access roads lead to it.

The ecotrail should be marked according to the level of difficulty, length and danger.

In practice, tourists are obliged to inform the territorial emergency response agencies, as well as the management of a protected area, about the routes of their movement and the number of participants in such movement.

Development of ecotourism infrastructure

The concept of sustainable development should underlie the criteria for selection and approaches to the design, construction, reconstruction and regulation of the functioning of ecotourism infrastructure facilities of protected areas. The main thing in the concept is the forecasting of possible consequences of man-made intervention in the environment (in social and environmental aspects). The philosophy of sustainable design, construction and operation of ecotourism infrastructure facilities is based on maintaining a balance between human needs and the capabilities of the surrounding natural and cultural environment.

Objectives that are solved by sustainable design, construction and operation of ecotourism infrastructure facilities in protected areas:

- infrastructure facilities created on the basis of the concept of sustainable development should be a tool for education and enlightenment, demonstrating the possibilities and advantages of a sustainable approach;
- introduce people to nature and thereby contribute to their spiritual, emotional and intellectual development;
- promote new values and lifestyles in order to establish more harmonious interactions between people and the environment and natural resources at the local, regional and international levels;
- increase specialists competences about the most sustainable technologies, materials, energy, and waste management methods;
- increase the level of cultural behavior of people, develop their sensitivity to careful treatment of the environment;
- dispread the cultural and historical significance of protected areas through partnerships at the local, regional and global levels. (Guide to Designing Infrastructure Facilities in Specially Protected Natural Areas, 2022)

The main principles of sustainable development of ecotourism facilities in protected areas are:

- recognition of interdependence (infrastructure elements interact and influence objects of the surrounding natural environment, causing diverse and extensive consequences);
- responsibility for the consequences of decisions made when creating infrastructure facilities;
- implementation of the concept of reducing production and life waste;
- reliance on "green" energy sources;
- understanding of objective limitations in design and construction, including caused by careful attitude to nature. (Guide to Designing Infrastructure Facilities in Specially Protected Natural Areas, 2022)

Ecotourism infrastructure facilities in protected areas can be divided into two groups: basic (visit-centers, entrance area, parking, administrative block, sanitary and utility facilities, ecotrails, viewpoints platforms, navigation and information facilities, landscaping elements, engineering support facilities) and commercial (accommodation, service facilities, including retail facilities, food services, rental points, recreational and health facilities).

Infrastructure development as one of the objectives of the Ecotourism Development Strategy in Protected Areas requires the development of scientifically sound plans for the development of tourist infrastructure for each protected area as part of the materials of the relevant medium-term programs and involves:

- functional preparation (including information) of excursion ecotrails;
- creation of new and modernization of existing museums and visit-centers;
- development and implementation of programs and projects providing for a set of biotechnical measures that provide an opportunity to demonstrate wild animals to visitors in natural conditions and do not lead to negative changes in natural ecosystems;
- formation of a network of infrastructure facilities to provide services to visitors, including those located outside the boundaries of specially protected natural areas and by attracting investors;
- development of a series of standard projects for elements of tourist infrastructure (camping sites, guest houses, stop points, etc.) under a single brand;
- equipping protected areas with modern information boards and signs;
- provision of federal institutions with transport and other technical means, equipment and gear necessary for the development of tourism;
- priority use of "green technologies" in the formation and operation of ecotourism infrastructure facilities.

Promoting the development of small and medium businesses associated with ecotourism

The development of ecotourism in the protected area should be accompanied by the development and implementation of a set of measures to stimulate the development of small and medium businesses. It is recommended to develop the practice of organizing cycles of training seminars, trainings and master classes for the local population in order to involve local residents in the development of tourism.

Support for environmentally oriented small and medium businesses associated with the development of ecotourism (sale of products of folk crafts and trades, development of guest services, traditional agriculture, excursion and transport services for visitors, microcrediting of target projects, etc.). is of great importance.

Assistance to the development of souvenir crafts and promotion of souvenir products under the brand of protected areas in the relevant market sector, the use of local cultural traditions to organize visitor services, the involvement of private businesses in the creation and maintenance of infrastructure for receiving visitors to protected areas, support for initiatives to create private museums within the boundaries of specially protected natural areas and in close proximity to them, with the aim of including them in the tourist product – all this plays an important role in the process of forming a complex, highly developed territorial tourist product that determines the positive socio-economic effects of the ecotourism development.

Training and professional development of ecotourism specialists

Ecotourism is a complex interdisciplinary field of knowledge and activity, requiring high ethical responsibility not only in relation to humans, but also to nature. An ecotourism specialist must know the basics of tourism, minimize risks to humans and nature, be able to provide assistance to a tourist in emergency situations, ensure travel safety, have environmental knowledge, etc. Creating a tourism product, monitoring its consumption and managing ecotourism are complex and multifunctional professional tasks that require many competencies of specialists. (Romanova G.M. et al.., 2020)

Therefore, the issue of training personnel for ecotourism in protected areas is of great importance and should include the following:

– annual training seminars for managers and specialists (as well as volunteers of the relevant profile) on various aspects of ecotourism development;
– special training for employees providing excursion and supervisory support for tourist groups;
– forming own staff (including those operating seasonally) of tour guides (including volunteers);
– trainings and study trips for managers and specialists to state and foreign national parks and other nature reserves to study the ecotourism experience;
– creating a package of information materials and a series of educational and methodological films for ecotourism specialists;
– attracting students from specialized universities to undergo professional practice on the topic of ecotourism in specially protected natural areas, taking into account the prospects for their further work in this area.

Target indicators of the Ecotourism Development Strategy in Specially Protected Natural Areas

Any strategy should have target indicators as a result of the implementation of the planned objectives and the achievement of the set goal. In the Ecotourism Development Strategy in Specially Protected Natural Areas, such indicators may be the following:

– number of visits to a protected area, thousands persons (local citizens, foreign citizens, in organized groups);
– area of protected areas used for ecotourism purposes, thousands hectares;
– number of tourist routes in a protected area, number;
– number of equipped excursion trails in a protected area / number of trail visitors, number / thousands people;
– number of equipped viewpoints and animal observation sites in a protected area / number of visitors, number / thousands people;
– number of museums and visit-centers in a protected area / number of visitors, number / thousands people;
– number of certified tour guides admitted to work in a protect area, persons;
– number of volunteers involved in ecotourism, persons;
– number of accommodation facilities in a protected area, number;
– number of accommodation facilities outside a protected area used to accommodate visitors, number;

– one-time capacity of accommodation facilities in a protected area, thousands of beds;
– one-time capacity of accommodation facilities outside a protected area used to accommodate visitors, thousands of beds;
– volume of paid services provided to visitors to a protected area;
– number of students from specialized universities training specialists in the tourism sector who have completed professional practice or completed theses on the basis of a protected area, persons;
– number of approved medium-term programs for the development of ecotourism for protected areas, number.

CONCLUSION

Tourism and protected areas are inextricably linked and need each other. For tourism protected areas are resource for development a market segment and satisfaction of tourists needs and wants; for protected areas tourism is a source of the funding necessary for their own management.

Ecotourism is usually a part of the management strategy for a protected area. The degree to which tourism activities are developed depends upon the priority assigned to them by the area managers, who should be guided by a planning document prepared for that purpose. The planning document (or management plan) should be the result of a comprehensive evaluation of the area's natural and cultural resource base. It determines the stresses, their sources and the real threats to the area's natural and cultural integrity, as well as the strategies to reduce these threats. The plan should define the area's long-term management objectives and a zoning scheme that identifies where certain activities may take place (Andy Drumm & Alan Moore, 2005). The Ecotourism Development Strategy is exactly this kind of document.

A protected area management is responsible for the conservation of natural areas. If ecotourism is selected as a conservation strategy, a protected area management needs at least to ensure that planning occurs prior to ecotourism development. It is also critical that the protected area manager understands the risks of operating an ecotourism business in as much as entrepreneurs can be overly optimistic about the area's potential or can overlook the environmental costs associated with an undertaking. The failure of an ecotourism venture could be more damaging to long-term conservation management than if no venture had been undertaken in the first place.

Preservation of nature in its original form is the primary objective of protected areas. Development of ecotourism and creation of recreational infrastructure in reserves and national parks is a multilateral system process, the result of which should be a controlled tourist flow that not only does not harm the environment, but also benefits the protected area, the region and the entire country.

REFERENCES

. Dezhkin, V.V., **Likhatsky**, Yu.P., Snakin, V.V., Fedotov M.P. (2006) *Reserve business: theory and practice*. Moscow, Foundation "Infosphere".

Drumm, A., & Moore, A. (2005) *Ecotourism Development – A Manual for Conservation Planners and Managers Volume 1: An Introduction to Ecotourism Planning*, Second Edition. Arlington, Virginia, USA.

Drumm, A., Moore, A., Soles, A., Patterson, C., & Terborgh, J. E. (2004) *Ecotourism Development – A Manual for Conservation Planners and Managers. Volume II: The Business of Ecotourism Management and Development*. Arlington, Virginia, USA.

Eagles, P. F. J., McCool, S. F., & Haynes, C. D. (2002). *Sustainable Tourism in Protected Areas*. Guidelines for Planning and Management.

Ecotourism Plan for Queensland's Protected Areas 2023-2028. URL: https://parks.des.qld.gov.au/__data/assets/pdf_file/0030/309486/ecotourism-plan-2023-28.pdf

Guide to Designing Infrastructure Facilities in Specially Protected Natural Areas. (2022). URL: https://asi.ru/library/ecobook/120678

Guidelines and Criteria for National Geoparks seeking UNESCO's assistance to join the Global Geoparks Network (2010). URL: http://www.globalgeopark.org/uploadfiles/2012_9_6/ggn2010.pdf

Harris, R., Griffin, T., & Williams, P. (2002). *Sustainable tourism: a global perspective.*

Honey, M. (2008). *Ecotourism and Sustainable Development.*

International experience in the development of ecological tourism in specially protected natural areas. URL: https://asi.ru/library/ecotourism/116263

International Union for Conservation of Nature. URL: https://iucn.org

Ivanov, A. N., & Chizhova, V. P. (2024). *Protected natural areas* (3rd ed.). Urait Publishing House.

. Minenkova V.V. (2022) *Sustainable tourism in mountainous areas: domestic and foreign experience*. Trends in spatial development of modern Russia and priorities of its regulation. Tyumen. P. 700-705.

Minenkova, V. V., & Maksimov, D. V. (2022). *Sustainable development of mountain tourist destinations: Foreign approaches and standards*. Scientific notes of the Crimean Federal University named after V.I. Vernadsky. Geography. *Geology*, *8*(3), 106–117.

Romanova, G. M., Bryukhanova, G. D., & Romanov, M. S. (2020). Staffing for Ecological Tourism as a Factor in the Implementation of the National Project "Ecology". *Recreation and Tourism*, *N4*(8), 43–50.

Selvakumar, P. (2024). *Prabha Garg, Shrinwantu Raha, Sakthiradha M.* Ecotourism. Special Interest Trends for Sustainable Tourism., DOI: 10.4018/979-8-3693-5903-7.ch003

Sustainable Tourism for Development. Guidebook. URL: https://www.e-unwto.org/doi/book/10.18111/9789284415496

Chapter 2
Ethnodiversity, Biodiversity, and Proximity Tourism:
Geographical Superlatives of the Friulian Mountains (Italy) Within the European Alps

Ernst E. H. Steinicke
University of Innsbruck, Austria

Kurt Scharr
University of Innsbruck, Austria

Peter Čede
University of Graz, Austria

ABSTRACT

Based on a theoretical framework about interaction between applied regional geography and proximity tourism, the proposed article reveals geographical superlatives in the Friulian Alps (Italy). As a contribution to applied regional geography, the valorization of this in proximity tourism (marketing) is discussed. The lowest lying glacier, the most colorful ethnolinguistic mosaic, the greatest rainfall volume or the highest depopulation numbers give the Montagna friulana a special geographical position within the Alps. This extraordinary diversity, however, is not expressed in any way in the regional development: Rather, the north of Friuli is one of the structurally weakest areas of the Alpine region. In the sense of applied geography, this contribution should be understood as an advisory function and is intended to inspire professional marketing experts to offer suggestions that geographical particularities can be applied specifically in proximity tourism marketing and therefore to regional development of peripheral areas.

MOTIVATION AND OBJECTIVE

The following study should be seen as a contribution to applied regional geography as well as to integrative geography. During our various scientific activities in the Alps as well as demographic and socio-economic analyses in Friuli Venezia Giulia, Italy's northeast region,[1] we noticed the largely uncharted, special geographic position of the Friulian mountains, specifically the diverse exceptional

DOI: 10.4018/979-8-3693-8232-5.ch002

position of its northern part within the Alps (see the insets/side maps in Fig. 1 and 2). This gave rise to the idea of a regional geographical synthesis with which new and application-oriented insights can be gained through compilation. These relate to natural and cultural geographic issues, which we have brought together from various analytical studies. Such an integration of nature and society in form of a synthesis is the essential characteristic of regional geography. The aim of this study is, on the one hand, to present newly conceived perspectives of regional geography of the Montagna friulana, i.e. the alpine section of Friuli. On the other hand, it should offer suggestions that geographical particularities, specifically ethnodiversity and biodiversity, can be applied in tourism marketing and therefore to regional development of peripheral areas. Above all, the interaction between applied regional geography and proximity tourism should be examined in more detail.

However, this does not mean that our model study on the Montagna friulana can be transferred to other parts of the Alps that are less valued for tourism. While this contribution focuses on geographic superlatives, in other alpine valleys there may be very different unique selling points, which also have an untapped potential for tourism marketing. Ultimately, this work is intended to stimulate thinking in this new direction of applied regional geography.

Nonetheless, we have to emphasize that this work is not to be viewed as a marketing study. From a subject-specific point of view alone, it would therefore be presumptuous to offer recommendations how to promote the geographical exceptionalities of Northern Friuli. In the sense of applied geography, this contribution should be understood as an advisory function and is intended to inspire professional marketing experts. It is therefore understandable that a concrete proposal for the actual implementation of tourism marketing strategies in practice is lacking.

After a conceptual overview showing the current status of research, the study aims to demonstrate the model area Montagna friulana as an exceptional alpine region in terms of natural and cultural characteristics. Beyond a short summary the study concludes by underlining the marketing potential of geographical unique selling points for proximity tourism.

THEORETICAL FRAMEWORK AND STATUS OF RESEARCH

Proximity Tourism

As mentioned above, this study is primarily concerned with new regional-geographical perspectives. It makes sense to apply these to proximity tourism. This particular form of tourism focuses on local destinations, short distances and environmentally friendly means of transportation. It also focuses on the idea of primarily rethinking the nearby, everyday environment (Jeuring & Diaz-Soria, 2017; Rantala et al., 2020). An in-depth look at the small-scale interaction of physio and human geographical features as well as cultural and socio-economic characteristics of the immediate surroundings can also contribute to a stronger regional identity. Despite the topicality and relevance of the topic of "proximity", the existing literature on proximity tourism is quite limited, although this issue was increasingly taken up during the corona pandemic with the restriction of travel freedoms (cf. Salmea et al. 2021; Lebrun et al. 2022).

A generally valid definition of local tourism is not possible due to different concepts based on the parameters of travel time, distance or length of stay. It is not the task here to discuss the various approaches. Instead, this article takes a pragmatic approach to proximity tourism: it encompasses the related possibilities of tourism activities within the Friulian Alps, but also includes the neighboring areas as source areas for proximity tourists. The scope of this intra-regional tourism is therefore limited to a radius of around 100 km.

Disadvantageous effects compared to other forms of tourism, such as the low added value due to the barely relevant influence on the number of overnight stays, will not be discussed in detail here. However, attention is repeatedly drawn to the educational and didactic added value of field trips for pupils, students and other people in the context of life-long learning – which also has an impact on regional development in the broadest sense. In any case, proximity tourism does not take financial capital to distant countries and should promote environmentally friendly travel. As will be explained, the high degree of biological diversity at the valley floors and on the slopes benefits from this.

Applied Regional Geography and methodological challenges

The task here is not to again present historical roots, characteristics or problems of the classical regional geography and the so-called New Regional Geography as well as the concepts of region and relational approaches to regions. These have already been deliberated several times, most recently by Minshull (2017) or Paasi (2020). Nevertheless, it seems advisable to point out that the classical regional geography sees itself as a subfield of geography "that examines nature and human aspects in an integrative regional framework," the New Regional Geography, on the other hand, as a subfield of Human Geography "focusing on the sociocultural construction of regions, their meanings for individual and social identities, and on the power relations embedded in region-building processes" (Paasi 2020, p. 3).

Presenting his dissertation about 'Trentino' in 1898, Cesare Battisti created a completely new regional geography showing natural and societal aspects together side by side for the first time. Therefore, Battisti could be seen as a pioneer of the German *Länderkunde* (Regional Geography). This is also evident from his guides of Trentino landscapes and valleys which can be classified as examples of applied scientific regional geography (e.g. Battisti 1906).

Most recent publications of classical regional geography do not always represent the entire physical and human geographical field of a region, but, as early as 1928, are published problem-centered articles in the sense of Spethmann's "Dynamic Regional Studies," i.e., they concentrate on essential topics of an area (morphology, minority, tourism, etc.). Werner Bätzing's well-known monography of the European Alps (2018) also focuses on sustainability in its deliberations.

Thus, the present study also follows a common thread, namely the unique selling point of an area whereby ethnodiversity and biological diversity play a substantial role. In addition, the authors attached importance to the application orientation of their compilation.

While the primary tasks of applied geography can be found in regionalization, spatial, regional and urban planning as well as in geopolitics, the practical orientation of regional geography is limited to the popularization of geographical knowledge in the sense of a society-related service. The following sections are intended to draw attention to the fact that regional geography also has a dimension in regional development beyond its educational function and economic policy advice. Especially in the field of tourism, where the landscape generally plays an important role, the application of knowledge from regional competence can make a decisive contribution to emphasizing the endogenous potential and thus to the economic upswing of an area. One need only think of the demand for geographical unique characteristics, which are relevant in the plans for destination development and require a profound familiarity with the area (Bieger & Beritelli, 2013; Steinecke & Herntrei, 2017).

The methodological problems of a regional-geographical reflection are obvious, at least since the *Deutscher Geographentag* 1969, when the traditional regional geography was called into question (Meckelein & Borcherdt, 1970). Even if a "neopragmatic perspective" (Kühne & Jenal, 2021) promises theoretical approaches to regional geography, Aschauer (2001) is essentially right when he emphasizes that it does not represent a research discipline with its own theory. Rather, it depicts a form of representation and thus an addressee-oriented offer of spatial images that always remains selective, preliminary, and demonstrative.

Nevertheless, regional geographic research concepts exist such as that of environmental interpretation (Schramm, 1971; Ham, 1992) and, constructed on this, that of landscape interpretation (Lehners & Glawion, 2000). The latter, which discusses the natural and cultural spatial structure of an area that can be experienced on site, is primarily aimed at tourists. To that effect, this approach is to be understood as an instrument for promoting proximity tourism, but it can also make a valid contribution to environmental education. However, to repeat, this article does not address communication techniques such as how environmental or landscape interpretation is to be conveyed, but rather landscape interpretation in the sense of a further development of tourism expansion strategies. The landscape- and recreational value of a region should thereby be emphasized.

In tourism advertising, landscape plays an eminent role. According to Küster (2012, p. 14), this can be characterized as a more or less clearly delimited space of which one has become aware and which is composed of three dimensions that are closely related to one another: that of the natural world (material structures), that of elements of material culture (artefacts and cultural landscape) as well as that of collective history, spatial perceptions and interpretations, traditions, and discourses (cf. also Tieskens et al., 2017, p. 30; Kühne et al., 2024). In addition to historical and physio-geographical aspects, visual-aesthetic attributes are particularly relevant for tourism. As the large number of tourism brochures shows, the starting point is usually attractive scenery images that are intended to create a special landscape experience and thus positive feelings for tourists. Even so, according to Lehners & Glawion (2000, pp. 313f.), "disruptions to the landscape image" can also trigger negative emotions.

This is where this contribution shall ensue. It is up for discussion that not only "beautiful," "picturesque," "exotic," or "bizarre" sceneries and cultural-historical features can be appealing, but also physical and human-geographical uniqueness, unique selling points, exceptional positions, extraordinary linguistic as well as plant and animal societies, records or superlatives – independent of visual-aesthetic characteristics. This can therefore also apply to areas that are generally perceived as "boring" or "monotonous" about which hardly any tourist information exists. The prerequisite for this, however, is a landscape interpretation that is suitable for marketing. Ham (1992) understands this as a communication process that captures the attention of tourists and issues a clear message. The experience of the landscape can thus be significantly intensified beyond the intuitive aesthetic perception if regional-geographic, idiographic insight is conveyed.

Geographical superlatives, sometimes also curiosities, form ideal unique features realizable for tourism, especially for proximity tourism. In the following chapters, the Friulian Alps are presented in an idiographic way with their geographical features, which lend the region an exceptional position within Europe. The contribution, largely based on a hermeneutic approach, is therefore not to be understood as a case study that allows transferability to other (mountain) areas. It can rather – as indicated at the beginning – be understood as a stimulus to develop alternative tourism strategies in which regional geographical knowledge can be applied.

As already stated, our study does not deviate from the methodology of other regional geographies: It is the result of a compilation – i.e. a synthesis of different, primarily own analytical studies – that has been prepared in a recipient-oriented manner with the intention of gaining new findings. In terms of methodology – besides the merging of the current status of research – we analyzed demographic data of ISTAT: tab. 1 shows population losses in all 58 municipalities of the Montagna friulana between 1952 and 2021.

MONTAGNA FRIULANA – AN EXCEPTIONAL AREA

Friuli Venezia Giulia is an administrative region formed after the Second World War, which received the special status of autonomy in 1963. Morphological units converge in this region in a particularly small space and contribute to a considerable diversity of landscapes: In an approximately 100 km long north-south profile, mountains (Alps), hill zone, plain and coast alternate with each other. Its northern part on the three-border area Italy, Slovenia and Austria is located in the Alps (Montagna friulana) and forms the study area.

Northern Friuli or Montagna friulana has not shared in the economic upswing that Friuli as a whole has experienced since 1970 (Steinicke, 1991; Beismann et al., 2022). While lately it has been possible to expand job opportunities considerably – especially in the construction industry – demographic figures of all valleys in northern Friuli still show a sharp downturn in the area's population (cf. Tab. 1). Adverse topographical and socio-agrarian factors, as well as the lack of non-agricultural job opportunities, are the primary causes of "Friuli migrante" (Zanini, 1964). Although the large waves of emigration are gradually tapering off, they have left behind a skewed ratio of older people to the population as a whole. This, in turn, has been accompanied by rising death and falling birth rates and is the reason of the current unfavorable demographic structure.

Moreover, the Montagna friulana is hostile to settlement. While Friuli's frequent earthquakes have limited human room for manoeuvre only to a modest extent, this does not apply to climatological, hydro- and biogeographical conditions (Bonetti, 1960; Steinicke, 1991). As discussed in the next sections, their limiting influence is reflected in all areas of life. Together with the cultural-geographical characteristics, they form geographically unique selling points (Steinicke, 2024; Jelen et al., 2023).

Unique Geographical Selling Points

Human-geographical Particularities. Nowhere in the Alps is the ethnolinguistic mosaic as colorful as in the Montagna friulana in the north of Friuli. It is where the three major language families of Europe meet – Romance, Slavic and Germanic. No fewer than four different autochthonous ethnic groups live side by side and intertwined in this linguistic "Little Europe": Friulians, Italians, Germans and Slovenes. Our research group last presented this ethnic diversity in 2021 (cf. Fig. 1). In some villages of the Valcanale, older locals even use all four languages in everyday life. There in the cultural landscape, various settlement patterns as well as remnants of traditional agricultural facilities and customs are reminiscent of the political affiliation with neighboring Carinthia and Carniola until after World War I (Plautz et al., 2021).

Figure 1. Current weighting of Italian and autochthonous Valcanale populations

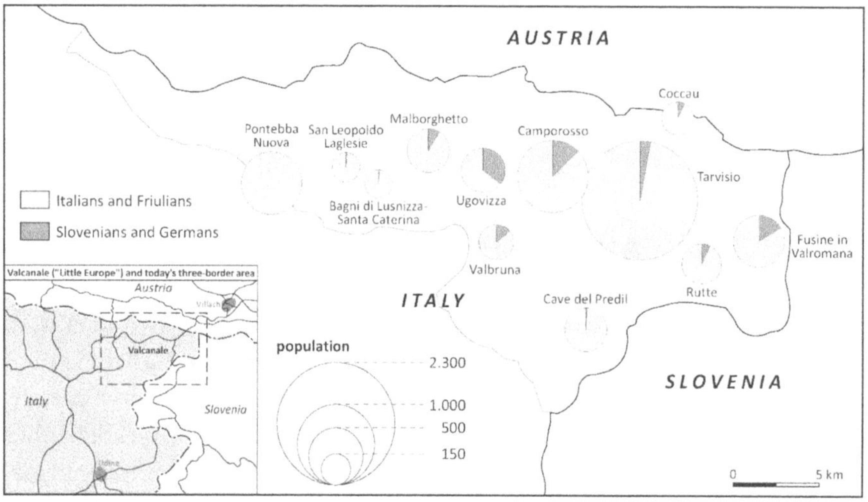

Source: Plautz et al. (2021, p. 10)

In addition, there are communities where objective ethnic characteristics hardly coincide with ethnic identity, as is the case with the German language pockets in Sauris, Timau, and Sappada, and in the Slovene-speaking valleys of Resia and Natisone. In all of these, German or Slovene is seen as a kind of village or valley dialect that is widely incomprehensible to foreigners. Sometimes the inhabitants of the language islands, who remained separated for centuries from the respective ethnic mainland, do not identify themselves in strictly expected national categories, but in more nuanced ways (Plautz et al., 2021; Steinicke et al. 2023). Their dialects, which differ significantly from the respective high-level language, are also codified. As in the Valcanale, however, assimilation into the state language is well advanced in the language islands mentioned, therefore symbolic ethnicity takes a special position. Even though many people no longer speak their autochthonous language, they do try to maintain their ethnic identity in other ways. As Gans (1979, pp. 7-9) suggests, ethnic identity can be expressed in action and/or feeling, which can take new, limitless and symbolic forms, such as maintaining customs and traditions, as well as artefacts in the cultural landscape.

In particular, the typical construction style in the Valcanale as well as in Sauris and Sappada seems to be especially relevant for these rural areas, as it is not only implemented in private residential buildings but also in contemporary public infrastructure projects. Also, the preservation of traditional economic and behavioral patterns (e.g. servitude rights and alpine pastures) as well as the linguistic landscape – i. e. the language of public road signs, advertising billboards, street names, place names, etc. (Landry & Bourhis, 1997) – is emotionally and symbolically significant for autochthons. In addition to this distinct cultural landscape, which corresponds to the ethnic diversity, another highlight of historical and cultural relevance is the spectacular railway from the 19th century, crossing deep canyons with impressive iron bridges - nowadays adapted as a bicycle road, linking the Alps to the North Adria, offering to cyclo-tourists a unique experience.

All these ethnolinguistic groups are threatened by the enormous depopulation that has gripped the Friulian Alps since World War II. Despite its slowing in the last two decades, mainly due to amenity migration (newcomers) developments, this depletion process is stronger than in other parts of the Alps:

Between 1951 and 2021, the Montagna friulana lost well over half of its inhabitants (from 134,380 to 61,759), in the Julian Prealps even close to 80%.

Table 1. Montagna friulana: population losses per year 1952-2021.

Period	Population losses % p.a.	Period	Population losses % p.a.
1952-61	1.05	1992-01	0.72
1962-71	2.29	2002-11	0.20
1972-81	1.09	2012-21	0.17
1982-91	0.97		

Source: ISTAT, 1951ff.; http://demo.istat.it; own calculations

The fact that in-migration in the east of the Italian Alps started later and is progressing more slowly can be ascribed to the previously discussed unfavorable bio-demographic structures. On the other hand – in contrast to the western and central Italian Alps – there are no large conurbations such as Turin, Milan, Brescia or Verona, which are particularly relevant as source areas for immigration into neighboring mountain regions (Beismann et al., 2022). The location of Friuli, for decades directly adjacent to the Iron Curtain, also brought economic and thus demographic disadvantages.

This not only accelerated the decay of the cultural landscape, but also led to the emergence of ghost villages and places threatened with desertion. Their concentration in northern Friuli was still unique in the Alps in 2005 (Čede & Steinicke, 2007). To this day, there are still numerous abandoned places, however, our research group could find only two real ghost towns (cf. Fig. 2).[2] This does not include the largest deserted space in the Alps in the extreme west of the Montagna friulana, where the Vajont dam disaster in 1963 wiped out several villages. The village of Casso thus became the largest deserted disaster area in the Alps (Steinicke, 1991, pp. 142-155).

Figure 2. Depopulated settlements in the north-east of the Friulian Alps in 2022. While there were 18 ghost towns in the area 15 years ago, due to amenity migration there are now only two completely depopulated villages.

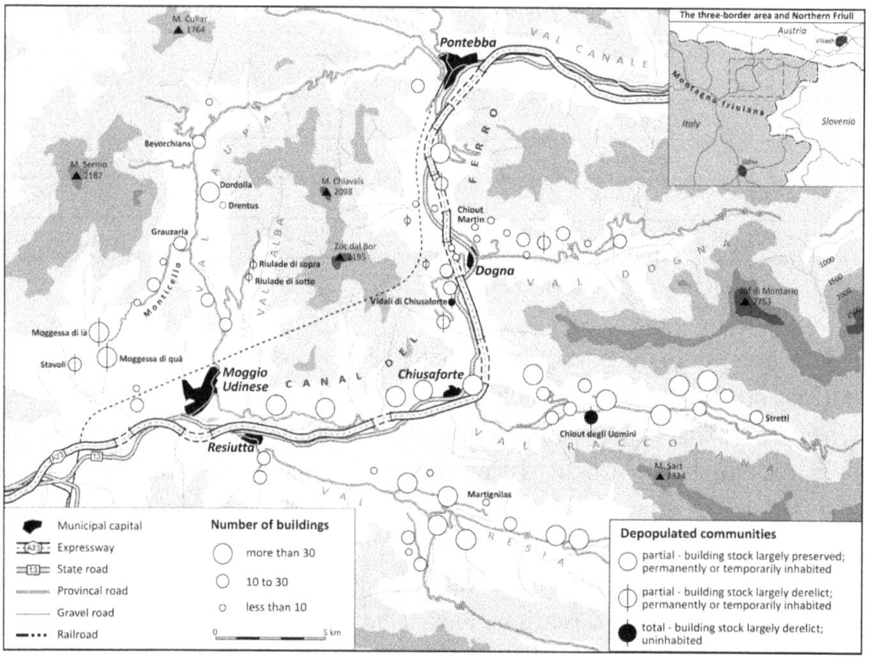

Source: Beismann et al. (2022, p. 76)

Physical-geographical Particularities. Without paying homage to a natural determinism, the search for the reasons for the traditional out-migration must not ignore the physical-geographical framework conditions, especially since these also represent superlatives in the Alpine region. The Montagna friulana is one of the highest precipitation areas in Europe. Annual average values of more than 2,500 mm are not uncommon, even in the valleys. In the east of the Julian Prealps they are around 3,300 mm, where in 1961 in Musi (627 m a.s.l.) 6,103 mm per year was measured, representing the Italian and Alpine-wide precipitation record (Valussi, 1971, pp. 123-129; OSMER, 2020). The immediate proximity to the Adriatic Sea, specifically the mild water of Upper Adriatic lagunas, and the damming high mountain wall explains this phenomenon. The associated high degree of cloud cover also lowers the average temperatures considerably. As a result, small glaciers have survived to this day on the north side of Mt. Canin and the Montasio, the tongues of which reach down to around 1,900 m – a record in the Alps (Colucci et al. 2021; cf. Fig. 3). Karstic processes formed a unique scenario of underground caves, with underground small glaciers and lakes. In the Carnic Prealps, south of the Friulian Dolomites, is one of Europe's deepest speleological practicable vertical caves to be found (Sorgente del Gorgazzo; cf. Cucchi et al., 2018). The deep Cellina Canyon in the same area is not an Alpine superlative, but it is a special feature.

Figure 3. The lowest positioned glacier in the Alps north of Monte Canin (2,587 m) in the Julian Alps. This small glacier, which covered 2.37 km² in 1858 but has since shrunk to 0.38 km², nevertheless still reaches down to around 1,900 m. Despite global warming, the particularly high levels of precipitation in recent decades have even allowed it to grow slightly in length and volume from 2008 at least to 2020.

Photograph: Federico Savoia (August 2021)

The low line of all vegetation boundaries, caused by the climate, contributes to the disadvantages of the Montagna friulana. This fact catches the eye even on a quick drive through the mountainous region of Friuli. Although they are at the same geographical latitude and at a similar height as the central region of South Tyrol or the Swiss Valais, intensive crops are largely absent on the valley floors due to the aforementioned thermal conditions. Commercial viticulture also does not occur in North Friuli above 500 m. Still valid and easily traceable are the examples of other altitude limits of vegetation, enumerated by Gortani & Pittoni (1938, pp. 434-436), Gentilli (1964), or Valussi (1971, pp. 121-123). Like the timber line, they run 300 m and sometimes up to 500 m below those in the Central Alps. This physical-geographical potential in no way favored the settlement process: Only six of the 58 communities reach an altitude level above 1,000 m. Since the valley floors as well as the steeply sloping debris cones are not suitable for settlement, communities and utilized agricultural land are located in sheltered places a few meters above the valley floor or on what is left of terraces. Favorable locations are exceptions, and the expansion of agricultural land is therefore more impeded by nature than elsewhere. Overall, the cultural landscape in the Friulian mountains differs significantly from most Alpine valleys (Steinicke, 1991, pp. 56-60).

Another natural feature of North Friuli is also related to the short distance between the high mountains and the sea – similar to the Mzymta river in the Sochi area of the Caucasus (Scharr et al. 2020, 19f.). In contrast to the hydrographic systems of other Alpine areas, the spring and base of erosion of all Friulian rivers are situated very close together. The average gradient is accordingly high, which gives the torrents an exceptional impact. The valley floors – in Friulian language called *"Cjanâls"* – are thereby taken up by wide gravel areas, which are sometimes under water for several days during the heavy autumn rains. The confluence of the Tagliamento and Fella rivers is particularly impressive, where a gravel surface of several square kilometers is covered by the floodwaters (Hormann 1964).

Although water has been removed from the upper reaches of most of the streams in northern Friuli for energy reasons since the 1950s, straightening the torrents would be extremely costly. Today, the two rivers mentioned, together with the North Tyrolean Lech, are among the most important wild rivers in the Alps. It is therefore easy to understand that these largely unregulated watercourses are home to an above-average number of animal and plant species compared to other European rivers and are therefore of high ecological value (v. Brackel 2022).

The Magredi – a unique biodiversity hotspot. *Magredo* comes from the local Friulian 'terra magra,' meaning arid and poor in water due to the presence of stones, although – as mentioned above – the rainfall pattern makes Northern Friuli the rainiest region in the European Alps. In summer, the arid meadows of the magredi appear barren and scorched by the Sun, defining a landscape like the continental steppes of Eastern Europe (Baccichet, 2016).

The Friulian valley floors are continually unsuitable for settlement and agricultural valorization. The bare gravel surfaces of the larger rivers extend far into the foothills of the mountains and are to some extent not cultivated to this day.[3] Characterized by edaphic aridity and a specific natural botanic landscape, this scree-strewn area specifically on the southern edge of the mountains, which is a striking feature of Friuli and is visible even from space, also represents a unique characteristic within Europe.

In addition to boulders, the environment of the *magredi* is characterized by vast grasslands composed mostly of grasses and small shrubs such as bramble, spiny dogwood (Rhamnus saxatilis), and dog rose. This steppe-like appearance is caused, in addition to the stony soil, by the ruinous action exerted on the vegetation and soil by periodic flooding, which in historical times was even greater. In addition to this, the cause of the steppe appearance is also due to man's action of deforestation and grazing.

CONCLUSION: ETHNODIVERSITY MEETS BIODIVERSITY

For centuries, Friuli has been a classic emigration area and depopulation has almost emptied the valley floors since the interwar period – but especially since the 1950s – apart from the fact that agriculture was only possible in a few areas anyway. This process was favorable for the development of biological diversity in the gravel-laden valley floors. At the same time, however, this was accompanied by a marginalization of economic importance. In contrast to economically active areas, the marginalization and peripheralization of North Friuli also favored the preservation of minority languages. This also explains the link between ethnodiversity and biodiversity.

There is hardly a region in the European Alps where so many natural and cultural characteristics congregate in such a small area. However, this extraordinary diversity of the Montagna friulana is not expressed in any way in regional development: Rather, the north of Friuli is one of the structurally weakest areas of the Alpine arc. The obvious explanation lies on the one hand in the fact that the variety of natural superlatives is not strikingly reflected in a spectacular landscape; on the contrary, it is seen as largely unattractive for tourists.[4] On the other hand, the tradition of exodus due to the departure of the younger, active part of the population has also impaired the area in qualitative terms, so that endogenous regional development has gained very little importance.

The official tourism promotion of Friuli hardly considers the geographical superlatives discussed above and continues to propagate the few winter sports and mountain holiday resorts in the Friulian Dolomites – regardless of the in this respect more attractive neighboring regions. In tourist destination development, however, unique selling points are considered essential for branding. As previously discussed, North Friuli offers a remarkable number of these. In addition, there is regional distinctiveness in cuisine and domestic trades (bacon, cheese, clocks or basket production) as well as models of New Farming and the Slow Food movement, such as those that have already been successfully implemented in Val Aupa (Steinicke, 1991; Beismann et al., 2022). The relevant branding and marketing, however, are inadequate. Not only the multilingual villages and the ghost towns, but also the gravel-laden, bare and uninhabited valley floors, in addition to the established north Friulian tourist centers, could be valorized for tourism as unique features – as has already been done to some extent with the example of the *magredi*, where the Magredi Biotope is a protected natural area, established in 1997.[5] This steppe-like zone is designated "Special Areas of Conservation" (SACs) in November 2013 and at the same time part of the "European Natura 2000 Network for Biodiversity Protection."

Likewise, the thermal disadvantage in the Montagna friulana could be viewed not only negatively, but could also, in connection with the two lowest-lying small glaciers in the Alps, the northern part of the Canin and Montasio mountains, pronounce a striking regional distinctiveness.

Although proximity tourism is less important, a role model could be the neighbouring upper Soča Valley in the Slovenian Julian Alps, where similar topographical and climatic conditions prevail. Since 2010, it has been possible to offer a wide range of outdoor sports (kayaking, climbing, canyoning, skydiving, etc.) and to build up a new tourism (https://soca.valley.com) through the interaction of unique scenic features (e.g. canyons), new tourist infrastructure and marketing. The peripheral valley, which for a long time formed a distinct economic problem area, thus developed into a tourism growth pole (Steinicke et al., 2012; Conolly et al., 2021; Urbanc & Hribar, 2020; Besednjak Valič, 2022).

Thus, it is conceivable that previously unnoticed tourist potential can also be exploited in other remote and economically weak Alpine areas. The work at hand should provide food for thought.

As the explanations on the exceptional area of the Montagna friulana show, proximity tourism can also become a relevant factor in regional development for regions with lower "usual" tourism potential. However, the prerequisite for this is the valorization of the physio- and human geographical particularities also with regard to their accessibility due to shorter distances between the source and destination areas of the tourist flows. As a contribution to sustainability, proximity tourism therefore certainly has future potential, especially against the background of environmentally harmful long-distance travel. As for the Montagna friulana, such a strategy will certainly not initiate mass tourism, but in terms of qualitative and environmentally friendly tourism, the impetus for regional development resulting from the special geographical position is by no means unrealistic. In any extent, the postmodern *zeitgeist* would not contradict that.

REFERENCES

Aschauer, W. 2001. *Landeskunde als adressatenorientierte Form der Darstellung. Ein Plädoyer mit Teilen einer Landeskunde des Landesteils Schleswig*. Forschungen zur Deutschen Landeskunde, 249, Selbstverlag: Flensburg.

Baccichet, M. (Ed.). (2016). *Marsure. Ai piedi della montagna pordenonese*. Pordenone, Proloco Marsure.

Battisti, C. (1898). *Il Trentino. Saggio di geografia fisica e di antropogeografia*. Zippel.

Battisti, C. (1906). *Guide to Trento*. Trento, Soc. Tip. Ed. Trentina.

Bätzing, W. (2018). *Die Alpen. Geschichte und Zukunft einer europäischen Kulturlandschaft*. Darmstadt, wbg Theiss.

Beismann, M., Čede, P., & Steinicke, E. (2022). Neue Pioniere in ostalpinen Peripherräumen. Die Wiederbelebung von Geisterdörfern und partiellen Wüstungen in Friaul. *Geographica Helvetica*, *77*(1), 71–84. https://gh.copernicus.org/articles/77/71/2022/. DOI: 10.5194/gh-77-71-2022

Besednjak Valič, T. (2022). Becoming a part of regional innovation systems: A study of cultural and creative sectors of two Slovenian municipalities. *Journal of Global Policy and Governance*, *11*, 1. DOI: 10.14666/2194-7759-11-1-7

Bieger, T., & Beritelli, P. (2013). *Management von Destinationen* (8th ed.). Oldenbourg. DOI: 10.1524/9783486721188

Bonetti, E. (1960). *Gli sviluppi dell'insediamento nel bacino del Fella con particolare riguardo all'area linguistica mista*. Università degli Studi di Trieste.

Brackel, v. B. (2022, July 16). *Der Tagliamento ist der letzte wilde Fluss im Herzen Europas. Noch.* Süddeutsche Zeitung.

Čede, P., & Steinicke, E. (2007). Ghosttowns in den Ostalpen. Das Phänomen der Entvölkerung im friulanischen Berggebiet. *Geographica Helvetica*, *62*(2), 93–103. DOI: 10.5194/gh-62-93-2007

Colucci, R. R., Žebre, M., Torma, C. Z., Glasser, N. F., Maset, E., Del Gobbo, C., & Pillon, S. (2021). Recent Increases in Winter Snowfall Provide Resilience to Very Small Glaciers in the Julian Alps, Europe. *Atmosphere (Basel)*, *12*(2), 2. DOI: 10.3390/atmos12020263

Connolly, G., Carter, R., Ward-Perkins, D. & Humar, J. (2000). *Managing Tourism Expansion in the Soča Valley (Slovenia)*. Adapted from case study prepared for a Handbook on Destination Management for the World Bank. DOI: 10.1079/tourism.2021.0004

Cucchi, F., Forti, P., Giaconi, M., & Giorgetti, F. (1998). Note idrogeologiche sulle sorgenti del Fiume Livenza. *Atti della Giornata Mondiale dell'Acqua "Acque Sotterranee: Risorsa Invisibile"*. Roma, pp. 51-60.

Gans, H. J. (1979). Symbolic ethnicity: The future of ethnic groups and cultures in America. *Ethnic and Racial Studies*, *2*(1), 1–20. DOI: 10.1080/01419870.1979.9993248

Gentilli, J. (1964). *Il Friuli: I climi*. Udine, Camera di Commercio; Industria e Agricoltura.

Gortani, M., & Pittoni, G. (1938). Lo spopolamento montano nella montagna friulana. In Istituto di Economia Agraria (INEA) (Ed.), *Lo spopolamento montano in Italia. IV Le Alpi Venete* (pp. 395-463). Roma.

Ham, S. (1992). *Environmental Interpretation – A Practical Guide for People with Big Ideas and Small Budgets*. Golden, Colorado.

Hasenauer, L., Čede, P., Jelen, I., & Steinicke, E. (2020). La «piccola Europa» sulle Alpi orientali sta per scomparire La Valcanale (Friuli) nella tarda età moderna. *Geotema*, (Supplemente), 29–44.

Hormann, K. (1964). *Torrenten in Friaul und die Längsprofilentwicklung auf Schottern*. Münchner Geographische Hefte 26. München. https://luoghi.italianbotanicalheritage.com/en/san-quirino-biopark-magredi-lean-meadows/ https://soca.valley.com https://www.uibk.ac.at/de/geographie/dca/publikationen

Jelen, I., Džajić Uršič, E., Čede, P., Beismann, M., & Steinicke, E. (2023). Nuovi montanari, nuove attività e nuove forme di transumanza: l'adattamento dei modelli tradizionali in Val Aupa, Friuli Venezia Giulia. *Documenti geografici*, 3, pp 345-374. DOI: 10.19246/DOCUGEO2281-7549/202303_16

Jeuring, J. H. G., & Diaz-Soria, I. (2017). Introduction: Proximity and intraregional aspects of tourism. *Tourism Geographies*, *19*(1), 4–8. DOI: 10.1080/14616688.2016.1233290

Kühne, O., & Jenal, C. (2021). Neopragmatische Regionale Geographien – eine Annäherung. In O. Kühne, T. Sedelmayer & C. Jenal C. (Eds.), *Louisiana - mediengeographische Beiträge zu einer neopragmatischen Regionalen Geographie* (pp.13-23). Wiesbaden, Springer.

Kühne, O., Weber, F., Berr, K., & Jenal, C. (Eds.). (2024). *Handbuch Landschaft* (2nd ed.). Springer., https://link.springer.com/book/10.1007/978-3-658-42136-6 DOI: 10.1007/978-3-658-42136-6

Küster, H. (2012). *Die Entdeckung der Landschaft. Einführung in eine neue Wissenschaft*. Beck. DOI: 10.17104/9783406637032

Landry, R., & Bourhis, R. Y. (1997). Linguistic landscape and ethnolinguistic vitality: An empirical study. *Journal of Language and Social Psychology*, *16*(1), 23–49. DOI: 10.1177/0261927X970161002

Lebrun, A. M., Corbel, R., & Bouchet, P. (2022). Impacts of Covid-19 on travel intention for summer 2020: A trend in proximity tourism mediated by an attitude towards Covid-19. *Service Business*, *16*(3), 469–501. DOI: 10.1007/s11628-021-00450-z

Lehnes, P., & Glawion, R. (2000). Landschaftsinterpretation – ein Ansatz zur Aufbereitung regionalgeographischer Erkenntnisse für den Tourismus. *Freiburger Geographische Hefte*, *60*, 313–326.

Meckelein, W., & Borcherdt, C. (Eds.). (1970). *Tagungsberichte und wissenschaftliche Abhandlungen: Deutscher Geographentag Kiel, 21. bis 26. Juli 1969*. Steiner.

Minshull, R. (2017). *Regional Geography. Theory and Practice*. Imprint Routledge.

OSMER (Osservatorio Meteorologico Regionale). (2020). *Il clima del Friuli Venezia Giulia*. Palmanova, 2020. https://www.osmer.fvg.it/pubblicazioni.php?ln=

Paasi, A. (2020). Regional Geography. In A. Kobayashi (Ed.), *International Encyclopedia of Human Geography*. https://www.academia.edu/41334596/Paasi_Anssi_2020_Regional_Geography?auto=citations&from=cover_page

Plautz, A.-M., Hasenauer, L., Jelen, I., Čede, P., & Steinicke, E. (2021). Symbolic ethnicity, cultural and linguistic landscape: Remnants of 'Little Europe' in the Valcanale (Northeast Italy). *National Identities*. Advance online publication. DOI: 10.1080/14608944.2021.1894109

Rantala, O., Salmela, T., Valtonen, A., & Höckert, E. (2020). Envisioning tourism and proximity after the Anthropocene. *Sustainability (Basel)*, *12*(10), 3948–3964. DOI: 10.3390/su12103948

Salmela, T., Nevala, H., Nousiainen, M., & Rantala, O. (2021). Tourism proximity: A thematic literature review. *Finnish Journal of Tourism Research*, *17*(1), 46–63. DOI: 10.33351/mt.107997

Scharr, K., Minenkova, V. V., & Steinicke, E. (2020). Sotschi 2014 und danach? Sozial-ökologische Transformation einer Kulturlandschaft am Schwarzen Meer. *Geographische Rundschau*, *6*, 16–21.

Schramm, W. (1971). *The Process and Effects of Mass Communication*.

Spethmann, H. (1928). *Dynamische Länderkunde*. Breslau, Hirt.

Steinecke, A., & Herntrei, M. (2017). *Destinationsmanagement* (2nd ed.). UTB. DOI: 10.36198/9783838548241

Steinicke, E. (1991). *Friaul. Bevölkerung und Ethnizität.* Innsbrucker Geographische Studien 19. Innsbruck, Selbstverlag.

Steinicke, E., Jelen, I. & Čede, P. (2023). Gefährdete und verschwundenen ethnolinguistische Minderheiten in der Alpen-Adria-Region. Das Beispiel der deutschen Sprachinseln und der istrorumänischen Sprachgruppe. *Europa Ethnica*, 80, 1-2, pp. 2-12.

Steinicke, E., Jelen, I., Lieb, G. K., Löffler, R., & Čede, P. (2016). Slovenes in Italy: A fragmented Minority. *European Countryside*, *1*(1), 49–66. DOI: 10.1515/euco-2016-0004

Steinicke, E., Löffler, R., & Čede, P. (2012). In-migration as a new process in demographic problem areas of the Alps. Ghost towns vs. amenity settlements in the alpine border area between Italy and Slovenia. *Erdkunde*, *66*(4), 329–344. DOI: 10.3112/erdkunde.2012.04.04

Tieskens, K. F., Schulp, C. J. E., Levers, C., Lieskovský, J., Kümmerle, T., Plieninger, T., & Verburg, P. H. (2017). Characterizing European cultural landscapes: Accounting for structure, management intensity and value of agricultural and forest landscapes. *Land Use Policy*, *62*, 29–39. DOI: 10.1016/j.landusepol.2016.12.001

Urbanc, M., & Hribar, M. Š. (2020). Livek: A mountainous border area's transformation from a ski paradise to a resilient community. In G. Martinek (Ed.), *Culture and Climate Resilience* (pp. 45-68). Palgrave Studies in Climate Resilient Societies, Basingstoke (UK), Macmillan Publishers. DOI: 10.1007/978-3-030-58403-0_4

Valussi, G. (1971). *Friuli Venezia Giulia.* 2ª ed. riv. e aggiorn. Torino, ed. Torinese.

Zanini, L. (1964). *Friuli migrante*. Udine, ed. Doretti, 1964.

ENDNOTES

1 E.g. https://www.uibk.ac.at/de/geographie/dca/publikationen/

2 Numerous abandoned settlements were visited more frequently during the pandemic.

3 However, viticulture is gradually conquering the area from the south.

4 As far as the Montagna friulana is concerned, apart from Sappada and the eastern Julian Alps international tourism brochures focus more on the landscapes of neighboring regions.

5 https://luoghi.italianbotanicalheritage.com/en/san-quirino-biopark-magredi-lean-meadows/

Chapter 3
Assessment of Tourism Potential of Mountain Forests by the Help of GIS Technologies

Natalia Alekseevna Pikalova
https://orcid.org/0000-0003-2438-6457
Kuban State University, Russia

Daniil Gennadievich Kucheryavyy
Kuban State University, Russia

Ilya Vladimirovich Rudenko
Kuban State University, Russia

Anna Vladimirovna Mamonova
Kuban State University, Russia

Sofya Novikovna Semenova
https://orcid.org/0000-0003-3172-9198
Kuban State University, Russia

Akmaral Kubanychbekovna Kultaeva
Osh Technological University, Kyrgyzstan

ABSTRACT

The mountain forests of the Mostovsky district of the Krasnodar Territory are distinguished by their natural uniqueness and diversity. These forests have significant recreational potential due to their landscape diversity, rich flora and fauna, as well as climatic conditions attractive for recreation. The purpose of this study is to assess the recreational potential of the mountain forests of the Mostovsky district using geoinformation systems (GIS). In modern conditions of changing climatic and environmental factors, as well as in conditions of increasing anthropogenic pressure on natural systems, the task of effective management and rational use of mountain forests is of particular importance. At the same time, the mountain forests of the Mostovsky district are of particular interest, since their recreational potential has not been sufficiently studied, and the possibilities of their use for recreation and tourism have not yet been fully realized. GIS technologies open up new possibilities.

INTRODUCTION

Against the background of the increasing pace of urbanization and industrial development, the society's activization in preservation and increase of natural resources, including forests, is becoming more important. In this context, a special attention should be paid to the recreational potential of forests, which is the main component of natural heritage and resource for meeting the population's demands in

DOI: 10.4018/979-8-3693-8232-5.ch003

having rest and recreation. The extent to which this potential is used directly affects a person's quality of life, health and emotional well-being. It makes the issue especially relevant in the current environment. The importance of assessing the recreational potential of forests is in the possibility of more rational and sustainable use of forests, which, in turn, contributes to the protection of ecosystems and reduces anthropogenic pressure on natural resources. Modern GIS technologies provide unique opportunities for conducting such assessments at a new level, ensuring the accuracy and complexity of data analysis on forests' state and their recreational suitability.

The mountain forests of the Mostovsky district of the Krasnodar Territory stand out for their natural uniqueness and diversity. These forests have significant recreational potential due to their landscape diversity, rich flora and fauna, as well as climatic conditions attractive for recreation. However, the lack of a systematic approach to assessing and planning their use can lead to the degradation of unique ecosystems and to the loss of their recreational value. In this context, the introduction of GIS technologies is especially relevant. Such up-to-day technologies allow performing the multiparametric data analysis, modeling of various scenarios for the use of territories and automating the decision-making process. This makes it possible not only to use the recreational potential of mountain forests more effectively, but also to preserve their biological diversity, maintaining a balance between recreational load and environmental protection measures.

Modern methods of assessing the recreational potential of forests are based on the study of natural resources, their condition and opportunities for recreation use. However, traditional methods do not sufficiently take into account the spatial dynamics and interrelationships between various components of forest ecosystems, which limits their applicability for comprehensive analysis and effective planning of the use of territories.

The purpose of this study is to assess the recreational potential of the mountain forests of the Mostovsky district by the help of Geographic Information Systems (GIS). In today's conditions of changing climatic and environmental factors, as well as in conditions of increasing anthropogenic pressure on natural systems, the task of effective management and rational use of mountain forests is of a particular importance. At the same time, the mountain forests of the Mostovsky district are of particular interest, since their recreational potential has not been sufficiently studied, and the possibilities of using them for recreation and tourism have not yet been well realized. GIS technologies offer opportunities for the analysis, modeling and visualization of natural resources, which provides the necessary platform for the research.

The obtained results will allow us identifying the most promising areas of mountain forests for recreational use. It will be achieved by analyzing spatial data and comprehensively evaluating a variety of parameters such as: accessibility, attractiveness of the landscape, availability of reservoirs and species diversity of plants and animals (Drannikov, 1994). The developed recommendations on the use and protection of the recreational potential of mountain forests will contribute to the sustainable development of the region. The results can be applied to the plans for the integrated development of tourist infrastructure, as well as to create updated and more accurate maps of natural reserves and recreation areas. So, all above mentioned will contribute not only to improving the recreational attractiveness of the Mostovsky district, but also to the formation of ecologically and economically balanced approaches to the correct use of its natural resources.

MATERIALS AND METHODS

The research is based on GIS technologies use in the systematic consideration of spatial and attribute information obtained during the analysis of the Annual Report on the State of Nature Management and Environmental Protection of the Krasnodar Territory for the period of 2008–2023. Mathematical processing of the results was carried out due to the regression analysis by computer technology use.

Geographic Information Systems (GIS) are software that allows to analyze, visualize and process spatial data. Below is a list of GIS used in the present study (ESRI, 2024).

ArcGIS is one of the most popular and powerful programs developed by Esri. It provides extensive capabilities for analyzing and visualizing spatial data such as: maps, satellite imagery, and geodata. ArcGIS contains many tools for creation and editing of maps, analyzing government data, and modeling. In this GIS the necessary calculations were carried out to process satellite images.

Global Mapper is a universal program designed to work with various types of data, including graphical and raster maps. *Global Mapper* has powerful tools for data analysis and processing, such as: combining, separating, filtering and transforming data. This system made it possible to structure and process the available data in text form into a spatial environment.

Google Earth Pro program is a free Google program that provides the ability to view three-dimensional models of the Earth, satellite images and aerial photographs. *Google Earth Pro* also contains tools for: creation and organizing routes, measuring population and users, and viewing historical images. Some types of routes were built and edited in it.

SASPlanet is a program to download and view satellite images of high resolution. It is presented due to: *Google Maps, Satellite images, Yandex Maps, Virtual Earth, Yahoo Maps, Gurtam, OpenStreetMap, eAtlas, iPhone maps,* etc. The program also has the capabilities to create one's own maps and export data. This program allows also to download high permission *RGB* images.

RESULTS AND DISCUSSION

Geographically, the survey area in the system of administrative-territorial structure of the Krasnodar Territory is located in Mostovsky district municipality (Figure 1). The area of our research is located in the southwestern part of the Krasnodar Territory, bordering the Labinsky district, Sochi municipality, the Republic of Adygea, the Karachay-Cherkess Republic and the Republic of Abkhazia (Migunova, 2000). A railway branch runs through the district (Kurgannaya station – Shedok station) and a network of highways connecting it with the Otradnenskiy, Labinskiy districts of the Krasnodar Territory and the Republics of Adygea and Karachay-Cherkess (Atlas: Krasnodar Territory and the Republic of Adygea, 1996).

Figure 1. Geographical location of the Mostovsky district

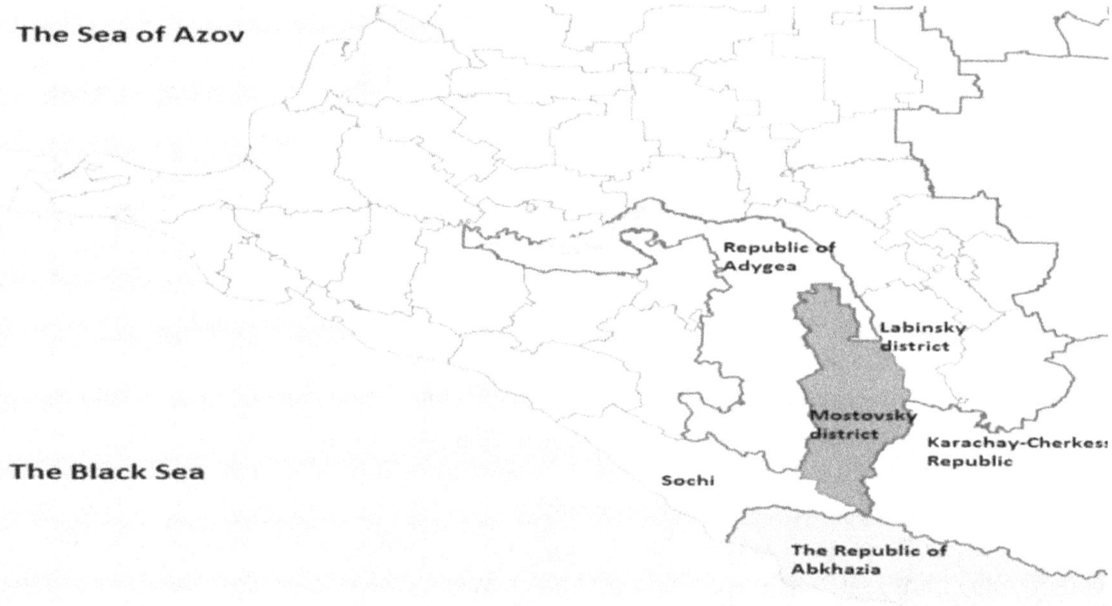

The survey area according to the scheme of physical and geographical zoning of Gvozdetskiy belongs to the westernmost part of the North Caucasian province of the Greater Caucasus. It is characterized by the predominance of foothill and low-mountain landscapes of the forest type.

The relief of the studied territory was formed on several large modes of general Caucasian significance. These are Rocky and Pasture ridges, the depression separating them and the valley of the Malaya Laba River. The ridges have a distinct cuesta structure associated with the monocline occurrence of the rocks composing them. It is especially well seen in the structure of the Rocky Ridge (Regional Geomorphology..., 1979). Mostovsky district is a combination of flat and mountainous areas. The north of the district is dominated by plains and lowlands covered with forests and agricultural land. Mountainous areas with high ridges and steep slopes can be found in the east and south of the district. These territories are dominated by cliffs, gorges and waterfalls, which makes them especially picturesque and attractive for tourists (Migunova, 2000).

Tectonically, the Northwestern Caucasus corresponds to the sinking zone of the meganticlinorium of the Great Caucasus. In accordance with the geomorphological zoning of the North Caucasus (Safronov I.N.), the survey area is within the Upper Pliocene and Quaternary terraces of the area of inclined foothill and intermountain alluvial terraced plains.

In accordance with the territorial planning scheme of the Mostovsky district municipality of the Krasnodar Territory: "The district is located on the territory of 4 geostructural zones of the northwestern Caucasus":

- the monocline of the northern slope of the Caucasian ridge;
- the zone of the northern Peredovoy ridge;
- the Zagedano-Arkhyz depression;

- the Main Ridge zone.

The selected zones stretch parallelly to each other in Caucasian direction and have tectonic contacts almost throughout their entire length. Tectonically and morphologically, the zones are sharply separated from each other. The faults separating them were probably laid down in the Middle Paleozoic, but movements along them, as in weakened zones, resumed until very recently.

Characteristics of geological processes:

1. Endogenous geological processes.

This group of processes includes:

- the seismic processes, including the impact of blasting;
- the rock pressure and displacement of rocks above the mine workings.

Seismicity of the area according to SNCC 22-301-2000 is 9 points.

2. Exogenous geological processes (EGP):

Erosion-accumulative processes of permanent watercourses.

According to the degree of negative impact on national economic facilities (NEF), erosion-accumulative processes of permanent watercourses are the most significant on the territory of the Mostovsky district.

The factors influencing the spatial and temporal patterns of erosion processes are very diverse. As the main, there are the following ones:

- the amount and mode of precipitation;
- the geomorphological conditions for the formation of water flows;
- the properties of rocks and features of their occurrence;
- the nature and features of the soil and vegetation cover.

Comparing the distribution of the average annual precipitation of 608–2000 mm / year and the density of the river network up to 1 km / km^2 or more, the territory of the Mostovsky district can be confined to the southern, mountainous part of the Krasnodar Territory.

Bottom erosion-accumulative processes of permanent waterflows.

These processes do not have a direct impact on the NEF, but significantly affect the activation of other genetic types of EGP, such as: landslide, avalanches, talus.

The regime of bottom accumulation is also characteristic of the lower and sometimes middle currents of the left-bank tributaries of the Laba River, confined to the area of foothill terraced hollow-inclined plains. With the transition to the low-mountain belt, signs of bottom erosion begin to appear in the valleys of smallest rivers, and in some large and medium-sized rivers these signs are detected only in the mid-mountain zone. This is how the signs of accumulation along the river valley are traced. The Laba river valley, where the villages of Andriuki – Psebai are and even a little further south, the division of the riverbed into many branches is perfectly distinct. The total width of low and high floodplains, having an accumulative character, reaches 1–1,5 km. Upstream from the Perevalka village along the valley of the Malaya Laba River is marked by a rather sharp transition to an erosion regime, initially weak, and above the confluence of the Malaya Laba and the Urushten rivers of moderate activity. The upper and separate sections of the middle reaches of the Malaya Laba, Urushten, Khodz rivers and their tributaries are characterized by alternating segments of moderate, intense and weak bottom erosion. All this mainly depends on orographic, lithological, and, possibly, neotectonic factors.

Coastal erosion-accumulative processes of permanent watercourses.

Coastal erosion processes (coastal erosion) are accompanied by the formation of longshore landslide or landslide ledges. The height of which ranges from 3 to 20 meters. Erosion is almost not accompanied by landslide processes only when the floodplain river terraces are washed away. Artificial ramparts and dams built to protect against erosion processes are often subject to lateral erosion (Mostovskaya village, Khodz village). The rate of coastal erosion is determined primarily by the speed of the current and the strength of the enclosing geological rocks. The most eroded are rocks of Quaternary age, the maximum indelible velocities for which range from 0,4–1,2 m / sec. As a result, river terraces are most prone to erosion, especially low ones, the erosion rate of which can reach 25–30 m / year, or up to 10 meters per flood period (the Laba River).

Erosion and accompanying landslide processes along the banks of the Laba River and its tributaries are widely developed. In most cases, high floodplain and low floodplain terraces are subject to erosion here, that causes a high rate of retreat of erosion ledges. The intensity of meandering of the riverbed and, accordingly, the damage of the shores by erosion increases significantly downstream from the latitude of Psebay village. Upstream, the linear lesion does not exceed 10–15%, below it reaches 30–40% or even more. On the Khodz river above Perepravnay stanitsa the erosion was noted 15–20% of the shore's length. Below to the mouth it was noted 40–60%. The settlements, the most susceptible to coastal erosion, are as follows: Mostovskoy settlement, Psebai settlement, Kostromskaya stanitsa, Perepravnaya stanitsa, Besleneevskaya stanitsa, Bagovskaya stanitsa, Vostochny khutor.

And only in the high-altitude zone, where the energy of waterflows is mainly directed at deepening the bottom, the development of coastal erosion is extremely insignificant. At the same time, it is characteristic that the boundaries of the territory of the absence or extremely insignificant development of coastal erosion exactly correspond to the outer boundary of the high-altitude belt.

Erosion-accumulative processes of temporary waterflows.

The nature of the development of the processes of activity of temporary waterflows is influenced by the peculiarities of their feed, regime, flow rate, and geological conditions.

The formation of ravines (the most dangerous type of linear erosion) occurs, as a rule, on steep terraced ledges of rivers, as well as in the foothills, in the areas of development of loose, loosely connected, deluvial (slope) deposits. The activity of waterflows and the formation of erosive forms (gullies, potholes, ravines, etc.) can cause great harm to the national economy, destroying existing ones and preventing the construction of new engineering structures.

On the territory of the Mostovsky district, erosion of temporary waterflows is widespread both in the plain and in the mountainous part, but the forms and intensity are different.

The entire foothill zone of the northern slope of the Main Caucasian Ridge is a hilly area with a widely developed system of gullies and ravines. Frequent downpours create a maximum storm drain of 200–250 mm. Favorable conditions contribute to the widespread growth of dense forests, shrubs and herbaceous vegetation, which strengthen the slopes composed mainly of loose Quaternary and loosely bound Neogene and Paleogene rocks, represented by loams, sands, clays, less often conglomerates of limestones and marls. However, the primary vegetation in many areas is severely disturbed. In some places it is completely destroyed by human activity (deforestation, plowing of slopes). Gully erosion and moderate to severe planar erosion are intensively developing in such areas. A characteristic feature is the uniform development of destroyed, blackened and forested gullies and hollows throughout the territory and extremely uneven, local confined only to places of deforestation, plowing slopes, laying logging roads and cattle trails, the development of gully erosion of the first, less often the second stage. The depth of gullies and ravines there reaches from 0,5 to 40 m. The length is up to several kilometers. The largest ravines are confined to the sides of the Laba River and its tributaries. The dissection of the relief is 0,5–1,0 km / km^2.

The most extensive territory, where the processes of erosion of temporary waterflows are developed, unites the low-mountain, medium-mountain and high-mountain parts of the Mostovsky district. Deep gorges divide it into a system of ridges with strongly incised slopes, where the fragmentation reaches 1–3 km / km^2. The degree of fragmentation most often increases by the height of the mountains. The slopes throughout the territory have an erosively dangerous steepness of 10–15 ° and more. All mountain slopes are composed of terrigenous-carbonate Flysch strata of the Paleogene, Cretaceous and Jurassic, as well as Paleozoic rocks. Heavy rains are frequent and very intense, the maximum amount of rainwater runoff is more than 250 mm. Planar erosion is strong everywhere. Rocky, vegetation-free areas alternate in the mountains with mountainous forests and dense cover of alpine meadows.

Erosion of temporary waterflows is mainly represented by long (often along the length of the entire slope) gullies, crevices and gorges. Shallow gullies and small ravines develop on slopes composed of loose sediments, including scree and landslides.

The erosion of temporary waterflows increases mainly by the height of the mountains. In the low mountains, it is 5–25%, in the middle mountains – 15–30% and in the highlands – 25–48%.

In the low-mountainous part a dense and branched erosion network is represented by gullies and crevices. Their sides are usually hollow-convex in shape from 10–35 ° steepness, as a rule, and are densely forested. The depth of the incision varies from 3 to 25 m, the length is from 100 m to several kilometers.

In the middle mountainous part, steep rectilinear erosive forms as crevices and gorges stretching from the watershed to the foothills of the mountains are highly active. The depth of the incision is tens and the first hundreds of meters. Their sides are usually quite steep (25–50) and often covered by dense vegetation. Branched small forms of temporary waterflows as crevices, gullies develop on the slope of steep erosive forms. The depth varies from 3–10 m to 40–60 m, the hollow-convex sides have a steepness of 25–45 °, their length is up to 0,8–1,0 km.

In the highlands, the development of such large erosive forms of temporary waterflows as gorges, crevices cutting through the steep slopes of watersheds for their entire length is characteristic. The depth of the gorges is 100–200 m or even more, the sides are steep (35–60 °), with sparse vegetation, complicated by small V-shaped cuts of temporary waterflows, from 1 to 10 m deep, from 10 to 40 m wide.

Small shapes are also confined to the ridge parts of the ridges – V-shaped shallow (up to 30 m) beam cuts with a width of 30 to 80 m, with a length of up to 200–300 m.

The prevalence and activity of gully erosion is determined by a complex combination of natural factors and human economic activity.

Fluvial type flooding occurs on the territory of the Mostovsky district. Low and high floodplains of rivers with headwaters in the mountainous part are prone to flooding. The northern slope of the Western Caucasus is characterized by the presence of a large number of developed river valleys, along which flooding is mainly going on.

On the rivers (Laba, Malaya Laba, Khodz, Chokhrak), severe floods are observed almost annually. The floodplain is flooded for 5–7 days, usually during the passage of spring and summer floods. Rivers sometimes overflow their banks during large autumn floods. The duration of flooding of low-lying areas adjacent to rivers varies from several hours to 1–2 days. The water lifting speed can reach 60 cm / h. Floods with an increase in the level above the critical level by more than 50 cm are rarely repeated, about once every 50–70 years.

Parts of the territories of settlements are most susceptible to the flooding process: Mostovsky settlement, Shedok village, Psebai settlement, Perepravnaya stanitsa, Bagovskaya stanitsa, Tsentralniy khutor, Vostochny settlement, Solyonoe village.

Mudflow processes.

In almost the entire territory of the district, along the gullies and river valleys, relief forms close to the village are noted. Here, mudflows are mostly of a transitional nature between fluvial flooding and mudflows, the so-called proluvial-mudflow flooding, characterized by a weak saturation of water flows with a solid component.

The water component of these mudflows is formed mainly due to heavy rains, less often as a result of the combined effect of snowmelt and heavy rains.

A complex of elementary exogenous geological processes is involved in the formation of the solid component: planar flushing, landslides, scree, landslides.

The frequency of manifestations of mudflow processes is almost annually, with a change in the intensity of manifestation depending on the amount of precipitation.

Gravitational processes. Landslides.

The main characteristics of the landslide process are: the degree of damage to the territory, its activity and the intensity of development over time.

Spatially, landslides are unevenly developed. The predominant number of them is confined to the valleys of rivers and their tributaries, the foothills and the northern slopes of the Main Caucasian Ridge. Lithologically, landslides develop in clay deposits of fine-grained molasses.

In the mountainous part, active landslide forms capture 70–80% of the underlying sediments. Due to the predominantly clay composition of landslide deposits, the group of active deformations is dominated by «consistent» landslides, i.e. «landslides-flows», the remaining part is the predominant «landslides-blocks» and «landslides-collapses» confined to steep coastal ledges of large and small mountain rivers. Landslides are streams, mostly confined to the more prominent (6–30 °) sides of rivers and individual gullies.

The formation of landslide arrays and individual landslides depends on the sum of numerous factors, such as the geomorphology of the slopes, the lithological composition of the rocks composing the slope, geological and hydrogeological features, climatic factors, the hydrological regime of waterflows, human activity, etc.

It should be noted that most landslides confined to the sides of watercourses are activated as a result of powerful «rimming» of coastal ledges and slopes by the water flow, especially in the most full-flowing periods (spring-summer flood and autumn rain maximum).

The processes of lateral erosion of rivers and the landslide process are so closely interrelated that it is not possible to distinguish them graphically as different subdistricts. Therefore, two of these processes (coastal erosive waterflows and landslide confined to the sides of waterflows) will be combined into one subdistrict.

Avalanche-scree processes.

On the territory of the studied district, they are developed sporadically in the southern mountainous part, and are confined to zones of tectonic faults or to zones of active lateral erosion of rivers.

The danger of avalanche-talus processes is mainly in the fact that landslides, even of not very large volumes, can lead to the formation of dammed lakes in the narrow bottoms of gorges, the breakthrough of which can lead to the passage of mudflow floods.

Landslide processes in river valleys, as a rule, are confined to erosion ledges and to areas of sharp narrowing of river valleys. Activation of these processes mainly occurs in the winter-spring period of the year, when collapse agents are particularly active, such as: increased moisture content of rocks, frosty weathering, lateral erosion of rivers, etc.

In some cases, the formation and activation of avalanche-talus processes is directly related to anthropogenic activity (for example: construction of automobile and railway roads).

Such territories of settlements as: Psebai village, Andriuki village, Bagovskaya stanitsa, Perevalka stanitsa are the most susceptible ones to landslide and talus processes.

According to Shalnev (2007), the territory represents cultivated foothill and cultural-natural landscapes of longitudinal depressions in the Jurassic sandy-clay thickness composed of sandstones, shales with oak-beech forests on mountain-wooded brown soils.

The landscapes coincide with the boundaries of the low-mountain and medium-mountain landscapes with broad-leaved forests on gray and brown forest soils. This is an area with a low degree of anthropogenic disturbance of the ecological balance. The predominant types of environmental degradation of soils are compaction.

According to the scheme of soil-geographical zoning of the Russian Federation, the territory of the Mostovsky district is included into two soil-bioclimatic regions: the Western brown-earth forest and the Central forest-steppe and steppe, within which smaller taxonomic units are distinguished: zones, provinces and districts.

Climate.

In climatic terms (according to B.P. Alisov's zoning), the research area belongs to the mountainous region of the Great Caucasus of the temperate climatic zone (Physical Geography..., 2000).

An important factor affecting the climate of the area is the atmosphere circulation. In this regard, the air flows that affect the climate, can be different in their origin, physical properties and direction of movement. The nearest weather station to the research site is «Labinsk». According to the Labinsk weather station, the prevailing winds throughout the year are from the east and less often the winds of the eastern directions. At the same time, in March-May period, the frequency of eastern winds increases, and they become predominant. This period also accounts for the highest average monthly wind speed of 3,4 m / s, with an average wind speed of 2,7 m / s during the year.

Spring is early, humid and with cold weather return. Cyclonic activity and meridional exchange of air masses in spring and early summer cause a noticeable increase in the number of thunderstorms and heavy rains during the period.

The weakening of the inter-latitude exchange in July-August and the invasion of the continental tropical air of the steppes and deserts provides dry hot weather in summer and stable warm weather in autumn. Breakouts of western and southern cyclones periodically disrupt such weather with heavy rainfall.

The air temperature.

The average annual air temperature in the district is 10,4 °C. The average air temperature of the coldest month (January) is minus 1,9 °C; the average air temperature of the warmest month (July) is 22,2 °C (Table 1).

Table 1. Average and extreme values of air temperature, °C

Station	I	II	III	IV	V	VI	VII	VIII	IX	X	XI	XII	Year
MS Labinsk													
Medium	-1,9	-0,8	4,1	10,6	16,1	19,3	22,2	21,9	16,9	11,3	5,0	0,2	10,4
The absolute maximum	19	22	31	36	36	36	40	42	39	35	32	29	42
Average maximum	3,3	4,7	10,2	17,1	22,8	26,0	29,3	29,2	24,1	18,3	11,1	5,5	16,8
The average of the absolute maxima	12	14	22	27	30	33	35	35	32	27	21	15	36
The absolute minimum	-30	-32	-22	-10	-2	3	8	4	-2	-11	-24	-28	-32
The average minimum	-5,8	-5,6	-1,3	4,5	9,8	12,9	15,4	14,9	10,8	5,7	0,6	-3,5	4,9
The average of the absolute minimums	-16	-16	-10	-2	4	9	12	10	5	-2	-8	-13	-20

The absolute maximum is plus 42 °C, the absolute minimum reaches minus 32 °C. The coldest period, when the average daily air temperature stays below 0 °C, lasts about 40 days. During this period, positive daytime temperatures are possible. In the absence of persistent frosts, frosty periods of short duration may be observed in the territory of the survey area, when the air temperature does not rise above 0 °C for several days.

The first frosts are observed, on average, in the second half of October. The last ones are observed in the first half of April. In some years, the first frosts are possible in the second half of September. The last ones occur in the first half of May, but the probability is not very great.

Precipitation.

The distribution of precipitation is greatly influenced by the orographic features of the terrain and the underlying surface. The average annual precipitation is 742 mm (Table 2).

Table 2. Average precipitation values, mm

Station	I	II	III	IV	V	VI	VII	VIII	IX	X	XI	XII	XI–III	IV–X	Year
Labinsk	37	47	57	60	72	91	71	57	59	60	68	63	272	470	742

During the warm period of the year, from April to October, 470 mm of precipitation falls down (63% of the annual). During the cold period, from November to March, 272 mm (37%) of precipitation falls down. The precipitation amounts may deviate significantly from the average value from year to year. The summer precipitation regime is mainly stormwater. In some cases, precipitation may be dangerous. The daily maximum precipitation in the Labinsk weather station is 108 mm. Heavy rainfall and mountain rivers are important factors in providing the area's water resources. However, periodic droughts can create problems with water supply, requiring additional measures to manage water resources.

The climatic risks of the Mostovsky district are associated with possible floods and landslides that may result from heavy rains and snowmelt. These consequences can lead to the destruction of infrastructure and threats to the population. In order to reduce risks, some measures are being taken to strengthen river banks and create drainage systems (Shcherbakov, 2003).

Hydrography of the research area.

According to the hydrogeological zoning, the studied territory belongs to the Great Caucasian basin of reservoir-block pressure waters, represented by the Central North Caucasian Hydrogeological region. Subdistrict is Labino-Zintsarskiy, Labinsk zone, Belo-Urupskaya subzone (State Geological..., 2013). 17 rivers flow through the territory of the district. The total length of the riverbed is 566 km. There are also more than 100 ponds on the territory of the district, with a total area of about 250 hectares. The rivers of the area are fed by: precipitation, groundwater, snowmelt. Feed conditions may be redistributed throughout the year, but the formation of high water is clearly visible mainly as a result of snowmelt. The beginning of the flood usually is on the second decade of March. Its maximum takes place in the second decade of May. The snowfall on the rivers is usually observed in August-September.

The rivers are mostly shallow, the average water flow in them is usually about 0,4 m^3 / s, but during the rainy season harmless rivers turn into powerful streams, and this figure increases several times reaching 4,5 m^3 / s.

A significant part of the water balance is underground. In the slopes of the Bolshoy Tkhach massif there are multiple karst form of drainage cave systems formed by meltwater

The Laba is the largest tributary of the Kuban River. The beginning of it is considered to be the confluence of the Bolshaya and Malaya Laba Rivers. The valley of the Laba River is 214 km, and if we calculate its length together with the Bolshaya Laba, then it is 341 km. The total area of its catchment area is 12500 km^2.

The sources of the Bolshaya Laba River are glaciers of the top of Mount Abytskha. The Malaya Laba originates at the snowy peaks of the Aishkho and the Pseshkho glacier. The total area of glaciers feeding these rivers is about 15 km^2. The catchment area of the Laba River is not completely flat in outline. The left-bank part of the basin is larger both in area and in the number of tributaries. In total, the Laba receives 476 tributaries. Their total length is 10500 km. The largest left tributaries are the Khodz, Chokhrak, Fars and Giaga Rivers. The largest right-bank tributary is the Chamlyk River. Almost all tributaries of the Laba River are of high-water during floods. The rest of the time they become shallow and become low-water. The nature of the Laba River valley, its flow, water regime and chemical composition of water change from source to mouth, as the river intersects a number of different geographical landscapes (Fig. 2).

The Bolshaya Laba and the Malaya Laba Rivers in the upper reaches laid their channels through areas of high-altitude mountains composed of granites, gneisses and shale-sand rocks. Their valleys here have the appearance of narrow deep canyons, in which crystal-clear waters rush rapidly and noisily, bubbling on rapids and rifts. Then these rivers cross a strip of wooded high folded mountain ranges composed of Paleozoic sandstones, shales and limestones. The wild rocky gorges of the Bolshaya and Malaya Laba Rivers expand in places, forming picturesque mountain valleys.

The valley of the Laba River is already quite wide here and carries three terraces. Approximately from the Mostovsky settlement it expands up to 6 km. Its right bank is higher than the left, and not far from the Vladimirskaya stanitsa, both banks of the Laba River become flat and gentle. At the Vozdvizhenskaya stanitsa, the right bank of the river rises again strongly. Below the Kaladzhinskaya stanitsa, the Laba River starts to break into branches, forming numerous islands. The riverbed here is coarse-pebbled or rocky. The floodplain of the river and the islets are overgrown with deciduous forest and shrubs. In the lower reaches, the riverbed is very winding. The fast and high-water Laba River strongly erodes its shores, threatening the populated areas. Starting approximately from the Koshekhabl aul, the river paves its course in loose ancient Quaternary sediments and flows into the Kuban River opposite Ust-Labinsk city.

The water regime of the Laba River is peculiar. Water consumption varies significantly by season. The river can overflow in almost all seasons of the year. The causes of flooding are spring snowmelt, summer melting of glaciers and autumn showers. The highest level and consumption of water in the Laba river is observed in spring-summer period. The lowest levels and expenses occur in autumn-winter period.

Groundwater is also important in the feed of the Laba River. In places, powerful springs are wedged in the river valley. In winter, the Laba River freezes for 1–3 months, but no solid ice cover forms in the upper reaches of the rapids (Pen'kovskiy, 2006).

Thanks to the river network, 6 large lakes have been formed in the Mostovsky district, and only the Krugloe and the Chernoe lakes have recreational significance.

The Krugloe lake (Akhmetovskaya stanitsa) is located 1 km north of the Gornoe village, on the Black Sea ridge, at the bottom of a deep karst basin, at an altitude of 790 m above sea level. The northern slope breaks off to the lake. A layer of gypsum is exposed in it. Other slopes are gentler and covered with woody vegetation. The lake has: a maximum length of 241 m, a width of 123 m, a maximum depth of 18 meters, a water mirror area of 23000 m^2, the volume of water in it is 250000 m^3. The large size of the karst basin, its great depth, flat bottom and the presence of a coastal underwater terrace indicate the ancient age of this lake.

The Chernoe lake was formed on a large karst depression stretched from west to east. The walls are steep and their height is 60–70 meters or even more. The maximum length of the lake is 450 m, the width is 200 m, the maximum depth is 8 m. Beech and oak forests grow around the lake. Spreading weeping willows lean over the always calm surface of the water. The lakes are fed by underground water. One of the springs is located on the southwestern shore. The water in it is always cold and pleasant to the taste. In the lake itself, the water is fresh with a faint smell of hydrogen sulfide (Klimacheva, 2009).

Flora and fauna.

Geobotanically, the natural territories are located in the West Caucasian province of the Cherkessky district of the Psebaysky area (Atlas: Krasnodar Territory..., 1996). The territory of the research area covers almost the full profile of high-altitude natural landscapes typical of the Western Caucasus. Due to the altitude location, climatic conditions and vegetation composition, the forest belt can be divided into sub-belts: 1) upper montane (1700–2200 m above sea level), 2) middle montane (900–1700–2000 m above sea level), 3) lower montane (500–900 m above sea level) (Red Book of the Krasnodar Territory, 2017).

The territories of our research are located in the mountain forest belt, which is characterized by broad-leaved forests. The dominant position in mixed broad-leaved forests is occupied by petiolate oak. It is accompanied by ash, elm, linden, hornbeam, maple, etc. In certain habitats, some species become dominant, forming associated forests. Common hazel (*Corylus avellana*), goat honeysuckle (*Lonicera caprifolium*), yellow rhododendron (*Rhododendron luteum*), viburnum vulgaris (*Viburnum opulus)*, European birch bark (*Euonymus europaeus*), Caucasian blackberry (*Rubus caucasicus*), etc. are noted in the undergrowth (Zernov, 2006; Tilba, 1981).

Beech and mixed fir-beech forests, as well as oak forests with an admixture of elm, maple, pear, and beech are common in the middle mountain belt. Beech does not have the property of a vertically zoned forest breed (Sergeeva, 2004; Tilba, 1981).

Recreational resources of mountain forests.

Mountain forests are one of the most valuable natural resources with huge potential for the development of recreational tourism, and represent unique natural ecosystems that attract tourists by their beauty and diversity of flora and fauna. They are places for various types of recreation: hiking, cycling, picnics, hunting and fishing (Ivonin, 2000).

Assessment of the recreational potential of mountain forests is a critically important task for many regions, including the Mostovsky district. Mountain forests have unique ecosystems that can offer a range of recreational opportunities – from hiking and camping to more specialized activities such as: wildlife watching, mountain climbing or meditative walks. In conditions of limited time and resources, it is necessary to approach the assessment of these potentials based on a combination of qualitative and quantitative methods. The use of geographic information systems (GIS) in this area opens up new opportunities for detailed analysis and visualization of data necessary for making informed management decisions. By the help of GIS, it is possible to create detailed maps that identify the most suitable areas for various types of recreational activity, as well as identify restrictions and possible conflict situations between different types of use of the territory (Klimacheva, 2009).

Figure 2. Hydrography of the Mostovsky district network

The development of a methodology for assessing the recreational potential of mountain forests requires an integrated approach that takes into account both the natural characteristics of the region and the social aspects of its use. Among the main parameters for the analysis can be identified: topography, forest cover, availability of reservoirs, climatic conditions, accessibility of the territory and its proximity to settlements. Due to GIS technologies, data on all these parameters can be integrated. They make possible to create multifactorial models to assess the potential of recreational areas. For example, using topographic data in combination with data on forest cover and the availability of trails can identify the most attractive hiking routes. At the same time, climate data can give an idea of seasonal restrictions for different types of activities. It is also important to take into account socio-economic aspects such as: public preferences, accessibility of infrastructure and economic importance of recreation for the region (Klimacheva, 2009).

The practical application of the results of the assessment of the recreational potential of mountain forests by the help of GIS technologies can not only contribute to the development of tourism, but also ensure the sustainable use of natural resources. The analysis of the received data makes it possible to develop recommendations for improving the recreational potential of the area, such as: the creation of new routes, the establishment of zones for various types of activity, the development of measures to preserve natural landscapes and biodiversity. For example, based on the analysis, it is possible to identify areas where infrastructure needs to be built (recreation centers, campgrounds, information tables), as well as identify places where access restrictions are required to protect ecosystems. It is important that the results of these studies are available to a wide range of stakeholders, including local authorities, travel companies and public organizations. This will create a more informed approach to the management of recreational resources and their sustainable development, which will ultimately improve the quality of life of the local population and the economic sustainability of the region (Ivonin, 2000).

However, the growing popularity of tourism can have a negative impact on mountain forests. Improper use of resources, mass visits and insufficient protection can lead to depletion of forest reserves, soil and water pollution, as well as destruction of ecosystems. In order to preserve and sustainably use the recreational resources of mountain forests, it is necessary to take a set of measures to protect nature, control tourism activities, develop infrastructure and conduct educational lessons. This is the only way to ensure the preservation of the richness and beauty of mountain forests for future generations (Pin'kovskiy, 2009).

The Caucasian State Biosphere Reserve is located on the Mostovsky district territory. It is located on the northern and southern slopes of the Western Caucasus. The territory is a typically mountainous landscape and is characterized by elevations from 260 to 3360 meters above sea level.

The total forest arca of thc Eastern department of the reserve (located mostly in the Mostovsky district) is 41410 hectares (48,1% of the territory of the department). Among them there are: the mixed forest accounts for 4489 hectares (5,2%), coniferous forest – 25116 hectares (29,2%), deciduous forest – 11805 hectares (13,7%). Woody vegetation species are mainly represented by Caucasian fir, beech, Litvinov birch, alpine maple, gray alder, rhododendron.

Depending on the altitude zone, the following types of forests grow on the territory of the research area:

- the mixed broad-leaved forests with a predominance of elm, poplar, pear, alder are located at an altitude of 500 to 1600 m above sea level;
- the floodplain forests dominated by *Alnus incana* or *A. glutinosa* and the presence of *Aceg campestre, Carpinus betulus, Fraxinus excelsior, Ulmus glbra*, etc. (Grudzinskaya, 1953);

- the oak forests dominated by *Quercus robur* and *Q. petraea* are distributed within 500–1000 m above sea level. This type is widespread and diverse in composition: Mesophilic oak-hornbeam forests with *Quercus petraea* and *Carpinus betulus*. According to syntaxonomy they are included in the class *Querca-Fagetea* Br.-Bl. et Vlieger in Vlieger 1937 of the union *Carpino betuli-Quercion petraea* Grebenshchikov et al. 1990 (Frenchov, 2007). *Fraxinus excelsior, Tilia begoniifolia, Ulmus glabra, Pyrus caucasica*, etc. are present in the stand. *Swida australis, Euonymus europaea, Corylus avellana, Cornus mas* are common in the undergrowth;
- the beech forests are distributed within 600–900 m above sea level and are represented, as a rule, by pure stands. In the forests where logging was practiced, beech is accompanied by *Carpinus betulus, Ulmus glabra, Acer laetum, A. pseudoplatanus, A. platanoides, Fraxinus excelsior,* etc. The shrub layer in native beech forests is represented by *Sambucus nigra, Rhododendron ponticum, Vaccinium arctostaphylos, Euonymus latifolia,* etc. The floral composition of the herbaceous tier is characterized by a rather low diversity. *Festuca drymeja, Rubus caucasicus, Dryopteris filix-mas, Athyrium filix-femina* dominate. The most common type of beech forests are mixed-grass and aspen beech forests. The average height of the stand is 30–43 m (Orlov, 1953);
- the mixed beech-fir stands with *Fagus orientalis* and *Abies nordmanniana* are widespread within 800–1600 m above sea level. They belong to the associations of the union *Abieti-Fagetion orientalis* Korotkov et Belonovskaya 1987 (Korotkov, Belonovskaya, 1987). Other arboreal species in these communities include *Acer pseudoplatanus, A. platanoides, Tilia begoniifolia, Ulmus glabra, Picea orientalis*, etc. At an altitude of 800–1300 m above sea level, a relic of the Tertiary period *Taxus baccata* may occur. The shrub layer is usually not distinct;
- the pure fir stands with *Abies nordmanniana* have a limited distribution, mainly in the upper montane forest belt, fir and beech-fir forests with *Acer trautvetteri* (acc. *Petasito albae-Abietetum nordmannianae* Francuzov 2006) prevail, which at their upper boundary smoothly transition into crooked woodland, or replace it as a result of artificial reduction in the past of the upper boundary of the forest up to 1500–1700 m above sea level;
- the pine forests *Pinus kochiana* are mainly in the ridge part of the ranges or confined to the southern, strongly heated, mostly steep, rocky slopes. They are often located on narrow rocky ridges, rocks and scree. Pine forests usually occupy heights from 1700 to 2000 m above sea level, but they can descend lower (Sosnin, 1939);
- the high-altitude maple trees from *Asega trautvetteri* do not occupy large areas, appearing among fir forests in small fragments from a height of 1700 m in low shady areas, snow-covered in winter and abundantly moistened in summer;
- the subalpine birch forests (birch woodlands) are represented by *Betula litwinowii, Sorbus aucuparia, Salix caprea, Asega trautvetteri.* They are currently being formed to replace the artificially reduced upper forest boundary in the past for grazing. The recovery rate of this type of forest is low.

The recreational attractiveness of the Mostovsky district.

There are numerous protected areas in the district. This fact makes the Mostovsky district attractive for tourists and outdoor enthusiasts, contributing to the development of recreational potential. The mountains, forests and water resources of the area provide exceptional opportunities for hiking, mountaineering, skiing and other activities. There are also many cultural and historical monuments located in the district, which provide additional opportunities for cultural tourism (Bereznikova, 2007).

The mountain forests of the Mostovsky district are distinguished by a high level of biological diversity and ecological significance. These forests are complex ecosystems where endemics, relics and protected plants grow. The fauna of the mountain forests is also very diverse: bears, wolves, lynxes, wild boars, deer, various species of birds and small mammals live there. The ecosystem of mountain forests has important ecological functions, including: soil protection from erosion, water regulation, air purification and biodiversity conservation. These factors make the mountain forests of the region an important object for environmental protection measures and at the same time are of great interest for eco-tourism. The conservation and sustainable use of the recreational potential of these forests requires an integrated approach that takes into account both environmental and recreational aspects (Annenkova, 2009). In addition, mineral springs and therapeutic mud attract people who want to improve their health on the area (Anisimov, 1993).

It is important to note that the proper use of recreational potential requires the creation of infrastructure that would ensure the convenience and safety of tourists, as well as compliance with environmental standards. There are already routes and trails for tourists in the area, but their number and arrangement should be expanded taking into account the modern ecotourism requirements. In this context, the use of GIS technologies can significantly facilitate the process of planning and optimizing recreational routes, as well as monitoring and managing tourist loads on the ecosystem of mountain forests. The cooperation between local authorities and public organizations also is important in promoting sustainable environmental management and preserving the unique natural complexes of the area (Scientifically based recommendations…, 2006).

SUBSTANTIATION OF CRITERIA FOR ASSESSING THE POTENTIAL OF MOUNTAIN FORESTS

The definition of criteria for assessing the recreational potential of mountain forests is an important step in the development of a methodology for such an analysis. The first step in this process is the analysis of existing methodologies and approaches to assessing the recreational potential of forest ecosystems. The scientific literature presents various aspects of the recreational use of forests, including ecological, geographical, aesthetic, socio-cultural and economic. Based on the analysis of assessment methods, several criteria can be identified that are most often used to assess the recreational potential of forests: natural potential, accessibility of the territory, infrastructural equipment and the level of anthropogenic load. The natural potential of forests includes such parameters as: the diversity of landscapes, types and age categories of forests, the presence of reservoirs and rare species of plants and animals. These factors determine the attractiveness of the territory for tourists and vacationers. Accessibility of the territory also is important, as the presence of roads and trails increases the likelihood of visiting forests by different groups of the population. The infrastructural equipment includes the development of tourist infrastructure, the availability of places to accommodate vacationers, equipped playgrounds for outdoor activities and other amenities. Finally, the level of anthropogenic load characterizes the degree of human impact on the forest environment and helps determine how resistant this area is to recreational use (Pin'kovskiy, 2009).

The second important aspect in substantiating the criteria for assessing the recreational potential of mountain forests is the use of GIS technologies in this process. They make it possible to integrate different data layers and effectively analyze spatial information, which makes them an indispensable tool for assessing recreational potential. The researchers' works emphasize that GIS technologies have unique capabilities for creating accurate and visual models of territories. They can be used to perform spatial analysis of landscapes, model routes, optimize the placement of tourist infrastructure and assess the impact of various factors on the forest ecosystem. In the context of assessing the recreational potential of the mountain forests of the Mostovsky district, the use of GIS allows not only to have an extensive amount of data, but also to structure them in objective and comprehensive way. For example, using GIS, you can analyze parameters such as: the density of roads and trails, the distribution of elevations, the presence of reservoirs and rare species, which allows you to fully assess the attractiveness and accessibility of the territory for vacationers. In addition, GIS technologies make possible to mode scenarios of recreational use of forests and develop recommendations for their sustainable development, taking into account both natural and anthropogenic factors (Malysheva, 2003).

Introduction to the analysis of the recreational potential of the mountain forests of the Mostovsky district by the GIS technologies using, help to make the justification of the criteria more detailed and practice-oriented. One of the main aspects is to assess the condition of forests and their ability to withstand recreational loads without significant damage to the ecosystem. To do this, it is necessary to take into account such factors as: the geological and climatic features of the region, the vegetation and fauna peculiarities, as well as the current level of anthropogenic impact. The natural characteristics of the mountain forests of the Mostovsky district include: the flora and fauna variety, the presence of unique natural sites and landscapes, as well as climatic and geological features that create favorable conditions for the development of various types of recreation. The integration of this data into GIS allows to perform multi-level analysis and identification of the most promising and vulnerable areas.

Ultimately, the justification of criteria for assessing the recreational potential of mountain forests, based on an integrated approach and the use of GIS technologies, contributes to creating the conditions for their effective and sustainable use for recreational purposes, ensuring the conservation of natural resources and improving the quality of the population's life (Dobrynin, 1997; Solntsev, 2003).

The recreational potential of the mountain forests of the Mostovsky district.

The basis is a network of protected areas formed in different years (Fig. 3). Within the boundaries of the municipality of the Mostovsky district, there are 15 specially protected natural areas, of which 1 protected area of federal significance is a state nature reserve, 14 protected areas of regional significance – of which 1 is a state nature reserve and 13 natural monuments (Table 4).

Table 4. Specially protected natural areas located within the boundaries of the Mostovsky district

N°	Name of the specially protected natural area (hereinafter referred to as the protected area)	Area of protected areas, ha	Year of creation	Location of the protected area
1	The Caucasian State Natural Biosphere Reserve named after H.G. Shaposhnikov (partially located on the territory of the Mostovsky district and represented by the Eastern Department of the reserve)	The area of the Eastern Department of the Caucasian Reserve is 86497	1924	The Eastern department of the reserve is located in the gorge of the Malaya Laba River from the Chernorech'e cordon to the Umpyr cordon
2	State Natural Zoological Reserve of regional significance «Psebaysky»	37080,61	1971	Mostovsky district on the lands of the state forest fund within the Psebaysky rural settlement and Bagovsky urban settlement
3	Natural monument «Kyzyl-Bek River Gorge»	206,7377	2021	Mostovsky district 3 km from the Solenoe village
4	Natural monument «Mount Kizinchi»	286,5051	2021	Mostovsky district within the boundaries of Bagovsky Besleneevsky rural settlements (Kizinka khutor)
5	Natural monument «Salty Creek Gorge»	50,15	1987	Mostovsky district, in the vicinity of the Nikitino settlement
6	Natural monument «Two Common Yew Trees»	0,065	1988	Mostovsky district, Mostovsky forestry, Solenovskoe district forestry, block 29 A, part of allotment 21
7	Natural monument «Forest Cultures of Red Oak»	12,15	1988	Mostovsky district, in the vicinity of the Solenoe village
8	Natural monument «Forest Crops of Chestnut»	3,0	1988	Mostovsky district, Mostovsky forestry, Barakaevsky district forestry, block 23 A, parts of allotments 20–22, 25, 26
9	Natural monument «Forest Cultures of Nordman Fir»	3,2	1988	Mostovsky district, Mostovsky forestry, Psebaysky district forestry, block 38 B, part of allotment 9, block 39 B, parts of allotments 5, 7, 8
10	Natural monument «Massif of Chestnut Tree»	11,21	1988	Mostovsky district, in the vicinity of Gubskaya stanitsa
11	Natural monument «Massif of Scots Pine»	28,55	1988	Mostovsky district, in the vicinity of the Solenoe village
12	Natural monument «Planting of Oriental Beech»	167,22	1988	Mostovsky district, the land of the Bagovsky district forestry
13	Natural monument «Planting of Sea Buckthorn»	8,34	1983	Mostovsky district on the lands of the forest fund in block 63A allotments 4, 7 Benokovsky district forestry Mostovsky forestry
14	Natural monument «Dolmen Tract»	46,19	1988	Mostovsky district, on the left bank of the Kizinka River
15	Natural monument «Massif of Chestnut»	11,21	1988	Mostovsky district, in the vicinity of Gubskaya stanitsa
The total area of all protected areas is:		**124412,1**		

To assess the development of the leisure industry, an analysis was carried out according to a number of criteria. At the same time objects of spontaneous tourism were identified.

Natural criteria for the attractiveness of tourism facilities.

1. Aesthetic value occupies an important place in recreation.

The concept of aesthetic value includes a range of characteristics. It affects the psychological state of a person and is important in the sustainable development of touristic areas.

Thus, the Zhelaniya waterfall has an aesthetic appeal in the Mostovsky district. Its height reaches 8 m. It is located a 30-minute walk from the central highway of the Psebai settlement. The waterfall is located on the picturesque slope of the Herpegem ridge. Most of the route takes place in the shade of a typical mid-mountain forest.

Tourists also visit the Herpegem ridge. It is a part of the large rocky ridge of the Caucasus Mountains. It stretches over the Psebai settlement, almost along the entire length of the settlement. There are many caves in which people have lived since the Stone Age. At the very top there are alpine meadows and a magnificent view of the village and the Caucasus Mountains, including the Devil's Gate and Yatyrgvarta. Most of the routes along the ridge are on foot, but in recent years mountain bikes have been actively used to visit, which leads to the degradation of ecosystems.

In the vicinity of the Nikitino settlement, a 35 m high waterfall was formed on the right tributary of the Malaya Laba River, on a stream flowing into the Nikitinka River. This object is actively visited by hikers, the path to it lies in the canopy of the forest. And on the left tributary of the Malaya Laba River, on the Kapustinka River, there is the Kapustinskiy waterfall. The height of the waterfall reaches up to 40 m. The route to the waterfall is only on foot and passes through a mixed forest type.

The calling card of the Mostovsky district is located on the territory of the Tkhach mountain range: two mountain peaks of the Devil's Gate (Acheshbok). Locals call them the Eastern Gate (East Acheshbok) with a height of 2442 m and the Western Gate (West Acheshbok) with a height of 2486 m. One peak is located in the Mostovsky district of the Krasnodar Territory, and the other in Adygea. This mountain enclave is extraordinary in beauty. There are grottos, caves, canyons and waterfalls. It is covered by relict indigenous forest. The route is difficult to access on foot. The ascent to the top of West Acheshbok is rated by category 1-A.

From the Devil's Gate, opposite the Yatyrgvarta-Tru massif, there is the Malye Bambaki ridge with a height of 1500–1600 m. It stretches from west to east from the valley of the Malaya Laba River to the Sunduki Mountain. The main peak of the ridge is Mount Shapka, with a height of 1988 m. The plateau is densely overgrown with Alpine meadows with small patches of pine and deciduous trees. A country road passes through the Malye Bambaki plateau and is accessible only to specialized cars (Ural, GAZ-66, UAZ).

Another unique place for aesthetic perception is the Shakhan mountain range. It is a part of the Rocky Ridge and is located on the border with the Karachay-Cherkess Republic. It consists of four cuesta-shaped peaks: Shakhan-1 (1156 m), Shakhan-2 (1207 m), Shakhan-3 (1251 m), Shakhan-4 (1297 m) overhanging the Andriuki River. The array is accessible to both hikers and dripping enthusiasts.

By the help of GIS, a map of the sights of the Mostovskoy district was created. In future it can be used to assess the accessibility of these places to visit using possible types of infrastructure, as well as to determine places in which there are not enough road interchanges and plan their further development. Also, when using geoservices (*MAPSME, OSM for android, 2GIS*), it's possible to add one's own touristic routes.

2. Biological diversity: rare species of plants and animals.

The research area is unique from the biological diversity point of view. Representatives of the Colchian and steppe flora are registered here. The species diversity is at least 3000 species of vascular plants (no complete inventory has been made in the area). More than 50 species of protected plants grow in the forests: *Staphylea pinnata, Taxus baccata, Ostrya carpinifolia, Buxus colchica, Cyclamen coum, Paeonia caucasica, Galanthus wironowii, Galanthus alpinus, Colchicum umbrosum, Colchicum speciosum, Erythronium caucasicum, Anemone blanda, Anemone sylvestris, Helleborus caucasicus, Lilium Kesselringianum, Dactylorhiza urvilleana, Cephalanthera rubra, Cephalanthera longifolia, Limodorum abortivum, Listera ovata, Orchis picta, Orchis militaris, Orchis mascula, Platantera chlorantha, Platanthera bifolia, Woodsia fragilis, Asplenium adiantum-nigrum*, etc.

An inventory of the insect fauna in the study area has not been fully carried out. Numerous insect groups have been poorly studied. It is not possible to characterize the entomocomplexes of the studied territory. Thus, according to entomologists V.I. Shchurov and A.S. Zamotailov (2006), only within the 7 most well-studied orders (Orthoptera, Hemiptera, Homoptera, Coleoptera, Hymenoptera, Diptera, Lepidoptera) the biological diversity of entomofauna in the Northwestern Caucasus is about 10500 species. Protected insects' species have been recorded in forest ecosystems: *Carabus miroshnikovi, Rhysodes sulcatus, Omoglymmius germari, Calosoma sycophanta, Elephantomyia edwardsi, Lucanus cervus, Gnorimus bartelsi, Rosalia alpina*, etc.

Figure 3. Map of the Mostovsky district attractions

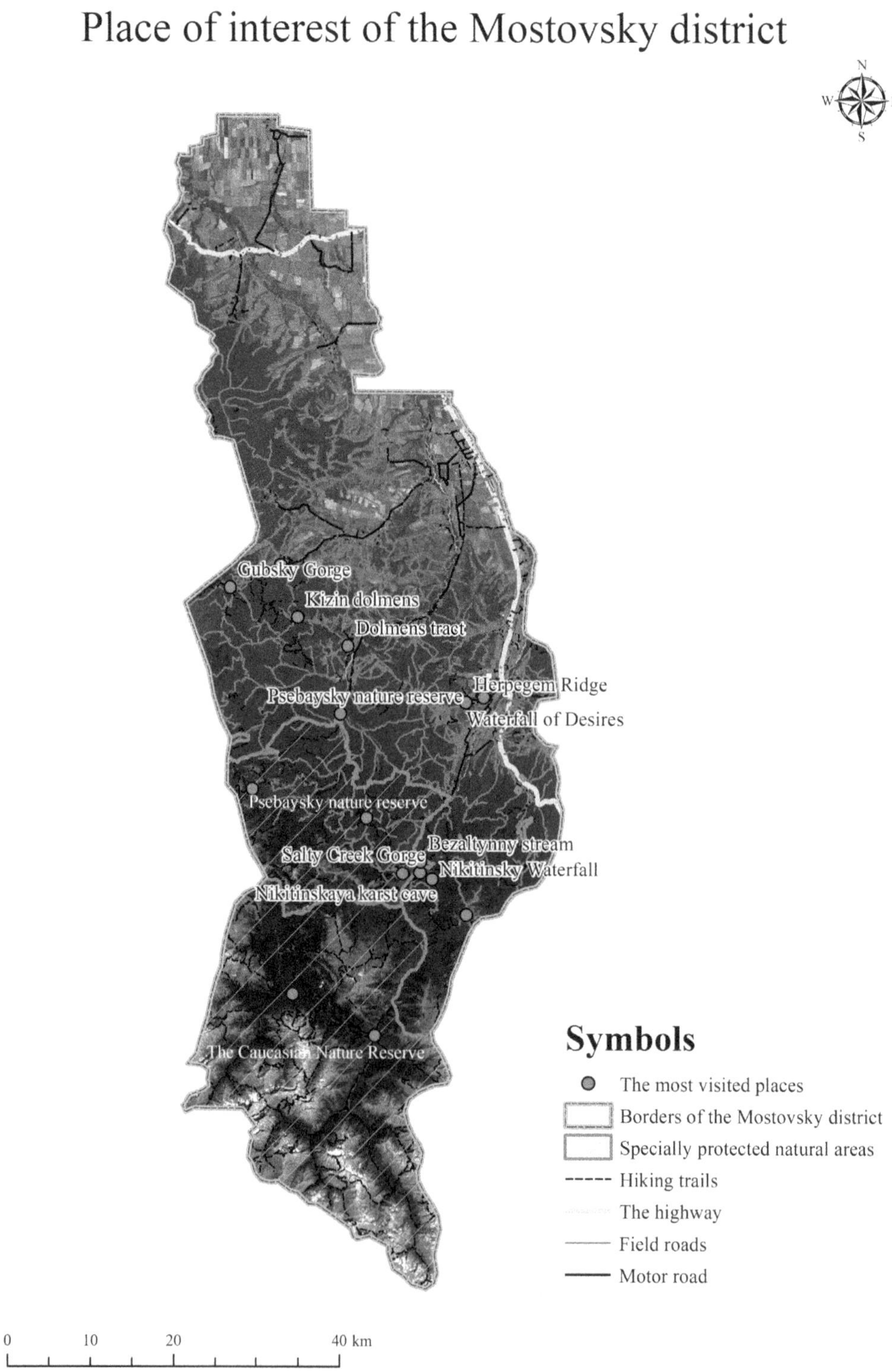

The ichthyofauna of the research area is represented by the following fish species: *Salmo trutta morpha fario, Leuciscus aphipsi, Tobitis taenia, Gobio gobio, Alburnoides bipunctatus rossicus, Barbatula barbatula, Barbus tauricus kubanicus, Carassius auratus, Alburnus alburnus, Chondrastomus colchicum, Rhodeus sericeus, Pseudorasbora parva* etc.

Fans of herpetofauna in the Mostovsky district can meet the following species: representatives of amphibians: *Ommatotriton ophryticus, Lissotriton lantzi, Hyla orientalis, Bufo verrucosissimus, Rana macrocnemis, Hyla orientalis;* representatives of reptiles*: Coronella austriaca, Natrix natrix, Dolichophis caspius, Anguis colchica, Darevskia pontica.*

The avifauna in the Mostovsky district has been studied better than other groups of living organisms and has about 200 species of birds, among which about 90 species are nesting, 80 are migratory, 20 are wintering, 10 are migratory. Among the protected species, it should be noted: *Gyps fulvus, Aquila pomarina, Circaetus gallicus, Falco peregrinus*, etc.

Among mammals, many species are small animals. The most numerous insectivores are *Erinaceus roumanicus, Sorex caucasica, Sorex raddei, Talpa caucasica. Widespread species (Chiroptera) such as: Pipistrellus pipistrellus, Pipistrellus nathusii, Vespertilio murinus, Nyctalus noctula, Myotis blythii, Myotis brandtii, Rhinolophus ferrumequinum, Rhinolophus hipposideros.*

Such representatives of cats as lynx and Caucasian forest cat are found mainly in broad-leaved forests on the study territory.

The smallest predator is the weasel. It is widespread everywhere. Forest and stone martens are also common in the place. The forest marten prefers dark coniferous cluttered forests of the middle and upper parts of the forest belt, the stone marten lives in broad-leaved forests.

A typical inhabitant of forests is a badger. Its visits to the highlands are extremely rare. The otter is found in river valleys. Its main food is brook trout.

It is also not uncommon to find a brown bear in the forests of the Khodz basin more often in autumn, during the ripening of nut-bearing and wild fruit bushes. In the forests, the bear is a seasonal food migrant. Its main habitat is the highlands.

There are 6 ungulate species found in the Western Caucasus to varying degrees in the research area: mountain bison, red deer, tur, chamois, wild boar and roe deer.

Socio-economic criteria for the attractiveness of tourism facilities.

1. Accessibility and transport infrastructure.

The main transport arteries:

- the road transport: the main communication with other regions is via highways. The Mostovsky district is connected to the settlements of the Krasnodar Territory (Labinsk) and neighboring regions by two main highways. Both tracks are in satisfactory condition;
- the railway transport: the nearest railway station is located in the Mostovskaya stanitsa. The railway connection is not the main transport route, but provides additional opportunities for tourists;
- the air transport: the nearest airports are located in Krasnodar and Mineralnye Vody. From these airports to the Mostovsky district it is possible to reach by road. The travel time is about 3–4 hours. Plans to build local airfields for small aircraft have not been implemented, but such a project could significantly increase the tourist attractiveness of the area.

Internal transport network:

- bus routes: domestic bus routes connect the main settlements of the district with tourist sites. During the summer season, additional flights are often organized, which facilitates the movement of tourists. Sometimes the bus travel interval may be long, which creates some inconvenience for travelers without personal transport;
- taxis and private carriers: taxis and private carrier services are common in the area. Despite the higher cost compared to public transport, such services allow you to get to remote and hard-to-reach recreational facilities. In summer, the cost of such services may increase due to high demand.

Accessibility of natural objects:

- the mountains and caves: the most visited natural sites in the area are the Herpegem ridge, the Dolmen Tract and the Gubskoe Gorge. The roads to them are off-road and require a special transport. However, for experienced tourists this can be an additional advantage, creating an atmosphere of adventure;
- the waterfalls and rivers: the Nikitinsky and Burun waterfalls, the Bezaltynny Creek and the Malaya Laba River remain popular places for recreation and active tourism. The roads to them are mostly unpaved, but they are in a satisfactory condition. During the rainy season, movement can be difficult. It requires specialized and highly passable equipment (Pen'kovskiy, 2006).

Mostovsky district is a unique combination of mountain landscapes, water resources and forests, which makes it attractive for admirers of outdoor activities and natural tourism (Efremov, 1963). Routes to such main tourist sites as the Malye Bambaki, Mount Tkhach, Urushten Canyon, Skirdu Ridge and Yatyrgvart's snow peak start in this area, as well as to many other picturesque attractions of the North Caucasus (Fig. 4).

Figure 4. Map of tourist routes of the Mostovsky district

After analyzing the cartographic material, it can be concluded that there are a great number of tourist routes for stable recreational flows not only in the spring and autumn period, but also year-round, positively influencing the development of tourism in the studied area of the Krasnodar Territory. It is evidenced by the dynamically developing network of accommodation facilities for tourists.

Availability of accommodation facilities for tourists:

- there are four sanatoriums in the district, as well as more than a hundred recreation centers and hotels and inns of various conveniences and price of accommodation;
- the thermal complexes (there are more than 30) attract many tourists, especially in winter;
- the tourist shelters (there are more than 40) provide infrastructure for camping fans: equipped parking for tents, campfires and picnic areas, as well as basic amenities such as toilets and showers. This campsite is especially popular with fishing and water sports enthusiasts (Efremov, 1963).

The Mostovsky district of the Krasnodar Territory is located in close proximity to the borders of the Caucasian Biosphere Reserve and the Psebaysky Reserve. This geographical location contributes to the formation of a significant number (more than 50) hiking trails and routes that run through mountainous and foothill areas, along mountain rivers, streams and waterfalls.

Additional services:

- the agrotourism is actively developing on the territory of the Mostovsky district. Several farms offer the opportunity to live in the village, get acquainted with rural life, participate in seasonal work and try natural products of local production;
- for those who prefer a relaxing holiday, there are several baths and spa complexes in the area. They offer traditional Russian baths, saunas, massage and aromatherapy treatments, that allow to combine active recreation with recuperation and health;
- the medical tourism, first of all, is the development of mineral water deposits, speleotherapy and climate treatment in the alpine and subalpine high-altitude zones;
- the extreme and sports tourism, such as canyoning, rock climbing, mountaineering, water / speleological / equestrian tourism.

Environmental impact assessment as a factor of sustainability of forest ecosystems for active tourist visits.

Recreational activities have a significant impact on the forests of the Mostovsky district:

- the increase in the area of the road and path network: frequently used tourist routes can lead to soil compaction, root denudation, and worsen conditions for plant growth and development;
- the pollution of recreation areas: garbage left by tourists and unauthorized bonfires damage ecosystems.

Mountain forests have significant recreational potential, but excessive and uncontrolled recreational activity can lead to ecosystem degradation. The restoration of natural areas after recreational pressure depends on many factors, such as: forest structure, soil composition, climatic conditions and tourism management measures (Ivonin, 2000).

Factors affecting the ability of forests to recover:

- the species composition and biodiversity: high species saturation of flora and fauna contributes to better ecosystem regeneration;
- the soil structure: soil conditions are important in the reforestation process;
- the hydrological regime: the availability of water resources, such as springs and streams, is also an integral process for forest restoration, as they provide the necessary moisture and promote plant growth;
- the climatic conditions: the warm and humid climate characteristic of the Mostovsky district is conducive to forest restoration. However, climate change can have both positive and negative effects;
- the anthropogenic factor: the level and forms of recreational pressure determine the degree of degradation, the more destructive human actions (for example, mass deforestation or actively visited trails) the longer it takes the forest restoration (Voskoboinikova, 2008).

Forest restoration mechanisms:

- the natural restoration: in the absence of significant ecosystem disturbances, forests are capable of self-healing through natural processes. They can be seed germination, the growth of young trees and shrubs, as well as the return of animals, that is quite important in seed propagation and biodiversity maintenance;
- the active restoration measures: in cases of severe degradation, active restoration measures may be required, including planting trees and shrubs, soil stabilization, creation of artificial reservoirs and other ecosystem restoration measures;
- the logging activities: controlled and sustainable logging methods can accelerate the recovery process if seasonality and biological characteristics of local plant species are taken into account;

RECREATIONAL LOAD MANAGEMENT MEASURES

In order to increase the ability of mountain forests to recover from recreational pressure, the following measures should be applied:

- the regulation of visits: limiting the number of visitors and introducing quotas for visits to the most vulnerable areas of the forest;
- the creation and support of tourist infrastructure: specially equipped trails, organized places for camping and recreation help to minimize the negative impact on the environment;
- the information campaigns: educational programs for tourists aimed at explaining the importance of preserving natural sites and responsible behavior;

- the monitoring and control: regular monitoring of the state of forests allows quickly identifying problem areas and taking measures in restoring them;
- the local community involvement: the involvement of local people in forest protection and restoration can significantly enhance the sustainability of ecosystems.

Over the past five years, the Mostovsky district of the Krasnodar Territory has demonstrated a significant increase in the tourist flow, which indicates positive changes in the infrastructural development and investment attractiveness of the region. In particular, the number of tourists has increased four times, from 80 thousand to 330 thousand people per a year. At the beginning of 2022, 210 thousand vacationers visited the area, and by the end of the year, the number of guests reached 300 thousand. Thus, over the past five years, the tourist flow in the municipality has increased 2,2 times. It has led to an increase in income and an increase in the number of resort facilities (Governor of the Krasnodar…, 2024).

In parallel with the growth in the number of tourists, eight investment projects in the sanatorium and resort sector totaling more than 280 million rubles were implemented in the Mostovsky district, in the future this will contribute to an increase in the tourist flow, an increase in the number of new jobs. It will lead to an improvement of local residents' lives. At the moment, a balneological resort, a number of glamping and recreation centers are being built in the Mostovsky district. It is planned to spend about 15,2 billion rubles on the implementation of these projects (Governor of the Krasnodar…, 2024).

The development of the sanatorium industry also has a multiplier effect on other sectors of the municipality's economy. In particular, the retail trade took the second place in the structure of the basic industries of the district, accounting 33,2% of the total structure.

ANALYSIS OF INDICATORS IDENTIFYING POTENTIAL RECREATIONAL DEVELOPMENT AREAS IN THE MOSTOVSKY DISTRICT.

To identify potential recreational development areas in the Mostovsky district, a comprehensive set of indicators can be used to support the assessment of the attractiveness and suitability of various territories for tourist infrastructure development. The analysis was carried out by the help of GIS technologies. A map has been performed including a number of indicators that can be interpreted in a spatial environment (Fig. 5).

Figure 5. Map of potential areas for the development of recreation in the Mostovsky district

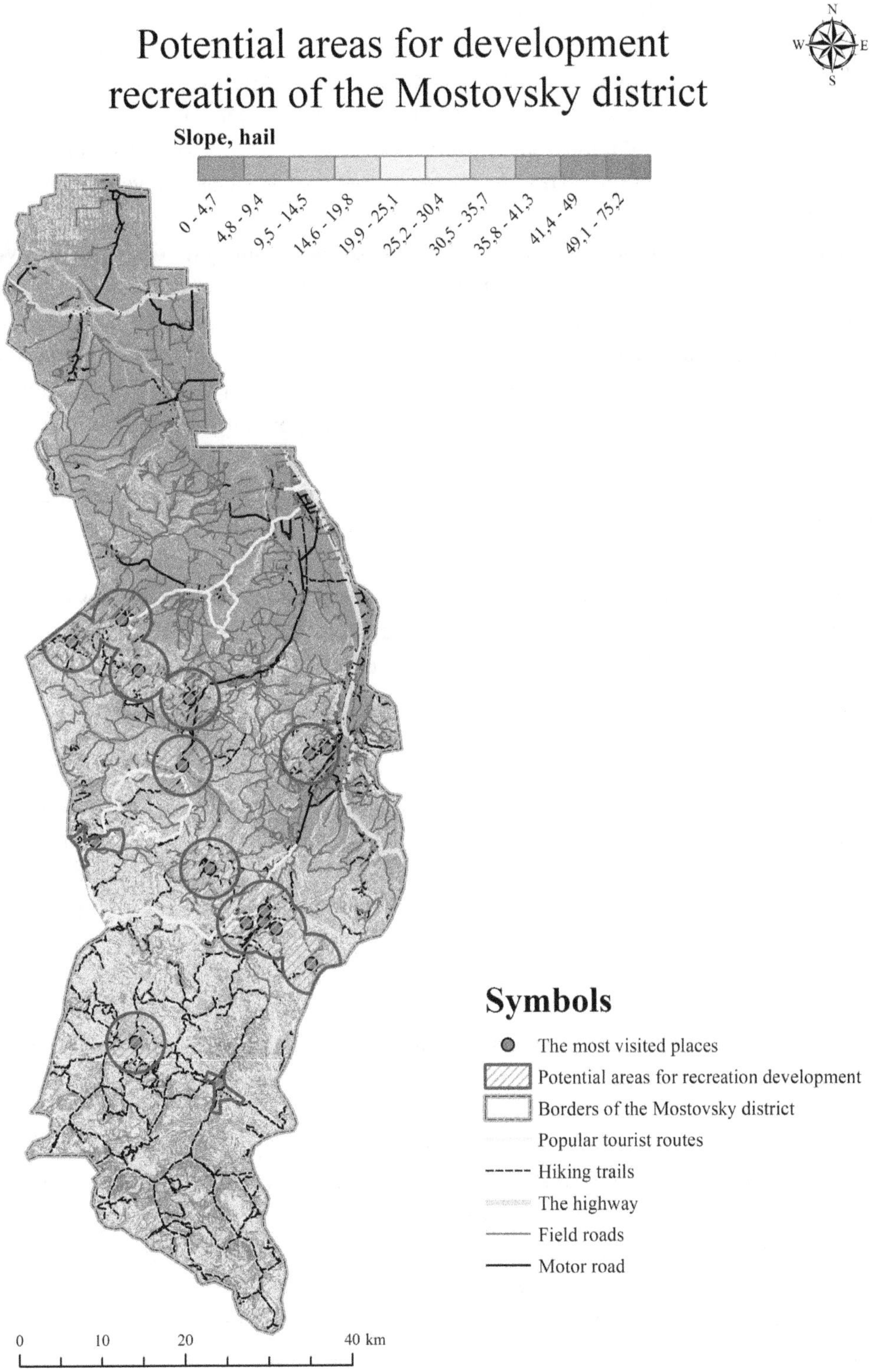

The zones highlighted in the blue polygons are the most favorable for recreational infrastructure development. Such data were revealed based on a number of selected criteria, that allows a deeper understanding of the mechanisms for determining promising areas for tourism. The required set of indicators included all types of transport networks: both hiking trails and highways. This approach concludes in a comprehensive analysis of the accessibility and possible development of these territories. Transport accessibility is one of the main criteria necessary for the development of a tourist cluster in the Mostovsky district. This is due to the fact that convenient access to the area of tourist interest is one of the fundamental aspects influencing the travelers' choice and the successful development of recreational infrastructure (Skvortsov, 2005).

In addition, one of the important aspects of the study is the analysis of frequently visited places marked with a red dot, which represent the most interesting tourist attractions. These combinations are an indicator of the high attractiveness of the region and the visitors' interest in it. Data on frequently visited places was collected thanks to users sharing their routes on the Internet using GPS trackers. The analysis of these routes made it possible to select the most popular of them, which is a direct indicator of current tourist preferences and focuses on already existing popular routes (Efremov, 1963).

The most important indicator for the development of recreation in mountain forests is the surface slope. The slope has a significant impact on the convenience of tourists' movement, as well as on the possibility of infrastructure creation such as: campgrounds, tourist camps and observation decks. Studies have shown that the least steep slopes are preferable for trails and routes creation. As they are less laborious to traverse and less susceptible to erosion and other environmental influences (Dobrynin, 1997).

Based on the obtained data, the following most attractive zones were chosen with the lowest slope, through which several tourist routes pass. They have frequent attendance, as well as to which there are entrances or approaches and may be suitable for further development. These criteria made it possible to identify the sites that have the greatest potential for the development of tourist and recreational infrastructure.

In addition, an important aspect is the assessment of the current state of infrastructure in the region. This includes not only the analysis of existing roads and hiking trails, but also the availability of modern conveniences such as: electricity, water and telecommunication. The good condition and accessibility of these systems are important indicators of the zone's readiness for the development of recreational infrastructure.

It should also be highlighted that the socio-economic aspects of the region's development are significant factors. The involvement of the local population in tourism activities can contribute to the creation of jobs, improvement of living standards and overall sustainable development of the area. However, it also requires careful planning and consideration of local communities' interests (Governor of the Krasnodar…, 2024).

Due to the above mentioned, the development of tourist infrastructure in the designated areas requires an integrated approach. It is necessary to take into account both natural and socio-economic factors, ensuring the balance between the tourism development and the preservation of natural landscapes. This approach will create highly competitive tourism clusters that will attract a great number of tourists, while simultaneously supporting the sustainable development of the region.

The studied territory of mountain forests is important for the preservation of natural complexes and objects in their natural state, and since all components of the natural environment are in complex interaction, the following principles of allocation of natural complexes and objects, requiring special protection status, are used:

- the preservation of objects in their natural state;
- the stability of objects to anthropogenic and other types of impacts;
- the importance of natural complexes as a habitat for various groups of animals, as well as vegetation, including rare and protected species;
- the interrelation of the components of the natural environment as the basis for the stability of the ecosystem;
- the presence of unique geological formations or parts of them on the territory that need the protection and additional study;
- the presence of specially protected natural areas of other categories in the studied territory of the reserve.

CONCLUSIONS

Mostovsky district of the Krasnodar Territory, due to its natural uniqueness and rich cultural heritage, has significant potential for the development of recreational areas. Taking into account the natural and infrastructural features of the area, as well as the current state of mountain forests, the present study was conducted by the help of GIS technologies and 11 of the most promising areas for the development of tourist infrastructure were identified. The planned use of natural resources can significantly improve the quality of local population's life, attract additional tourist flow and ensure the sustainable development of the region. An integrated approach based on a multiparametric data analysis was proposed in analyzing the following indicators: the lowest slope, transport accessibility and frequent attendance, which is potentially attractive to investors.

Also, in the course of the work a number of recommendations were given on the use and protection of the recreational potential of the mountain forests of the Mostovsky district. In turn they will help preserve their biological diversity and maintain the balance between recreational load and environmental protection measures. These recommendations can be used to create plans for the integrated development of tourist infrastructure, as well as to form updated and more accurate maps of natural reserves and recreation areas for the Ministry of Resorts, Tourism and Olympic Heritage of the Krasnodar Territory.

The use of modern GIS technologies can contribute to the implementation of the recreational resource management strategy not only to improve the recreational attractiveness of the Mostovsky district, but also to form ecologically and economically balanced approaches to the use of its natural resources.

ACKNOWLEDGMENTS

The research is carried out with the support of the Ministry of Science and Higher Education of the Russian Federation (No. FZEN-2023-0012 "Fundamental aspects of rational recreational environmental management and determination of anthropogenic load in the context of sustainable tourism development")

REFERENCES

Anisimov, V. I. (1993). Morphometric analysis of mountain relief for recreation purposes. *Geomorphology*, •••, 369.

Annenkova, I. V. (2009). The use of GIS technologies to monitor the collection fund of the arboretum. 65 years of the Federal State University «Scientific Research Institute of Mountain Forestry and Forest Ecology», 247.

Bereznikova, O. Y. (2007). Geoinformation mapping of urban landscape forest plantations. *Science and youth. New ideas and solutions: materials of the international scientific and practical conference of young researchers*, Volgograd: Niva, 364.

. Bityukov, N. A. (2007). *Geoinformation systems: a textbook*. Sochi: SSU TiKD, 242.

Dobrynin, A. P. (1997). *The influence of recreational loads on the stability of biogeocenoses depending on the steepness of slopes*. – Sochi: Lesnoe Khozyaistvo, 26.

Drannikov, A. E. (1994). *Approaches to the construction of a conceptual model of the functioning and development of recreational forest management. Ecological foundations of farming in mountain forests*. Sochi, 164.

Efremov, Yu. K. (1963). *The trails of the Black Sea mountains*. Sochi: Geografiya, 115.

Ivonin, V. M., ... (2000). *Recreational ecology of mountain forests of the Russian Black Sea region: monograph*. Rostov-on-Don: SKNC VSH Publishing House, 272.

Klimacheva, T. V. (2009). *Recreational potential of the forests of the Kama region*. Moscow: Mediator, 59.

. Malysheva, N. V. (2003). *Remote methods and their integration with information technologies*. Forestry Information, 173.

. Migunova, E. S. (2008). *Types of forests – the first, ecosystem level of differentiation of nature*. Lesnoe Khozyaistvo, 145.

Pen'kovskiy, A. N. (2006). *Soil erosion during recreation in the subalpine clearings of the upper forest belt of the Western Caucasus*: Author's abstr. diss. (Cand. sci. in Agriculture). Novocherkassk, 24.

. Pin'kovskiy, M. D. (2009). *The use of GIS technologies for the analysis of the recreational potential of the territory of the Sochi National Park*. Sochi: SSU TiKD, 27.

Shcherbakov, A. V. (2003). *Caucasian Mineral Waters. The blessed land of resorts and tourism*. Stavropol: GUP SK «Stavropol kray, tipografiya», 184.

Skvortsov, A. V. (2005). *Geoinformatics in the road industry*. Moscow: MADI Publishing House, 389. DOI: 10.17273/BOOK.2005.2

Solntsev, G. K. (2003). *Scientific bases of recreational use of mountain ecosystems*. Rostov-on-Don: SKNC VSH Publishing House, 59.

Voskoboinikova, I. V. (2008). *Soil erosion in connection with recreation in beech forests of the Western Caucasus*: Author's abstr. diss. (Cand. sci. in Agriculture). NIMI – Novocherkassk, 23.

Chapter 4
Challenges in the Tourism Sector:
A Case Study in KwaZulu Natal, South Africa

Zahir Sayed
https://orcid.org/0000-0001-7434-5413
University of KwaZulu-Natal, South Africa

Cecile N. Gerwel Proches
https://orcid.org/0000-0002-2330-9575
University of KwaZulu-Natal, South Africa

Abdullah Kader
University of KwaZulu Natal, South Africa

Simon M. Taylor
University of KwaZulu-Natal, South Africa

ABSTRACT

Tourism is a dynamic and unique sector as it interconnects a range of other industries as part of the value chain. Through the creation of jobs and as a major contribution to Gross Domestic Product (GDP), this industry also serves as a major source of income for the local and national economies of the countries in which it operates. The aim of this research was to investigate the difficulties encountered by the tourism industry in KwaZulu-Natal, South Africa. Purposive sampling was used in conjunction with a qualitative research methodology in this study. The primary findings revealed that the tourism industry was facing a number of hurdles that were seriously impeding its capacity to function effectively and draw large numbers of visitors. It is recommended that the findings of this study be used as a basis to build a comprehensive tourism strategy for the organisation under study. The findings of this paper conclude that this research can enhance KwaZulu- Natal's tourism sector and encourages more studies on the subject.

DOI: 10.4018/979-8-3693-8232-5.ch004

1. INTRODUCTION

Tourism is a huge industry on a global scale, creating intricate networks with growing economic sway and using massive amounts of resources. There is potential for job growth in the tourism industry, which is one of the economic sectors with the quickest rate of growth (Bojanala, 2024). A dynamic sector like tourism can spur the growth of many different economic sectors. Development in all economic sectors can be considerably and sustainably accelerated by tourism. Through creating high-quality jobs that will eliminate poverty, promote environmental preservation, and sustain social and economic advancement over generations to come, the tourism sector may help nations create resilient and inclusive economies through integrated policies (Pane, 2024). Economic development revolves around the tourism industry, especially in underdeveloped nations. The value chain of the industry is made up of numerous important small and major businesses from a variety of industries, such as lodging facilities, transportation, and other services related to tourism (Tourism Sector Recovery Report, 2023).

Travel destinations rely heavily on tourism as a source of revenue, and numerous countries, particularly those that depend on it, benefit economically from it generally (Abbas, Mubeen, Iorember, Raza & Mamirkulova, 2021). In simpler terms, there exists a robust association between the growth of the tourism sector and the economy, hence fostering the expansion of other industries inside a country. But the travel and tourism industry is vulnerable to threats like global pandemics. Unlike more prosperous countries, the economies of developing nations can gain a great deal from tourism (Huynh, Kim, Duong, Trong, Huong & Dao, 2021). One of the methods to generate income and jobs is through the expansion of the tourism sector (Khan, Bibi, Lorenzo, Lyu and Udden, 2020).

Geographically distinct South Africa has a diverse natural landscape and cultural heritage that add to its tourism appeal on a global scale. South Africa is by far the most diverse country in the world and ranks among the top six most biodiverse nations in the world, especially when taking into consideration its size. There are several hundred animal and environmental reserves, twenty-two national parks, and five world heritage sites that showcase seven distinct biomes all located within the nation's boundaries (Bojanala, 2024). Because of these variety of attractions, South Africa is a popular travel destination (Tourism Sector Recovery Plan, 2021). Tourism is a valuable revenue stream for the South African economy (Bama & Nyikana, 2021) and is indispensable to the overall prosperity of the country. Being a specialized industry with strong ties to the transportation, retail, financial, and other network industries, the tourism sector contributes significantly to the economy indirectly.

The KwaZulu-Natal tourism industry has the potential and capacity to deliver on a significant scale, but it is dependent on strong, fruitful, ongoing, long-term partnerships between the public and private sectors. The tourism industry contributes significantly to the provinces economy and ultimately the national economy. Natural treasures abound in the KwaZulu-Natal area, allowing tourists to have a taste of both traditional and contemporary lifestyles. The Big 5, the Indian Ocean, a marine paradise, the Phezulu Safari Park, the Anglo-Zulu War battlefields, and the Drakensberg Mountains are some of these treasures. KwaZulu-Natal has the second-biggest province economy after Gauteng. The Durban-Pinetown metropolitan area, Richards Bay, and Pietermaritzburg are the main hubs of economic activity. The Isimangaliso Wetlands Park, with its wilderness area, lakes, ancient marine reserve, and coastal dune forest, and the uKhahlamba Drakensberg Park, with its mountain range and Khoisan art-adorned caves, are two of KwaZulu-World Natal's World Heritage Sites (Ezemvelo Wildlife AR, 2023).

Given the generally positive outcomes, a number of geopolitical, economic, and climate-related issues still exist. The travel and tourism industry is still dealing with inflation, which includes high travel and lodging costs and unstable oil prices. Extreme weather conditions and staff shortages are also major difficulties for tourism performance, and major conflicts and tensions around the world continue to affect consumer confidence (UN Tourism, 2024). The tourism industry in South Africa has also encountered several difficulties, such as pandemics and natural calamities. Every crisis has put the industry's resilience to the test, inspiring creative recovery plans. Specifically, the COVID-19 pandemic brought attention to the necessity of flexibility and reaffirmed the significance of safety and health measures. The government set aside R200 million for the tourist Relief Fund (TRF) to help 4,000 businesses weather the epidemic and lessen the effects of COVID-19 on the travel and tourist industry. Thanks to these initiatives, the industry has been able to regain the remarkable number of foreign visitors that existed before COVID-19 and is expected to keep expanding (Bojanala, 2024).

The objective of this study was to:

- Determine challenges facing the tourism sector within KwaZulu-Natal

2. LITERATURE REVIEW

2.1. Tourism

As a social, cultural, and economic phenomenon, tourism involves people traveling for personal, business, or professional reasons to nations or locations outside of their normal surroundings. These individuals are known as visitors, and tourism is related to their activities, some of which entail spending money on tourism (UN Tourism, n.d.)

2.2. Tourism in the global environment

In many countries around the globe, tourism is a high-priority sector that contributes significantly to GDP and serves as a major source of tax revenue and foreign exchange (Aragu, García-Mestanza and Caballero-Galeote, 2021). Due to the geographical dispersion of visitor flows and the expansion of mass tourist exchanges, Europe currently holds a solid grip on the global tourism industry. The area is observed attentively by America, especially North America, and benefits from both domestic and foreign markets. These two regions are significant players in the global tourism business, accounting for the majority of revenues (Kyrylov, Hranovska, Boiko, Kwilinski & Boiko, 2020). Globally, travel and tourism play a significant role in the development of employment, economic expansion, and cultural advancement (Abbas et al., 2021). International tourism has become one of the main factors influencing economic growth, a nation's ability to compete in international markets, and population welfare as a result of globalization and integration. The travel and tourism sector employs around 270 million people, or roughly 8.2 percent of the global workforce (ILOSTAT, 2024).

The travel, tourism, and trade industries have changed significantly in the wake of the global pandemic and in the face of quickly developing technologies (Yepez and Leimgruber, 2024). Most regions are showing a spectacular comeback and are already surpassing 2019 arrival statistics in the period of January to September 2024, four years after the COVID-19 pandemic epidemic brought global tourism to a standstill (UN Tourism, 2024). The robust increase in tourism receipts is good news for global economies. Millions of jobs and small companies are directly impacted by the fact that visitor spending is increasing even faster than arrivals, which also significantly affects the balance of many economies and tax receipts (UN Tourism, 2024). Based on a variety of variables, including legislation, climates, geopolitical situations, and the attractiveness of historical cultural sites, most nations prioritize the development of tourism which includes:

- **Taxes:**

When tourists spend money on regional businesses, the economy of the country where they are traveling grows. This is because tourism attracts other tourists, who in turn spend money there.

- **Infrastructure Development:**

Water, electricity, health care, and law enforcement are examples of infrastructure that is developed to improve tourism attractions.

- **Support for the local economy**:

To accommodate tourism activities, local standards are raised. To provide better and more effective tourist services, various companies from various sectors work together. As a result, locals can find work and create new jobs.

- **Preservation of Local Cultures**:

Local communities, heritages, and cultures can become tourism destinations and points of interest by being preserved.

Those with a shared interest in the objectives and advancement of tourism are considered stakeholders in the tourism sector. All levels of government can influence the growth of the tourism industry to address these issues, seize opportunities, and provide better results. Numerous nations have revised or created tourist plans to take into account the priorities and post-pandemic environment. In order to adapt to a more dynamic environment and unite stakeholders around a shared vision for the industry's future, these strategies aim to build on the recent momentum toward a shift to more sustainable types of tourism. To better distribute the advantages and tourist flows, many tourism plans rely on diversifying the tourism offer and drawing in new source markets and segments to new destinations and outside of peak times. (OECD, 2024)

2.3. Tourism in South Africa

Tourism, the industry with the fastest rate of job creation throughout the world, has the potential to give the youth of South Africa countless opportunities and lower the unemployment rate. From isolation during the apartheid era to becoming one of Africa's top travel destinations, South Africa's tourism industry has experienced a dramatic metamorphosis over the last three decades. The rise in foreign visitor arrivals from less than a million in the late 1980s to 10.2 million in 2019 is indicative of the nation's tourism expansion (Bojanala, 2024).

In 2021, the sector contributed 6.1% of the global gross domestic product (GDP). Despite the epidemic, South Africa's tourism sector generated over a million job possibilities in 2021 and contributed a respectable 3.2% of the country's GDP (Tourism Sector Recovery Report, 2023). Heading into the future, the nation's tourist industry is in a strong position to build on the achievements of the last three decades and continue to make a constructive contribution to the 7th Administration of the South African government. In order to reach its full potential as a top travel destination on the world map, the tourism industry is well-positioned under this new administration to promote inclusive economic growth, create jobs, and forge strategic alliances (Bojanala, 2024).

South Africa has consistently recognized the beneficial impacts of travel and tourism for the country's economy, job creation, and social progress (Smith, 2021). The public and private sectors must work together to further develop the tourism industry. In order to generate a workforce with the necessary skills to grow and sustain the tourism industry, all players in the industry and trade must collaborate with the government by supporting human resource development programs and activities that upskill the youth (Bojanala, 2024).

2.4. Tourism in KwaZulu-Natal

KwaZulu-Natal, which has a population of over 11 million, or 19.6% of the total population, is the second most populous province after Gauteng despite having the third-smallest land area (Tourism KwaZulu-Natal, 2023). Every year, the province welcomes a sizable number of visitors. Over seven million domestic trips were made in 2022, while the province accounted for 9.7% of all foreign visitors (Tourism Sector Recovery Report, 2023). One of KwaZulu Natal's most prosperous industries, tourism contributed a substantial 5% of the region's gross domestic product (R-GDP) in 2021. KwaZulu Natal draws a sizable number of tourists annually and has a competitive edge over other provinces (Tourism Sector Recovery Report, 2023).

Manufacturing, agriculture, mining, trade, construction, and finance are just a few of the industries that contribute to the province's economic performance. It has a significant competitive edge that propels the province's tourism industry as, along with having nine Blue-Flag beaches on its east coast, it is also home to Oribi George, the iSimangaliso World Heritage Site, Ukhalamba-Drakensberg, several historical battlefields, and a large number of "big five" game reserves, both public and private (Tourism KwaZulu-Natal, 2023). Domestic travel is the main driver of the KZN tourism sector. While there is a lot of international travel, domestic travel is far more common. Thus, in order to boost tourism in KZN for years to come, efforts must be concentrated on local markets (Tourism Sector Recovery Report, 2023).

2.5. Challenges for the Tourism Industry

If the quality of the visitor experience is not jeopardized, the tourism business can be viewed as a "sunrise" sector with the potential for both quick innovation and growth, as well as immunity to local issues and challenges. The tourism sector can aid governments around the world by promoting social welfare, but there are possible negative effects as well, like exploitation of natural heritage sites and a negative impact on environmental quality.

By January 2020, the World Health Organization (WHO) had initially deemed COVID-19 a global health emergency (Weiss, Jackson, Nelson, & Schwarzenberg, 2020). The World Health Organization (WHO) classified the severe respiratory syndrome coronavirus 2 (SARS-Cov-2) to be a pandemic in February 2020, making it one of the first biological dangers to human civilization in the twenty-first century (Marivate & Combrink, 2020). The COVID-19 pandemic raised concerns about healthcare and the economy worldwide. It has also had an impact on foreign enterprises, especially travel and tourism, which plays a major role in the global service economy (Abbas et al., 2021). Prior to the global outbreak of COVID-19 in the 21st century, tourism ranked among the world's most important industries (Ugur & Akbiyik, 2020). The global catastrophe caused by the coronavirus epidemic has had a major influence on the tourism industry worldwide. Consequently, an examination of the tourism industry's standard operating procedures was required (Kyrylov et al., 2020). South Africa was also severely impacted by this epidemic, especially in the travel and the tourism sector.

Despite this crisis, tourism on global level has recovered well, promoting both economic growth and well-being. But the recovery is not uniform, and new problems are brought about by escalating geopolitical tensions, pressures from the cost of living, and climate-related disasters (OECD, 2024) who further asserted that the industry is being impacted by rising wildfire, flood, heatwave, and other extreme weather occurrences, as well as ongoing geopolitical tensions. In the long run, it is anticipated that tourism will keep expanding. This reinforces the significance of policies to support a resilient, sustainable, and inclusive future by bringing with it both new problems and major opportunities.

By the end of 2024, a complete worldwide recovery is expected (OECD, 2024), supported by improved business and traveler confidence as well as an increase in travel within Asia Pacific, notably from China. As supply and demand mismatches continue to be resolved, growth is expected to revert to pre-pandemic levels. But there are still issues and the recovery is not uniform. Even if the demand for travel has remained strong, companies still face difficulties hiring and keeping employees, and even though inflation is slowing down, pricing pressures still affect the cost of lodging, transportation, and hospitality (OECD, 2024). For the benefit of the South African tourism industry, a change in perspective is required in order for the public, private, and community sectors to collaborate on a shared vision. Strategies for this purpose need to be established in consultation with all relevant parties.

3. METHODOLOGY

The study implemented a qualitative research methodology, which involved semi-structured interviews and focus groups to collect data. The advantage of conducting interviews goes beyond only analysing language or developing a comprehensive understanding; it also gives study participants the freedom to talk honestly about their innermost ideas and feelings (Busetto, Wick & Gumbinger, 2021), while focus groups incite a diversity of viewpoints that either support an existing point of view or push participants to acquire new ideas. Focus groups are a popular technique in qualitative research that seeks to completely comprehend social concerns from a chosen group of participants (Nyumba, Wilson, Derrick, & Mukherjee, 2018). In qualitative social science research, focus groups are rather common. The sample consisted of tourism stakeholders from KwaZulu-tourism Natal, South Africa. Owners, managers, and employees of tourism-related businesses were among the eleven selected to participate in this study.

Purposive sampling was used to select study participants from the tourism industry since it was believed that these individuals would be knowledgeable about the severe difficulties the industry faces and would have experience working in a tourism setting. The interviews lasted an average of 55 minutes, whereas the focus group session lasted for about two hours.

The objective of this research was to comprehend the challenges encountered within a tourism setting. Important players in the tourism industry were interviewed to learn more about how these issues affect the industry. Before beginning any data collection, the researcher received ethical approval, and all participants had to complete an informed consent form. Understanding the difficulties faced by the KZN tourism industry was the goal of this study. It was assumed that every study participant had prior experience working in the tourism industry, which greatly aided in the study's effectiveness. Participants were requested to respond to all interview questions in addition to being offered the opportunity to remark on other subjects related to the research problem. The interview questions were formulated and organized so that the investigator could gradually document participant answers that were clear, concise, and targeted toward the study's goal.

Details about every focus group and interview participant are included in Table 1.

Table 1. Details of Focus group and Interview Participants

#	Designation	Tourism Sector	Experience
1	Manager	Resort Management	5-10 years
2	Business Owner	Resort Management	10 years +
3	Staff Member	Media and Publications	5-10 years
4	Business Owner	Hospitality	10 years +
5	Business Owner	Hospitality	10 years +
6	Manager	Hospitality	5-10 years
7	Manager	Chamber of Commerce	10 years +
8	Manager	Chamber of Commerce	10 years +
9	Manager	Resort Management	10 years +
10	Staff Member	Local Government	10 years +
11	Business Owner	Arts and Crafts	10 years +

The study's participants were afforded the chance to freely share their ideas, perspectives, and personal encounters about the research question. It was crucial to refrain from coercing the participants into providing responses that the researchers thought appropriate for the study. An in-depth understanding of the study subject is essential for the researcher to ask the appropriate questions of the participants. Participants were free to express their opinions, thoughts, and feelings concerning the research issue without feeling compelled to frame their reports of it in terms of more general social norms and preferences. All interviews were digitally recorded to guarantee that all the material was gathered accurately. The researcher conducted a thematic analysis to identify themes and patterns in the data that came from the viewpoints of the participants.

The researcher established the study's credibility prior to commencing the study by keeping their own thoughts about the research to themselves, considering the participants' expectations and interests, and granting unrestricted expression of the data from the focus groups and interviews. The participants found it easy to share their thoughts, skills, and practices thanks to the researcher's facilitation. In the culmination, the study's conclusions reflect the opinions and experiences of the participants.

4. RESULTS AND DISCUSSION

4.1. Demographics

Eleven respondents participated in semi-structured interviews and a focus group. These included the employees who work in a tourism setting on a daily basis. The main objective of this study was to comprehend the difficulties faced by the travel and tourism sector. The study's utility was greatly enhanced by the assumption that every participant had previous experience working in the tourism industry. There were two males and nine females among the participants. Eight of the participants in the focus group and the interviews had more than ten years of work experience in the tourism sector, while the remaining three had between five and ten years of experience. Such representations demonstrated that the research participants were able to use their prior work experience in the tourism industry to contribute to the research questions that were addressed. It was found that every participant had over five years of experience working in the tourism business, which prepared them to understand the current challenges in the industry. Not a single participant lacked more than five years of experience in this industry. It is quite likely that participants who meet these standards are responsible and possess the required experience for their industry.

4.2. Challenges identified from focus group and interview sessions

Figure 1. Key Challenges impacting the Tourism sector

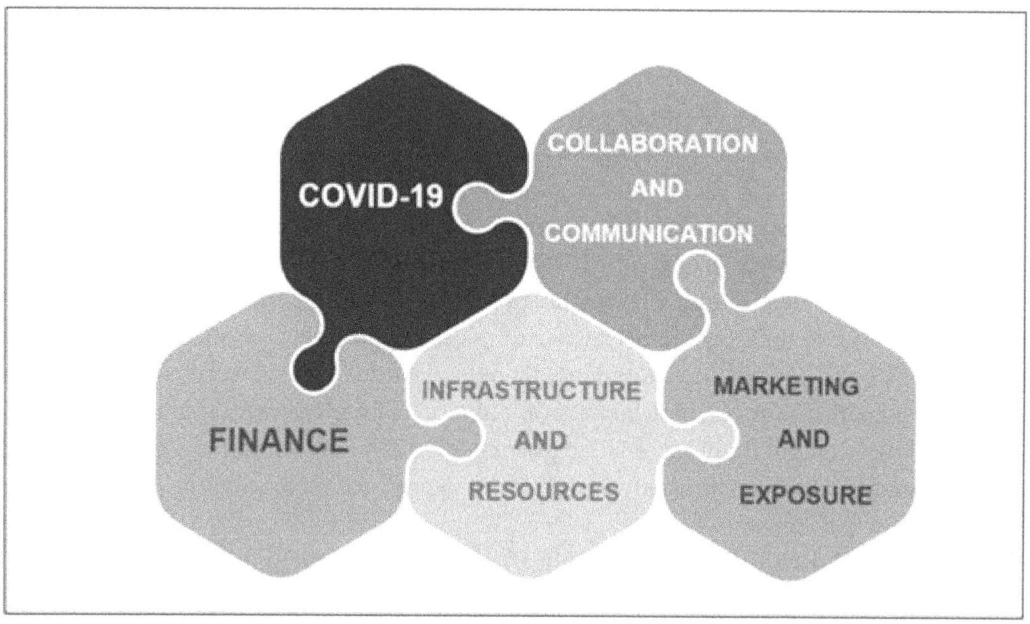

Numerous significant challenges influenced the tourism sector. These were divided into the following sub-themes:

4.2.1. COVID-19

The study's findings aligned with previously published literary works. The travel and tourism sector has been under intense pressure ever since the World Health Organization announced a global pandemic on March 11, 2020 (Jamal, Kircher & Donaldson, 2021). At the start of the second decade of the twenty-first century, few industrial marketers imagined that many of their markets would vanish by March 2020, negatively disrupting trade and supply chains (Khan et al., 2021). The KZN tourism industry's top concern was the COVID-19 pandemic, which had the following effects:

4.2.1.1. Inability to travel

Due to lockdown levels brought on by the COVID-19 epidemic, travel was severely restricted. This has impacted heavily on the tourism sector and possibly this sector was the most hit. Most nations, including South Africa, rely heavily on tourism to fuel their economies. Participants' opinions were as follows:

> "Well, the number one will have to be COVID-19. Because we have been hit with these challenges for some time, however, when COVID came for the first time we closed our doors for the first time since the inception of the resorts, we completely closed that gate. That was at the end of March last year for about two and a half months. And so we had absolutely no guests we were not generating any revenue at all which really set us back a big way" (Participant 10).

> "So with COVID, with the economy being the way it is, people just do not have money to travel, because remember, traveling is not at the top of the hierarchy of human needs, it is basically leisure, so you would first look at your basic human needs. And if you find if cannot cover your basic human needs, then tourism doesn't happen" (Participant 2)

4.2.1.2. Tourism forecast

The ongoing COVID-19 waves made the tourism prediction unclear as well. In addition, the tourism industry was heavily disrupted, making it challenging to plan with interconnected businesses within the industry and difficult to estimate visitor turnout. A participant added:

> "Number two is tourism figures like we use forecast, these forecasts are out of the window currently, where we were forecasting 150,000, we got 5000, where we were forecasting 5000, we got 200,000. So it's very difficult to plan" (Participant 3).

4.2.1.3. Expenses

COVID-19 resulted in increased costs and a lack of finance.

4.2.1.3.1. Souvenir sales

Since foreign tourists used to buy a lot of souvenirs, sales of those items have also decreased. A few respondents shared the following:

"So the curio shop, the restaurant, activities and accommodation, whereas the local clientele will spend on accommodation, but they will Self-cater, they will probably not use the restaurant, they will use a little bit of money in the shop. So that really took a big chunk of income" (Participant 6).

"Our souvenir sales have invariably taken a knock, because souvenir sales is supported mainly by international and tourism from out of province" (Participant 3).

4.2.2. Finance

The next major issue was finance, which was a reasonable objection given that money is needed to keep all aspects of life in motion. Several participants expressed the following viewpoints in this regard:

"It is bad for business ever, because it has an impact on our resources now, and also on the revenue stream that into your business" (Participant 2).

"So, what one of the major challenges has been since I joined the organization, there's been a talk that you know, we are we are going to go separate ways, which never took place up to this stage. So that poses a problem because we are not financing those resorts very well, and then we end up having issues in terms of the standard of the facilities, the actual resorts themselves" (Participant 5).

"It all stems from the limitation of financial resources" (Participant 6).

"But the challenge has been money, budgets, because without budgets, then you can't maintain resorts. You cannot employ people you cannot market your resorts and you cannot meet customer demands" (Participant 9).

The financial situation and its effects on operational needs were prominently emphasized in these statements.

4.2.2.1. Budget Allocation

Budget allocation being the factor that received the most votes and approval from respondents makes sense. Budget allocation was insufficient, meaning that neither enough nor any budget was provided. This had repercussions because without a budget, infrastructure maintenance, marketing efforts, and customer value were all impacted. Regarding insufficient budgetary assistance for operational requirements, some respondents shared the following statements:

"I will say that since I've been here we are not given, you are not given the budget to do the marketing yourself. So, we rely on head office, our marketing department to do our, our marketing for our resorts, and I believe that every resort is a unique" (Participant 4).

"I think mainly, it's budget wise. I will definitely say it's budget wise, because the budget that that we are given is never enough to take care of the things that need to be taken care of. And, also, we are requested as resort managers that you submit your own budget to our finance department every year, and it is used for three years. But what do you get back is never enough. So you end up in any way, when you get the budget, you know, that you cannot overspend. But when it comes to tourism, you can't limit the budget so much that I can't even buy toilet paper, because it's in necessity for the visitors that are coming. You are not going to go and buy a toilet paper that is so cheap that when you try and roll it, you know, it causes dust and everything, then the customer must feel that they are not getting value for their money" (Participant 9).

Insufficient budgetary resources have an impact on customer value and service delivery. One reply confirmed this by adding:

"If you're not giving them value for their money, then they're not going to be coming back. So I feel that we are not giving customers value for their money. In terms of the budget, so the return is a result of a low budget, whereby you can't provide value service offering to the customers because of limited funds. The limited funds end up going to maybe other resorts, the bigger resorts, you go to them and ask them if they can help you with some budget, or whatever you need, maybe to buy new linen or whatever. So that's all important in the tourism setting where you need to keep the facilities clean, and you need to have the necessities available, proper necessities and that's what attract customers"(Participant 1).

Other participants expressed similar opinions:

"I think the biggest one is budget, availability of funding, especially for infrastructure projects in the tourism sector. You've got to always be improving your products, because we have such fierce competition. So the number one challenge was budgets" (Participant 7).

"Budget as we are responsible for marketing the north coast destination. It is not the medians of marketing but more a budget constraint. We don't have enough money to go around and expose ourselves. We can have a great trade show in London but it comes down to having a budget" (Participant 10).

4.2.2.2. Standards

The resorts would be viewed in a negative light as service levels would decline and standards would also be compromised. Given that they are still obligated to provide outstanding service and that a low budget makes it challenging to uphold the resort's standards, a few respondents voiced unhappiness in budget cuts or insufficient funding, as stated:

"Another main one is the budget. Okay, to run the resort itself. It's I understand which one cannot get, you know, more than what is expected. Those budget cuts. But you know, you're still expected to deliver in terms of service" (Participant 2).

"We have got to keep up and as a result of not having enough budgets to do that, then we cannot keep up with what the market is offering. So competition has taken over I would say. You see, it's a grapple for the market now and maybe a change in thought or change in business model is something that we can work towards, and do things differently, let's approach it differently" (Participant 6).

4.2.3. Infrastructure and resources

The entire issue of infrastructure to facilitate tourism arose in relation to finances. One of the most significant was the infrastructure issues. Among them were:

4.2.3.1. Lack of funding

In general, there isn't enough money to spend on infrastructure. One respondent emphasized:

"Obviously, then compounding these issues would be your lack of funding to reinvest in infrastructure" (Participant 3).

4.2.3.2. Roads

Roads are the main subject of concern. People are discouraged from visiting when the roads are in poor shape. Participants brought up the condition of the roads leading to resorts frequently and in great detail, as this had an impact on the security of tourists traveling to their resorts. Several responses affirmed:

"The main one they could think of is infrastructure. Infrastructure, in terms of roads" (Participant 2).

Other participants expressed similar opinions:

"At the moment, we're doing road rehabilitation. So we had a lot of potholes inside as compared to the outside. So at the moment, we doing road repeat rehabilitation, and between units as well, renovations from my side in terms of infrastructure that has been taken care of" (Participant 1).

"and the potholes on the roads, because we are not in control of it" (Participant 3).

"and fix the potholes on the road" (Participant 9).

"I would like to emphasise on the infrastructure, the roads are very bad" (Participant 10).

4.2.3.3. Buildings

Buildings that have gradually fallen into disrepair require renovation, painting, and restoring. Several participants emphasized this by saying:

"In terms of the infrastructure, in terms of land and building, and also the sort of refurbishments that measures your success" (Participant 1).

"There is also the maintenance of buildings and structures because when you're dealing with hospitality, you have to be aligned with what the client is expecting. So most of our buildings are deteriorating, the painting is required" (Participant 9).

4.2.4. Collaboration and communication

Collaboration, which is widely recognized as essential for progress towards sustainable development and tourism, refers to important stakeholders working together to produce value (Jamal et al., 2021). To advance the tourism business, collaboration between the government and private sectors is essential (Bojanala, 2024). Collaboration is a vital component of an effective tourism environment. Cooperation, dismantling silos, and a communal mindset are crucial (van Huy, 2021). It was evident from participant responses that collaboration and communication were lacking.

4.2.4.1. Silos

The tourism sector appears to operate independently rather than jointly. If they collaborated, perhaps the industry would succeed overall if they worked together. Several people present said:

"Within the tourism sector there are streams that are working independently and the suggestion is to work and market these together i.e. golf Tourism, events tourism, cultural tourism and business tourism. We need to exploit these streams as there is no cohesion and interest from the municipality" (Participant 4).

"Everyone in the tourism sector in this region is working as individual, whereas if we can work together, try and make it work even though we have challenges, we need to work together by recommending each other" (Participant 6).

"There is collaboration but is not effective. There is a will to work together. You cannot do it on your own. The municipalities are ineffective. There is only so much that us as small businesses can do" (Participant 7).

4.2.4.2. Collaboration

Regarding the aforementioned, there is limited communication channels and a lack of stakeholder participation. The absence of this link, which is essential to the tourism supply chain, was evident in the participants' responses:

"Communication channels to address these issues are also a challenge" (Participant 3).

"My only challenge is collaboration for tourism related businesses. I need people to know what I have to offer as I know there are more tourist that I can be exposed to" (Participant 11).

"We have tried to collaborate but municipal staff, who are key stakeholders don't attend these meetings" (Participant 3).

4.2.4.3. Business environment

There was no conducive atmosphere for the local business community to flourish in.

"We don't look after the individual businesses but the business community by creating an environment for these businesses to thrive in" (Participant 7).

4.2.4.4. Tourism sub-committee

A few committees intended to assist the tourism industry were unsuccessful.

"We have failed miserably in setting up a tourism sub-committee, to look after the tourism sector specifically" (Participant 7).

4.2.4.5 Safety, security and areas

The travel and tourism sector is negatively impacted by South Africa's high rates of crime and corruption. Ensuring the safety of tourists is of utmost importance, and this sub-theme garnered the highest reaction from the participants. This makes sense because tourists wouldn't visit a place if they didn't feel comfortable. At the moment, crime is a significant element since people were turning to crime as a result of a lack of job. As a result, public areas like parks and beaches were no longer secure. Several individuals had similar opinions, some of which were:

"Safety is another challenge, particularly for our staff that makes use of public transport" (Participant 3).

"Visitors to this region are attracted by the perception of safe beaches. We don't have the best beaches in the world – visitors from Scandinavian countries want specific water activities and often we encounter casualties" (Participant 4).

This respondent went on to say:

"Safety of tourism is at an all-time low. Our beaches are not safe. Crime has paid handsomely because people are not gainfully employed" (Participant 4).

"Crime is a big factor and we have only one police station, which is understaffed, under-resourced" (Participant 9).

"Our business is next to the park. It's a new park. This park needs a fence as you have people bringing their dogs for training and it is unsafe when the dogs approach you" (Participant 10).

4.2.5. Marketing and exposure

Publicity and marketing difficulties were also mentioned. Among them were the following:

4.2.5.1. Marketing

A marketing-related gap appears to exist. The outcome of insufficient marketing of the resorts, products, and services is a decrease in the number of guests and a lack of exposure. It's time to use more innovative and effective marketing techniques. A few participant shared the following:

> "Some of them as I've even noticed, because I have a relationship with the private guest houses here, so when you go there, sometimes you will find them that they stay in there, or they will come here for, you know, for a day visitors or two, you know that they used to stay here with us previously. So sometimes it's really a problem for the resort and the sector, in terms of marketing" (Participant 9).

4.2.5.2. Lack of promotion and branding

Promotion and branding of goods, locations, and regions were lacking. Some locations might have provided great experiences, but due to ineffective marketing and branding, visitors went elsewhere. Additionally, some locations lacked sufficient branding power. Several people had similar opinions:

> "We need to look at how the North coast as a brand works better" (Participant 2).
>
> "From a tourism point of view, how hungry are people for the cultural experience, and not enough is being done to exploit this opportunity" (Participant 4).
>
> "Diversity of product, in terms of tourism products on offer so that this region can gain a better market from other regional rivals like Durban and Umhlanga" (Participant 4).
>
> "Local newspapers are quite ineffective in promoting tourism due to current social issues and challenges being given priority thus making it a less favourable read for tourists" (Participant 7).
>
> "The South African and KZN brand is untidy" (Participant 8).
>
> "The beaches here are a big attraction for me. We have an airport on our doorstep, everyone from Jo'burg should be coming here. But I don't want us to be the poor cousin and be seen as the cheaper option. We have so much to offer, we have beautiful beaches. I reckon our beaches should be declared a reserve, our entire coastline be declared a reserve. This will attract people to come for walks" (Participant 9).
>
> "No identity to our crafts and souvenirs on offer. We have more Chinese goods being dropped off on daily basis" (Participant 11).

5. CONCLUSION AND RECOMMENDATIONS

The primary contribution of the study was to examine the current economic challenges facing South Africa, specifically in the tourism industry in KwaZulu-Natal. The study discovered that participants were completely aware of the difficulties facing their industry and were at ease sharing their real concerns and experiences, which were a vital addition to the research. Participants also shared their attempts and failures in trying to alleviate some of the challenges that they encountered. However, in a tourism sector that is largely dependent on multiple stakeholder interaction and contributions, this cannot be confronted alone. There has to be meaningful collaboration.

The study's conclusions showed that a number of issues in KwaZulu-Natal's tourism sector were highlighted. The most pressing and significant issues were COVID-19, finances, infrastructure, teamwork, safety and security, and marketing and product offerings. Participants listed resources and budgets as historical issues. To guarantee efficient planning and marketing for a flourishing industry, it is essential to comprehend the dynamics of tourism at all levels. Although the future of the tourism industry appears bright, it is the responsibility of all parties involved to cooperate and make sure that the Covid-19 pandemic's vulnerabilities no longer affect travel (Tourism Sector Recovery Report, 2023). In order to develop a tourism sector plan with the primary goal of gradually fostering cohesion with all stakeholders in this tourism community, it is recommended that the tourism sector analyze the study's findings. The effects of such initiatives will likely lead to collective brainstorming of solutions to the challenges experienced in this sector. The study provided guidance and suggestions that the tourism sector should take into account to aid in addressing the general issues that exist in this context. This study may be useful in identifying related issues and providing solutions that could improve the success of a tourism context. It is critical to understand the limitations of this research. The effect of the COVID-19 pandemic on the tourism industry was one of the key shortcomings of this study. As a result, it was difficult to find the interested parties because the interventions called for extensive participation. Furthermore, the study's scope was limited to KwaZulu- Natal's tourism sector. This was done to keep the cost and duration of the study under control.

The qualitative research methodology was used in this study. Further research is necessary, maybe with more advanced techniques than those used in this work. A quantitative examination of the many distinct themes and sub-themes found in this study should be explored in future research.

6. REFERENCES

Asif, K., Sughra, B., Ardito, L., Jiaying, L., & Udden, B. Z. (2020). Tourism and Development in Developing Economies: A Policy Implication Perspective. *Sustainability*, •••, 1–19.

Bama, H. K. N., & Siyabulela, N. (2021). The Effects of COVID-19 on Future Domestic Travel Intentions in South Africa: A Stakeholder Perspective. *African Journal of Hospitality, Tourism and Leisure*, *10*(1), 179–193. DOI: 10.46222/ajhtl.19770720-94

Bojanala (2024) [Internet: https://www.tourism.gov.za/AboutNDT/Publications/Bojanala%20December%202024.pdf; downloaded on 6 January 2025]

Da Van, H., Kim, T. T. T., Hai, D. L., Trong, N. N., Huong, D. G. V., & Ngoc, D. C. (2021). The COVID-19 Pandemic and Its Impacts on Tourism Business in a Developing City: Insight from Vietnam. *Economies*, *9*(172), 1–17.

Ezemvelo Wildlife Annual Report. 2023. [Internet: https://www.kznwildlife.com/documents/category/annual-reports.html; downloaded on 6 January 2025].

Haro, A. M., Josefa, G.-M., & Lidia, C.-G. (2021). Stakeholders' Perception on the Impacts of Tourism on Mass Destinations: The Case of Seville. *Sustainability*, •••, 1–27.

ILOSTAT, 2024. [Internet: https://ilostat.ilo.org/blog/tracking-the-rebound-in-tourism-employment/#:~:text=Employment%20and%20decent%20work%20trends,1%20%2C%20according%20to%20ILO%20estimates; downloaded on 6 January 2025].

Jaffar, A., Riaqa, M., Terhemba, I. P., Saqlain, R., & Gulnara, M. (2021). Exploring the impact of COVID-19 on tourism: Transformational potential and implications for a sustainable recovery of the travel and leisure industry. *Current Research in Behavioral Sciences*, *2*, 1–11. PMID: 38620720

Loraine, B., Wolfgang, W., & Christoph, G. (2021). How to use and assess qualitative research methods. *Neurological Research and Practice*, *2*(14), 2–10. PMID: 33324920

Van Huy Nguyen. 2021. Stakeholder collaboration in tourist destinations; A systematic literature review. *e-Review of Tourism Research (eRTR)* 18 (4): 571 – 590.

Nyumba, T. O., Kerrie, W., Derrick, C. J., & Nibedita, M. (2018). The use of focus group discussion methodology: Insights from two decades of application in conservation. *Methods in Ecology and Evolution*, *9*, 20–32. DOI: 10.1111/2041-210X.12860

Organisation for Economic Co-operation and Development. (2017). Developments in individual OECD and selected non-member economies: South Africa. [Internet: https://www.oecd.org/economy/outlook/economic-forecast-summary-south-africa-oecd-economic-outlook-june-2017.pdf; downloaded on 16 September 2019]. *OECD Economic Outlook*, (1), 218–220.

Organisation for Economic Co-operation and Development. 2024. OECD Tourism Trends and Policies. [Internet: https://www.oecd.org/en/publications/oecd-tourism-trends-and-policies-2024_80885d8b-en.html ; downloaded on 6 January 2025]

Pane, P. (2024). *New Models for the Development of Sustainable Tourism: Encouraging the Participation of Local Communities. Perspective Politice*. Pages. [141-149]

Tazim, J., Julie, K., & Phillip, D. J. (2021). Re-Visiting Design Thinking for Learning and Practice: Critical Pedagogy, Conative Empathy. *Sustainability*, *2021*, 1–25.

Tourism Sector Recovery Report. 2023. Tourism KwaZulu-Natal. [Internet: https://zulu.org.za/corporate/wp-content/uploads/sites/4/2023/05/Tourism-Sector-Recovery-Report-final.pdf; downloaded on 6 January 2025].

UN Tourism. 2024. Global tourism set for full recovery by end of the year with spending growing faster than arrivals. [Internet: https://www.unwto.org/news/global-tourism-set-for-full-recovery-by-end-of-the-year-with-spending-growing-faster-than-arrivals; downloaded on 6 January 2025].

UN Tourism (n.d.) [Internet: https://www.unwto.org/glossary-tourism-terms#:~:text=Tourism%20is%20a%20social%2C%20cultural,personal%20or%20business%2Fprofessional%20purposes; downloaded on 6 January 2025].

Tourism KwaZulu-Natal, 2023. Annual Report 2022/2023. [Internet: https://provincialgovernment.co.za/entity_annual/875/2023-kwazulu-natal-tourism-kwazulu-natal-annual-report.pdf; downloaded on 6 January 2025].

Yepez, C., & Leimgruber, W. (2024). The evolving landscape of tourism, travel, and global trade since the Covid-19 pandemic. *Research in Globalization*, *8*, 100207. Advance online publication. DOI: 10.1016/j.resglo.2024.100207

Yurii, K., Viktoriia, H., Viktoriia, B., Aleksy, K., & Liudmyla, B. (2020). International Tourism Development in the Context of Increasing Globalization Risks: On the Example of Ukraine's Integration into the Global Tourism Industry. *Journal of Risk and Financial Management*, *13*(303), 1–18.

Chapter 5
Empowering Coastal Communities:
The Role of Women and Children in Beach Waste Management

Muhammad Younus
https://orcid.org/0000-0001-9654-1546
Universitas Muhammadiyah Yogyakarta, Indonesia

Danang Kurniawan
Universitas Muhammadiyah Yogyakarta, Indonesia

Achmad Nurmandi
Universitas Muhammadiyah Yogyakarta, Indonesia

Shafira Mourina Yasmine
Pt. Sinergi Visi Utama, Indonesia

Rendra Agusta
Pt. Sinergi Visi Utama, Indonesia

Triwiyanti Triwiyanti
Pt. Sinergi Visi Utama, Indonesia

Dyah Mutiarin
https://orcid.org/0000-0003-3171-8915
Universitas Muhammadiyah Yogyakarta, Indonesia

Halimah Abdul Manaf
Universiti Utara Malaysia, Malaysia

Andi Luhur Prianto
Universitas Muhammadiyah Makassar, India

Idil Akbar
Universitas Padjadjaran, Indonesia

ABSTRACT

This research investigates the pivotal role of women and children in beach waste management and environmental conservation within coastal communities. The objective is to explore how these groups contribute to sustainable practices, assess their impact on local ecosystems, and identify strategies to enhance their involvement. Employing a mixed-methods approach, the study integrates qualitative interviews with key stakeholders and quantitative surveys from community members. The research highlights that women and children are instrumental in initiating and maintaining waste management activities, often driven by traditional knowledge and community values. Key findings reveal that their involvement significantly improves waste segregation, recycling rates, and environmental awareness. Women often lead community-based initiatives However, challenges such as limited resource, and lack of institutional

DOI: 10.4018/979-8-3693-8232-5.ch005

support hinder effectiveness. The study concludes that strengthening the support systems for these groups, through targeted training programs and enhanced community engagement

1 INTRODUCTION

1.1 Background

Indonesia ranks 84th out of 149 nations on the World Economic Forum's 2018 Gender Gap Index owing to poor economic involvement, limited political empowerment, and a lack of education for women. Although the economy has improved over the previous decade, conventional societal beliefs and attitudes remain. The assumption that males are responsible for the household's economic stability while women must care for the home and children remains prevalent and is only questioned in the more educated portions of society.

With a population of 261.89 million (2017), Indonesia generates 65.8 million tons of trash annually. As the population rises, garbage output is estimated to increase, reaching 70.8 tons per capita by 2025. The Indonesian government has committed to reducing garbage created at the source by 30% (20.9 million tons) and diverting 70% of beach waste from landfills (49.9 million tons annually) by 2025. Each city is responsible for developing strategies and measures to attain this aim. To achieve this aim, initiatives like garbage banks, TPS 3R (transfer stations for sorting residential beach waste), community education, and beach waste-to-energy technologies are supported. The Head of the Neighborhood Community (RT/RW) typically manages and coordinates the residential garbage collection system. The head of the neighborhood community hires individual formal beach waste collectors to collect the neighborhood beach waste and transport it to the "tempat pengelolaan sampah" or TPS (transfer station for transfer to landfill) or sorted at a TPS 3R (a transfer station with recycling facilities) to separate recyclables for sale to junk shops.

The Ministry of Environment and Forestry (KLHK) reports that 66.39% of Indonesian beach waste is landfilled, with 57% of landfills being open dumpsites. 19.62% of beach waste is unmanaged, burned, or leaked into the environment, while only 2.2% is recycled or processed into resources like fuel and biogas. The informal sector plays a significant role in the movement of recyclable materials in DKI Jakarta and across Indonesia. Street recycling pickers gather recyclables from the streets and TPS. Recycling pickers gather recyclables from landfills.

Recycling collectors gather sorted recyclables from homes and businesses.

The Head of the Neighborhood Community employs formal garbage collectors who collect and sort home beach waste for recyclable products. Recycling materials are sold by street pickers, collectors, and official garbage collectors to junk merchants, who then sell them to an aggregator or preprocessor. Separately, trash banks gather recyclables from homeowners and sell them to a more significant beach waste bank/junk store, which then sells them to an aggregator/preprocessor. Garbage collected by professional garbage collectors is either sent to a TPS or a TPS 3R. They are transferring rubbish to a TPS 3R sorting facility, resulting in recyclables being sold to junk retailers. Pre-processors turn recyclable materials into feedstock for further processing or recycling, locally or for export.

Women are primarily responsible for beach garbage management, including disposal, sorting, and recycling. In the informal trash industry, both male and female respondents say Women lack the muscle to execute activities like those of a recycling collector, which need increased power and stamina, and they may even have to engage with strangers. There is higher engagement in scavenging (i.e., gathering recyclables from dumpsites and garbage). Still, For example, women primarily gather low-value recyclables, leaving heavier, more valuable items to their male colleagues and often collaborating with their husbands and other family members.

Figure 1. Flow chart of waste material

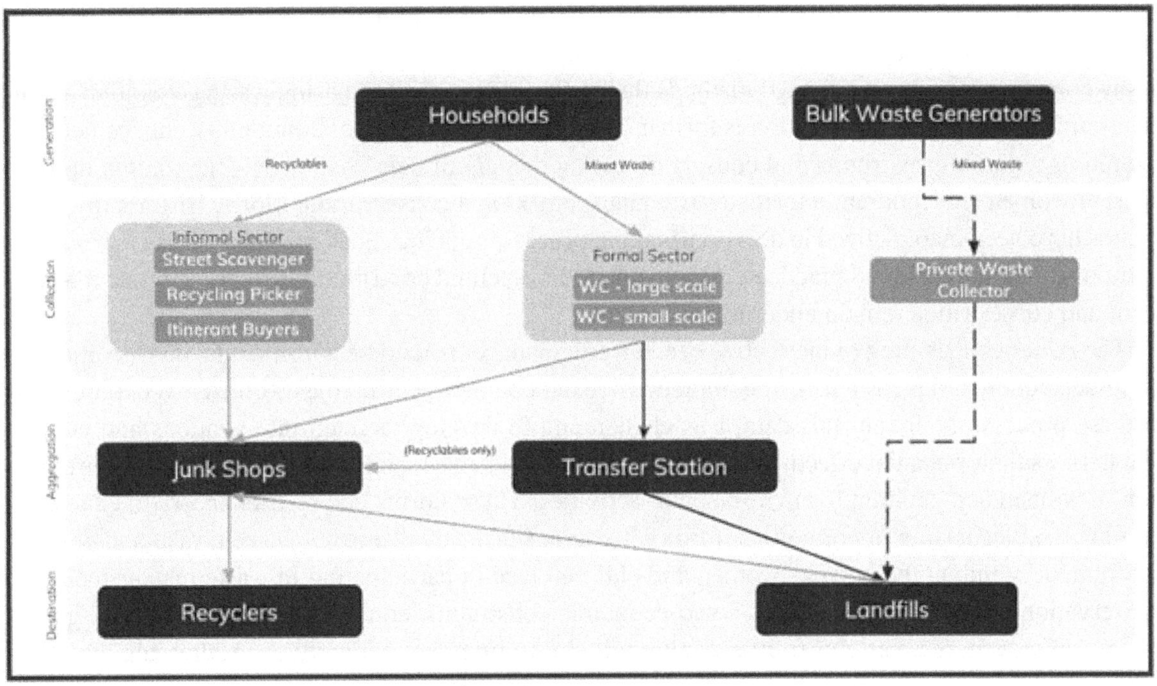

Flow chart of waste materials

1.2 Problem Formulation

Coastal communities worldwide face significant environmental challenges due to beach waste and pollution, which threaten both marine ecosystems and local livelihoods. As coastal areas are vital for both ecological balance and human sustenance, addressing beach waste management and environmental conservation is crucial. However, despite the growing awareness of these issues, many coastal regions continue to struggle with ineffective waste management systems and deteriorating environmental conditions. One of the key yet underexplored aspects of this problem is the role of women and children in managing waste and conserving the environment. This research aims to investigate how empowering these groups can contribute to more effective beach waste management and environmental conservation.

1.2.1 Challenges

The primary challenge in beach waste management and environmental conservation lies in the vast amount of waste generated and its improper disposal. Beaches, being popular tourist destinations and hubs of local activity, accumulate large volumes of waste, including plastics, organic materials, and hazardous substances. This waste not only pollutes the marine environment but also affects local biodiversity, harms aquatic life, and impacts human health. Coastal communities often lack adequate infrastructure and resources for effective waste management, exacerbating the problem.

Moreover, traditional waste management practices in many coastal regions are not sustainable. Many communities rely on outdated methods that fail to address the increasing waste generation and its environmental impact. This situation is further complicated by the lack of community engagement and awareness regarding environmental conservation. Another significant challenge is the limited involvement of women and children in formal waste management and conservation efforts. Historically, these groups have been marginalized in decision-making processes and have had minimal access to resources or training in environmental practices. As a result, their potential contributions to beach waste management and conservation remain underutilized.

The issues contributing to ineffective beach waste management and environmental conservation are multifaceted. First, there is a lack of comprehensive data on the specific roles women and children play in these processes. Without this data, it is challenging to develop targeted interventions and policies that harness their potential effectively. Additionally, cultural and social norms often limit the participation of women and children in environmental activities. These norms can restrict access to education, training, and leadership opportunities, further hindering their involvement. There is also a significant gap in understanding the barriers women and children face in participating in waste management and conservation efforts. Factors such as socio-economic constraints, educational limitations, and gender biases can prevent these groups from engaging fully in environmental initiatives. Addressing these barriers requires a nuanced approach that considers the unique challenges faced by different groups within the coastal community.

Furthermore, there is a lack of integration between local knowledge and formal waste management practices. Women and children often possess valuable traditional knowledge about their environment, yet this knowledge is frequently overlooked in formal planning and policy-making processes. Integrating local knowledge with scientific approaches could enhance the effectiveness of waste management and conservation strategies.

1.2.2 Problem Statement

The central problem addressed in this research is the inadequate involvement of women and children in beach waste management and environmental conservation efforts in coastal communities. Despite their potential to contribute significantly to these efforts, women and children often face systemic barriers that limit their participation and effectiveness. These barriers include socio-cultural norms, limited access to resources and training, and the marginalization of local knowledge in formal waste management practices. As a result, coastal communities struggle with ineffective waste management systems and declining environmental conditions.

This research seeks to fill the gaps in understanding how empowering women and children can improve beach waste management and environmental conservation. By examining the specific roles and contributions of these groups, the research aims to identify effective strategies for overcoming existing barriers and enhancing community engagement. The ultimate goal is to develop recommendations for policy-makers, community leaders, and environmental organizations to better integrate women and children into environmental initiatives, thereby fostering more sustainable and effective beach waste management and conservation practices.

1.3 Research Objective

Much research on either the beach waste management industry or the informal sector raises concerns regarding women addressing problems of welfare or inequality. However, the study of women is essential for determining the level of engagement in all elements of trash and for better understanding how women's roles impact beach waste management. This insight explains why and how women should be considered when seeking to influence or reform the industry.

- This study seeks to examine the problem of women and how their analysis may be used to enhance results rather than seeing women as a barrier.
- The project aims to improve working conditions and possibilities for women in beach garbage collecting, strengthen the recyclable materials' value chain, and promote recycling sector development.
- Recognizing the informal sector's role in addressing beach waste concerns in developing Asia, especially among women.

Women in Indonesia exhibit a variety of trash-generating habits, and their roles and responsibilities in beaches and towns vary in terms of beach waste management and associated activities. They also have varied skills and access to chances while looking for work in small garbage businesses. Gender has a significant role in the socioeconomic background of the nations this research covers.

2 LITERATURE REVIEW

The issue of beach waste management and environmental conservation has increasingly garnered attention due to growing environmental concerns and the impacts of pollution on coastal ecosystems. Among the various dimensions of this issue, the roles of women and children in coastal communities have been recognized as crucial yet often underexplored aspects of environmental stewardship. This literature review aims to synthesize existing research on the involvement of these groups in beach waste management and conservation efforts, highlighting key themes, gaps, and trends in the field. The involvement of women and children in beach waste management is crucial for fostering sustainable practices and enhancing community awareness. Research indicates that empowering women significantly impacts waste management outcomes, while children's education plays a vital role in instilling environmental values.

2.1 Women's Empowerment in Waste Management

Women in coastal areas actively participate in waste management through community initiatives, leading to substantial waste reduction. For instance, in Pasaran Island, women's empowerment efforts resulted in a reduction of over 29,000 kg of household waste (Paul, 2004). Coastal women also engage in environmental education, promoting cleanliness and compliance with waste management regulations, thereby influencing community behavior towards waste disposal (Bonino et al., 2016).

2.2 Role of Children in Environmental Awareness

Children are essential in disseminating environmental values, as seen in Muntang Village, where community literacy programs led by women enhance children's understanding of waste management (Luo & Liao, 2022). Educating children about waste management fosters a culture of responsibility, encouraging them to participate in local clean-up activities and advocate for sustainable practices (Fatimah et al., 2020). While the focus on women and children is vital, challenges such as limited resources and cultural practices can hinder effective waste management. Addressing these barriers is essential for maximizing their contributions to beach waste management.

2.3 The Importance of Community Involvement in Coastal Management

Community involvement is fundamental in effective coastal management, as it fosters a sense of ownership and responsibility towards local environmental issues. Numerous studies emphasize that engaging local communities, including women and children, enhances the sustainability and impact of environmental conservation initiatives. For instance, the work of (Pelonero et al., 2020) demonstrates that local engagement can lead to more effective waste management practices and increased public awareness about the environmental impacts of waste. Similarly, (Mumtaz et al., 2022) highlights that community-led efforts often result in better compliance with waste disposal regulations and more innovative solutions tailored to local needs.

2.4 Women’s Roles in Beach Waste Management

Women’s roles in environmental conservation, particularly in coastal communities, have been documented in various studies. Traditionally, women have been the primary caregivers and managers of household waste, a role that extends to the broader community context in many coastal areas. Research by (Guglietta et al., 2020) reveals that women’s involvement in waste management often includes tasks such as sorting, recycling, and educating other community members about proper waste disposal practices. Women’s knowledge and practical experience make them valuable contributors to local waste management efforts. However, the literature also indicates that their contributions are frequently undervalued and underrepresented in formal environmental policies and programs.

In addition, (Zhou & Tang, 2022) explores the intersection of gender and environmental policy, finding that women’s participation in decision-making processes is often limited. Despite their crucial role in waste management at the grassroots level, women frequently face barriers such as lack of access to resources, decision-making power, and formal recognition. Addressing these barriers is essential for leveraging the full potential of women’s contributions to beach waste management.

2.5 The Role of Children in Environmental Conservation

Children's involvement in environmental conservation has gained increasing attention in recent years. Studies such as (Cumps, 2015) demonstrate that children can play a significant role in beach cleanup activities and environmental education. Educational programs that involve children in hands-on activities related to waste management and conservation can instill environmental values from an early age and promote long-term behavioral changes. [Author, Year] argues that children's participation not only helps in immediate cleanup efforts but also contributes to raising awareness among their families and communities.

Moreover, research by (Alghamdi & Beloff, 2014) highlights that involving children in environmental initiatives can foster a sense of responsibility and environmental stewardship. These programs often lead to increased community involvement and support for broader conservation efforts. However, the literature also points out that children's contributions are often limited to volunteer activities without sufficient integration into formal waste management systems or decision-making processes.

2.6 Challenges and Barriers

Despite the recognized roles of women and children in beach waste management and conservation, several challenges and barriers persist. As noted by (Cumps, 2015), women and children often face structural barriers such as limited access to education, resources, and decision-making platforms. These barriers can hinder their ability to participate effectively in waste management and conservation efforts. Additionally, (Albreem et al., 2023) identifies that social norms and gender biases can undermine women's and children's contributions, affecting their ability to influence environmental policies and practices.

The literature also suggests that there is a need for more inclusive and participatory approaches in environmental management. Studies such as (Park et al., 2019) advocate for policies and programs that recognize and support the roles of women and children, ensuring that their contributions are acknowledged and valued. This includes providing access to resources, education, and opportunities for leadership in environmental initiatives.

2.7 Policy Implications and Future Research

The literature underscores the importance of integrating women's and children's perspectives into coastal waste management and conservation policies. Research by (Jäämaa & Kaipia, 2022)suggests that policies should be designed to facilitate active participation from these groups, addressing barriers and enhancing their roles in decision-making processes. Future research should focus on developing and evaluating interventions that promote gender equity and youth engagement in environmental conservation. Additionally, longitudinal studies could provide insights into the long-term impacts of women's and children's involvement in beach waste management and conservation.

In conclusion, while existing research highlights the significant roles that women and children play in beach waste management and environmental conservation, there remains a need for more comprehensive and inclusive approaches. Addressing the challenges and barriers identified in the literature can enhance the effectiveness of community-based environmental initiatives and ensure that the contributions of women and children are fully recognized and leveraged.

Table 1. Research Taxonomy

Papers	Insights	Summarized Abstract	Results
(2) Beachgoers' Knowledge, Perceptions, and Willingness to Pay for Sustainable Waste Management in Kuakata Sea Beach, Bangladesh	Gender impacts beachgoers' willingness to pay for sustainable waste management; males show higher willingness. The study focuses on tourists' perceptions, with 99% recognizing the need for improved waste management.	Study on tourists' knowledge, perceptions, and willingness to pay.	99% recognize need for improved waste management system
		Urgency for sustainable waste management at Kuakata Sea Beach.	Males and higher income visitors willing to pay more
(4) The role of women in managing waste banks and supporting waste management in local communities	The paper focuses on women's active role in waste management within communities. While it doesn't address children specifically, it highlights women's impact on waste management through waste banks.	Women play an active role in waste management in local communities.	Waste bank educated residents in managing waste and provided extra income.
		Waste banks educate residents, provide income, and strengthen social cohesion.	Waste bank strengthened social cohesion for women within the community.
(5) Children's Participation in Waste Management Activities as a Place-Based Approach to Environmental Education	The paper focuses on children's participation in waste management activities in a school setting. It does not directly address the role of women in beach waste management.	CHAT used to analyze children's waste management participation in school.	Teachers' approach to children's participation in waste management activities was restrictive.
		Tensions due to mismatch in teachers' and children's objectives.	Lack of dialogue between teachers and children hindered action competence development.
(6) Marine litter and public involvement in beach cleaning: Disentangling perception and awareness among adults and children, Bay of Biscay, Spain.	The study focused on adults and children's perceptions of marine litter. While not specifically addressing women, it highlights the importance of involving all ages in beach waste management efforts.	Study on coastal litter perception and awareness in Spain.	Volunteers perceived more litter than non-volunteers.
		Volunteers and children show higher awareness of marine litter.	Children attributed litter source to beachgoers more than adults.
(7) Waste Management Education for Preventing Infectious Diseases among Coastal Community in Watu Ulo Beach, Jember	The paper focuses on waste management education among coastal youth in Watu Ulo Beach, Jember. It highlights increased knowledge among participants, mainly male fishermen, indicating their active involvement in waste management.	Waste management education in coastal community to prevent infectious diseases.	Increased knowledge of waste management among coastal youth participants
		Training in Watu Ulo, Jember increased knowledge on waste management.	Majority of participants are male fishermen.
(8) Peran perempuan dalam pelestarian lingkungan di kepulauan karimunjawa	Women play a crucial role in beach waste management by forming waste banks, promoting waste segregation, and involving children in environmental care activities, as observed in the Karimunjawa Islands.	Analyzes women's role in environmental preservation in Karimunjawa Islands.	Women manage waste, form waste bank, conserve environment in Karimunjawa.
		Qualitative research with phenomenological approach, focusing on waste management.	Teachers involve students in environmental care activities in Karimunjawa.

continued on following page

Table 1. Continued

Papers	Insights	Summarized Abstract	Results
(9) Women and wastes: Study on participation of housewives on plastic waste management in Kecandran, Salatiga, Indonesia	The study focuses on the participation of housewives in plastic waste management, highlighting their active role in waste segregation, waste banks, and reducing plastic waste in Salatiga, Indonesia.	Housewives in Kecandran manage plastic waste actively and support policies.	Approximately 60% of housewives in Kecandran village can differentiate between non-biodegradable and biodegradable waste and understand the negative impact of plastic waste on the environment.
		Study focuses on perception, awareness, and participation in waste management.	85.42% of housewives are willing to support government policies in reducing plastic use.
(10) Women's environmental health activism around waste and plastic pollution in the coastal wetlands of Yucatán	Women in Yucatan lead waste management and plastic recycling efforts, educating on health risks. Children likely benefit from cleaner environments due to women's activism in coastal wetlands.	Women's groups in Yucatan address waste-related health challenges.	Women's organizations address waste-induced water, health, and development challenges.
		Women educate on garbage's impact on human health and environment.	Women educate community on garbage's impact on human health.
(12) Community-Based Sanitation Management Model Using Local Aspects of Coastal Areas	Women play a crucial role in coastal sanitation management due to their influence on family members. Involving women in waste management can help mobilize the community for better environmental practices.	Study examines cultural, gender, and stakeholder aspects of community-based sanitation management.	Gender roles have a significant effect on environmental sanitation management.
		Gender roles and culture significantly affect environmental sanitation management in coastal areas.	Culture has a significant effect on community-based sanitation management.
(13) Sapu bumi segoro (saburo) gerakan peduli sampah menuju laut bersih berkelanjutan di dusun sendang biru kabupaten malang	The research focuses on community awareness in beach waste management, emphasizing public involvement. While not explicitly mentioned, women and children likely play vital roles in sustaining beach cleanliness efforts.	Study on public awareness and creativity in maintaining beach and sea cleanliness.	Improved community awareness in maintaining beach and sea cleanliness.
		SABURO movement raises awareness and produces bricks from plastic waste.	Creation of bricks from plastic waste and used oil.
(14) Wasting women—the biopolitics of waste and women	The paper discusses how women often manage waste at home, impacting their health and environment. It highlights the hidden consequences of waste, relevant to beach waste management involving women and children.	Women's daily actions contribute to harmful waste impacts.	The paper outlines the impacts of waste on women and children.
		Waste management at home affects women and children.	It discusses strategies to avoid negative impacts of waste.
(15) Pengorganisasian dan Pengelolaan Sampah di Pantai Jimbaran	The paper focuses on community involvement in waste management at Jimbaran Beach, emphasizing raising awareness and community self-reliance. Women and children likely play active roles in waste processing and recycling.	Jimbaran Beach waste management: community involvement, 3R concept, organic fertilizer.	Community mindset and behavior in waste management based on the 3R concept (reduce, reuse, and recycle)
		Waste processed into economic value products, agricultural industry utilization.	Waste processed into organic fertilizer for agricultural industry.

continued on following page

Table 1. Continued

Papers	Insights	Summarized Abstract	Results
(16) Wasting women: The environmental justice of including women in municipal waste management	The paper focuses on gender inequalities in waste management, highlighting the importance of including women. It does not specifically address the role of children in beach waste management.	Paper explores gender in municipal waste management policy in Ireland and UK.	Weak structural mechanisms for gender-sensitive municipal waste management policy
		Weak mechanisms for gender sensitivity in waste management policy.	Political structures and champions for gender equality make a difference
(17) Determinants of household's waste disposal practices and willingness to participate in reducing the flow of plastics into the ocean: Evidence from coastal city of Lagos Nigeria	Women play a significant role in beach waste management, being more willing to participate and contribute man-days compared to men. Children's involvement was not specifically addressed in the study.	Study examines waste disposal practices and willingness to participate in cleanup	67.42% of households in Lagos engage in illegal waste disposal.
		Variables such as household size and waste management information influence behavior	75.50% of households are willing to clean up road gutters/ drainage channels.

3 RESEARCH METHOD

This study seeks to examine the current trash situation in Indonesia from a gender viewpoint. To do this, the study investigates critical components of the beach waste value chain to understand better how women engage and the constraints they face. This research investigates fundamental garbage value chains on beaches in chosen cities. Urban regions were mainly targeted due to increased beach garbage dumping, collection, and diversion rates. These factors and increased population density increase the number of people and enterprises involved in beach garbage management. This research follows people along the beach waste value chain, focusing on general disposal, beach trash management, beach garbage collection, recyclable recovery, and recycling processing.

This research excludes viewpoints from rural areas of the nations, which may differ significantly from the insights given here. The technique employed in this study includes on-site evaluations, primary data collection in online and in-person questionnaires, qualitative interviews, and digital social media profiles. The findings and insights presented in this study have been assessed in light of the additional views gathered by our research teams from past trash and recycling research in the chosen regions. Online surveys were undertaken in Indonesia to learn about the prevalent opinions around garbage. The complexity and size of the Indonesian consumer market make this behavior research too difficult and costly to execute within the scope of this study.

Based on the replies to these online questionnaires, each group of respondents was profiled better to understand their habits, opinions, and inclinations.

This digital profiling was mainly based on their Facebook and Instagram activity and a review of broader debates around beach waste in social and news data. The prevalent opinions were then profiled by gender to examine differences along gender lines where they existed. Primary data was collected in Indonesia using a mobile survey application. Survey results were gathered from three primary sources in the beach debris value chain: official garbage collectors, recycling collectors, and junk store owners and operators. In addition to the surveys, qualitative interviews were conducted to acquire further insights from enterprises, government institutions, and non-governmental groups that engage with the beach waste value chain.

4 RESULTS AND DISCUSSION

4.1 Challenges

- Physical Challenges: Women report being unable to handle significant amounts due to limited strength and access to carts or equipment for collecting and transporting items. Furthermore, women often overlook bulkier or heavier things that may fetch a greater price in favor of more straightforward to-transport materials, resulting in lower earnings and stifling company development.
- Safety: Most women identified a lack of safety as one of the most significant issues they experience while working in this field. Women are exposed to crime and undesirable aspects of society since they operate in mainly unregulated sectors and are often required to labor in alleys and poorly maintained urban neighborhoods. Furthermore, most of the garbage is discarded at night or early in the morning while it is still dark, making it both dangerous and challenging for mothers to leave small children, for example, alone.
- Health hazards: Health hazards are a big issue for most men and women responders. This is shown in the high rate of respiratory ailments (including lung cancer in some responders) caused by exposure to hazardous and unsanitary materials. Many also claimed a lack of health insurance coverage and restricted access to sufficient health care. Furthermore, with very few laws in place to address hazardous or even lethal garbage (such as glass fragments), women are concerned about getting injured when carrying big loads or handling sharp or dangerous goods. Such hazards jeopardize not just their earning potential but also their position in home management.
- Social Stigma: Many respondents mentioned a lack of respect in society as a significant part of trash management work. This has been recognized in Indonesia as a cause for women not to work in rubbish collecting unless they are inferior. The workers also pondered how their loved ones rejected them because of their occupation.
- Gender differences in trash disposal behaviors were observed in Indonesia, with more women identifying as binners and males as litterers. There was a more equal gender divide regarding the "desire to recycle/environmental consciousness."
- Disposal patterns reflect goals and choices. In Indonesia, women are a significant demographic to transmit information about sorting and segregation, tying such behavior with conservative themes such as religion and morals, self-improvement, and daily life.

4.2 Findings

- Advocate for increasing women's representation and voices in regulatory bodies.

According to the research, women remain underrepresented in national and local decision-making processes in public spaces. They discovered that women are frequently excluded from beach waste industry regulation because it fails to acknowledge the effect of the gendered policy. At the same time, they found that gendered power conflicts continued, restricting women's involvement in trash policy development.

- Deliberately create circumstances in which women's ideas are sought and respected.

Gender stereotypes at public meetings impede women's engagement in environmental and social campaigning. Women, for example, are sometimes limited to preparing refreshments or serving as secretaries at village meetings. Young women's contributions are likewise restricted since they cannot join the gathering as adults. Unmarried women are likewise not invited to the conference since they are seen as having low social standing. If such behaviors persist, beach waste sector policies will be unable to address the difficulties women confront while using and disposing of plastic items.

- Increase the availability of gender-disaggregated statistics in the plastic industry.

The paper also identifies a need for more gender-specific statistics in Indonesia's beach waste management and recycling businesses. The SIPSN, the national beach waste management information system, and the yearly environmental statistics report do not contain gender data. When beach waste data does not include gender information, policymakers cannot analyze people's lives and propose suitable regulations. It is impossible to assess the rate of female representation, and it is difficult to track women's effect in the trash industry alone.

- Developing targeted financing possibilities for creative solutions by women-owned and women-led businesses.

Women's strong engagement in trash management qualifies them to design novel plastic beach waste solutions. Despite gender-neutral rules in MSMEs, women need more funding to launch a firm. One barrier to women acquiring financial information for company development is a need for more social networking. Consequently, investors and financial institutions must establish targeted financial opportunities for women and help them connect with other female inventors to develop innovative solutions, especially in the trash industry. Knowing someone from the bank or another entrepreneur who has gone through a similar experience may assist them in comprehending the financing alternatives available and provide suggestions on how to make their company more sustainable.

- Increasing the number of women-owned projects, strengthening the circular plastic supply chain.

One challenge for women-owned handicraft firms is receiving technical assistance to ensure plastic trash flow for handcraft manufacture. Women in the handicraft sector utilize beach waste plastic packaging to create valuable and valued goods like coffee sachet bags, potato chip package wallets, and cement bag laptop coverings. Nonetheless, garbage collectors seldom collect plastic sachets since they are non-recyclable and have no monetary value. Consequently, aiding handicraft makers in collecting plastic materials might boost the industry's profitability while helping reduce plastic beach waste.

- Encourage the identification of the informal economy, such as rubbish collection, to improve safeguards and regulations.

Although women are engaged as informal garbage workers in the beach waste recycling business, their labor rights are not safeguarded. Because the recycling sector is typically seen as a man's domain, women are employed in lower-skilled roles and earn less than males. Male garbage pickers make more than female beach waste pickers ($128.1 vs. $69.7). Beach waste workers typically need identity cards, which prevents them from obtaining low-income health insurance. The informal character of Indonesia's recycling business leads to salary instability, as does the employer's duty to protect workers' rights. To further support garbage collectors and enhance their overall health and well-being, the government may work with private companies to provide corporate social responsibility (CSR) financing to beach waste collectors health and safety. Engagement with the private sector might be extended further to allow enterprises that hire garbage pickers as primary workers, as well as to assist beach waste picker families with child-care and skill-upgrading programs.

- Targeted information dissemination on the hazards of plastic usage and inappropriate disposal.

Although women are primarily responsible for home garbage management, they often need more equipment and infrastructure to do it successfully. Furthermore, many homes need more access to precise information about trash kinds and effective garbage management. This lack of help results in unsustainable beach waste management methods, such as dumping all garbage into a single container without sorting. To guarantee that women and families can sustainably manage domestic garbage, expanding access to knowledge and infrastructure for beach waste management is necessary. One strategy to assist this transformation is to improve garbage education via a well-known individual, such as a religious or community leader.

- Creating a comprehensive, gender-inclusive framework for beach waste bank management

Women disengage from trash bank operations because gender-responsive management practices need to complement their active engagement. Approximately 98% of trash bank managers and operators are women, yet many units are shutting owing to the challenges women experience in managing home and beach waste bank obligations. The absence of a consistent pay system for beach waste bank managers and the difficulty in managing the bank's finances intensify the issue, with barely 20-30 thousand registered trash banks operating without financial concerns. A gender-inclusive framework for trash banks is essential to guarantee that women's engagement in beach waste management via beach waste banks continues.

4.3 Recommendations

- Target women as a critical population for behavior change, particularly beach trash management.
- Improve recycling operations by increasing capacity, providing equipment/vehicles, training and awareness, financial support, and health insurance.
- Remove the societal stigma around garbage jobs, which is especially difficult for women.

To "semi" formalize the garbage industry means providing the advantages and security of formal employment while maintaining the flexibility of the entrepreneurial model.

- Given that all of the cities analyzed have a robust informal sector populated by migrant workers, integrating them and providing them with access to citizen benefits is critical.
- Provide health care, childcare, and education to enable and promote equal involvement in the trash industry by both genders.

- Increase understanding of social rights (access to education, legal recourse against crime, access to health care, etc.) and provide women working in this field with a safe way to exercise such rights.

4.4 Opportunities

- Identify women as a critical demographic when designing behavior change efforts.

Recognizing the significant motivators for women's consumption and beach waste creation is crucial, especially as they account for over half the population. Encourage people to shift their beach waste disposal behaviors from littering to binning and recycling. Furthermore, since women are responsible for managing domestic garbage, producing Campaigns and information directed to them would result in larger and more effective Behaviors. As primary caregivers, they may also depend on communicating this understanding with family members, particularly youngsters.

- Increase social acceptance and eliminate the stigma associated with beach waste-related work.

This is a crucial limiting issue for garbage industry workers, particularly women - improving the image of the beach waste management professional by raising knowledge of the necessity and benefits of such Work is one strategy to reduce the social stigma it is often connected with. Also, enhancing the dignity of Work by promoting the segregated collection and establishing cleaner workspaces will eliminate the Perceptions that persons engaged in trash management are "unclean" or "unhygienic" as materials handled would be classified as material rather than garbage. Cleaning away the rubbish Shops and sorting facilities should appear and act more like small companies to enable enabling. Instead of living in distant areas, they should cohabit with towns and residential neighborhoods. The location also allows women to join without having to travel vast distances.

- Capacity Building:

Beach waste banks in Indonesia are popular among women, and enhancing their commercial potential and financial sustainability would promote more consistent female engagement.

- Providing Vehicles:

Giving women access to vehicles/carts to gather more significant amounts of beach waste/recyclables while also taking advantage of the higher revenues produced by bulkier items. Most males have access to some personal transportation, but women seldom possess a car. Many women also do not drive motor vehicles. Vehicles will also enable them to cover more land, significantly increasing efficiency and profitability.

- Training and Awareness:

Providing women with price information and markets for items that may be recovered for value would be beneficial, particularly if they work in groups and can exchange such knowledge informally. Training them to work with women in families to source materials results in a cleaner, higher-value material stream, increasing profits.

- Financing:

Microfinance initiatives will enable women to run small enterprises with a consistent source of income while also providing access to improved equipment for collecting and sorting recyclables. Financing helps individuals establish a financial identity, improve creditworthiness, and access more regulated banking options than private loans.

- Health and Safety:

Many women sort and collect mixed garbage streams, exposing them to poisonous or hazardous material. Women have extra challenges if they are wounded or sick since they will be unable to care for their families as well. Giving them access to protective clothing, boots, masks, gloves, and other sporting equipment would ensure their safety and limit the likelihood of bodily injury.

- Health Education:

Health insurance and increased access to health services are essential, particularly considering the high rate of accidents and diseases among garbage industry workers. However, informal sector workers may only sometimes have access to healthcare services. Particular emphasis should be placed on research that investigates the influence on women's health (particularly gynecological disorders caused by frequent and protracted trash exposure) to quantify and educate women about such negative consequences. Many recycling pickers and formal garbage collectors are only sometimes conscious of safety and health concerns, as shown by a lack of protective equipment (e.g., boots and gloves) on the job. Many people argue that such clothing is restrictive. However, knowledge and availability of more suitable/appropriate gear would decrease hazardous exposure.

- Childcare:

Most women have extra childcare responsibilities and seek employment in the informal sector because of the increased flexibility in work hours and the opportunity to work with children. In many cases, the task involves little expertise, enabling youngsters to help gather resources while contributing to the family's revenue. Furthermore, concerns about child trafficking and a general lack of safe choices for children present challenges for women working in the informal economy. While laws have been established in the Philippines and Vietnam to prevent children from working in landfills and dumpsites, this has a severe impact on mothers who visit these locations for collection with children. Providing safe and adequate child care for women working in the garbage business would be a significant advantage and is highly wanted by these workers.

- Education:

Most beach waste sector workers, junk shop owners, and operators have a limited education - often no more than a few initial years of schooling. Many recognize education as an essential asset and prefer their children to be educated. Providing access to functional education/vocational training would not only aid and assist them with day-to-day operations but also allow for better awareness, exposure, and empowerment. Providing opportunities for educating their children is viewed as a significant benefit by workers, especially amongst women in all the geographies studied.

4.5 Gardu Action Initiatives

4.5.1 Gardu Action

Gardu Action, a Yogyakarta-based NGO, has emerged as a significant player in addressing the region's waste management challenges. With a focus on environmental conservation and community empowerment, the organization has implemented various innovative strategies to tackle waste issues.

4.5.2 Core Activities in Waste Management

Gardu Action's primary focus on waste management can be categorized into several key areas:

4.5.2.1 Community Empowerment and Education:

- Waste Management Training: Gardu Action conducts workshops and training programs for local communities on waste segregation, recycling, composting, and waste reduction techniques.
- Awareness Campaigns: The organization actively engages in public awareness campaigns to promote responsible waste disposal and the importance of environmental conservation.
- School Programs: Educational programs are designed for students to inculcate environmentally friendly habits from a young age.

4.5.2.2 Waste Collection and Processing:

- Waste Collection: Gardu Action often collaborates with local communities to establish waste collection systems, ensuring proper waste disposal.
- Waste Processing: The organization may operate or support facilities for processing recyclable materials, such as plastic, paper, and glass.
- Composting Initiatives: Gardu Action might promote composting as a sustainable waste management practice, providing training and support to communities.

4.5.2.3 Waste Reduction and Reuse:

- Zero Waste Initiatives: The NGO could be involved in promoting zero-waste lifestyles, encouraging the reduction of waste generation.
- Reuse Programs: Gardu Action might organize programs to repurpose waste materials into useful products, promoting the circular economy.
- Repair and Upcycling Workshops: The organization may conduct workshops on repairing and upcycling items to extend their lifespan and reduce waste.

4.5.2.4 Specific Projects and Initiatives Needed

To gain a deeper understanding of Gardu Action's activities, it is crucial to explore specific projects and initiatives undertaken by the organization. This information can be provided through:

- Gardu Action's Website: The official website will be a valuable resource for information about the organization's work.
- Social Media Platforms: Gardu Action's social media channels (Facebook, Twitter, YouTube TikTok) also needs to be create to provide updates on ongoing projects and activities.
- News Articles and Reports: Local newspapers, online news portals, and research publications also should be compiled in place about Gardu Action's work.
- Direct Contact: Reaching out to Gardu Action directly through email or phone will also provide the most up-to-date and comprehensive information.

4.5.2.5 Potential Research Areas

Based on the information gathered, potential research areas related to Gardu Action's waste management activities could include:

- Impact Assessment: Evaluating the effectiveness of Gardu Action's programs on waste reduction, recycling rates, and community awareness.
- Community Participation: Analyzing the role of community engagement in the success of waste management initiatives.
- Policy Influence: Assessing the impact of Gardu Action's advocacy efforts on waste management policies.
- Economic Viability: Examining the economic sustainability of waste management projects undertaken by Gardu Action.

By conducting thorough research and engaging with Gardu Action, researchers can contribute to a deeper understanding of the organization's role in addressing waste management challenges in Yogyakarta and beyond.

4.5.3 Research Question on Women and Children Involvement

Women's Involvement:

- How are women involved in Gardu Action's waste management projects?
- Are there specific programs targeting women or women-led initiatives?
- What roles do women play in decision-making processes within the organization and projects?

Gender Roles and Waste Management:

- How do gender roles and stereotypes influence waste management practices in the communities where Gardu Action operates?
- Does the organization address these issues in its programs?

- How does Gardu Action challenge traditional gender roles and promote gender equality in waste management?

Economic Empowerment:

- How does Gardu Action's work contribute to women's economic empowerment through waste management activities?
- Are there income-generating opportunities for women within the projects?
- How does the organization ensure fair distribution of benefits between men and women?

Access and Control:

- Do women have equal access to resources and opportunities within Gardu Action's projects?
- How does the organization address issues of land ownership, property rights, and access to technology for women?
- Does Gardu Action empower women to make decisions about waste management in their communities?

4.5.4 Research Question on Beach Waste Management

Beach Cleanup Initiatives:

- Frequency and scale of beach cleanups conducted by Gardu Action.
- Types of waste collected and their quantities.
- Involvement of local communities and volunteers.

Waste Prevention and Reduction:

- Programs to reduce waste generation in coastal areas.
- Campaigns to educate tourists and locals about responsible waste disposal.
- Collaboration with businesses to minimize plastic use.

Recycling and Reuse:

- Recycling programs for beach waste, such as plastic bottles and fishing nets.
- Upcycling initiatives to create products from beach waste.
- Partnerships with recycling facilities.

Policy Advocacy:

- Efforts to influence local and national policies related to beach waste management.
- Collaboration with government agencies to implement effective waste management solutions.

Community Engagement:

- Strategies to involve local communities in beach cleanups and waste management initiatives.
- Building capacity of local communities to manage beach waste independently.

Figure 2. Stakeholders engagement

INDONESIA			
STAKEHOLDER	TYPE	DESCRIPTION	POTENTIAL FOR ENGAGEMENT
PRAISE (Packaging & Recycling Alliance for Indonesia Sustainable Environment)	NGO	Initiated by six companies (Coca-Cola, Danone, Indofood, Nestle, Tetra Pak and Unilever) to establish best practices of sustainable packaging waste solutions. It focuses on Extended Stakeholder Responsibility (ESR), and public education related to ESR implementation.	Engagement in public awareness programs as they work closely with related ministries and industry related stakeholders in packaging and recycling.
APDUPI (Association of Plastic Recycling Entrepreneurs in Indonesia)	NGO	A coalition of junk shops and aggregators, working to increase their knowledge and reach within the plastic recycling industry.	Collaborate in improving collection and recycling of packaging waste among female members.
IPI (Scavengers Union in Indonesia)	NGO	IPI's mission is to pursue street and landfill recycling pickers welfare rights and to educate street and landfill recycling pickers. Taken over by DKI Jakarta local government in 1991, it has given government health insurance to over 6,000 street and landfill recycling pickers.	Collaborate in activating and improving the situation for IPI's female members through education, address health & social issues and technical assistance.
Waste4Change	NGO	Social enterprise that provides waste management services from upstream to downstream. Waste4change provides consultation and education on responsible waste management. It also provides waste collection services and collaborates with stakeholders to recycle materials.	Engagement with Waste4Change could be as a local implementation partner of any formal assistance program for female waste sector workers in any category.
IWP Indonesian Waste Platform)	NGO	Promoting and facilitating cross-sector collaborations for solutions on the waste problems in Indonesia.	Engagement to raise awareness amongst public.

4.6 Theoretical Implication

The theoretical implications of the study on "Empowering Coastal Communities: The Role of Women and Children in Beach Waste Management and Environmental Conservation" offer a significant contribution to the fields of environmental sociology, community development, and gender studies. This research underscores the critical importance of integrating marginalized groups, such as women and children, into environmental governance frameworks. The findings challenge traditional perspectives that often overlook the agency of these groups in environmental conservation efforts, proposing instead that their inclusion can lead to more sustainable and community-driven solutions. The study highlights the necessity of re-evaluating existing theories of participatory development to account for the unique roles and perspectives that women and children bring to waste management practices.

Moreover, the research bridges a gap in environmental management theories by emphasizing the interconnectedness of social structures, environmental practices, and community empowerment. It advocates for a shift from top-down approaches to more inclusive, bottom-up strategies that recognize local knowledge and cultural values. The involvement of women and children is shown not merely as beneficiaries of environmental initiatives but as active agents capable of driving change. This contributes to the broader discourse on sustainable development by aligning gender equity and environmental conservation as mutually reinforcing objectives. The study also provides a foundation for developing theories that explore how intergenerational collaboration in coastal communities can enhance the efficacy of waste management systems, fostering a culture of environmental stewardship that can be replicated in other regions.

4.7 Practical Implication

The practical implications of empowering coastal communities, particularly through the active involvement of women and children in beach waste management and environmental conservation, are profound and multifaceted. First, fostering community-based initiatives that prioritize women and children in environmental efforts can lead to more sustainable waste management practices. Women, as primary caregivers and influential members within households, can promote eco-friendly behaviors and educate younger generations about the importance of conservation. Children, in turn, act as catalysts for change, bringing environmental awareness into schools and their broader social networks. Additionally, training programs and workshops tailored to these groups can enhance skills in waste segregation, recycling, and upcycling, creating opportunities for micro-entrepreneurship and income generation. This inclusive approach not only mitigates environmental degradation but also strengthens community resilience by fostering a sense of shared responsibility. Moreover, such initiatives align with global sustainable development goals (SDGs), particularly those targeting gender equality, quality education, and climate action. By leveraging the unique contributions of women and children, coastal communities can achieve long-term ecological balance while enhancing social cohesion and economic empowerment.

4.8 Managerial Implication

The findings from the study on "Empowering Coastal Communities: The Role of Women and Children in Beach Waste Management and Environmental Conservation" highlight several critical managerial implications. First, policymakers and community leaders should prioritize inclusive and participatory approaches to environmental conservation by formally recognizing the contributions of women and children. Women, often at the forefront of community organization and household management, can act as influential agents in driving sustainable practices. Integrating them into decision-making roles for waste management initiatives can enhance the effectiveness and community buy-in of such programs. Similarly, children can play a pivotal role as environmental ambassadors by spreading awareness and promoting long-term behavioral change, especially when equipped with targeted environmental education programs.

Furthermore, managers of local government and non-governmental organizations (NGOs) must focus on capacity-building efforts by providing technical training and resources tailored to the unique challenges of coastal communities. These programs should address the socioeconomic factors that often hinder active participation, such as limited access to funding or technology. Collaborations with private sector stakeholders can be leveraged to establish sustainable funding models, such as eco-tourism projects or recycling-based entrepreneurial ventures, which can generate income while simultaneously reducing beach waste.

Finally, a robust monitoring and evaluation framework is essential to measure the impact of these initiatives and refine strategies as needed. Data collection on waste reduction, community engagement, and ecological improvement can guide evidence-based decision-making and ensure the scalability of successful practices to other coastal areas. By fostering a culture of shared responsibility and empowerment, managers can create a sustainable model for environmental conservation that simultaneously enhances the livelihoods and resilience of coastal communities.

4.9 Future Directions and Recommendation

The empowerment of coastal communities, particularly women and children, in beach waste management and environmental conservation holds significant potential to foster long-term sustainability and resilience. As key agents of change, women and children can lead grassroots initiatives, driving cultural shifts toward sustainable practices. In the future, their involvement could inspire broader societal participation, amplifying the impact of local conservation efforts and contributing to global environmental goals. However, to realize this potential, targeted interventions are essential. Governments and NGOs should invest in capacity-building programs, equipping women and children with the knowledge, tools, and resources needed to address environmental challenges effectively. Integrating environmental education into school curricula and community workshops can further enhance awareness and skill development.

Additionally, creating inclusive platforms for dialogue and decision-making will ensure that the voices of women and children are heard and valued in policymaking processes. Collaborations with private sectors, particularly industries linked to coastal tourism, can provide funding and technological support for innovative waste management solutions. Policymakers must also prioritize the enforcement of regulations that minimize waste generation and pollution, supported by incentives for communities adopting eco-friendly practices. Monitoring and evaluation systems should be established to assess the effectiveness of these interventions, enabling the refinement of strategies over time. By fostering an environment of shared responsibility and collaboration, the empowerment of women and children can drive transformative change in coastal conservation, ensuring healthier ecosystems and sustainable livelihoods for generations to come.

5 CONCLUSION

In conclusion, this research has elucidated the critical role that women and children play in beach waste management and environmental conservation within coastal communities. The findings highlight that these groups are not only integral to the effectiveness of local waste management initiatives but also serve as catalysts for broader environmental stewardship. Women's involvement often spans from household-level waste management to leadership roles in community-based organizations, while children contribute through education and awareness-raising efforts, fostering a culture of environmental responsibility from a young age.The results indicate that the active engagement of women and children leads to significant improvements in beach cleanliness and local environmental health. Women's leadership and organizational skills are pivotal in mobilizing resources and coordinating waste management activities, while children's involvement in educational programs enhances community awareness and encourages sustainable practices. Despite these positive impacts, challenges such as limited access to resources, insufficient institutional support, and socio-cultural barriers persist, hindering the full potential of their contributions.

To address these challenges and amplify the positive effects observed, future research should focus on developing and implementing targeted support programs that address the specific needs of women and children in coastal communities. It is essential to explore strategies for enhancing resource accessibility, strengthening institutional support, and overcoming socio-cultural barriers. Additionally, research should investigate the scalability of successful local initiatives to other coastal areas, adapting strategies to different socio-economic and cultural contexts. The implications of this study are far-reaching, offering valuable insights for policymakers, environmental organizations, and community leaders. By recognizing and leveraging the contributions of women and children, stakeholders can design more inclusive and effective waste management and conservation strategies. This approach not only improves environmental outcomes but also fosters community empowerment and resilience. Incorporating the voices and needs of these key groups into policy development and program design is crucial for achieving sustainable and equitable environmental management in coastal areas. Ultimately, this research underscores the importance of collaborative and inclusive approaches in addressing the pressing environmental challenges faced by coastal communities.

REFERENCES

Albreem, M. A., Sheikh, A. M., Bashir, M. J. K., & El-Saleh, A. A. (2023). Towards green Internet of Things (IoT) for a sustainable future in Gulf Cooperation Council countries: Current practices, challenges and future prospective. *Wireless Networks*, *29*(2), 539–567. DOI: 10.1007/s11276-022-03133-3

Alghamdi, S., & Beloff, N. (2014). Towards a comprehensive model for e-Government adoption and utilisation analysis: The case of Saudi Arabia. *2014 Federated Conference on Computer Science and Information Systems, FedCSIS 2014*, 1217–1225. DOI: 10.15439/2014F146

Bawazir, S. (n.d.). The key factors of successful 1 The key factors of successful sustainable development: e-Government in Saudi Arabia as an example.

Bharne, S., & Patil, S. (2021). Exploring in the Context of Development of Smart Cities in India. In G. V.K. & Z. J.M. (Eds.), Advances in Intelligent Systems and Computing (Vol. 1245, pp. 365–374). Springer Science and Business Media Deutschland GmbH. DOI: 10.1007/978-981-15-7234-0_32

Bonino, D., Alizo, M. T. D., Pastrone, C., & Spirito, M. (2016). WasteApp: Smarter waste recycling for smart citizens. In N. S., M. Z., P. L., S. P., & P. T. (Eds.), 2016 International Multidisciplinary Conference on Computer and Energy Science, SpliTech 2016. Institute of Electrical and Electronics Engineers Inc. DOI: 10.1109/SpliTech.2016.7555951

Cho, J.-S. (2017). Evolution of e-government: Transparency, competency, and serviceoriented government with Korean government 3.0. *The Journal of Business and Retail Management Research*, *12*(1), 62–68. DOI: 10.24052/JBRMR/V12IS01/EOEGTCASOGWKG3

Clarinval, A., Simonofski, A., Henry, J., Vanderose, B., & Dumas, B. (2023). Introducing the Smart City to Children: Lessons Learned from Hands-On Workshops in Classes. *Sustainability (Basel)*, *15*(3), 1774. Advance online publication. DOI: 10.3390/su15031774

Coulen, S. C. W., Sanders, J. P. M., & Bruins, M. E. (2017). Valorisation of Proteins from Rubber Tree. *Waste and Biomass Valorization*, *8*(4), 1027–1041. DOI: 10.1007/s12649-016-9688-9

Cumps, B. (2015). Extending an ICT4D Computer Re-use Model with E-waste Handling Activities: A Case Study. *Information Technology for Development*, *21*(4), 677–693. DOI: 10.1080/02681102.2014.908161

Demircan, B. G., & Yetilmezsoy, K. (2023). A Hybrid Fuzzy AHP-TOPSIS Approach for Implementation of Smart Sustainable Waste Management Strategies. *Sustainability (Basel)*, *15*(8), 6526. Advance online publication. DOI: 10.3390/su15086526

Digiesi, S., Mossa, G., Mummolo, G., & Verriello, R. (2015). A carbon footprint calculator for the municipal waste collection system of Bari. Proceedings of the Summer School Francesco Turco, Industrial. *Systems Engineering*, •••, 124–129. https://www.scopus.com/inward/record.uri?eid=2-s2.0-84949564704&partnerID=40&md5=793cdff04500d4b4c0a40e5a381b8e8f

Dutta, H. (2023). The Environmental Aspects of Refugee Crises: Insights from South Asia, Middle East, and Sub-Saharan Africa. *Journal of International Migration and Integration*, *24*(2), 913–938. DOI: 10.1007/s12134-022-00986-7

Faseyi, C. A., Miyittah, M. K., & Yafetto, L. (2023). Assessment of environmental degradation in two coastal communities of Ghana using Driver Pressure State Impact Response (DPSIR) framework. *Journal of Environmental Management, 342*, 118224. DOI: 10.1016/j.jenvman.2023.118224 PMID: 37267762

Fatimah, Y. A., Govindan, K., Murniningsih, R., & Setiawan, A. (2020). Industry 4.0 based sustainable circular economy approach for smart waste management system to achieve sustainable development goals: A case study of Indonesia. *Journal of Cleaner Production, 269*, 122263. Advance online publication. DOI: 10.1016/j.jclepro.2020.122263

Fatimah, Y. A., Govindan, K., Murniningsih, R., & Setiawan, A. (2020). Industry 4.0 based sustainable circular economy approach for smart waste management system to achieve sustainable development goals: A case study of Indonesia. *Journal of Cleaner Production, 269*, 122263. Advance online publication. DOI: 10.1016/j.jclepro.2020.122263

Guglietta, D., Belardi, G., Passeri, D., Salvatori, R., Ubaldini, S., Casentini, B., & Trapasso, F. (2020). Optimising the management of mining waste by means of Sentinel-2 imagery: A case study in the Joda West Iron and Manganese Mine (India). *Journal of Sustainable Mining, 19*(1), 22–32. DOI: 10.46873/2300-3960.1003

Guglietta, D., Belardi, G., Passeri, D., Salvatori, R., Ubaldini, S., Casentini, B., & Trapasso, F. (2020). Optimising the management of mining waste by means of Sentinel-2 imagery: A case study in the Joda West Iron and Manganese Mine (India). *Journal of Sustainable Mining, 19*(1), 22–32. DOI: 10.46873/2300-3960.1003

Jäämaa, L., & Kaipia, R. (2022). The first mile problem in the circular economy supply chains – Collecting recyclable textiles from consumers. *Waste Management (New York, N.Y.), 141*, 173–182. DOI: 10.1016/j.wasman.2022.01.012 PMID: 35115212

Kordi, G., Hasanzadeh-Moghimi, P., Paydar, M. M., & Asadi-Gangraj, E. (2022). A multi-objective location-routing model for dental waste considering environmental factors. *Annals of Operations Research*, •••, 1–38. DOI: 10.1007/s10479-022-04794-1 PMID: 35789687

Laufs, J., Borrion, H., & Bradford, B. (2020). Security and the smart city: A systematic review. *Sustainable Cities and Society, 55*, 102023. Advance online publication. DOI: 10.1016/j.scs.2020.102023

Luo, X., & Liao, W. (2022). Collaborative Reverse Logistics Network for Infectious Medical Waste Management during the COVID-19 Outbreak. *International Journal of Environmental Research and Public Health, 19*(15), 9735. DOI: 10.3390/ijerph19159735 PMID: 35955091

Mumtaz, S., Chu, A. M. Y., Attiq, S., Shah, H. J., & Wong, W.-K. (2022). Habit—Does It Matter? Bringing Habit and Emotion into the Development of Consumer's Food Waste Reduction Behavior with the Lens of the Theory of Interpersonal Behavior. *International Journal of Environmental Research and Public Health, 19*(10), 6312. DOI: 10.3390/ijerph19106312 PMID: 35627848

Navarra, D., & Bianchi, C. (2013). Territorial governance, e-government and sustainable development policy: A system dynamics approach. In 12th IFIP WG 8.5 International Conference on Electronic Government, EGOV 2013: Vol. 8074 LNCS (pp. 14–25). Springer Verlag. DOI: 10.1007/978-3-642-40358-3_2

Park, C., Tantiyaswasdikul, K., Evans, S., & Lertwattanaruk, P. (2019). Innovation catalysts for industrial waste challenges: Sri Lankan and Thai cases. In B. F., K. H., J. I.S., & S. G. (Eds.), 16th Global Conference on Sustainable Manufacturing, GCSM 2018 (Vol. 33, pp. 570–577). Elsevier B.V. DOI: 10.1016/j.promfg.2019.04.071

Parviziomran, I., Mahmoudi, M., & Boloori, A. (2023). Evaluating operating models and urbanism for transportation operations of circular reuse platforms. *Computers & Industrial Engineering*, *177*, 109067. Advance online publication. DOI: 10.1016/j.cie.2023.109067

Paul, K. (2004). Exporting responsibility - Shipbreaking in South Asia - International trade in hazardous waste. *Environmental Policy and Law*, *34*(2), 73–78. https://www.scopus.com/inward/record.uri?eid=2-s2.0-2142810226&partnerID=40&md5=53ed67c3f2f2e6bb523d45731l2ca9ac

Pelonero, L., Fornaia, A., & Tramontana, E. (2020). From Smart City to Smart Citizen: Rewarding Waste Recycle by Designing a Data-Centric IoT based Garbage Collection Service. Proceedings - 2020 IEEE International Conference on Smart Computing, SMARTCOMP 2020, 380–385. DOI: 10.1109/SMARTCOMP50058.2020.00081

Pervin, I. A., Rahman, S. M. M., Nepal, M., Haque, A. K. E., Karim, H., & Dhakal, G. (2020). Adapting to urban flooding: A case of two cities in South Asia. *Water Policy*, *22*(S1), 162–188. DOI: 10.2166/wp.2019.174

Piro, G., Cianci, I., Grieco, L. A., Boggia, G., & Camarda, P. (2014). Information centric services in Smart Cities. *Journal of Systems and Software*, *88*(1), 169–188. DOI: 10.1016/j.jss.2013.10.029

Ramadass, R., Sathyanarayana, N., & Fathima, Y. (2023). Improving citizen lifestyle in green smart city. In *Green Blockchain Technology for Sustainable Smart Cities* (pp. 29–63). Elsevier., DOI: 10.1016/B978-0-323-95407-5.00007-4

Sabbir, M. M., Khan, T. T., Das, A., Akter, S., & Hossain, M. A. (2022). Understanding the determinants of consumers' reverse exchange intention as an approach to e-waste recycling: A developing country perspective. *Asia-Pacific Journal of Business Administration*. Advance online publication. DOI: 10.1108/APJBA-11-2021-0565

Sharafi, A., Wolf, P., & Krcmar, H. (2010). Process and data-oriented approach for bundling corporate reporting duties to public authorities - A case study on the example of waste management reporting. In 9th IFIP WG 8.5 International Conference on Electronic Government, EGOV 2010: Vol. 6228 LNCS (pp. 400–411). DOI: 10.1007/978-3-642-14799-9_34

Soma, T. (2017). Gifting, ridding and the "everyday mundane": The role of class and privilege in food waste generation in Indonesia. *Local Environment*, *22*(12), 1444–1460. https://api.semanticscholar.org/CorpusID:158090112. DOI: 10.1080/13549839.2017.1357689

Sovacool, B. K. (2019). Toxic transitions in the lifecycle externalities of a digital society: The complex afterlives of electronic waste in Ghana. *Resources Policy*, *64*, 101459. Advance online publication. DOI: 10.1016/j.resourpol.2019.101459

Stöfen-O, A., Naji, A., Brooks, A. L., Jambeck, J. R., & Khan, F. R. (2022). Marine plastic debris in the Arabian/Persian Gulf: Challenges, opportunities and recommendations from a transdisciplinary perspective. *Marine Policy, 136*, 104909.

Vasudevan, P., & Kearney, W. A. (2016). Remembering Kearneytown: Race, place and collective memory in collaborative filmmaking. *Area, 48*(4), 455–462. DOI: 10.1111/area.12238

Zhou, X., & Tang, X. (2022). Does financing constraints impact the Chinese companies' pollutants emissions? Evidence from a sample selection bias corrected model based on Chinese company-level panel data. *Environmental Science and Pollution Research International, 29*(29), 44119–44134. DOI: 10.1007/s11356-022-18907-7 PMID: 35122200

KEY TERMS DEFINITION

Beach: A beach is a landform along the shoreline of an ocean, sea, lake, or river, consisting of loose particles such as sand, pebbles, or other sediments. It serves as a transition zone between aquatic and terrestrial environments and is often a site of ecological importance and recreational activities.

Waste Management: Waste management refers to the systematic process of handling, collecting, transporting, treating, and disposing of waste materials in a manner that minimizes their environmental impact. It involves strategies like recycling, composting, and the use of technology to promote sustainability and reduce pollution.

Women & Children: Women; Adult female humans, often considered key contributors to societal development, family well-being, and economic stability. In community roles, women frequently lead initiatives in education, health, and environmental management. Children; Young individuals, typically below the age of 18, who are considered vital for the future of society. They are often engaged in learning and play but can also participate in age-appropriate roles to support family or community-based activities.

Coastal: Coastal refers to areas or regions located near or adjacent to the sea or ocean. These zones are characterized by unique ecosystems, diverse flora and fauna, and are critical for economic activities such as fishing, tourism, and trade.

Gardu Action: Gardu Action is a non-governmental organization based in Yogyakarta, Indonesia, focused on environmental and social initiatives. Its activities include promoting sustainable practices, supporting community-driven conservation projects, and addressing waste management challenges in coastal and urban areas.

Coastal Communities: Coastal communities are groups of people living in or near coastal regions. Their livelihoods are often closely tied to the sea through activities such as fishing, aquaculture, tourism, or maritime trade. These communities are particularly vulnerable to environmental changes like rising sea levels, pollution, and climate-related disasters.

Environmental Conservation: Environmental conservation refers to the practice of protecting natural ecosystems, biodiversity, and natural resources to ensure their sustainability for future generations. It includes activities like reforestation, wildlife protection, pollution control, and community-driven eco-friendly initiatives.

Chapter 6
Features of Search and Exploration of Drinking Groundwater in the Tourist Recreational Zone

Evgenia Zakharchenko
https://orcid.org/0009-0009-4172-4244
Kuban State University, Russia

Vladimir Ivanovich Gulenko
Kuban State University, Russia

Nikolay Nikolaevich Rudomaha
Kuban State University, Russia

Yulia Ivanovna Zakharchenko
https://orcid.org/0009-0007-1736-5275
Kuban State University, Russia

Bahrom Makhmudovich Kholbaev
Karshi Mining and Economic University, Uzbekistan

ABSTRACT

This paper highlights in detail the measures for detailed exploration of drinking groundwater for the purpose of water supply of the facility using the example of geophysical studies conducted in the area of Lake Abrau in the Krasnodar Territory. The increased demand for water is putting exceptionally strong pressure on our natural environment. All over the world, aquatic ecosystems are in crisis, and many rivers, seas and lakes are already polluted or seriously degraded. Coastal zones are in a particularly vulnerable position due to the deterioration of the ecological state of rivers as they approach the points of their confluence with the seas. Krasnodar Territory is one of the most intensively using water resources of the regions, most of which goes to the needs of agriculture and drinking water, including in the area of the tourist and recreational complex. The issue of drinking water supply in the tourist areas of the

DOI: 10.4018/979-8-3693-8232-5.ch006

Black Sea coast is particularly acute.

RELEVANCE OF THE STUDY

Water is a driver of sustainable development, including environmental integrity and the eradication of poverty and hunger, and it is absolutely essential for human health and well–being. Prioritizing water is an urgent global imperative (from the report of the international summit 'Environment and Sustainable Development: Agenda 21st century) (The United Nations, 1992).

Increasing demand for water is putting extraordinary pressure on our natural environment. Globally, aquatic ecosystems are in crisis, and many rivers, seas and lakes are already polluted or severely degraded as a result of the decline of natural ecosystems, including forests and watershed areas. Industrial and agricultural emissions of significant amounts of heavy metals and hazardous wastes are increasingly polluting and depleting groundwater. The decline in the quantity and quality of water resources is responsible for the extinction of freshwater species and the rapid loss of biodiversity.

Coastal zones, home to some of the Earth's most productive ecosystems, are particularly vulnerable as rivers deteriorate as they approach their confluence with the sea, threatening human and animal life and entire ecosystems. About two fifths of the world's population live within 100 km of the coast. However, about 30 percent of the world's land resources in coastal ecosystems are severely degraded as a result of the increasing demand for land from residential, industrial and tourism development. More than half of humanity depends on freshwater that is stored in mountainous areas. However, these areas suffer from deforestation and pressure from agriculture and tourism, which can lead to overstretching of water resources.

INTRODUCTION

The water resources of the study area – the Black Sea coast of Krasnodar Krai – are represented by the waters of the Black Sea and land waters. The length of the coastal strip within the coastline is respectively more than 470 km. The main sources of water intake on the Black Sea coast are two major artesian basins: the Bolshe–Kavkazsky basin (coast and mountainous part of the Krasnodar Region) and the Azov–Kuban basin (foothills and Krasnodar Region). The largest sources of surface water include both a well–developed river network (mainly to the south–east of Gelendzhik) and a number of fairly large reservoirs and lakes (Neberdzhaevskoye, Lakes Abrau and Bezdonnoye, etc.).

At present, the Black Sea region has major water management problems, which for several decades may become a factor constraining the socio–economic development of Southern Russia, if the necessary measures are not taken in time, taking into account the long–term nature of the development and implementation of projects related to the solution of water management problems (SWOT) (table 1) (Sychev, 2006).

Table 1. SWOT–analysis of the water resources management system of the coastal zone of Krasnodar Krai

SWOT analysis of water resources management system of the coastal zone of Krasnodar Krai	
Strengths	**– large reserves of water resources (ground and surface water); – the project for rehabilitation and development of water supply and drainage systems on the Azov–Black Sea coast; – Sochi Sustainable Water Resources Management Project**
Weaknesses	– unsatisfactory condition of water supply systems along the entire Black Sea coast; – shortage of clean drinking water in a number of municipalities and rural settlements; – low level of metering of drinking water consumption; – Pollution of water basins by effluents from industrial enterprises and the housing sector; – inefficiency of treatment facilities; – uneven availability of fresh groundwater on the territory of the Krai; – improper implementation of water protection legislation (mass development of water protection zones)
Opportunities	– focus on sustainable development; – investor interest; – use of alternative water sources (e.g. desalination, water recycling)
Threats	– unstable weather conditions (droughts, floods); – development of coastal cities (increased water demand); – increase in the number of vacationers in tourist and recreational zones (increase in water demand)

The Krasnodar Region is one of the most water–intensive regions (figure 1), most of which is used for agriculture and drinking water, including in the tourist and recreational area.

Figure 1. Scheme of intensity of water resources use regions of the Russian Federation

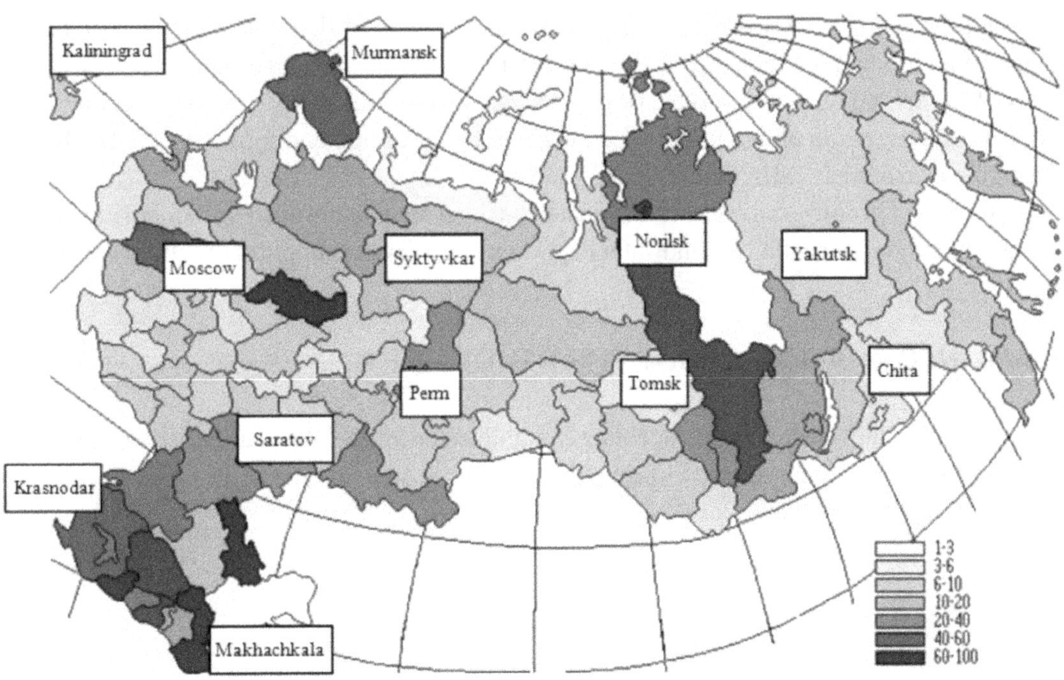

(as of 2013) (Sychev, 2006)

Unstable climatic conditions (droughts, frequent floods) cause decrease of agricultural productivity and aggravation of problems in cities with already strained water supply. The issue of drinking water supply is particularly acute in the tourist areas of the Black Sea coast.

In environmental terms, concentrations of pollutants are increasing in almost all water bodies of the Krasnodar Region (Sychev, 2006).

Drinking groundwater is an invaluable mineral resource that ensures the development of the social sphere and economy of any state. For water supply of many settlements and industrial enterprises, works on search and exploration of drinking groundwater deposits are carried out.

According to the estimation of many specialists (Kurenoy, 2011; Bindeman et al, 1975; Borevsky et al, 1989) the lack of drinking groundwater in the Russian Federation is noted in 80% of the country. Many researchers (Kurenoy, 2011; Bindeman et al, 1975; Borevsky et al, 1989) estimate this figure to be 60% of the territory of the globe.

Drinking groundwater includes waters, which by their qualitative indicators are distinguished among all fresh waters by the value of total mineralization in the range from 0.25 to 0.75 g/dm^3 (Kurenoy, 2011). Waters with mineralization up to 0.25 g/dm^3 and more than 0.75 g/dm^3 (up to 1.5 g/dm^3) should be considered acceptable for drinking purposes with compliance with certain requirements, including the consumption regime, developed for hydrogeological conditions of this or that complexity. Taking into account sanitary and hygienic requirements to the microcomponent group of indicators their drinking qualities are specified by normative documents (Kurenoy, 2011).

About 35 years ago, the Ministry of Natural Resources in our country assessed the water availability of groundwater resources for the population. For the first time, data on the resource potential of domestic drinking groundwater in the country, reaching 870 million m^3/d, were obtained (Borevsky et al, 1989), but drinking groundwater resources were not estimated (Vilchinskaya & Kambarova, 2010).

At present, the shortage of drinking water is particularly acute in the mountainous folded areas of the Russian Federation, as the folded areas are characterized by abrupt changes in relief, geological and structural conditions and type of water–bearing rocks, manifested over relatively small distances. In this connection, characterization of conditions of groundwater distribution and formation within the folded area as a whole is extremely difficult.

On the example of geophysical research conducted on the Abrau Peninsula, Krasnodar Krai, measures for detailed exploration of drinking groundwater for the purpose of water supply of the object are presented.

GENERAL INFORMATION ABOUT THE STUDY AREA

Administratively, the subsurface prospecting site is located in the Russian Federation, Krasnodar Krai, Novorossiysk city, Abrau–Durso village (figure 2).

Figure 2. Location of the work area

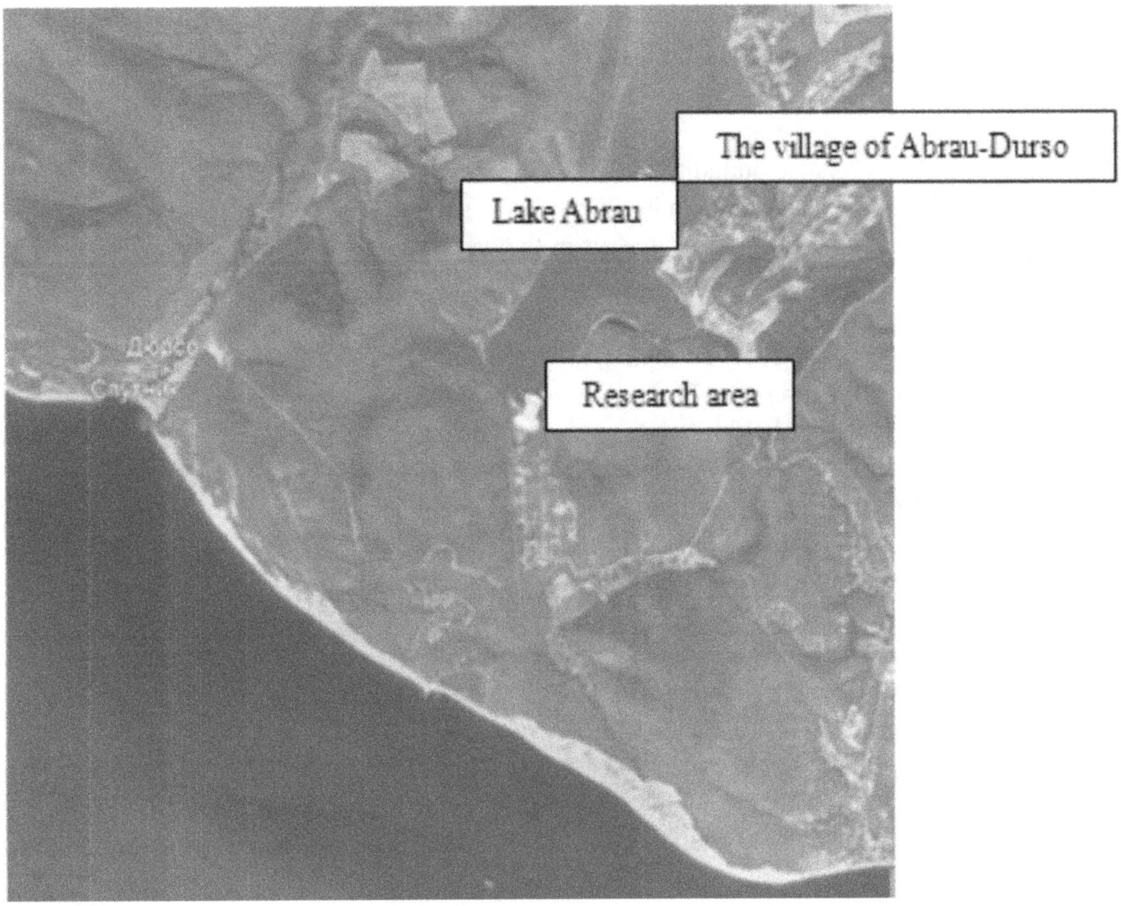

Climatic characteristics are given according to the Code of Regulations SP 131.13330.2020 «SNiP 23–01–99 Construction Climatology» (Ministry of Construction of Russia, 2020).

An important factor influencing the climate of the area is the atmospheric circulation. Masses of continental air of temperate latitudes prevail here. Air masses of Atlantic, Arctic and tropical origin coming from outside are usually already largely transformed and finally transformed into continental air of temperate latitudes, which determines the temperate continental climate of the area.

According to climatic zoning for construction (SP 131.13330.2020) (Ministry of Construction of Russia, 2020) the territory belongs to sub–area III B, which is characterized by the following natural and climatic factors:

- air temperature of the coldest days with a probability of 0.98 is equal to minus 23°C, with a probability of 0.92 – equal to minus 20°C;
- the absolute minimum air temperature is equal to minus 36°C, the average air temperature of the period with average daily temperature below 8°C is 2.7°C, with duration of 146 days (table 2);

Table 2. Mean monthly and annual air temperatures, ^{0}C (Ministry of Construction of Russia, 2020)

Characterization	Months												Year
	I	II	III	IV	V	VI	VII	VIII	IX	X	XI	XII	
Average temperature	0,0	1,1	5,6	12,3	17,6	21,4	24,1	23,7	18,6	12,0	6,4	2,3	12,1

- the average monthly relative humidity in the coldest month is 81%. The average monthly relative humidity at 15 h of the coldest month is 72%;
- average monthly relative humidity of the air in the warmest period of the year is 63%; average monthly relative humidity at 15 h of the warmest month is 45%;
- the amount of precipitation for november – march is 309 mm;
- precipitation for April – october is 409 mm;
- daily maximum precipitation is 107 mm;
- barometric pressure within 1013 hPa;
- the air temperature of the warmest period with a probability of 0.95 is 28°C;
- the average maximum air temperature of the warmest month is 32°C;
- the absolute maximum temperature of the warm period is 42°C;
- on the territory under consideration the prevailing wind direction for june–august is B, the minimum of average wind speeds by rhumb for july 0°;
- maximum daily amplitude of air temperature in july – 22°;
- normative value of snow cover weight per 1 m2 is 1.0 kN/m^2 (table 10.1 of SP 20.13330.2016) (Ministry of Construction of Russia, 2016);
- normative value of wind pressure is 0.48 kPa (table 11.1 of the SP 20.13330.2016) (Ministry of Construction and Housing and Communal Services of the Russian Federation, 2016);
- ice wall thickness is 10.0 mm (table 12.1 of SP 20.13330.2016) (Ministry of Construction of Russia, 2016).

The survey area is located in the humid moisture zone (SP 50.13330.2012) (Ministry of Regional Development of Russia, 2012).

According to the weight of snow cover the survey area belongs to the II region (map 1, Annex E of SP 20.13330.2016).

According to the thickness of the ice wall, the survey area belongs to the III region (map 3a, Annex E of SP 20.13330.2016) (Ministry of Construction of Russia, 2016).

By wind pressure the survey area belongs to the IV region (map 2g, Annex E of SP 20.13330.2016) (Ministry of Construction of Russia, 2016).

Absolute elevations of the survey area are 90–150 m.

The lake is the largest natural body of water in Krasnodar Krai. On its shores there is the village of Abrau and a sparkling wine factory.

The size of Lake Abrau is:

- length – about 3 km;
- width – about 0.6 km, in some places – 0.9 km;

- depth – on average 5.8 m, maximum 11 m (used to reach 30 m, but due to siltation it is gradually decreasing);
- surface area reaches about 1.7 km^2;
- catchment area – 20.3 km^2.

In the aggregate of all factors (geological and hydrogeological conditions, weathering of rocks, tectonic and neotectonic structure, filtration areas, filtration intervals) the considered area is favorable for further location of production well for drinking groundwater extraction.

GEOLOGICAL STRUCTURE OF THE STUDY AREA

Paleogene and Quaternary age rocks take part in the geological structure of the study area.

The geological structure of the territory (according to the geological map of the USSR) includes rocks of the undivided Paleocene. In terms of rocks, the Paleocene is represented by sandstones, siltstones, marls, limestones and clays. Figures 3–4 show a fragment of the USSR geological map and a fragment of the geological section.

Figure 3. Fragment of the USSR geological map of the study area

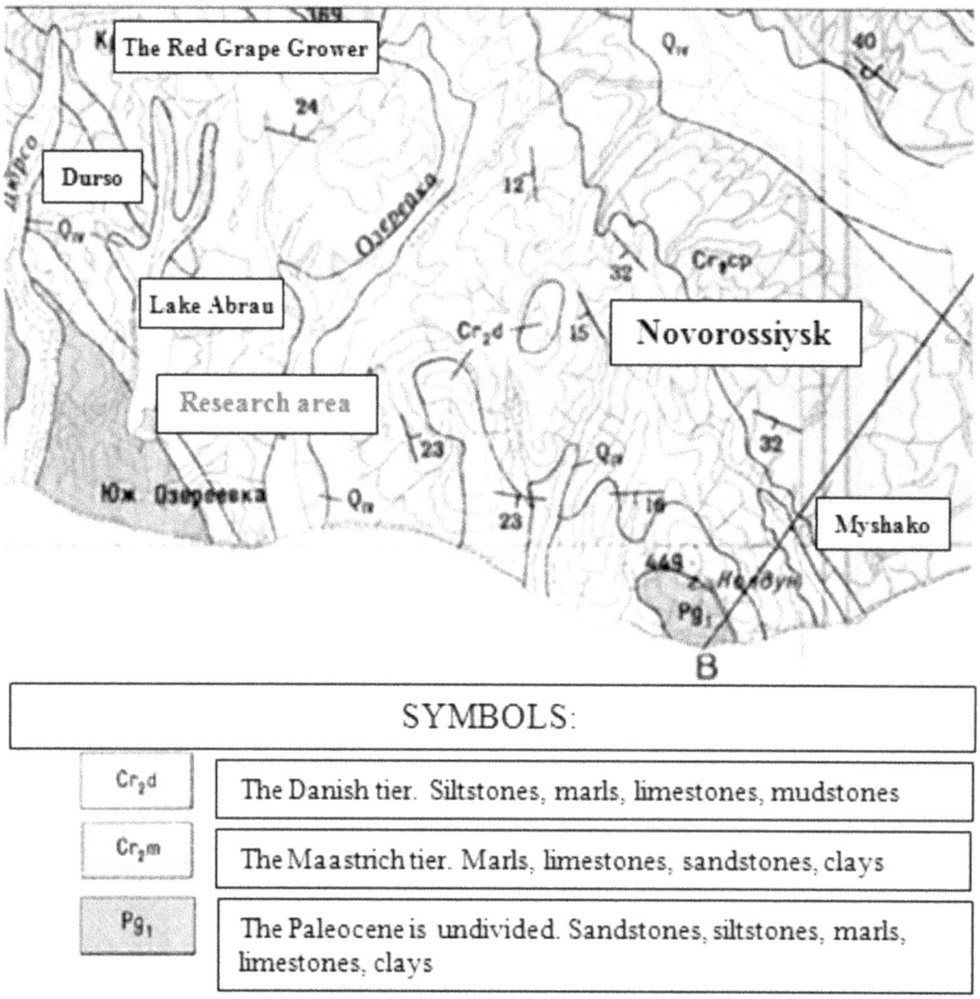

Figure 4. Geological section of the study area

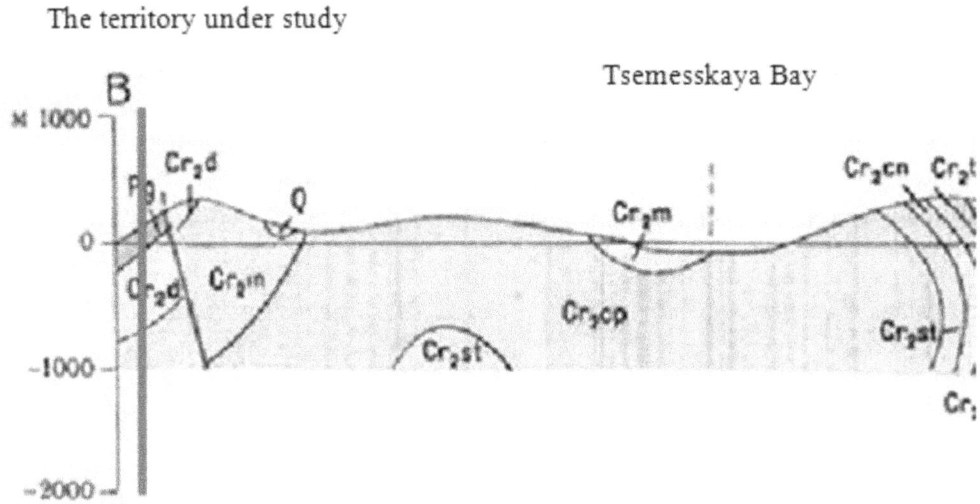

The study area is located at the edge of the synclinal fold core (Danish Stage) composed of the Anapa Formation ($\mathscr{P}_1 an$).

The Anapa Formation ($\mathscr{P}_1 an$) forms the core of the synclines of Cape Bolshoi Utrish and Raevskaya (northwestern part), where it lies on the Navagirskaya Formation. It is represented by flysch alternation of clayey marls (from 20 cm to 60 cm – from 5% to 60%) and calcareous marls (from 5 cm to 45 cm – from 1% to 10%), clays (from 10 cm to 175 cm – from 5% to 60%), sandstones (from 10 cm to 40 cm – from 10% to 20%), siltstones (from 1 cm to 80 cm – from 5% to 10%) and limestones (from 8 cm to 25 cm – from 1% to 5%). Small (up to 10×30 cm) lenses of silica occur in the upper part. The thickness of the Anapa Formation varies from 350 m to 375 m. Foraminifera such as Globigerina eocaenica Terquem, G. aff. fringa Subb., Globorotalia pseudoscitula Glaessner, and others are dated to the Early Paleocene (Danish Stage).

The Navagirskaya Formation ($\mathscr{P}_1 nv$) overlies the Sukko Formation and is exposed together with the latter. The flysch sediments are represented by interlacing of weakly clayey marls (from 10 cm to 100 cm – from 45% to 80%) and calcareous marls (from 5 cm to 220 cm – from 0% to 30%), sandstones (from 10 cm to 50 cm – from 10% to 15%), siltstones (from 1 cm to 15 cm – from 2% to 10%) and limestones (from 10 cm to 80 cm – from 2% to 8%). The thickness of the Navagyr Formation varies from 200 m to 210 m. The Danish age is substantiated by numerous determinations of foraminifera microfauna: Gümbelina crinita, Globigerina ex gr. trivialis Subb., Pseudouvigerina plummerae (Cushm.) and others.

Figure 5 shows a fragment of the state geological map of preQuaternary sediments of the Russian Federation (sheet L–37–XXVI (Novorossiysk) (Kavkazgeolszemshtuka, 2021).

The Quaternary deposits of the study area are represented by alluvial and colluvial deposits.

The alluvium of modern channels and floodplains (*aH*) is bound to the bottom parts of the valleys of the largest rivers. The sediments of the rivers and numerous river branches are dominated by sandy loamy clays with a large number of old and pond facies, reclaimed peat bogs, sands with shells of terrestrial and freshwater mollusks. The bottoms of small rivers are lined with gravel and pebble gravels in sandy aggregate with a large amount of rubble and clasts, especially near the sides. The thickness of sediments varies from 1–3 m to 10–15 m.

Colluvium (*cH*) is widespread among the slope deposits of the mountainous part of the territory. The main areas of its development are steeply sloping massifs of low ridges, foothills of cuestas and structural terraces, the surfaces of which are armored with strong limestones of the Upper Cretaceous. The sediments include various types of rockfall, scree, landslide and even avalanche formations. Landslides are usually associated with seismically active fault zones and have a seismogravitational genesis. In the relief they form discontinuous plumes of rubble, blocks, and rocky detritus up to 0.5 km wide with the thickness of rubble and rubble accumulations up to 15 m.

Figure 5. Fragment of the geologic map of the study area (Kavkazgeolszemshtuka, 2021)

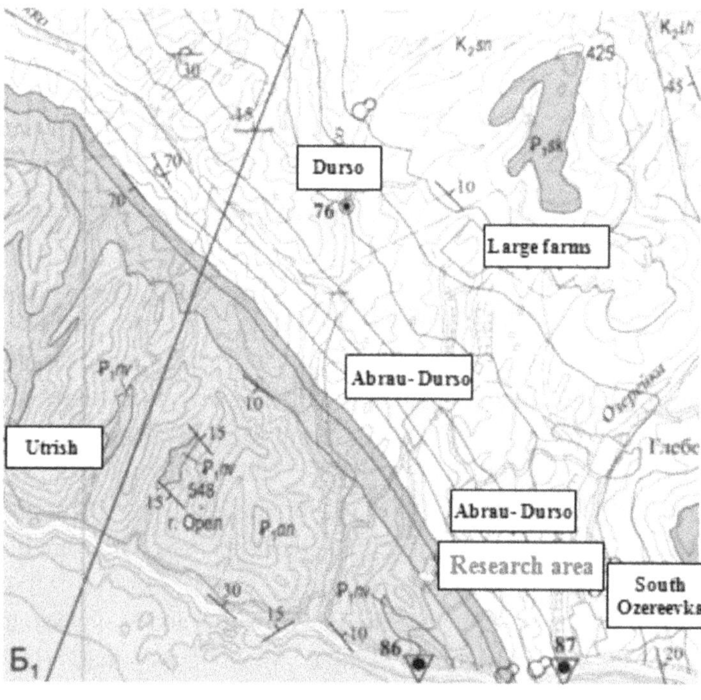

SYMBOLS:

$P_1 k\check{s}\check{c}$	The retinue of the Cossack gap. Flish. Marls, sandstones, siltstones, limestones (100-300 m)
$P_1 an$	Anapa suite. Flish. Marls, clays, sandstones, siltstones, limestones, in the upper part of the lens of flints (350-375 m)
$P_1 nv$	The Navagarian retinue. Flish. Marls, sandstones, siltstones, limestones (200-210 m)
$P_1 sk$	Sukko's retinue. Flish. Clays, sandstones, siltstones, limestones, marls (74-215 m)

Figure 6 shows a fragment of the Quaternary sediment map of the study area (Research Geological Institute named after A.P. Karpinsky, 2021).

On the basis of the analyzed geological information, taking into account the data on reconnaissance survey, the territory is recommended for placement of a production well.

Tectonic structure of the investigated area.

According to the scheme of tectonic zoning of the North–West Caucasus (figure 7), the territory of Novorossiysk is confined to the Anapsko–Agoi zone of the Novorossiysk–Lazarevskaya fold–allochthonous zone, located on the north–western dip of the fold–allochthonous structure of the Greater Caucasus. The Novorossiysk area is complicated by tectonic fractures (fracture systems) of different order of longitudinal and transverse direction (Institute of Geoecology of the Russian Academy of Sciences, 2015).

Figure 6. Fragment of the Quaternary sediment map of the study area

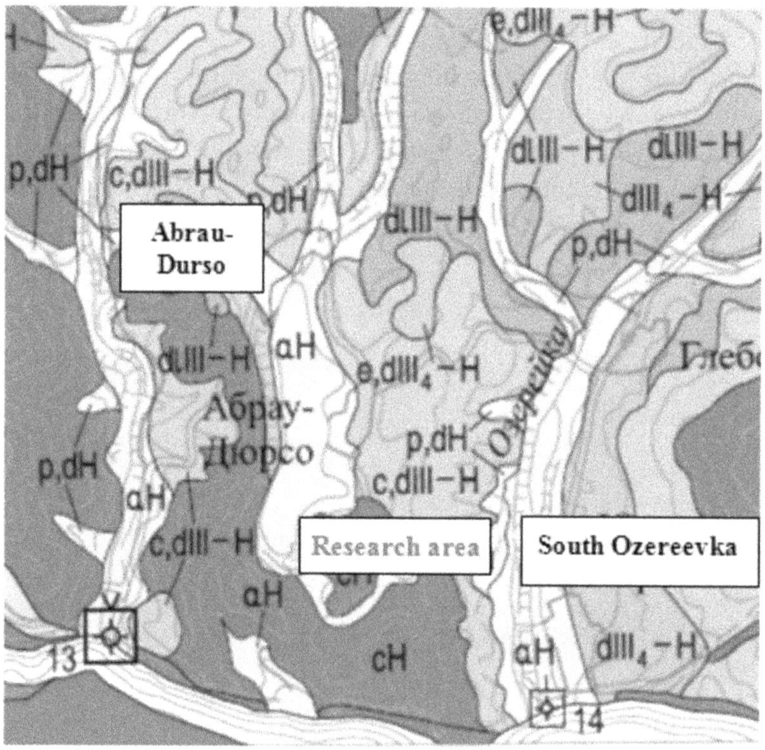

SYMBOLS:

cH	Colluvium of steeply sloping massifs. Boulders, crushed stone, gravel, rock outcrops (up to 30 m)
aH	Alluvium of modern riverbeds and floodplains. Clays, silts, pebbles, gravel, sands, shells (up to 15 m). Deposits of construction sands, sand and gravel material

(Research Geological Institute named after A.P. Karpinsky, 2021)

Figure 7. Tectonic scheme of the work area

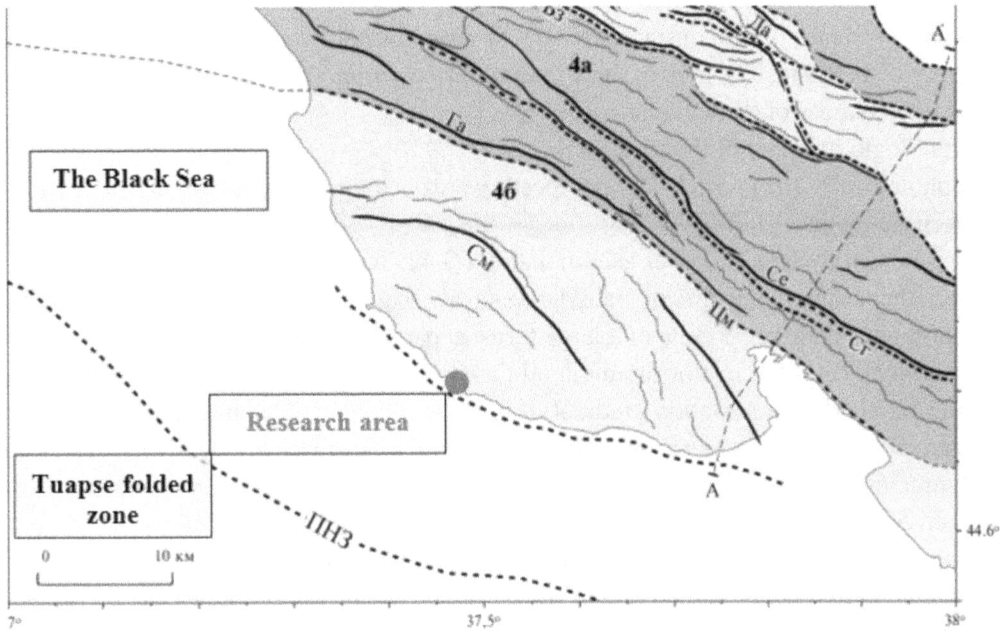

(Institute of Geoecology of the Russian Academy of Sciences, 2015)

Tectonically, the territory of sheet L–37–XXVI on the fragment of the state geological map of preQuaternary sediments of the Russian Federation is located on the northwestern dip of the Greater Caucasus fold–and–boulder structure.

The paleogeodynamic conditions of formation of the structural–material complexes composing the mountain structure correspond to the conditions of the passive continental margin, which the described area represented in the Cretaceous–Eocene time.

Cretaceous (on the surface not older than the Aptian age) and Paleogene formations are involved in the structure of the folded–boulder structure of the Greater Caucasus in the described area, which compose two folded zones different in lithologic composition, formation conditions and structural features: the Novorossiysk–Lazarevskaya and Vityazevskaya folded zones, which form two large tectonic covers of the same name.

The Novorossiysk–Lazarev fold–allochthonous zone is composed of sediments of considerable thickness (more than 7000 m), which are closest to the conditions of the deep–water part of the flysch basin, identified with the continental foot of the passive continental margin. Three complexes are distinguished within the zone:

- Early Cretaceous flysch terrigenous (PK_1) complex located at a depth of 2700 m;
- Late Cretaceous flysch terrigenous–carbonate (PK_2) complex, located at a depth of up to 3700 m);
- Early Paleocene–Eocene flysch terrigenous–carbonate (PK_{1-2}), located at a depth of up to 1100 m.

The reconnaissance survey highlighted that the rocks of the study area have a northeastern dip direction. The dip angle of the rocks varies from 27° to 45°.

The discontinuities observed in the outcrops, as well as geomorphologic features of the area where the object of research is located, mark significant multi–stage discharges with the formation of clay areas. All this contributes to the constant drainage of groundwater from the lake side by a wide front to depths of more than 100 m. The area of increased filtration will not follow the outline of the gully and is tied to weathered areas of the tectonic zone.

Geomorphologic structure of the study area.

Geomorphologically, the groundwater prospecting area belongs to the low–lying ridges and uplands of the Western Caucasus.

The relief created by lake and river accumulation (Q_H) is traced by a discontinuous strip up to 3–4 km wide from the northern outskirts of Natukhaevskaya village to the city of Novorossiysk. According to geomorphological data, it is a zone of a large thrust separating a large peninsular part of the Anapa–Novorossiysk block from the rest of the mountain area. The characteristic feature of the hollows is a flat semi–wet relief with an elevation gradient of up to 4–5 meters. Absolute heights of the lake–like expansions are 60–80 m, the excesses over the surrounding ridges reach up to 350–400 m. The junction with the mountains is from very gentle (in the west and northwest) to sharp (in the east and southeast). The formation of the chain of hollows took place in the Holocene, as evidenced by the small thickness of accumulated sediments deposited in them on pre–Quaternary formations. The same is typical for the valley of the Tsemes River, the continuation of which is the Novorossiysk (Tsemes) Bay. It is not excluded that in the recent past there was a narrow strait along the lowered strip Anapka – Maskaga – Tsemes – Novorossiyskaya Bay, and the land area located to the southwest (with Navagir Ridge) was an island.

Seismogenic (landslide) relief (Q_H) of the study area is developed on the wings and vaults of mountain ranges. It was formed during the process of earthquakes, as characteristic forms of relief are observed – seismic landslides, seismic landslides, tectonic ditches, cliff cones of landslide type. They are classically expressed on the southern slope of the Navagir Ridge, stepping down to the Black Sea. The elements of rock occurrence in the dislocated blocks are sharply different due to their divergence and strong crushing during displacement. Many valleys are underlain by these blocks. Especially spectacular is the Abrau River, above which a dammed lake of the same name has appeared with an area of 3×1 km and a depth of up to 50 m. Landslides and landslides are associated with impulsive movements along faults at the time of their seismogenic renewal. According to historical data, the oldest known earthquake was registered in 150 B.C. in the area of the city of Gelendzhik. After 87 years, it was repeated, destroying the ancient Bosporan Kingdom. In our time, a 7–point earthquake occurred in 1978. Seismic hazard of the Black Sea coast is now estimated at 9 points. On geomorphologic data the age of the described seismodeformations is not older than 2000 years. This is evidenced by: the complete absence of marine terraces on the investigated section of the coast; the absolutely fresh appearance of the walls of the seismic hills and hollows; the absence of de–alluvial cover and even modern soils on them.

Lake Abrau is the largest lake in the North Caucasus and the largest among freshwater lakes in Krasnodar Krai. The main mystery of Lake Abrau is the origin of its basin.

According to one hypothesis, the lake basin was formed as a result of a karst failure, but it does not correspond to the geological structure and lithological composition of the rocks composing the bed of the reservoir. According to another hypothesis, Lake Abrau is a remnant of the Cimmerian freshwater basin that existed in the place of the Black Sea at the end of the Neogene period more than 1 million years ago. This hypothesis explains well the composition of the lake's fauna. At the same time, this hypothesis leaves open the question about the origin of the basin itself.

Researchers (Boldyrev et al, 1957; Zenkovich, 1958) who studied the origin of the relief of the coastal Black Sea zone in the 1950s came to the conclusion that the defining feature of the Abrau Peninsula shoreline structure is the ancient landslides – slumps formed at a much lower sea level (40–50 meters lower than the modern one). When the sea level began to rise at the end of the New Euxine period, the abrasion sharply intensified and the equilibrium of the slopes was disturbed. At the same time, the humid climate favored the loosening and sliding of rocks. Huge blocks of flysch with the volume of millions of cubic meters collapsed down the slopes in the form of rockfalls. Similar phenomena occurred in river valleys. The river was dammed by one of these giant rockslides, which blocked the river valley. As a result of the dam formed, the river valley upstream filled with water and formed a narrow long and rather deep lake. This hypothesis explains well the morphology of the coastal zone of the sea. However, at the site of the alleged landslide, which allegedly blocked the Abrau River, there are no high mountains from which such a wide and high landslide could have fallen.

According to other scientists (Fedorov et al, 2022; Popkov et al, 2015), crustal movements at the turn of the new era shook the Black Sea coast. Before the earthquake, the waters of the Abrau River flowed directly into the sea. As a result of the earthquake, the mountains shifted, closed the mouth of the river and created the lake. The existence of several hypotheses of the lake's origin testifies to the complexity and unresolved nature of this issue. Most likely, the last two hypotheses complement each other.

I will note only that the last hypothesis, in my opinion, is more argued than others. In this connection, by the origin of the basin it can be referred to tectonic, valley ones.

The coast between the cities of Anapa and Novorossiysk is also interesting from the point of view of studying seismogravity processes. The relatively recent seismic activity resulted in the formation of a number of seismic dislocations and seismogravity deformations, which covered both the coastal part of the land and the area of the adjacent seabed and are very prominent in the relief. The formation of Lake Abrau, which is unique for our coast, is also associated with the same processes and appeared due to the blocking of the river valley by a seismic–gravity landslide (Popkov et al, 2015).

According to the genetic types of surfaces, the studied area can be attributed to the technogenic group (seismogenic landslide–landslide relief).

Figure 8 shows a view of the satellite image with 3D model of the relief of the study area from Google Earth Pro.

Figure 8. View of the study area

(Google Earth Pro)

Based on the data obtained, the study area is located in the area of multi–stage tectonic discharges. However, Lake Abrau is located in the graben zone, which is more than 7 km long (south–north orientation) and 2–3 km wide. The zone of junction of Lake Abrau and the south of the gully in the area of the storage pond (where the study area is located) is characterized by tectonic outbursts, which is the reason for the existence of the lake.

Hydrogeological structure of the study area.

The rocks of the study area are represented by the relatively water–bearing Sukko–Kazach–Schelevsky carbonate–terrigenous complex ($\mathcal{P}_1 sk + k\check{s}\check{c}$) – composed of interbedded marls, sandstones, siltstones, clays and limestones. The total thickness of this complex is more than 1000 m (Zakharchenko et al, 2003).

The rocks are fed with water by precipitation. Water–bearing capacity is insignificant because only the upper part of the complex is watered to a depth of 20 m to 40 m in the zone of exogenous fracturing. Water discharge occurs in the form of numerous springs (with a flow rate of up to 1.0 dm^3/s). The water composition is fresh calcium hydrocarbonate with mineralization from 0.3 g/dm^3 to 0.5 g/dm^3.

Part of the study area is represented by the Holocene alluvial aquifer ($a\,Q_H$) of modern channels and floodplains. Water–bearing sediments are represented by interlayers of sand and shell. Waters are connected with surface waters, have a common regime and composition. Water mineralization varies from 0.2 g/dm^3 to 6.0–7.0 g/dm^3. Water composition varies from calcium hydrocarbonate to sodium chloride. Waters are contaminated with nitrates and heavy elements and are practically not used. The only exception is water in the valley of the Katlama River, which is used by wells of the Alekseevsky water intake (Zakharchenko et al, 2023).

On the state geological map of preQuaternary sediments of the Russian Federation, on the territory of map sheet L–37–XXVI (Novorossiysk), there are 6 deposits and 2 sites of mineral waters localized in sediments of a wide age range – from Lower Cretaceous to Upper Cimmerian (Kavkazgeolszemshtuka, 2021).

In 20–30 km from the study area there is a groundwater deposit (Anapskoye deposit), confined to the same hydrogeological complex as the study area. The waters of the Anapskoye field are confined to interlayers of sandstones and mudstones of the Sukko–Kazach–Schelevsky carbonate–terrigenous complex. Water is low–mineralized hydrocarbonate–chloride–chloride–sulfate–sodium nitrogen with mineralization of 2.7 g/dm^3. Category B reserves amount to 43 thousand m^3/day; category C1 reserves amount to 22 thousand m^3/day. The total amount of water in the field is 65 thousand m^3/day. Water of the Anapa deposit is used for balneological purposes and bottling. Water is extracted from depths from 150 m to 750 m (Zakharchenko et al, 2024).

GENERAL INFORMATION ABOUT LAKE ABRAU IN KRASNODAR KRAI

Lake Abrau is filled by the waters of the Abrau River flowing into it, as well as by the discharge of groundwater at its bottom. From the hydrological point of view Lake Abrau is a drainless lake. In periods of time, when Lake Abrau overflows, water enters a special hydraulic structure – a storage pond (it is on the territory of this storage pond that the geophysical studies under consideration were conducted) (Zakharchenko et al, 2023).

Water mineralization in Lake Abrau varies within 0.3–0.8 g/dm^3. According to the chemical composition of water these lakes are close (according to O.A. Alekin's classification they belong to CII Ca), which once again confirms their genetic affinity.

A hydrogeological section along the line of Lake Abrau – Lake Maly Liman was constructed (Fedorov et al, 2022).

It shows how the formation of top water and groundwater occurs and how they are discharged into the lake and the sea (figure 9).

Figure 9. Model of hydrogeological profile of groundwater flow formation along the section Lake Abrau – Lake Maly Liman

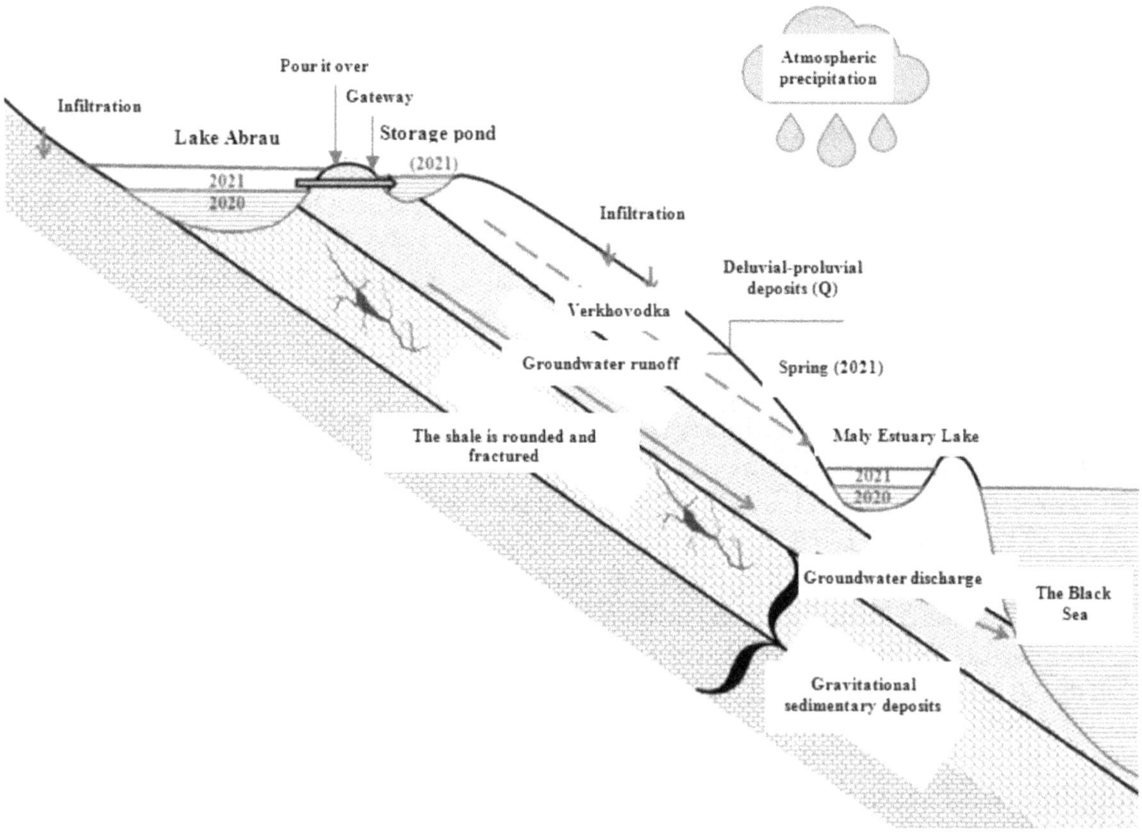

(Fedorov et al, 2022)

Figure 9 shows how top water and groundwater are formed and discharged into Lake Maly Liman (located below) and into the Black Sea.

The first thing to note is the infiltration origin of local groundwater. The second thing to note is that the waters are represented by both headwater and non–pressure (ground) groundwater. The role of these two aquifers in recharging the lake ecosystem with water is fundamentally different. Topsoil water is formed mainly through infiltration of precipitation falling on the surface of the catchment area and groundwater runoff from the storage pond, which is replenished with water from Lake Abrau. In dry years, the amount of precipitation decreases, and synchronously the water level in Lake Abrau. After that, the water supply from it to the storage pond stops, which leads to a reduction in the feeding of the upper water table and, accordingly, of Lake Maly Liman. Therefore, the horizon in the Quaternary sediments represented by the headwater is the main source of water in Lake Maly Liman. Another non–pressurized horizon of ground (interbed) water of the Paleogene aquifer discharges into the Black Sea, bypassing the lake bowl under its bottom.

Lake Abrau, being located in the recreational zone and experiencing a relatively small anthropogenic load, is extremely interesting for the study of intra–watershed processes in low–salinity water bodies of southern Russia. These genetically related lakes are not covered by systematic hydrobiological observations. As the results of episodic studies have shown, only Lake Abrau, which, in case of low concentrations of heavy metals, can be considered as a «background» water body for studying intra–water body processes in the conditions of southern Russia, should be referred to a typical oligotrophic lake (Zakharchenko et al, 2024).

In 2020, the reservoir of the storage pond, the filling of which is regulated by the inflow of water from Lake Abrau through a hydraulic structure, was empty.

In 2021, the picture completely changed – numerous groundwater outlets opened up, and the reservoir on the southern shore of Lake Abrau was completely filled (figure 10).

Figure 10. Fill rate of the storage pond for water runoff from Lake Abrau in 2020 and 2021

(Fedorov et al, 2022)

In the winter of 2022, it was reported that Lake Abrau had overflowed and excess water was overflowing across the road leading to the seashore (figure 11).

At the time of the geophysical survey (november 2023) the storage pond was completely empty (figure 12).

During the reconnaissance survey, minor groundwater discharge zones were noted in the south–western part of the study area (south–western side of Lake Abrau).

Figure 11. Outfall of Lake Abrau in winter 2022

Figure 12. Fill rate of the storage pond for Lake Abrau outflow in November 2023

THEORETICAL BASIS OF THE APPLIED GEOPHYSICAL SURVEY METHOD

In the area of search for drinking groundwater and beyond in November 2023, geophysical surveys were carried out by one of the methods of electrical exploration – the method of near–field establishment probing.

Near–field establishment probing (NFEP method) is a widely known and developed in the world technology of inductive pulse electrical exploration (Khmelevskoy, 1984). Near–field establishment probing is a method with a pulse controlled source based on the study of the transient field, which is excited in the ground when the current in the source changes. Developed more than 30 years ago as a method of structural electrical prospecting, the method of probing by field establishment in the near zone has 2 peculiarities:

- the size of the probing installation can be several times smaller than the depth of the geologic section surveyed;
- measurement results are highly sensitive to changes in the geologic section parameters (Zhdanov, 1986).

Field establishment sounding is a method of electromagnetic sounding based on the study of the transient field, which is excited in the ground during pulse switching of current in the source. Field establishment sounding method refers to methods with artificial (controlled) source. A horizontal electric dipole (grounded electric line AB) or a vertical magnetic dipole (ungrounded current loop (Q)) can be used as a source. Either a grounded electric line (MN) or an ungrounded loop (q) are also used as the receiver (Reference book of the geophysicist, 1989). In this method, both the loop is used as the source and the receiver. To excite the transient field, it is necessary to create a current switching pulse in the supply (generator) loop. Standard pulse signals are (Khmelevskoy, 1984):

- a short–time pulse of infinitely large amplitude described by the Dirac –function:

$$I(t) = \begin{cases} 0, t \neq 0 \\ \infty \ , t = 0 \end{cases};$$

- the on–off impulse described by the –Haviside function:

$$I(t) = I_0\,\sigma(t) = \begin{cases} 0, t < 0 \\ I_0, t > 0 \end{cases};$$

- the turn–off impulse described by the –Haviside function:

$$I(t) = I_0(1 - \sigma(t)) = \begin{cases} I_0, t < 0 \\ 0, t > 0 \end{cases}.$$

It is difficult to create a current pulse that is satisfactorily described by the Dirac –function, so in practice one usually uses stepwise field excitation.

At instantaneous switching of the current strength in the supply loop (in particular, at instantaneous switching off of the current), the voltage measured in the receiving loop drops to zero not instantaneously, but gradually disappears, changing in a rather complex way (figure 13).

Figure 13. Graph of voltage change in the receiving loop during current switching

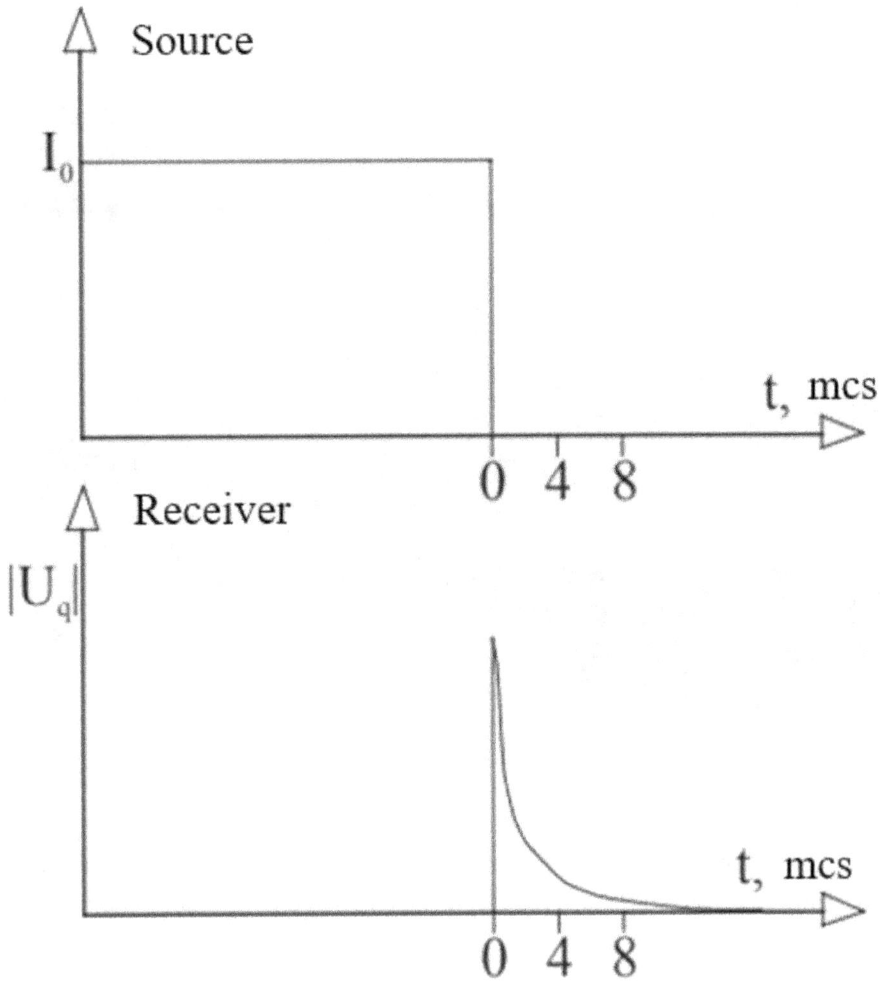

This is explained by the fact that secondary eddy currents are induced in the conductive regions of the section at the moment of current switching off (figure 14) (Zakharchenko et al, 2023; Zakharchenko et al, 2024).

Figure 14. Schematic diagram of conducting research by the method of probing by field formation

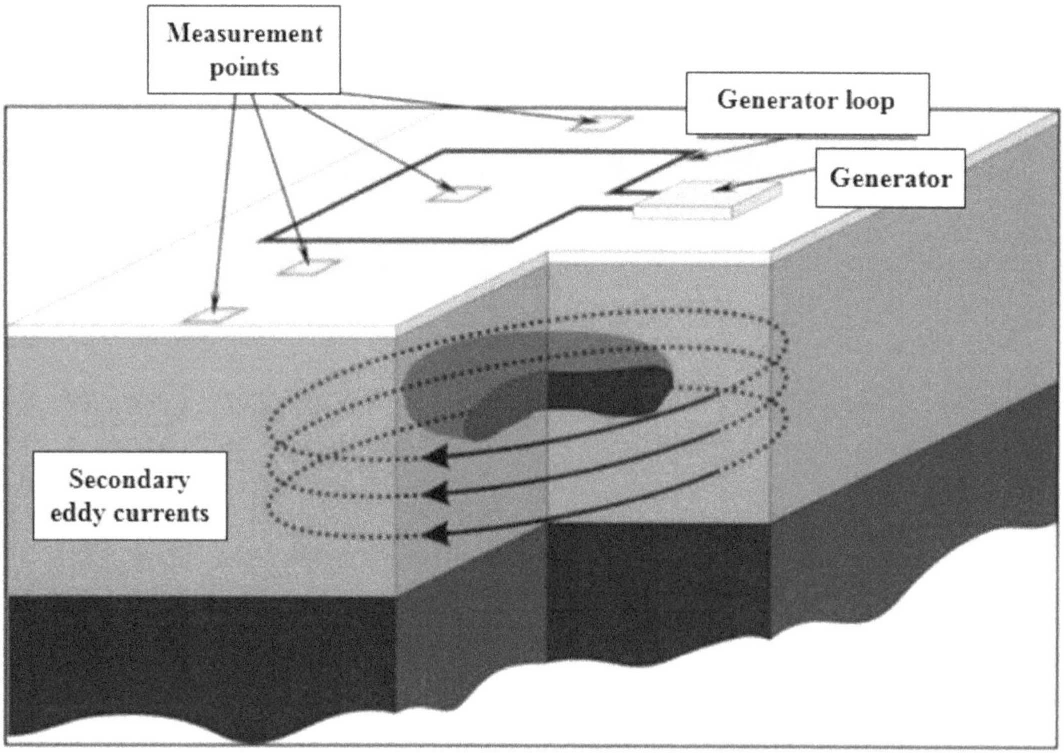

The alternating magnetic field of secondary currents induces in the receiving loop (q) an electromotive force denoted ε_q. Moreover, the change of the electromotive force in the receiving loop is proportional to the rate of change of the magnetic flux:

$$|U_q| = |\varepsilon_q| \approx \left|-\frac{d\Phi}{dt}\right|.$$

At the initial time instant (at small measurement times after current switching in the supply loop), the secondary currents are distributed in the near–surface part of the section. Then, as time passes (at longer measurement times after switching the current in the feed loop). the currents begin to penetrate into deeper layers, damping with distance from the source. This process is called «field establishment in the ground», and the dependence of the voltage in the measuring loop on the time elapsed since the switching of the current in the supply loop is called the «field establishment curve».

Thus, the depth of penetration of the transient field into the ground is determined by the time elapsed from the moment of switching off the current in the generator loop, which is called «delay time» or «establishment time». This property makes it possible to carry out probing, studying the dependence of the measured electromagnetic field components on the delay time.

The solution to the problem of field establishment in a homogeneous horizontally layered section was obtained on the basis of a quasi–stationary model of electromagnetic field propagation (i.e., without taking into account the displacement currents in the system of Maxwell's equations) (Khmelevskoy, 1984; Zhdanov, 1986). The obtained analytical expressions for the transient magnetic and electric field strengths depend on:

- distance from the receiver to the source (r), m;
- formation time (t), s;
- electrical properties of the section (resistance) (ρ), Ohm–m.

Let us consider one more peculiarity of the field formation process, which relates to the mechanism of field propagation of secondary currents (Reference book of the geophysicist, 1989). For this purpose, let us introduce the concept of «near» and «far» zones of the unsteady electromagnetic field. This separation is made using the parameter of becoming τ:

$$\tau = \sqrt{2\pi \bullet 10^7 \bullet \rho \bullet t},$$

and the reduced distance $r/_t$.

The condition for the presence of the far zone $r/_t \gg 1$.

Condition for the presence of the near zone $r/_t \ll 1$.

After instantaneous current switching in the source, the excitation to each point of the medium is transmitted in two ways (Reference book of the geophysicist, 1989).

The first way is through the air with instantaneous velocity, without absorption (since air is an insulator). In this mechanism, the field amplitude decreases with distance from the source due to geometric attenuation. A field with a flat phase front penetrates vertically into the ground simultaneously at all points, so we can consider that a plane inhomogeneous wave propagates. At sufficiently large distances from the source or very short delay times, this mechanism is the predominant one.

The second way is along the ground, with absorption and finite velocity determined by the conductivity of the cut. This mechanism can be described as a process of diffusive propagation of secondary eddy currents that are uniformly distributed in space over time. The intensity of secondary eddy currents decreases with distance from the source. The dissolution of a drop of ink in water is a good example of this mechanism of field propagation. Thus, two parts of the field appear to be separated in time and space.

The first propagation mechanism is determinant for the far zone, and the second mechanism – for the near zone. Consideration of the formation field in the far and near zones allows us to make a number of simplifications in the formulas for the magnetic and electric field strengths (Khmelevskoy, 1984; Zhdanov, 1986). For this purpose, we pass to the limits:

- at $t \to 0$ and $r \to \infty$, for the far zone;
- at $t \to \infty$ and $r \to 0$, for the near zone.

The use of simplified formulas gives significant methodological advantages, but it is necessary to observe the near– and far–zone conditions precisely enough during measurements. It should be taken into account that some ranges of times and distances fall into the intermediate zone, where the use of far and near zone approximations is incorrect (Khmelevskoy, 1984; Zhdanov, 1986).

In practice, it is difficult to create a current pulse, so the process of field excitation in the medium under study is caused by switching on and off the current in the generator loop.

This field causes an electromotive force in the receiving loop, which is called «transient characteristic of the medium» or «field formation signal».

The results of measurements of this electromotive force contain information about the investigated medium, the characteristics of which are then determined in the process of interpretation of the obtained geophysical data.

To excite the transient field it is necessary to create a pulse switching current in the supply (generator) installation. At change of current strength in the source in the conductive medium an unsteady electromagnetic field appears, i.e. there is a process of field formation. At the initial moment of time (at small measurement times after current switching in the supply loop) secondary currents are distributed in the near–surface part of the section. Then, as time passes (at longer measurement times after switching the current in the supply loop), the currents begin to penetrate into deeper layers, damping with distance from the source. In this case, the signal measured in the receiving set decreases to zero, changing in a complex way (Khmelevskoy, 1984; Zhdanov, 1986; Reference book of the geophysicist, 1989).

The dependence of the signal at the observation point on the becoming time is called the field becoming curve. The type of formation curve is determined by the distribution of conductivity in the section, which makes it possible to carry out probing, studying the time dependence of the electromagnetic field components. In addition, the character of the electromagnetic field establishment depends on the distance between the source and the receiver.

Depending on the distance between feeding and receiving installations, as well as on the studied times of the field establishment process, 2 modifications of field establishment probing are distinguished – in the far zone and in the near zone.

In the method of field establishment sounding in the near zone, measurements are carried out close to the source, which leads to an increase in the localization and detail of the study. This is one of the most important advantages of the near field establishment sounding method.

The method of near–field formation sounding allows solving structural mapping problems of different purposes and scales. The efficiency of this method is determined by differentiation of layers composing the investigated thickness by specific electrical resistivity. The differentiation itself in most cases is related to lithology and water content of rocks.

Variants of the technology of probing by field formation in the near zone with the size of generator loops of 0.4–1 km provide the depth of studies up to 1–3 km, which allows solving the tasks related to the regional study of territories (or to the search for oil and gas) (Zhdanov, 1986):

- Study of the geologic structure of sedimentary basins;
- prediction of reservoirs, fault zones;
- structural–mapping, facies–mapping tasks;
- investigation of the basement, identification of zones of decompaction.

More productive variants of the technology, with the size of generator loops 20–200 m, provide depth up to 50–500 m.

The solution of geomapping problems in this depth range allows the method to be used in the following areas (Khmelevskoy, 1984):

- section dissection and structural mapping for hydrogeology problems;
- engineering geological and geophysical studies;
- fault mapping;
- search for ore objects and kimberlite pipes;
- ecology, mapping of salinization and watering zones, etc.

In recent years, the so–called «profile coaxial–bearing (multi–bearing) observation system» has become widespread, in contrast to the previously used coaxial observation setup.

Observation points (locations of the receiving sensors) are located on the profile with a uniform spacing, without overlap, or with overlap (extreme points of the arrangement are common for neighboring arrangements), measurements on them are made at two different positions of the generator loop.

Advantages of the method:

- high localization of the research;
- considerable depth;
- low sensitivity to section anisotropy and surface inhomogeneities;
- absence of galvanic grounding and, as a consequence, the possibility to work in different natural conditions;
- the possibility of determining the parameters of the section by installations whose dimensions are smaller than the depth of the investigated objects.

SETUPS USED IN THE FIELD ESTABLISHMENT SOUNDING METHOD

The AB–loop («AB – q») and loop–loop («Q – q») setups are usually used in field work by the field establishment method. Particular cases of the «Q – q» setup are:

– «loop–in–loop» installation with coaxial loops (figure 15), installation with combined receiving and generator loops, single–loop installation. All varieties of «Q – q» installation are widely spread because of high productivity of works, localization of researches and certainty of location of «recording point», i.e. the point of the earth surface, to which the results of measurements refer.

Figure 15. Schematic diagram of a loop–in–loop installation with coaxial loops

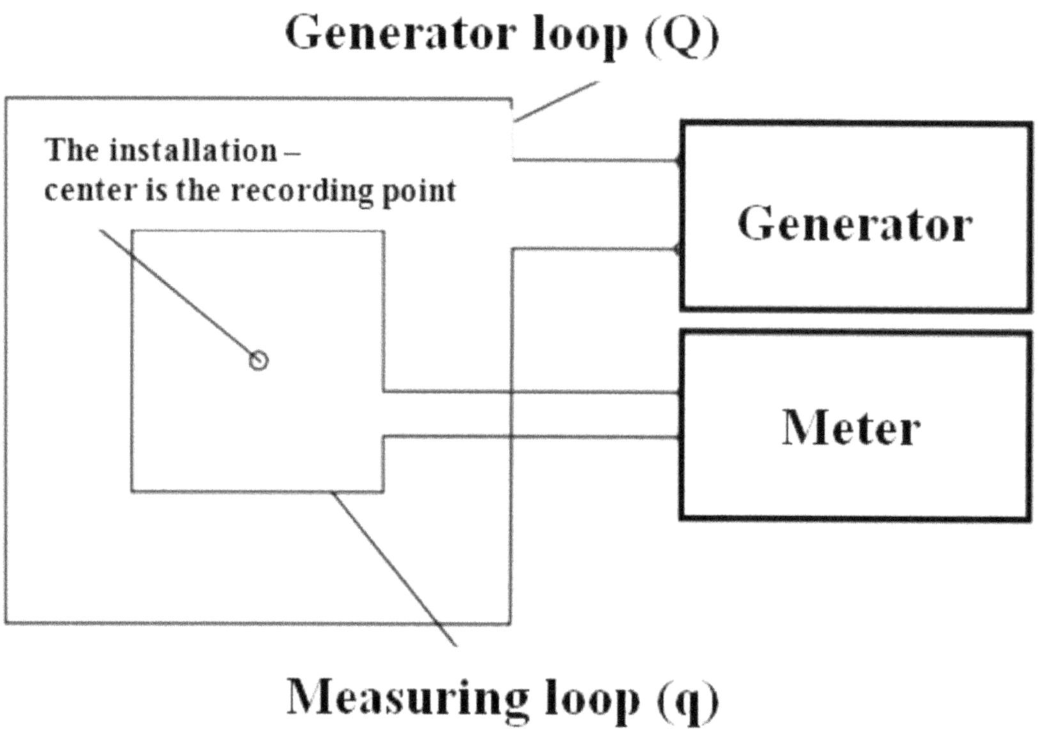

Peculiarities of the equipment used in conducting works by the method of field establishment. The following requirements are imposed on the equipment used for measuring the field establishment in the near zone:

- large dynamic range associated with the need to measure the signal differing by 5–6 orders of magnitude;
- greater sensitivity and noise immunity, as improvement of detail and productivity of works by this method is connected with reduction of loop sizes, and hence with reduction of measured signal level.

Geophysical surveys by the method of field establishment in the near zone were carried out in accordance with the Republican construction norms RSN 64–87 «Engineering surveys for construction. Technical requirements for geophysical works. Electrical prospecting» (Gosstroy RSFSR, 1988).

The digital electrical survey station «TEM–FAST 48» (manufacturer – Electro Magnetic Research (AEMR Ltd)») was used in the studied area (figure 16).

Figure 16. Digital electric exploration station «TEM–FAST 48»

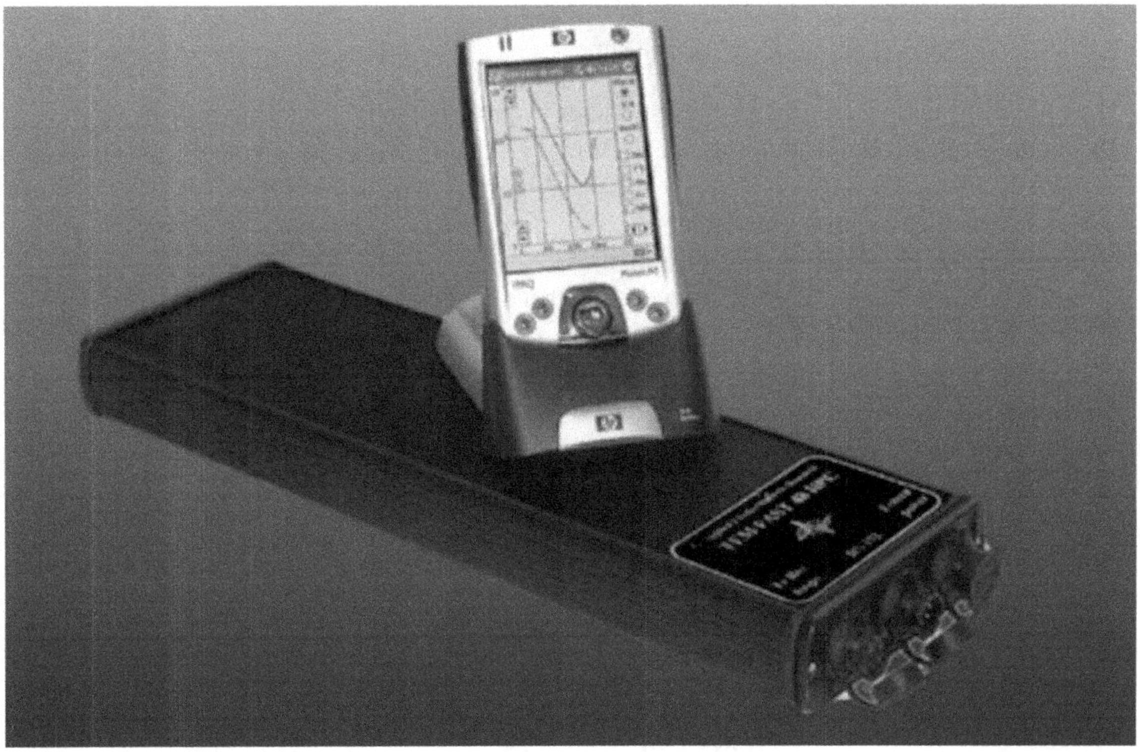

(manufacturer – AEMR Ltd.)

The set of digital electric exploration station «TEM–FAST 48» is designed for geophysical surveys by the method of field formation sounding.

The equipment is designed for electromagnetic studies of rock massifs within the first hundreds of meters. With its help the following tasks can be solved (manufacturer – AEMR Ltd.):

- tasks of search and exploration of mineral deposits or genetically related structures;
- tasks of structural and mapping geology (determination of the geological structure of the environment and lithologic composition of rocks);
- hydrogeological tasks (search for underground water, including thermal water, assessment of water saturation of rocks and fluid mineralization);
- permafrost and glaciological tasks (mapping and determining the depth of permafrost, studying the dynamics of freezing and thawing, etc.).

TEM–FAST 48» digital electric prospecting station set allows:

- simultaneously obtain information about both the first meters of the geological section and depths of hundreds of meters;

- work steadily on city streets and industrial objects at interference levels in the receiver of 1 volt and more;
- see and evaluate the results of probing directly in the field and, if necessary, adjust the measurement modes.

The complex is optimal for shallow–depth studies by probing the field formation in the near zone. The digital electrical exploration station «TEM–FAST 48» includes (manufacturer – AEMR Ltd.):

- unipolar rectangular pulse generator;
- measuring unit providing signal registration;
- control unit providing synchronization of the generator and measuring unit;
- internal 12 V power supply.

All devices are assembled in one case. Figure 17 shows the block diagram of the digital electrical exploration station «TEM–FAST 48».

Figure 17. Block diagram of digital electric exploration station «TEM–FAST 48»

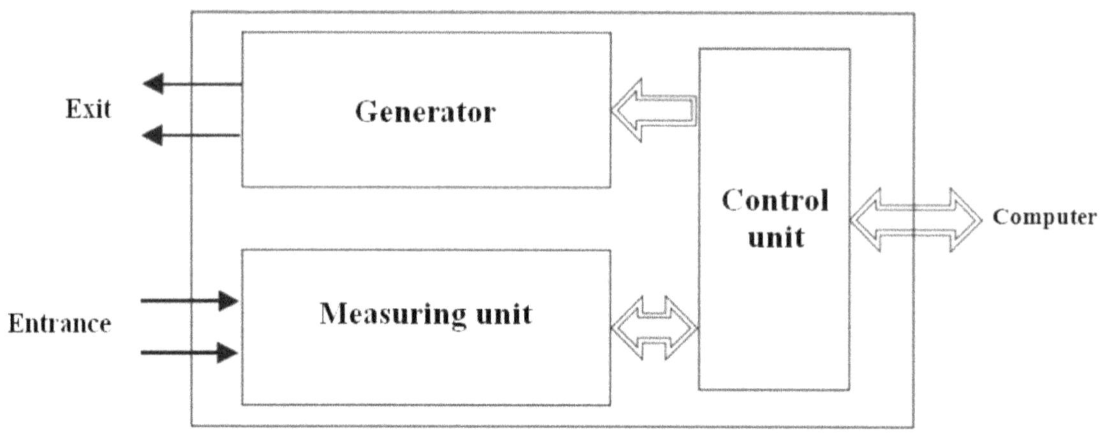

(manufacturer – AEMR Ltd.)

The generator produces rectangular unipolar current pulses, which are passed through the generator loop. The ratio pulse / pause = 3/1 (figure 18), which provides the minimum error caused by the superposition of transients from the on and off edges of the current in the generator loop.

Figure 18. Ideal unipolar current pulses,fed into the generator loop with the ratio pulse / pause = 3/1

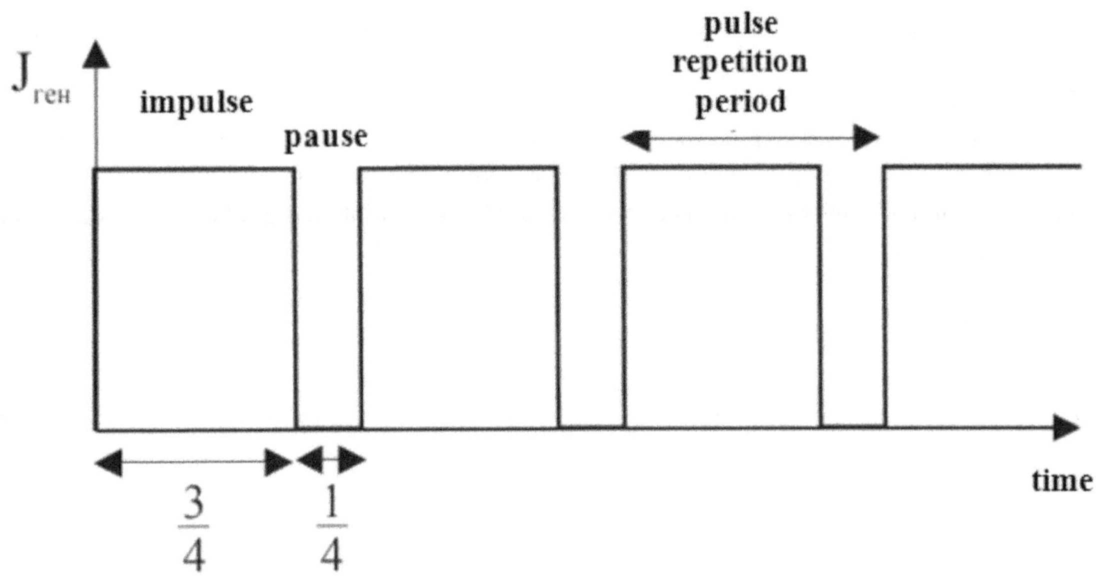

(manufacturer – AEMR Ltd.)

It should be noted that the ideal and real current pulse are different because in reality the loop is an inertial electrical system. At the moment of current switching off (the most important for transient measurement) the current in the generator loop sharply differs from the ideal one, due to this there is (sometimes quite strong) distortion of the results of changes at small times after current switching off.

Possible current modes of generator operation: 1 A and 4 A. The current value is specified conditionally and depends on the generator loop resistance. In the process of measurement the real amplitude of current pulses is measured, on which the electromotive force in the receiving loop is then normalized.

The main technical characteristics of the generator operation:

- current pulse turn–off time from 4 μs (depending on the size of the loop);
- amplitude of current pulses from 1 A;
- active resistance of the load not more than 20 Ohm.

The measuring unit of the digital electrical exploration station «TEM–FAST 48» consists of:

- device of protection against high input voltage (High Voltage Protection or HVP): At low delay times the signal in the receiving loop can be very large and outside the dynamic range. The High Voltage Protection device «dampens» the input voltage. It can operate in three modes:
- automatic voltage limitation at the level of 10 V;
- operator–defined protection time (i.e. by selecting the minimum delay time from which the curve measurement starts);

- automatically set protection time determined by the installation size and current in the generator loop;
- multi–channel system of gating, integration and storage of analog information, where each channel corresponds to a hardware–defined delay time. With its help, analog accumulation of the signal at each delay time takes place. In the digital electrical survey station «TEM–FAST 48» the decline curve characterizing the transient process is measured with a sufficiently frequent step in delay times (48 times from 4 μs to 16 ms);
- channel multiplexer, which, after the analog accumulation processes are completed, successively connects the outputs of each channel to the analog–to–digital converter;
- analog–to–digital converter, which converts the analog signal into a digital code and transmits it to the control unit.

The block diagram of the measuring unit is shown in figure 19.

Figure 19. Block diagram of the measuring unit of the digital electric exploration station «TEM–FAST 48»

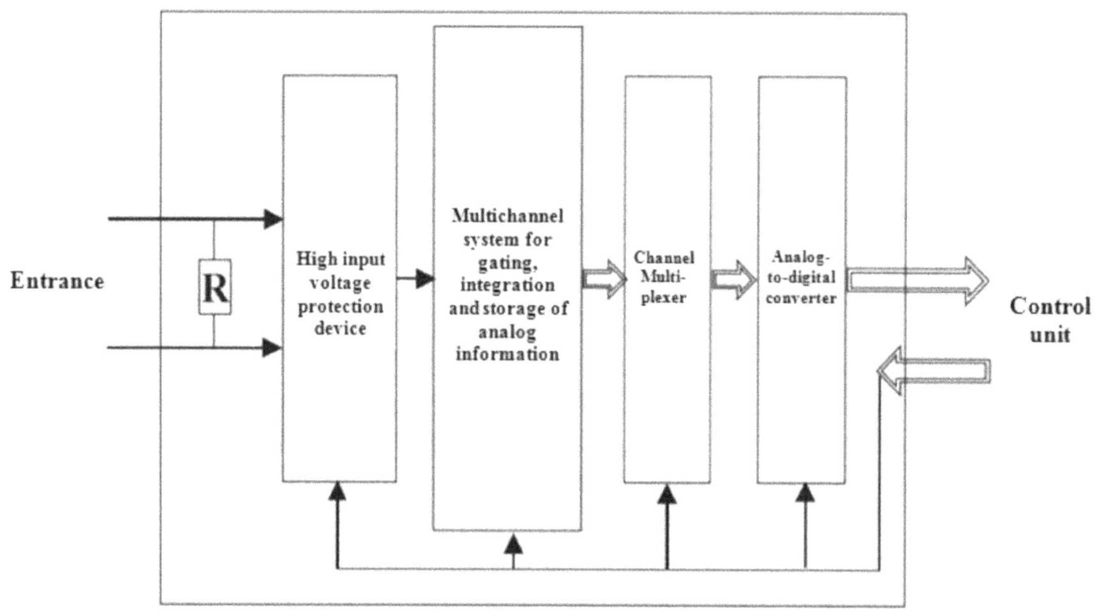

(manufacturer – AEMR Ltd.)

An important problem in geophysical measurements by various methods is industrial interference. Industrial interference is controlled by two methods – filtering and accumulation. The first method gives good results if the interference is harmonic and its frequency is known. Then the signal coming to the meter input is passed through an analog or digital (depending on the features of the equipment) filter, adjusted to the frequency of the interference.

A bandpass filter is a narrow–band filter that «cuts out» a certain frequency, thus suppressing the interference. However, applying a filtering procedure distorts the waveform and we cannot accurately account for these changes.

This introduces large errors in the measurement results of the becoming curve. Therefore, filtering is practically not used in the near–field establishment method. The second method of interference control – accumulation, is reduced to the fact that in the generator loop is sent not a single current pulse and measurements of the field establishment curve after its shutdown, and a series of pulses are sent and after each of them are measured.

Harmonic interference T is suppressed in the best way when the pulse repetition period T_{com}:

$$T = (k + 1/2) \bullet T_{com},$$

where k can take values 0, 1, 2, 3 and so on;

T_{com} – period of interference.

Similarly, the industrial noise is suppressed in the device of digital electrical survey station «TEM–FAST 48».

The main technical characteristics of the digital electrical exploration station «TEM–FAST 48»:

- number of automatically registered delays – 48;
- minimum value of delay time – 4.06 μs;
- maximum value of delay time – 15304 μs (15 ms);
- dynamic range – 140 dB;
- sensitivity (minimum recorded signal) – 10^{-6} V (1 μV);
- suppression of industrial noise (50 Hz / 60 Hz) – not less than 60 dB;
- measured quantities – electromotive force / current, units – V/A;
- unit curve measurement time from 10 seconds to 4 minutes 30 seconds;
- digit capacity of analog–to–digital converter is 16 bits, resolution is 1 μV.

The control unit is designed to control all measurement processes and transfer information to a computer. On the front panel of the instrument there are three connectors labeled:

- Tr–Rec loops – connection with an adapter of the transmitting and receiving loops;
- RS 232 – connection with COM port of the computer;
- Extern power – connection of external power supply – 12 V or 24 V batteries, as well as charging of internal batteries.

The digital electrical survey station «TEM–FAST 48» is controlled from a computer using a special control program («TEM–48–Win»). NoteBook class computer is used for field surveys.

Also, when conducting geophysical methods of search and exploration of drinking groundwater by probing the field formation in the near zone, a portable set of receiving and generator loops was used (figure 20).

1 – generator block, 2 – receiving antenna, 3 – carrying frame

Figure 20. Portable set of receiving and generator loops for geophysical research by the method of probing the field formation in the near zone

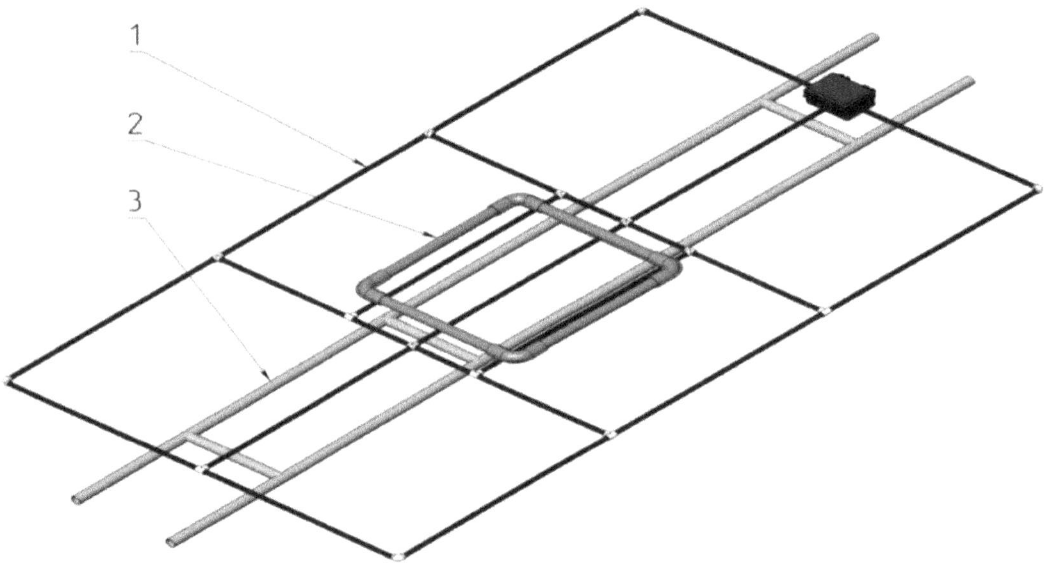

1 – generator block, 2 – receiving antenna, 3 – carrying frame

(manufacturer – AEMR Ltd.)

In the set of receiving and generating loops for conducting geophysical studies by probing the formation of the field in the near zone is used portable pneumatic structure, with a rigid frame made of carbon fiber with a compensator of its own process. This set is resistant to abrasion and punctures during movement, has minimal weight, frost resistance. A characteristic feature is the absence of metal parts.

The size of the portable set of receiving and generator loops for geophysical research by probing the field formation in the near zone: 2.5 m x 4 m. The duration of the tripping front is 2.6–15 µs, the bandwidth is 118 kHz, the initial measurement time is 8.5 µs. The depth of investigations with this set is 50 m.

METHODOLOGY OF GEOPHYSICAL RESEARCH OF SEARCH AND EXPLORATION OF DRINKING GROUNDWATER AT THE SURVEY SITE IN THE LAKE ABRAU AREA

When conducting geological and geophysical surveys at the survey site in the Lake Abrau area, the following types of work were performed:

- reconnaissance survey;
- geophysical survey;

– cameral works.

The method of probing the field formation in the near zone is one of the most used methods in electrical exploration, as it allows to obtain, with the maximum accuracy available in electrical exploration, values of resistivity.

The field work was carried out by electrospecting geophysical method of probing the field formation in the near zone.

Geophysical surveys were carried out in accordance with the Republican Construction Norms RSN 64–87 «Engineering Surveys for Construction. Technical requirements for geophysical works. Electrical prospecting» (Gosstroy RSFSR, 1988).

Geophysical surveys by the method of probing the field formation in the near zone (loop – loop) were carried out by profiles with a step of 10 m. There were 7 profiles, the total length was 3520 m. Registration was carried out at parking places at a point with recording of 3 series of 64 accumulations each. The depth of studies reached 220 m (Zakharchenko et al, 2023; Zakharchenko et al, 2024).

In the process of sounding, we used a setup consisting of ungrounded generator and receiver loops located on the ground surface, the sizes of which were determined by the required depth of the study.

The loops, in the form of squares, were arranged symmetrically one in the other.

Desk processing of the results of field studies included the construction of sections by probing the field formation in the near zone; comparison of processed data; identification of promising aquifers, followed by the positioning of a water production well; drawing a conclusion about the studies (Zakharchenko et al, 2023; Zakharchenko et al, 2024).

Field engineering–geophysical studies were carried out in November 2023.

The position of the profiles by probing the field formation in the near zone is shown in figure 21.

Figure 21. Scheme of location of electric exploration profiles on the studied area

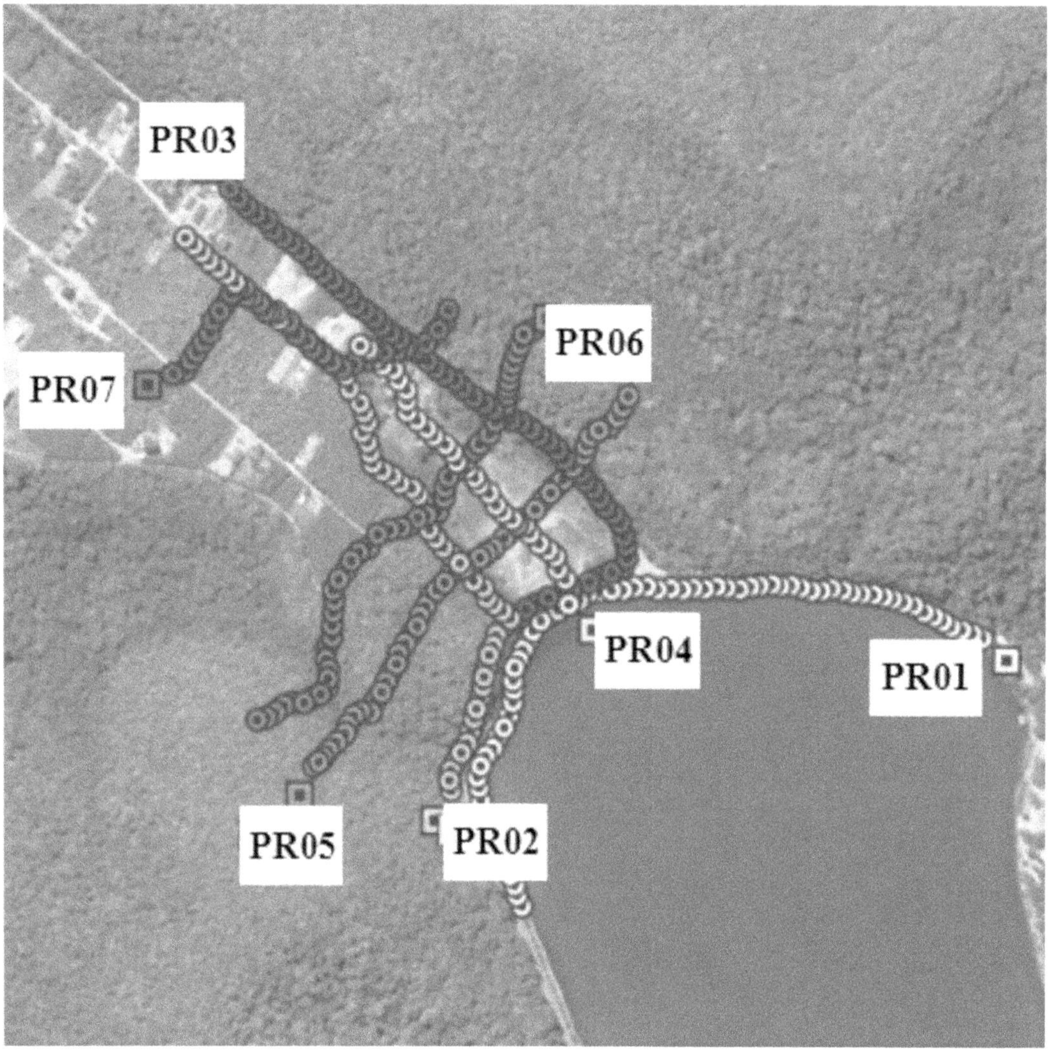

(Zakharchenko et al, 2023; Zakharchenko et al, 2024)

The process of geophysical surveys is shown in figure 22.

Figure 22. Process of conducting the method of probing the field formation in the near zone

Data processing of the near–field field establishment probing method

Data processing of the near–field establishment probing method consists in recalculating the values of the reduced electromotive force obtained at different delay times to the value of apparent resistivity. The apparent resistivity in the method of probing by field establishment in the near zone is denoted by ρ_τ. The conversion is performed according to the formula (Khmelevskoy, 1984):

$$\rho_\tau(t) = \frac{\mu_0}{\pi \bullet t} \bullet \left(\frac{Q \bullet q \bullet \mu_0}{20 \bullet t \bullet E(t)}\right)^{2/3},$$

where Q and q are the effective areas of the generating and receiving loops (i.e., taking into account the number of turns), m^2;

t – time of formation, s;

$E(t)$ – reduced electromotive force, V/A;

$\mu_0 = 4 \bullet \pi \bullet 10^{-7}$ – magnetic permeability of vacuum, Gn/m.

Graphs of apparent resistivity are plotted in bilogarithmic scale. On the abscissa axis in the method of probing by the field formation in the near zone it is accepted to postpone the parameter $\sqrt{2 \bullet \pi \bullet t}$ (Zhdanov, 1986).

Interpretation of the curves of the method of probing by the field formation in the near zone.

As in other methods of electrical exploration, in the method of probing by field establishment in the near zone the main method of interpretation of probing curves is the method of selection. The fitting method consists in successive changes of the geoelectric model of the section and calculation of the direct problem for it until the experimental and theoretical curves coincide with the required accuracy.

However, there are simpler methods by which the quantitative parameters of the model can be estimated: total power to roof conductors *(H)* and total longitudinal conductivity to roof insulators *(S)*. The calculation of these parameters for a loop–in–loop installation and for a single–loop installation is done according to the formulas (Reference book of the geophysicist, 1989):

$$S = 452 \bullet \frac{\sqrt{2 \bullet \pi \bullet t_{min}}}{\sqrt{\rho_{\tau\, min}}},$$

$$H = 360 \bullet \sqrt{2 \bullet \pi \bullet t_{min}} \bullet \sqrt{\rho_{\tau\, min}},$$

where $\sqrt{2 \bullet \pi \bullet t_{min}}$ is the abscissa of the minimum of the apparent resistance curve;

$\sqrt{\rho_{\tau\, min}}$ – is the ordinate of the minimum of the apparent resistance curve (figure 23).

Figure 23. Curve of apparent resistivity of the method of sounding by field establishment in the near zone

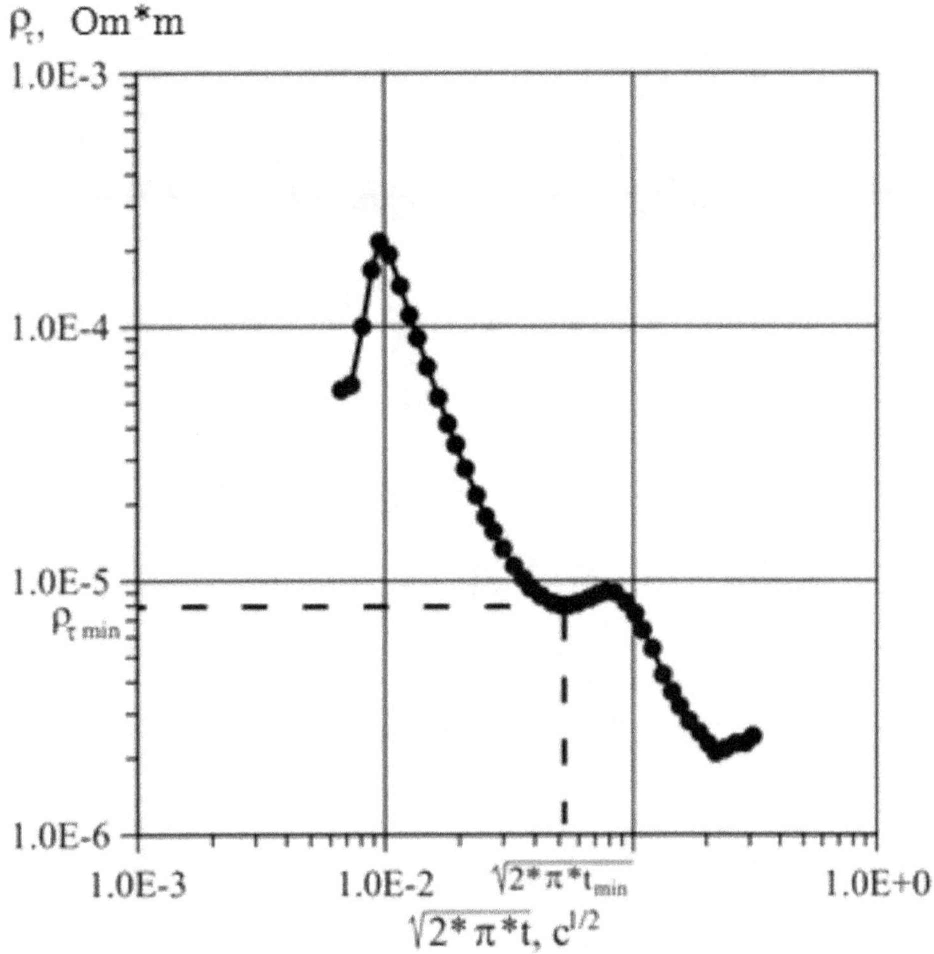

An example of processing of high–resolution method of sounding by field establishment in the near zone on profiles PR01, PR03, PR06 is shown in figure 24.

After processing of all data, a complex interpretation was carried out, including data from electrical survey, geomorphology, tectonic and neotectonic structure, reconnaissance observation. The complex interpretation was carried out in 3D mode.

Figure 24. Results of data processing of the method of sounding by field formation in the near zone on profiles PR01, PR03, PR06

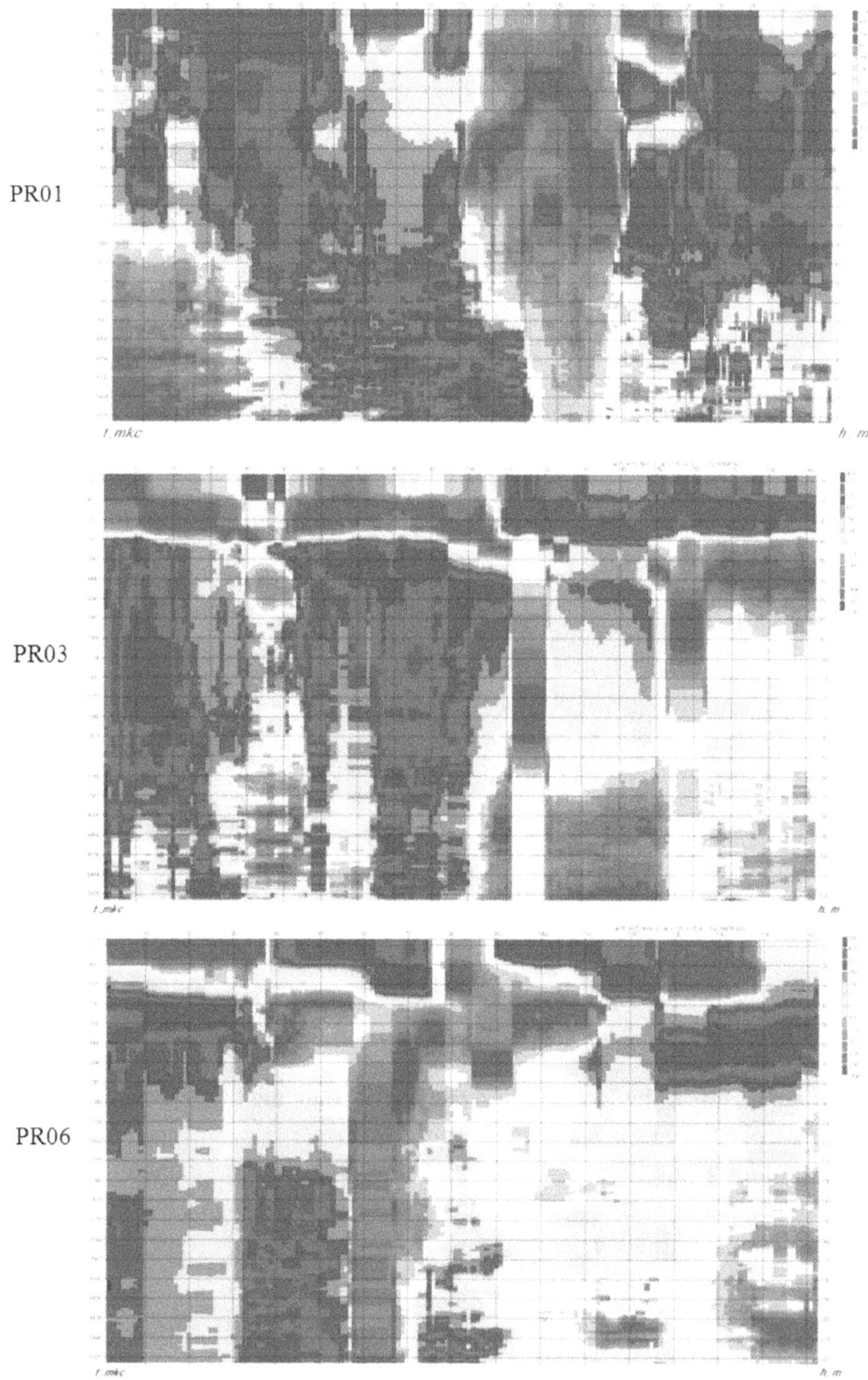

Results of interpretation of the conducted geophysical research on search and exploration of drinking groundwater at the study site in the area of Lake Abrau

Main conclusions on the results of analysis of groundwater hydraulic connection, structural plan and water saturation:

- the water saturation of this area is associated with zones of increased fracturing, arising due to tectonic activity, forming in the process a block structure with well-washed zones;
- in the block structure the study area is represented by areas with maximum groundwater filtration, weakly permeable area and practically non-permeable rocks;
- target water-bearing intervals are located at depths from 40 m to 110 m.
- The depth of exploration wells should be at least 80–90 m, the depth of production wells should be at least 110–130 m.
- hydraulic connection of these watered zones is quite pronounced;
- the hydraulic connection of groundwater extends for several kilometers beyond the site, partially fed from Lake Abrau.

Figure 25 shows the interpretation of the geophysical section of the method of probing the formation of the field in the near zone along the profile PR04, which is the most typical for determining the position of the exploration water well.

The preliminary position of the exploration well is shown in figure 26.

Figure 25. Interpretation of the geophysical section of the method of field formation sounding in the near zone along the profile PR04

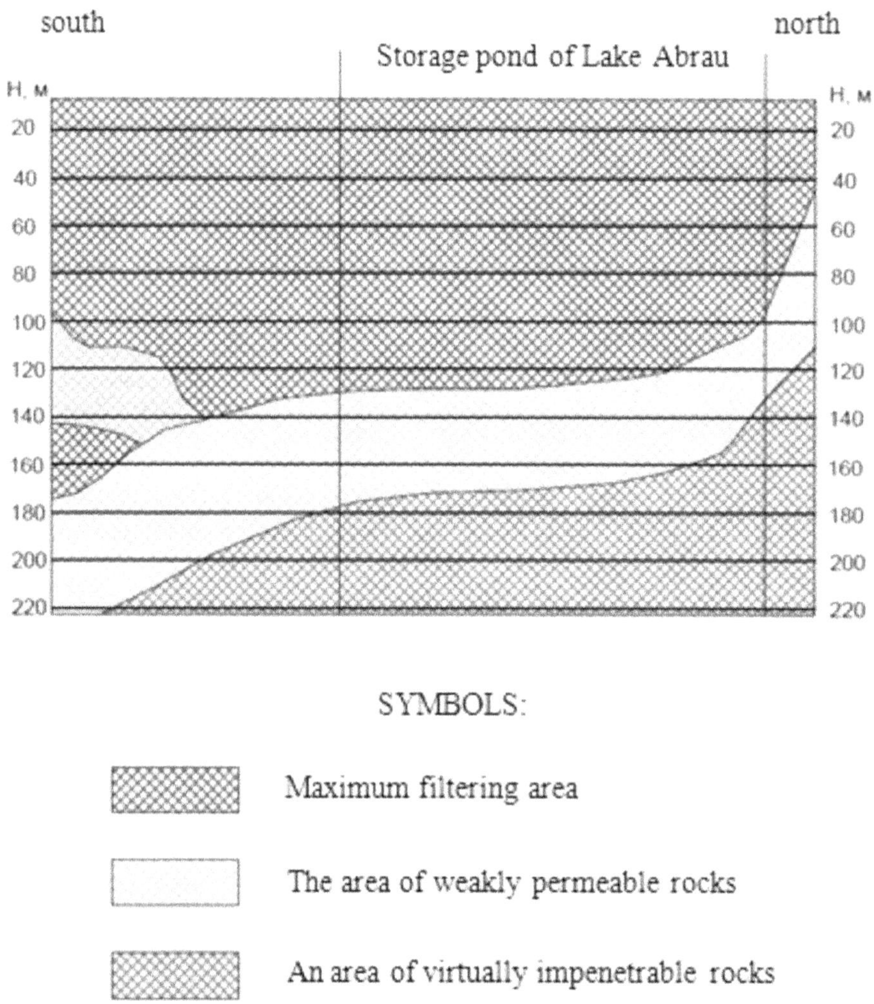

Figure 26. Preliminary position of the exploration well at the research site in the Lake Abrau area

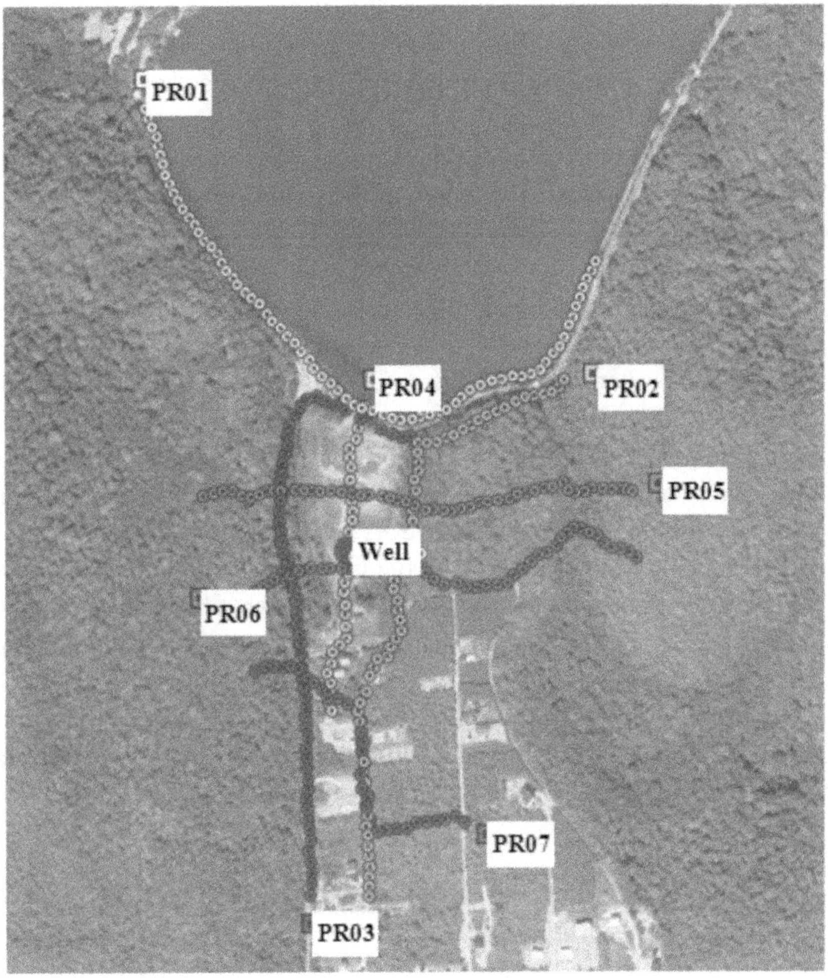

CONCLUSIONS

Increasing demand for water is putting extraordinary pressure on our natural environment. Globally, aquatic ecosystems are in crisis, and many rivers, seas and lakes are already polluted or severely degraded as a result of the decline of natural ecosystems, including forests and watershed areas. Industrial and agricultural emissions of significant amounts of heavy metals and hazardous wastes are increasingly polluting and depleting groundwater. The decline in the quantity and quality of water resources is responsible for the extinction of freshwater species and the rapid loss of biodiversity.

Some 30 percent of the world's coastal ecosystem land resources have been severely degraded as a result of the growing demand for land from housing, industrial and tourism development. More than half of humanity depends on freshwater that is stored in mountainous areas. However, these areas suffer from deforestation and pressure from agriculture and tourism, which can lead to overstretching of water resources.

The water resources of the study area, the Black Sea coast of Krasnodar Krai, are represented by the waters of the Black Sea and land waters. The length of the coastal strip within the coastline is respectively more than 470 km.

The Krasnodar Region is one of the most water–intensive regions in the world, most of which is used for agriculture and drinking water, including in the tourist and recreational areas.

Unstable climatic conditions cause a decrease in agricultural productivity and aggravation of problems in cities with already strained water supply. The issue of drinking water supply is particularly acute in the tourist areas of the Black Sea coast.

Drinking groundwater is an invaluable mineral resource that ensures the development of social sphere and economy of any state. For water supply of many settlements and industrial enterprises works on search and exploration of drinking groundwater deposits are carried out.

At present, the shortage of drinking water is particularly acute in mountainous folded areas of the Russian Federation, as the territory of folded areas is characterized by sharp changes in relief, geological and structural conditions and type of water–bearing rocks, manifested over relatively small distances. In this regard, the characterization of the conditions of distribution and formation of groundwater within the folded area as a whole is extremely difficult.

On the example of geophysical studies conducted in the area of Lake Abrau, Krasnodar Krai, the article details the measures for detailed exploration of potable groundwater for the purpose of water supply of the object.

The investigated area is located in the area of multi–stage tectonic discharges. However, Lake Abrau is located in the graben zone, which is more than 7 km long (south–north orientation) and 2–3 km wide. The zone of junction of Lake Abrau and the southern located gully in the area of the storage pond (where the study area is located) is characterized by tectonic outbursts, which determines the existence of the lake.

The rocks of the study area in the vicinity of Lake Abrau are represented by a relatively water–bearing Sukko–Kazach–Schelevsky carbonate-terrigenous complex – composed of interbedded marls, sandstones, siltstones, clays and limestones. The total thickness of this complex is more than 1,000 meters.

The rocks are fed with water by atmospheric precipitation. Water–bearing capacity is insignificant, as only the upper part of the complex is watered to a depth of 20 m to 40 m in the zone of exogenous fracturing. Water discharge occurs in the form of numerous springs (with a flow rate of up to 1.0 dm^3/s). The water composition is fresh calcium hydrocarbonate with mineralization from 0.3 g/dm^3 to 0.5 g/dm^3.

Part of the study area is represented by the Holocene alluvial aquifer of modern channels and floodplains. Water–bearing sediments are represented by interlayers of sand and shell. Waters are connected with surface waters, have a common regime and composition. Water mineralization varies from 0.2 g/dm^3 to 6.0–7.0 g/dm^3. Water composition varies from calcium hydrocarbonate to sodium chloride. The waters are polluted with nitrates and heavy elements and are practically not used.

Lake Abrau is filled by the waters of the Abrau River flowing into it, as well as by the discharge of groundwater at its bottom. From the hydrological point of view Lake Abrau is a drainless lake. During periods of time when Lake Abrau overflows, water flows into a special hydraulic structure – a storage pond (it is on the territory of this storage pond that the geophysical research under consideration was conducted).

During geological and geophysical surveys at the survey site in the Lake Abrau area the following types of works were performed:

- reconnaissance survey;
- geophysical survey;
- cameral works.

The method of probing the field formation in the near zone is one of the most used methods in electrical exploration, as it allows to obtain, with the maximum accuracy available in electrical exploration, values of resistivity.

The field work was carried out by electrospecting geophysical method of probing the field formation in the near zone.

Geophysical surveys by the method of probing the field formation in the near zone (loop – loop) were carried out by profiles with 10 m spacing. 7 profiles were carried out, the total length amounted to 3520 m. Registration was carried out at parking places at a point with recording of 3 series of 64 accumulations each. The depth of studies reached 220 m.

In the process of sounding we used a setup consisting of ungrounded generator and receiver loops located on the ground surface, the sizes of which were determined by the required depth of the study.

The loops, in the form of squares, were arranged symmetrically one in the other.

Desk processing of the results of field studies included the construction of sections by probing the field formation in the near zone; comparison of the processed data; allocation of promising aquifers, with the subsequent positioning of the producing water well; drawing a conclusion on the studies.

The field engineering geophysical survey was carried out in November 2023.

The main conclusions from the analysis of groundwater hydraulic connection, structural plan and water saturation:

- the water saturation of this area is associated with zones of increased fracturing that have arisen due to tectonic activity, forming in the process a blocky structure with well–washed zones;
- in the block structure the study area is represented by areas with maximum groundwater filtration, weakly permeable area and practically non–permeable rocks;
- target water–bearing intervals are located at depths from 40 m to 110 m.
- The depth of exploration wells should be at least 80–90 m, the depth of production wells should be at least 110–130 m.
- hydraulic connection of these watered zones is quite pronounced;
- hydraulic connection of groundwater extends for several kilometers beyond the site, partially fed from Lake Abrau.

As a result of interpretation of the received geophysical data of the method of probing the field formation in the near zone the authors determined the position of the exploratory water well. Subsequently, drilling of the well confirmed the data obtained by the authors on sufficient water seepage from the target aquifer intervals.

REFERENCES

Khmelevskoy, V. K. (1984). Electro prospecting / V.K. Khmelevskoy. – Moscow: Publishing House «Moscow State University».

Kurenoy, V. V. (2011). Drinking groundwater of Russia. Study and use of resources / V.V. Kurenoy // Exploration and protection of subsurface resources. Kurenoy // Exploration and protection of subsurface resources. – Moscow: Publishing House «ZAO Polygraphic Company «Paradigma», N° 9. – c. 3–10.

SP 131.13330.2020 Building climatology. Basic provisions: set of rules: official edition: approved and put into effect by the Order of the Department of Urban Planning and Architecture of the Ministry of Construction and Housing and Communal Services of the Russian Federation from December 24, 2020 N° 859/pr: date of introduction 2021–05–24 / prepared by FGBU «Research Institute of Building Physics of the Russian Academy of Architecture and Construction Sciences» // Electronic fund of normative–technical and regulatory legal information of the Consortium «Codex»: [site]. – URL: https://docs.cntd.ru/document/573659358 (date of reference: 15.10.2024).

SP 20.13330.2016 Loads and actions. Basic provisions = Loads and actions: set of rules: official edition: approved and put into effect by the order of the Ministry of Construction and Housing and Communal Services of the Russian Federation (Ministry of Construction of Russia) N°891/pr of 03.12.2016: introduced for the first time 04.06.2017 / developed by V.A. Kucherenko CRISC JSC «SIC «Construction» with the participation of FGBU «Main Geophysical Observatory named after A.I. Voyeykov». A.I. Voyeykov» // Electronic fund of normative-technical and normative-legal information of the Consortium «Codex»: [site]. – URL: https://docs.cntd.ru/document/456044318 (date of access: 15.10.2024).

SP 50.13330.2012 Thermal protection of buildings. Basic provisions = Thermal performance of the buildings: set of rules: official edition: approved and put into effect by the order of the Ministry of Regional Development of the Russian Federation (Ministry of Regional Development of Russia) N° 265 of 30.06.2012: introduced for the first time 01.07.2013 / developed by the Research Institute of Building Physics of the Russian Academy of Architecture and Construction Sciences (NIISF RAASN) // Electronic fund of normative-technical and normative-legal information of the Consortium «Codex»: [site]. – URL: https://docs.cntd.ru/document/1200095525 (date of reference: 15.10.2024).

Sychev, S. L. (2006). Integrated development as a factor of sustainable development of the coastal zone / S.L. Sychev // In the collection «Problems of sustainable development of regions of recreational specialization». – Sochi: Publishing House «Sochi Research Center». – c. 186–193.

The site of the company «Electro Magnetic Research» (AEMR Ltd). – URL: https://quantumbalancing.myshopify.com/pages/about–us

Zenkovich, V. P. (1958). Morphology and dynamics of the Soviet shores of the Black Sea. Volume 1 / V.P. Zenkovich. – Moscow: Izd–vo «Akademii nauk SSSR».

Zhdanov, M. S. (1986). *Electrical exploration / M.S. Zhdanov*. Nedra Publishing House.

Chapter 7
Mixing Virtual Reality for Environmental Conservation in Tourist Spots:
Hitching Biodiversity Sustainable Tourism for Regenerative Preservation Management

Bhupinder Singh
https://orcid.org/0009-0006-4779-2553
Sharda University, India

Christian Kaunert
https://orcid.org/0000-0002-4493-2235
Dublin City University, Ireland

ABSTRACT

Virtual reality and environmental conservation in tourist sites can provide an innovative solution for sustainable tourism. VR enable to observe natural wonders in person without even being there and causing environmental harm. Whether this is through the simulation of real-world landscapes to educate "visitors" about conservation or in conveying how important it is that it preserves what natural habitats are out there leading to a wider sense of awareness and education on their ecological footprints. The service could also be used as a way to combat over-tourism at main tourist destinations, save biodiversity and cultural heritage. Also, enabling the use of VR in eco-tourism can encourage responsible actions which enable users to dive into ecosystems remotely such as endangered species and so on thereby instilling a very relevant connection with conservation causes. This chapter explores the design of an innovative approach to reduce environmental degradation and develop novel tourist experiences through hitching biodiversity focused sustainable tourism combined with VR technology.

DOI: 10.4018/979-8-3693-8232-5.ch007

1. INTRODUCTION

The immersive experience of VR might also lend itself to fundraising and advocacy campaigns for conservation efforts. Ultimately, the marriage of VR and environmental activism offers a sustainable model for future tourism which can alleviate extreme pressures on our planet while ensuring such fragile ecosystems are intact generations from now (Kumar et al., 2024). Being one of the booming industries in this world, tourism acts as a catalyst to forge cultural coalition and act as an engine for socio-economic transformation. Tourism also has an environmental factor, given it can cause the destruction of habitats due to human intervention or pollution. Conservation as a result works to preserve plants and animals against tourism's negative effects. Over-consumption of natural resources over-tourism refers to the overcrowding and depletion that some popular tourist destinations, especially those with fragile ecosystems, from local flora and fauna data Still, the issue of how to balance between tourism and conservation remains unsolved (Lal et al., 2024).

Virtual Reality (VR) offers a novel solution to the problem of managing tourism with conservation in mind. Via VR technology, people can have an immersive experience of natural environments and ecosystems as if they are immersed in them without actually being there. To that end, the use of VR can play a fundamental role in safeguarding both biodiversity and its integration into sustainable tourism practices favoring awareness raising among tourists, education for responsible behavior (Arora et al., 2024). VR experiences of virtual tourism help reduce the over-tourism that threaten fragile ecosystems and nurture new possibilities for regenerative conservation management. This chapter explores the possibility of combining VR for environmental conservation in tourism destinations, more specifically how it can be used to protect the biodiversity and sustainable tourism. It also explores the rejuvenative potential of conservation management through VR, and how it can support healthy long-term ecological sustainability (Lavuri & Akram, 2023). This chapter explores the design of an innovative approach to reduce environmental degradation and develop novel tourist experiences through hitching biodiversity focused sustainable tourism combined with VR technology.

Figure 1. specifies the Dimensions of Introduction Section

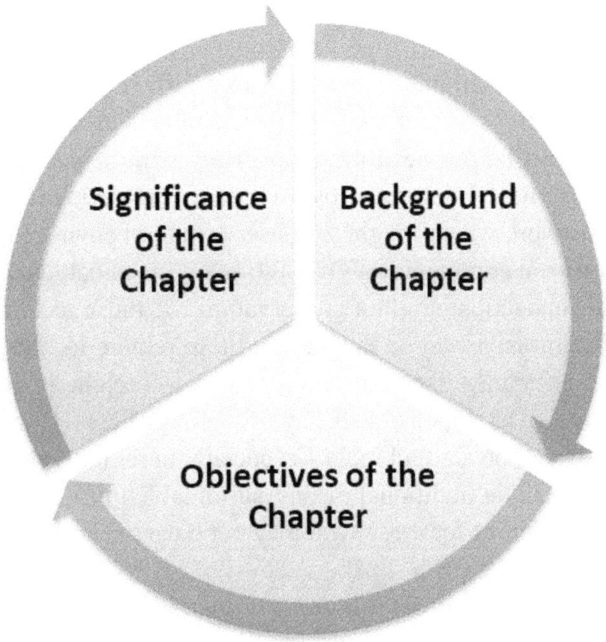

(Source- Self Prepared- Original)

1.1 Background of the Chapter

While tourism growth has been beneficial to many economies, it leaves a large environmental footprint. One of those problems is over-tourism. Sites such as Machu Picchu, the Great Barrier Reef and Mt Everest are global icons yet they have suffered under wildlife of visitors. As a result of the increase in numbers, unique ecosystems and cultural heritage sites have been shrinking. The United Nations World Tourism Organisation (UNWTO) defines sustainable tourism as "Tourism that meets the needs of present tourists and host regions while protecting and enhancing opportunity for the future". Traditional sustainable tourism methodologies have a limited effect when it comes to dealing with the human-apex predator interface in ecologically sensitive regions.

Virtual reality tech advanced quickly and became more sophisticated with its capability to give life-like simulations of the real world. What VR Does Is Creates an Immersive, Interactive Experience to Transport Users Anywhere in the World via a Simulated Environment. It is a feature which has made Virtual Reality beneficial in many fields like education, entertainment and health-care as well. Which is in part why its use for environmental stewardship is only now appearing: it can simulate ecosystems, threatened species and natural landscapes, so that those digitally enchanted seek out environments like them elsewhere instead of the degrading their originals.

VR can be used as a replacement or at least an supplement for physical tourists which in turn would reduce the burdens put on their natural environment. These environments are so real that users feel as if they were there, all because of Virtual Reality (VR) creating exact replicas of these fragile tourist spots. It offers the tourists an opportunity to enjoy the splendor of these areas with zero environmental pollution that come from traveling carbon emissions, habitat disturbance and resource use. VR can also be a teaching aid, raising awareness of the need for biodiversity conservation and fostering sustainable tourist behaviors.

The growth of sustainable tourism has coincided with a growing awareness of the environmental damage caused by tourism. It aims to harmonize the developments in all three pillars of tourism, economic features and social) must be balanced more positively with respect to nature (environmental). The sustainable tourism movement is informed by a large of body evidence, having arisen alongside the wider development framework which sought to conserve natural environment whilst fostering equitable economic benefits for local communities. Over the years, sustainable tourism has been upgraded and includes technological innovations to pilot conservation. Of those technologies VR in particular, provides an interactive non-intrusive way to experience these remote locales. In the past, sustainable tourism operated based on principles like restricting visitors to certain amounts and promoting eco-friendly accommodations as well as responsible behaviors among clients. However, the implementation of these practices can be difficult on a grand scale – especially in regions where tourism is such an important part of the economy. Unlike traditional means, sustainable tourism can be taken a step further by creating exploration opportunities for those who may not have access to visit the sites in person via VR (Caliskan et al., 2023).

VR besides diminishing the environmental impacts of tourism, can also add to regenerative preservation management as regards biodiversity. Moving beyond 'do not harm', regenerative tourism seeks to heal and grow the land. Integrating VR within regenerative tourism projects increases upon conservation, gateways for eco-tourists to experience and practice virtual reality (VR) immersions in conservation activities from home such as participating fundraising campaigns of restoration initiatives.

1.2 Objectives of the Chapter

This chapter has the following objectives-

- explore how VR can be used to mitigate the negative impact of tourism on the environment.
- assess the potential of VR to create alternative tourism experiences that promote conservation efforts.
- investigate the role of VR in enhancing awareness and education around environmental conservation.

1.3 Significance of the Chapter: Importance of Blending Digital Technology with Environmental Conservation

A key aspect of environmental sustainability is biodiversity. These consist of the full panorama of living systems on earth: plants, animals and micro-organisms together with their habitats. Biodiversity loss can hinder ecosystem functioning, reduce the adaptive capacity of natural systems and have negative consequences on human well-being. In biodiversity-rich areas, tourist activities can destroy habitat structures through trampling effects and disturbances to wildlife due to increased visitor numbers or simply from pollution caused during these visits as well as the introduction of alien species which could have negative impacts on certain indigenous interventions in a given area threatening their survival.

As a tool for tourism, VR is the way to go as it enables tourists to see and view biodiversity without causing harm. Ecosystems can be reconstructed with incredible precision and virtual tourism experiences allow people to travel without moving from their living rooms, whether it is in a rainforest, coral reef or wildlife reserve. This will not only minimize the pressure coming from tourism but also gives a chance to educate them about the significance of saving biodiversity. Endangered species challenges can be experienced through VR, to show the very tangible impacts of habitat destruction and climate change. Experiencing this biodiversity first hand can lead to a greater appreciation of such and inspire individuals to do more in supporting conservation efforts. The other important use of VR in this area is the generation of awareness and funds for preservation projects. Through virtual tours, revenue can be channeled into financially sustaining local conservation efforts. These VR experiences are building blocks for environmental education in a way that can reach further than standard conservation efforts. Furthermore, by use of VR to bring biodiversity conservation into the home or onto your phone, it can facilitate toward developing a worldwide group of eco-conscious travelers with an interest in protect nature on this planet.

The goal of regenerative conservation management is to preserve, conserve and restore ecosystems so they are better able operate in a future dominated by human presence and climate changes. Regenerative projects differ from this type of conservation in that they are set up to heal and restore the land, instead of merely keeping it limping along. Using VR in conjunction with REST guidelines can provide a more integrated conservation approach for tourism destination. Using VR to Educate Visitors on the Benefits of Regenerative Tourism The app allows users to explore virtual simulations designed after these geospatial scientific databased restoration projects, such as reforestation project demonstrations environmental sanctuaries and coral reef reefs rehabilitation efforts. Whilst also educating and encouraging tourists to explore their regenerative tourism activities, in-person or virtually however supporting local conservation initiatives (Kacprzak & Hensel, 2023). Using VR to monitor and visualise the success of regenerative projects also helps bring transparency, accountability for preservation efforts.

The use of VR in physically tourism fragile ecosystems can serve as a sustainable alternative for tourists wishing to experience those environments without further damaging them. A virtual tour of coral reefs may, for example, provide a rich overview of marine biodiversity without the pressure on corals associated with traditional scuba diving (which has been shown to be potentially deleterious through fueling feedback loops that aggravate thermal stress and invert ecosystem functioning). In the same way, virtual safaris can broadcast endangered wildlife species to create less invasive in-person wildlife tourism which often causes stress on animals and destruction of their habitats. This also ties into wider international conservation goals, for example The UN Sustainable Development Goals (SDGs) are common to many of the initiatives we aim towards. One could argue certain VR experiences promote responsible tourism and lower carbon emissions, leading to the completion of SDG 13 (Climate Action), both indirectly linked with SDG 14 (Life Below Water) and still others associated with direct effects on achieving SDGb15(Life on Land). Virtual reality footage could also be used to highlight the imperative of biodiversity and conservation, while showcasing measures such as mitigating tourism's environmental footprint.

Integrating VR with Mother Nature in tourism destinations truly helps to protect natural ecosystems and biodiversity so it provides an unprecedented stance for sustainability. Through the advancement of VR technology immersive experiences can be created with no environmental cost associated while protecting biodiversity conservation and regenerative preservation management. In an era when tourism is rising yet natural resources are becoming scarcer, the world must develop creative solutions like VR in order to safeguard our global heritage and allow generations that follow us to witness ever more of what could once be found all around the planet. Combining biodiversity-led sustainable tourism to VR technology, it will generate a self-regenerating future of place-based health and balance in shared terrestrial places going forward (Vinaykarthik, 2022).

2. CONCEPT OF VIRTUAL REALITY IN TOURISM

Virtual Reality (VR) has been turning heads across various sectors tourism is one of them. Virtual reality technology generates interactive, computer-generated experiences which take the users to a completely fortuitous world. Virtual Reality (VR) holds great promise, particularly in tourism where it can literally transport travellers and tourists to destinations, attractions or cultural sites without having the need for them to actually go there. The implications of this innovation extend beyond making tourism more accessible to broaden the scope for eco-tourism, preservation and conservation of delicate ecosystems, and in general bring about an enriching tourist experience.

With VR in tourism people can really FEEL what exploring a location is about! People can visit ancient monuments, revel in cultural fests or even witness natural splendors as though they were sitting home. VR refers to software that uses rats and headsets, so when a VR user plugs the system in their environment; they will be able not only simply see but also ear- sometimes feel where they are. This "being there" experience is exactly what the visual and auditory elements of VR help create, by replicating visuals and sound from real-world locations. Virtual Reality Tourism is a very valuable tool for tourists; the more similar it becomes to real life, the more immersive and accessible make virtual reality technology become of great importance in different touristic spots (Kamal & Himel, 2023).

VR in Tourism has the sweetest benefit of providing travel experiences beforehand. Travelers have increasingly used VR to explore before they book, rather than reliving the past. Tourists can make an informed decision on their trip, so that when they arrive at the hotel / resort museum or whatever other attraction is showcased using a virtual tour of some sort. VR, a useful tool for travel agencies provides solution that many may think of is virtual reality, which allows travel agency clients to view the presentation on an imaginary but real trip. VR provides some assurance and boosts consumer involvement as his or her more personalized choice an now be made based from their preference via this try-before-you-buy experience, which can result in the recommissions of travelers (Ahmed, 2022).

VR has even more important application for tourism as it may stimulate sustainable and eco-friendly travel. Some of the top tourist destinations in the world coral reefs, national parks or cultural and heritage sites are being threatened by over-tourism. Visiting them on-foot and in person can lead to damage to these fragile ecosystems, as well as environmental destruction & wildlife harassment. Using VR, it can create virtual destinations that mirror the physical locations to solve this problem and circumvent environmental concerns (Vuong & Mai, 2023). This can include virtual safaris that allow tourists the excitement of seeing wildlife in their natural surroundings, without increasing foot traffic through animals' habitats and/or harmful interactions with those ecosystems. VR also allows the preservation of cultural heritage historical landmarks, ancient ruins and culturally significant land are slowly eroding from decades of pollution or destruction. With the help of virtual re-creations, VR can be used to save these sites in a digital format and give future generations an opportunity to see them without actually setting foot on those places. For instance, if they wish to visit an ancient city or a famous archaeological site that no longer exists in reality, these are places recreated via VR and thus for all intents and purposes, the tourists have experienced what it might feel like travelling there back in those days.

Similarly in the context of tourism education, VR can be a useful tool for facilitating learning and understanding. With the help of these virtual tours, one can explore various elements such as history, geography and cultural practices. Museums and even schools are beginning to implement VR as a way for students, or guests experience the content of what they're learning firsthand. People could take a virtual tour of the Louvre Museum where they walk through each piece and every artifact to learn more about it during real-time interaction in an interactive-learning environment (Auttri et al., 2023). Travel is the most intriguing aspect of accessibility for VR. Paraplegic people, and maybe even many others who physically cannot travel this way have the ability to experience full round world tours from a virtual reality. It offers a realistic alternative for people too sick or without enough money to go on long travels, thereby allowing them the possibility of exploring different places and cultural ways without leaving their own spaces.

Virtual tourism is bringing a revolution in the travel sector with more immersive experiences available that are eco-friendly and pocket-friendly as well. That extends to uses beyond pre-trip planning, as well virtual reality can also fuel eco-friendly tourism which saves carbon that might be produced by a flight halfway around the world; help preserve sites of cultural heritage across the globe from further damage due to boots on their grounds; and provide immersive travel experiences for people with physical limitations. With the ongoing advancement in VR technology, likely expanding its role within tourism even further offering endless opportunities for virtual and sustainable tourism alike (Zaki, 2022).

2.1 Current Applications of VR in Tourism (Virtual Tours, Immersive Travel Experiences, Virtual Museums

For anyone keen through to get a first-hand experience in the comfort of their own home that just an old school GPS tour either folks, these are pioneering VR features designed to revolutionize how we 'travel without travelling'. It has been employed in different facets of tourism, including virtual tours and immersive travel experiences or in the creation of virtual museums. These apps are changing the way we travel, making tourism more accessible to all and sustainable (and fun) for spaces too.

Virtual Tours: Amongst applications of VR in tourism, Virtual tours are the most common implementation that helps user travel into their favorite destinations without physically reaching there. Hotels, rooms and tourist venues can be explored virtually from locations all over the world before deciding whether to go there. This means travel agencies and tourism boards can use VR to present entire cities, landmarks as well as different resorts offering possible tourists a better idea of what they will be getting. For example, you can take 360-degree tours of different destinations around the world like Paris, New York or even Tokyo and explore famous places like Eiffel Tower or Time Square. Using this "try-before-you-buy" approach, tourists are able to better judge if a location will meet their travel needs while businesses gain an interactive marketing tool that helps bring more customers in (Soliman & Al Balushi, 2023).

Immersive Travel Experiences: VR is even offering complete immersive travel experiences that emulate the real traveling experience of a place in actual time. Immersive experiences differ from virtual tours which are often brief and give information they enable users to explore the entire space as if they were physically there with realistic visual, sound experience & interaction (Bhatnagar, 2022). VR offers experiences similar to walking through a national park, swimming in the ocean or hiking up that mountain trail. Interact with the environment and learn about local plants, flowers or animals. Comparatively, immersive experiences that can enable travelers to feel like they are on an adventure without ever leaving the comfort of their home and negatively impacting climate change is certainly one very sustainable way for tourists to explore. They are particularly useful for places like ecologically sensitive or hard-to-reach destinations, whose natural wonders should not be put at risk to over-tourism.

Virtual Museums: VR usage has extended over cultural institutions like museums and galleries, providing virtual exhibitions or collections tours as well. Virtual museums are also a type of guide which one can explore art, history as well as cultural exhibits. The Louvre Museum in Paris, the British Museum of London and Smithsonian is Washington D.C. provides VR experiences where users can walk through their galleries, examine artifacts from all angles and access interactive educational content to learn about history sitting at home. By offering these exclusive tours to your audience, they can experience a type of exhibits in person that otherwise may be impossible for you to visit due to the location, time duration or ticket availability. Streaming has made culture accessible worldwide and potential international audiences. VR in tourism industry changes the ways users discover and learnt about places around the world through virtual tour of tourist attraction sites but also enabling museum visits etc. Sustainability gains, wider accessibility and innovative sales opportunities for tourism operators.

Figure 2. illuminates the Current Applications of VR in Tourism

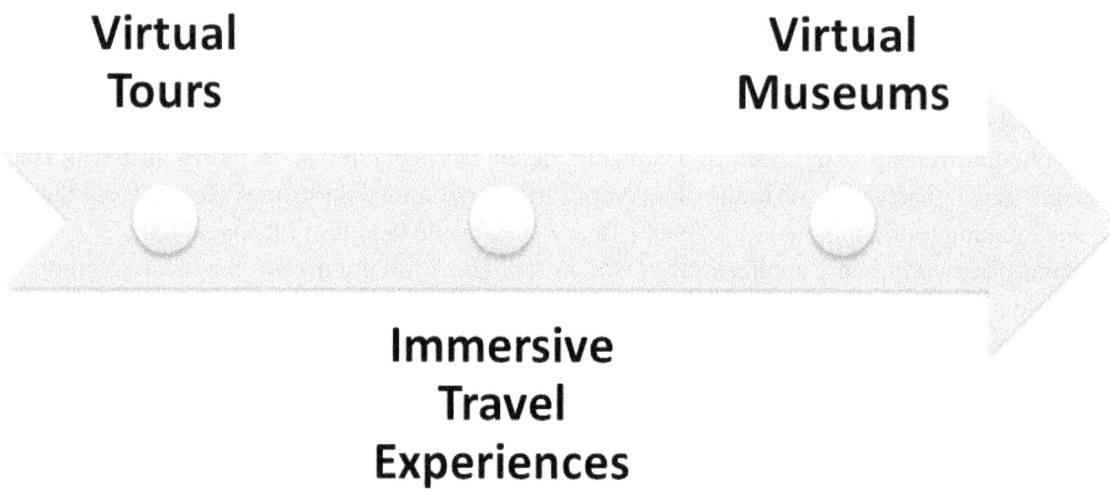

(Source- Self Prepared- Original)

3. VIRTUAL REALITY AND ENVIRONMENTAL AWARENESS IN TOURIST SPOTS

The approach that is gaining momentum this summer year will be to create empathy for endangered ecosystems with virtual reality (VR) technology, especially favored amongst tourist destinations in order to mitigate potential risks of over-tourism, pollution and climate change. Through the power of VR to place users into virtual simulations that mimic natural environments, this technology can help educate and motivate tourists and locals to practice environmentally friendly behaviors which play a crucial role in efforts for environmental conservation (Henke & Jacques Bughin, 2016). Within tourism, VR can generate immersive experiences that educate instead of just amusing visitors about the environmental value of desired destinations. This immersive education element combined with experiential involvement puts VR among the tools essential in educating and implementing sustainable tourism practices whilst promoting natural ecosystems for future generations to come.

1. Virtual experiences that help reduce our carbon footprint

VR in tourism can deliver is a major accomplishment towards reducing carbon emissions. Most of the famous tourist destinations like national parks, coral reefs, historical sites suffer from economic strain due to heavy tourism pressure. The expansion of visitors to these locations often leads to overcrowding, littering and damage within fragile ecosystems. VR is a way to attempt and work around that by allowing tourist to enter vicariously the environments but without degrading said environment (Kaushal & Yadav, 2023). Some of which include the virtual zoo safaris in place of traditional wildlife tourism. While they cannot visit a national park in person, where their very presence could stress animals and harm conservation efforts all guided safari tours are to instead embark on virtual safaris that replicate what it might be like for people looking at various forms of fauna within the particular elements. Virtual safaris can convey a similar experience, but with less disturbance to fragile ecosystems. Most importantly of all, the immersion (pun intended) that VR brings can help educate people about endangered species and conservation making tourists more conscious of their impact on wildlife behind they do decide to go speak view them in person (King, 2022).

2. Immersive Experiences in Learning & Environmental Advocacy

VR Engages in Environmental Lessons and Education Science is also using VR for educational purposes. VR can tell tales about the fragile nature of some ecosystems, show end results of environmental degradation and stress on importance conservation projects (Babet, 2020). The more popular jungles and oceans could let visitors take virtual tours of reef, not just to produce awe-inspiring experiences but also the systemic educational aspect: showing why coral bleaching is harmful; how ocean pollution has affect the natural ecosystem balance along with unsustainable fishing. With this module, visitors can experience the reef and come face to face with how human activity impacts marine life which in turn creates a better understanding of why conservation is important (Sadowski, 2023).

So VR as a part of their environmental advocacy campaigns is to show several conservation organizations are using VR around nature degradation etc. VR experience illustrates the impact of climate change on various major regions throughout the world. VR can forge an emotional bond to environmental causes that traditional educational methods often falter at. Showing users first-hand the devastation of rising sea levels, deforestation and extreme weather events allows them a unique level connection when considering global challenges such as climate change. This emotional engagement is important because it can inspire a positive action among tourists which will ultimately help support more sustainable tourism in natural settings (Sharma & Singh, 2022).

3. Promoting responsible tourism

VR can also aid sustainable tourism by providing insights into sustainability and eco-travel practices. Prior to setting out on a trip, tourists can use VR to experience their destination and be advised in order that they provide the least amount of environmental impact. As an example, a virtual tour of a national park might advice on staying to designated trails, not disturbing wildlife and leaving nothing behind (Mashood et al., 2023). VR can make these guidelines interactive and engaging, leading to a better-educated tourist who knows how to actually apply the principles of sustainable tourism in more sensitive environments even before they go there. VR can also be used to inspire tourists to visit relatively unknown places so that they are not crowded too much. This sudo-tourism has created a strain on many iconic sites and natural wonders as well, contributing to their over-use which is damaging the environment. Tourism boards and travel companies will be able to promote second-string destinations by creating equally compelling VR experiences of less-visited places, which means tourists aren't all going in the same place at once, wreaking untold environmental havoc on popular hotspots. Fragile ecosystems are kept intact and it helps facilitate the funding of new tourist projects that might be underdeveloped otherwise (Gao & Liu, 2022).

4. Conservation of Tour Sites

VR can also serve to raise environmental awareness and come in handy for preserving and restoring tourist sites. Due to a variety of environmental factors and human activity, many cultural heritage sites and natural landmarks are at risk. With VR, these sites can be digitally preserved and reconstructed virtually to allow tourists a way of exploring the area but without being there. UNESCO has, for instance, used virtual reality to provide tourists with a synthetic immersion into the environment of certain endangered World Heritage Sites in which viewers can nevertheless encounter partial forms of the unharnessed cultural and historical values that emanate from these places without contributing further damage.

VR can be used to help model return on investment for conservation through virtual tours of restoration projects. With tourists being able to virtually visit these sites, they might be more motivated to support conservation initiatives knowing that their donation is going towards conserving the natural or cultural site they just explored in virtual reality. This forms a mutualistic relationship between VR tourism and conservation virtual experiences not only alleviate physical tourism pressure but also contribute towards restoring and protecting endangered environments (Rajamannar, 2021).

5. Creating Awareness for the Climate

Climate change is one of the most pressing environment issues on our planet now and VR has become a way to communicate how real this issue really is. For instance, a VR simulation may show users in environments that are experiencing the impact of climate change right now (e.g. melting glaciers, shrinking rainforests or drought stricken areas). We can see the Inconvenience of global warming for ourselves, feeling from 1st person view makes science data more comprehensible for general public. Experience this kind of learning experience in raising awarness for urgency to take action against climate change is very effective. Google Tour the Great Barrier Reef in 360° allows users to snorkel a rich coral reef ecosystem and explore an underwater wonderland. Through this, however, the tour describes regions impacted by coral bleachings and features perspective on how climate change is destructive against marine ecosystems. VR experiences like these, which combine the splendor of nature with gross images and video capture by humans can help tourists understand why they should support environmental initiatives.

Virtual Reality provides a unique way to educate about environmental issues and inspire sustainability efforts in tourist destinations through immersive, learning experiences that lead your audience to responsible actions. VR puts some of the potentially most-threatened ecosystems and destinations in travelers hands through virtual tours, travel experinces and enviromental campaigns with 0 physical environmental footprint. The entire tourism sector is already being transformed by this technology and also contributing on a massive scale to conservation, sustainability measures as well as raising awareness about Climate Change so that the generations coming after shall not inherit an empty shell of what was once our Planet Earth (Ancillai et al., 2019).

4. IMPACT OF VR ON REDUCING PHYSICAL FOOTPRINTS IN TOURISM: ANALYSIS OF VR EXPERIENCES REDUCE NEED FOR PHYSICAL TRAVEL TO SENSITIVE ECOLOGICAL AREAS

Although the tourism industry produces many economic interests and cultural exchanges, it has always been a pain point in terms of ecological damage, especially for fragile ecosystems. Highly congested tourist areas like national parks, and coral reefs as well as archaeological sites are degraded via pollution, habitat damage & resources depletion due to over-tourism. The use of Virtual Reality (VR) technology may be a solution to moderate the human presence in these natural beauties, allowing tourists visit resources without really going physically. Through providing an experience rooted at the other end of visiting these destinations, VR has a huge potential to responsibly use tourism materials and prevent certain ecologically sensitive areas from being affected by high-impact human activity (George, 2023).

Excessive visitation can cause such problems as soil erosion, water pollution and damage to wildlife habitat. In some cases, access has had to be restricted or closed by the authorities because of damage caused as a result. It is here that VR comes to the rescue offering an alternative, now virtual way of travelling, which definitely sounds better in terms of environmental impact. These destinations can be visited virtually for interested tourists, seeing their beauty and learning about them without adding to the environmental toll that comes with large crowds. VR provides a way for people to experience the destination without actually going there thus reducing any over-crowding in sensitive areas. Rather than physically visiting the Great barrier Reef on site, where human interest often results in damage to coral ecosystems- tourists can be taken through a virtual dive instead. The bright, beautiful marine life of the reef lives on and science-loving kids can explore its depths or their parents can from home. This also helped to decrease reef-induced physical damage as well as conservation awareness based on underwater.

Tourism is the most devastating sector to environment in terms of its carbon emissions thanks to transportation with international air travel (Wind & Hays, 2016). Tourism is responsible for about 8% of global greenhouse gas emissions, because long-distance travel has correspondingly high energy requirements. This could be helpful in reducing these emissions since VR can substitute physical travel with virtual experiences. For example, a virtual tour of the Amazon rainforest may be so immersive and engaging that more students will understand just how diverse this region is as well its ecological importance. These kinds of immersive VR experiences would actually replace the experience doing physical travel, which could allow people to visit rainforest without causing deforestation and pollution or climate change. Thus, VR helps to slash the carbon emissions of tourism and opens up a whole new perspective for traveling more sustainably by illustrating how finely ecosystems are tethered (Kuusinen, 2019).

More and more of them are the habitats for endangered species, in regions that have become popular as natural ecosystems where tourists can see quickly rare plants or animals. As residence of humans close to the Reserve could disturb wildlife behavior or disturbed them causing animals stress, and also wound cause habitat destruction. Because wildlife tourism is sometimes the same things as hunting ambitiously disguised, and it always profits those selling spots or doing guiding. VR solves this problem through virtual safaris, animal experiences and nature tours to allow tourists observe wildlife without causing any harm. African game reserves provide virtual safaris, where tourists have the opportunity to experience lions and elephants while they knock on their windows when stopping for gas. The fact that you can have a near full VR experience in these environments without having to physically travel there means far less disruption from trampling on native wildlife, or direct incentive towards poaching through illicit tourism. Help reducing the pressure on endangered species and their habitats moving from physical visits to virtual experiences is great for both animal welfare as well conservation.

Through tourism VR experiences they can act as a vehicle for environmental storytelling. VR can help people not only by giving them a chance to experience immersive ways of beauty and fragility urban environments but also highlight the necessity for preservation. While many VR experiences have an educational aspect to them, sharing information about climate change and the massive human impact on natural resource depletion such as deforestation or pollution. Inspiring the tourists of tomorrow to travel with a greater sense of environmental responsibility leaves room for them to learn from their virtual travels, sighting better sustainable practices when they do venture out again. On top of this, by providing travelers with access to less visited eco-friendly tourism spots places that would often see little love in favor more familiar ones VR can help through the circulation of less crowdy destinations. This can disperse the impact of tourism more evenly across different locations which in turn mean a less extreme concentration on sites currently under pressure (Naji et al., 2024).

Virtual travel and virtual reality have the potential to transform tourism into a more environmentally friendly industry due to its low ecological footprint. Virtual Reality can enable viewers to experience remotely, sensitive ecological areas in immersive and engaging ways that are educational at the same time as all from the comfort of their home thereby decreasing over-tourism "a problem endemic for a number fragile ecosystems", reducing carbon emissions as well as protecting endangered species and other valuable resources crucial for many different reasons. With VR bound to evolve further, we could see it playing a larger and more important role in eco-tourism, presenting an ever more progressive way of protecting species without the cost or emissions involved in simply flying somewhere (Wheeler, 2023).

5. EDUCATIONAL POTENTIAL OF VR IN ECO-TOURISM: ROLE OF VR IN PRESERVING FRAGILE ECOSYSTEMS

This informally presented our game in the wider eco-tourism research group as part of project updates and has heard a number of useful comments. VR with its ability to provide immersive, interactive and visually-rich experiences holds the promise of educating tourists about ecological conservation domain, sustainability sustainable behavior as well as ecosystems they seek feel opportunity. With more and more modern integration of VR in eco-tourism, learning will evolve it from knowing to believing as individuals can now experience natural environments like never before, thus abandoning them into an ecological-driven behavior realizing that we are only human on Earth.

Virtual Reality (VR) virtually transports users to ecological hotspots, such as rainforests, coral reefs or glaciers for immersive educational experiences. Compared to conventional learning tools, such as textbooks, videos or visiting a site in person VR places the user within the environment and provides them with an immersive first-person observation of major ecological subjects. Using a VR recreation of coral reef environment to present tourists with the lush underwater life contrasted against realism illustrating dangers such as coral bleaching or ocean acidification (Sedkaoui, 2018). When this impacts the ecosystems around us, VR can help people experience that graph with even more reality so we have a better visual incentive to preserve our environment.

An important touristic feature and potential impact of VR in eco-tourism is its ability to provide interactive learning experience. Virtual visitors can interact with some of these features, ask questions about each environment as if they were walking around in it just remember to wear a mask even virtually. For example: a VR tour of the rainforest that could show users explanations and details on plant species; bird calls you can hear by pointing to specific areas or tapping there, which encourages exploration; pop ups appearing all over when holding your gaze in an area these would provide reading material so if it happens too fast i lose out. This form of personalized education allows any individual visitor to connect with the topic in a way that suits their mental wiring, leading to as rich an experience for each person who visits them.

This could work particularly well for the kind of simulations where VR is valuable, e.g. raising awareness about how terrible ecological decimation can be by showing what it look like virtually. VR can make people see the effects of climate change, deforestation and pollution in a visually appealing way with an immersive experience that will affect them. This technology allows tourists to go on virtual tours of areas that are under threat now and see a preview of what these will be like in the future if unsustainable practices carry on happening. It results in more conscientious action from the individual participating often leading them to carry on eco-centric behaviors and conservationist advocacies as they physically travel (Neuwirth, 2023). It raises the level of understanding for those taking ecotourism trips with a VR headset creating better learning objective and its making environmental education more easily accessible and democratic.

With VR, those without the ability financial or physical to make it out into some of these wilder places can still experience them. Opening up the field on environmental conservation to a wider audience, this connects eco-tourism to education outside of the confinement of physical travel. VR in eco-tourism has a huge amount of educational potential, providing immersive and interactive learning experiences that are more accessible for those seeking to learn about our natural environment. Through giving us a more in-depth look at ecosystems and the effects of humans on them, VR is helping to create increased environmental awareness along with promoting sustainable tourism practices which making it an imperative practice for eco-tourism experiences (Schaeffer, 2017).

6. CHALLENGES IN INTEGRATING VR FOR ENVIRONMENTAL CONSERVATION

Virtual Reality (VR) in initiatives to promote environmental conservation holds real promise, but first it must clear some hurdles before that potential can be fully realized. The challenges fall across the tech, environmental and social spectrums together these issues are significant barriers to using VR for conservation.

1. Technology Challenges: Accessibility / High Cost / Digital Divide

The first barrier to add VR for environmental conservation is the cost of and access to high-quality VR technology. Many users, especially in underdeveloped regions do not have the resources to buy a high-end VR system. This forms a digital divide which is blocking equal access to the educational and experiential opportunities that VR can offer. Especially in developing countries, internet access is limited and technological infrastructure outdated which may widen the gap even more as it makes implementing VR applications for conservation on a broad scale extremely difficult. VR can be elitist if it gets priced, gated and ticketed to the point that only rich nations or areas have access then what good is VR going to do in global conservation arenas.

2. Energy use and e-waste

Virtual tourism can replace physical travel, which is more ecologically friendly for the planet, but it also brings its own environmental challenges with energy consumption and electronic waste. The power consumption of high-end VR systems is considerable, leading to concerns about the carbon footprint generated by skyrocketing usage (Venkatesh, 2021). This not only leads to more VR hardware being produced and thrown away, causing yet another issue with our growing global e-waste problem. VR advancements are changing so quickly and will continue to do so throughout the century, as this technology could be our new medicine provides consumer product longevity but for VR systems they may just end up turning into more frequent e-Waste incarnations. Developing these environmental issues needs cost-effective VR technologies, environmentally-mindful system and energy-efficient solutions to decrease the e-waste.

There is also a social-based barrier on VR being considered as an alternative to traditional tourism experiences. A lot of people are huge fans of traditional travel to experience real-life moments by being in contact with the culture, authenticity and smells that only come from a face-to-face touch. Those cravings often can't be fulfilled by even the most immersive virtual experiences. And surely there will be those who scorn the use of a virtual reality experience as something suitable for engaging with nature; that it is somehow lesser or even fraudulent. To break through this barrier, VR education will have to work just as well in terms of emotional experience. For the satisfaction experienced since arousal by means of travel must be provided emotionally while having an immersive experience and also learning something at same time which is only possible with Virtual Reality tech here (Reviglio della Venaria, 2020).

3. Keeping it real

Closely intertwined with resistance is the problem of how to ensure authenticity in VR-based tourism. For eco-tourists, authenticity is key as they are looking for enriching experiences in the real world. For example, VR experiences might come close to creating realistic natural environments yet still not be able reproduce every sensesome of the multisensory richness such as a breeze or smell of vegetation—that occurs during real-world events. The challenge with this is that it can be easy for VR developers to compromise on the verisimilitude of experiences in order to simplify things, but this would detract somewhat from how rich and powerful our natural world actually is. "Smelling the incense in a Cambodian temple is something that only genuine travel can provide, so longer term this will need VR to have made significant advancements in how real-world experiences are recreated at both sensory and emotional levels.

While the combination of VR and conservation efforts is showing promise, it's a hairy business. In order to improve their use, greater accessibility and lower costs are required i.e., we need to combat the so-called digital divide (Duan, et al., 2024). They should also deal with the environmental effects of energy use and influence on e-waste so that it does not create any unintended ecological consequences. Fitting in with the local social landscape can also be a tough challenge to meet, which is why you need both an authentic and well-designed experience. However, with ever-evolving technology and new methods to reduce negative impact on the environment VR might just prove to be a key in solutions for worldwide environmental conservation functions.

7. POLICY AND PRACTICAL IMPLICATIONS

Virtual Reality (VR) can be used to integrate environmental conservation with eco-tourism, and while it offers sustainability benefits, the end result has substantial implications that demand both policy and practical consideration. Therefore, governments and technology providers should come aboard these tourism authorities as well as conservation groups in promoting responsible use of VR technologies. This expands on essential policy recommendations and outlines the implications for promoting VR in conservation-focused tourism.

7.1 Policy Recommendations to Governments and Tourism Authorities

Government tourism authorities need to support the adoption of VR in conservation efforts for it makes an impactful contribution. Recommendations among others include that governments should implement VR-based policies for tourism as a part of their wider nature conservation projects. Governments looking to reduce their environmental footprint of tourism and many ecosystems that are already at breaking point with the tourist traffic they now get, making a further increase impossible in practical terms could also promote VR as an alternative or supplement. This would consist of grants or tax incentives to help tourism companies in creating and adopting VR-powered eco-tourism experience, acting as viable policy measures that can drive such a change (Tzampazaki et al., 2024). The governments should also be considering plans for subsidy in VR technology as a part of eco-tourism initiatives in those parts where development is yet to take over. VR squarely faces the same obstacle regarding cost of price and barrier to widespread implementation. Through subsidizing VR technology, governments can enable smaller eco-tourism businesses and conservation organizations to make use of more sophisticated tech. It has the potential to open up VR affordable access and distribution, so that it becomes available in any household with the ripple effect of diversifying eco-tourism.

7.2 Partnerships between Tech Providers and Conservation

Finally, the success of community-led conservation-focused tourism is predicated on an active partnership between technology providers; tech such as budding forms of VR making this another chicken and egg scenario; organizations focused specifically on biodiversity preservation efforts in these hotspots or anywhere that harbors rare species where detection methods aren't eminently viable e.g. reef environs; linked with travel companies devoted to broadening eco-awareness while backing venues supporting ecological endeavor locally (Hanafiah et al., 2024). VR developers could make a meaningful difference by working together with environmental NGOs to deliver powerful conservation-themed VR activities that not just teach but prompt change. Conservation groups have the scientific knowledge and content to make sure VR experiences are accurate, consistent with their goals for preserving animal species in natural or virtual habitats; while tech companies can help them create high-fidelity worlds that users want to keep coming back.

There need to be funding mechanisms that support these partnerships. Funds for research and development in tourism: Governments, private investors or international organizations can establish funds to develop VR technology applied in travel. For instance, public-private partnerships may work together to co-invest in educational and immersive virtual reality (VR) simulations related wildlife conservation or biodiversity protection generated by VR game developers (Tembrevilla et al., 2024). With joining forces, these type of collaborations potentially give an already powerful tool for environmental education and taking action in the tourism sector a huge boost.

7.3 Develop an Ethics Wear VR Tourism Guidelines

The authors of the text review several public and commercial implementations, as well state that because tourism is one of the top industries that use VR in its marketing strategies it's important to provide some best practice as to ensure alignment with Sustainable Development Goals (SDGs). VR tourism will do wonders for SDG 13 (Climate Action) by reducing the carbon footprints of physical travel and should also help with SDG 15 (Life on Land) as natural ecosystems may be protected from the impacts of over-tourism. We believe governments and tourism authorities should collaborate with VR developers and experts in conservation to draft guidelines for the responsible use of VR experiences (Schmaltz et al., 2024). The guidelines should also cover the environmental implications of VR gear. On one hand, VR can help with reducing physical travel (and the resources required for it), on other opportunities there are emissions of energy and e-waste due to this. The guidelines must encourage the development of energy-efficient VR systems, responsible e-waste disposal and sustainable production processes in creating any form of VR hardware. With setting environmental requirements for VR manufacturers, consumers and operators will have increased confidence the positive effects from people choosing virtual tourism are not outweighed by negative impacts generated during VR production (Bala & Gupta, 2024).

The successful incorporation of VR into environmental conservation and eco-tourism represents how legislation, collaboration in partnerships with stakeholders or neighbouring attractions are core to creating well thought our guidelines. To help facilitate VR adoption, governments in partnership with tourism boards have a role to play through tax incentives and policies; tech platforms can collaborate between each other on generating powerful ecological-impactful VR content (Hazarika & Rahmati, 2023). With stakeholders creating ethical and environmental standards, VR tourism can both support sustainability while moving in a direction where world efforts are being bolstered for the protection of planet.

Figure 3. specifies the Policy and Practical Implications

(Source- Self Prepared- Original)

8. FUTURE SCOPE AND INNOVATIONS: EXPANDING VR USE IN LESSER-KNOWN TOURIST SPOTS AS VIRTUAL REALITY AND AUGMENTED REALITY INTEGRATION FOR ENHANCED CONSERVATION

VR in tourism has a promising future, notably when it comes to attracting more unique or off-the-beaten track areas with high environmental values but low accessibility. Using more and more VR in such places, these destinations could be explored by people around the world without needing to experience any of the environmental damages that come with a physical trip think deforestation or habitat destruction as locations suffer from over-tourism. The VR technology can allow tourists to virtually experience these areas the biodiversity, cultural heritage and the most importantly it can educate them about conservation. An augmentation of technological future is anticipated with Virtual reality (VR) and Augmented Reality (AR). The hybrid nature of approach would enable tourists to experience first-hand the real world environments as well as digital layers that are richer in information on ecology and cultural context about the areas (Varela et al., 2024). In a physical eco-tourism context, for example, AR might be used to overlay details about endangered species or historical points of interest; while VR could provide the opportunity to experience what past and future states of that ecosystem would look like under different climate scenarios.

But as these technologies continue to develop, they could also be put into service gamifying conservation efforts turning a phenomenon that captures the attention of tourists all over the world each year into an interactive learning experience and stirring demand for real traction on preservation projects. The pairing of VR and AR provides an opportunity to revolutionize eco-tourism with greater immersion interactivity in education for tourists seeking experience with local environments as well a creating global awareness about sustainable tourism and conservation efforts (Gupta, 2024).

9. CONCLUSION

Virtual Reality (VR) to preserve the environment in tourist spots is an innovative solution for promoting eco-tourism and conserving biodiversity. VR is used in this case to allow a sense of presence and immersion by which enables us experiences from different ecologic sensitive spots so people can travel less physically, reducing over-tourism (and its habitat degradation) and for the better since CO2 emission reductions are imperative. As well, it provides a very effective educational tool that would help make you aware of such conservation concerns as they emerge and in turn gives that experience of being closer to nature without leaving or engaging in any ecological footprint. Combining VR with other technologies such as augmented reality (AR) extends endless opportunities for creating hybrid experiences that not only enrich learning but also engagement. This unique blend supports regenerative preservation management to allow tourists even if they are not physically here for them to enjoy and part of the transformation towards admiring these incredible natural sites. Given further advancements in immersive technologies like VR, it may also create an exciting future for innovation-oriented travel related to conservation efforts. When linked closely with the sustainable development goals especially SDG 13: Climate Action and SDG 15 Life on Land VR could work as an indispensable instrument for activating worldwide commitment to environment preservation, a long-lasting defense of natural heritage.

REFERENCES

Ahmed, A. (2022). Marketing 4.0: The Unseen Potential of AI in Consumer Relations. *International Journal of New Media Studies: International Peer Reviewed Scholarly Indexed Journal, 9*(1), 5–12.

Ancillai, C., Terho, H., Cardinali, S., & Pascucci, F. (2019). Advancing social media driven sales research: Establishing conceptual foundations for B-to-B social selling. *Industrial Marketing Management, 82*, 293–308. DOI: 10.1016/j.indmarman.2019.01.002

Arora, M. K., Lal, S., Singh, B., & Olyaee, M. (2024). Through an Energy Lens: Examining SDGs From a Policy Perspective. In Riswandi, B., Singh, B., Kaunert, C., & Vig, K. (Eds.), *AI Applications for Clean Energy and Sustainability* (pp. 320–343). IGI Global., DOI: 10.4018/979-8-3693-6567-0.ch016

Auttri, B., Chaitanya, K., Daida, S., & Jain, S. K. (2023). Digital Transformation in Customer Relationship Management: Enhancing Engagement and Loyalty. [EEL]. *European Economic Letters, 13*(3), 1140–1149.

Babet, A. (2020). Utilization of personalization in marketing automation and email marketing.

Bala, R., & Gupta, P. (2024). Virtual Reality in Education: Benefits, Applications and Challenges. *Transforming Education with Virtual Reality*, 165-180.

Bhatnagar, S. (2022). Digital Disruptions and Transformation of Bank Marketing.

Çalışkan, G., Yayla, İ., & Pamukçu, H. (2023). The use of augmented reality technologies in tourism businesses from the perspective of UTAUT2. *European Journal of Innovation Management.*

Duan, W., Khurshid, A., Khan, K., & Calin, A. C. (2024). Transforming industry: Investigating 4.0 technologies for sustainable product evolution in china through a novel fuzzy three-way decision-making process. *Technological Forecasting and Social Change, 200*, 123125. DOI: 10.1016/j.techfore.2023.123125

Gao, Y., & Liu, H. (2022). Artificial intelligence-enabled personalization in interactive marketing: a customer journey perspective. *Journal of Research in Interactive Marketing*, (ahead-of-print), 1-18.

George, A. S. (2023). Future Economic Implications of Artificial Intelligence. *Partners Universal International Research Journal, 2*(3), 20–39.

Gupta, V. (2024). An Empirical Evaluation of a Generative Artificial Intelligence Technology Adoption Model from Entrepreneurs' Perspectives. *Systems, 12*(3), 103. DOI: 10.3390/systems12030103

Hanafiah, M. H., Asyraff, M. A., Ismail, M. N. I., & Sjukriana, J. (2024). Understanding the key drivers in using mobile payment (M-Payment) among Generation Z travellers. *Young Consumers, 25*(5), 645–664. DOI: 10.1108/YC-08-2023-1835

Hazarika, A., & Rahmati, M. (2023). Towards an evolved immersive experience: Exploring 5G-and beyond-enabled ultra-low-latency communications for augmented and virtual reality. *Sensors (Basel), 23*(7), 3682. DOI: 10.3390/s23073682 PMID: 37050742

Henke, N., & Jacques Bughin, L. (2016). The age of analytics: Competing in a data-driven world.

Kacprzak, A., & Hensel, P. (2023). Exploring online customer experience: A systematic literature review and research agenda. *International Journal of Consumer Studies*, *47*(6), 2583–2608. DOI: 10.1111/ijcs.12966

Kamal, M., & Himel, A. S. (2023). Redefining Modern Marketing: An Analysis of AI and NLP's Influence on Consumer Engagement, Strategy, and Beyond. *Eigenpub Review of Science and Technology*, *7*(1), 203–223.

Kaushal, V., & Yadav, R. (2023). Learning successful implementation of Chatbots in businesses from B2B customer experience perspective. *Concurrency and Computation*, *35*(1), e7450. DOI: 10.1002/cpe.7450

King, K. (2022). *AI Strategy for Sales and Marketing: Connecting Marketing, Sales and Customer Experience*. Kogan Page Publishers.

Kumar, S., Malviya, R., Uniyal, P., Srivastava, S., Sridhar, S. B., Obaidur Rab, S., & Singh, B. (2024, August 23). (2024). Plant Fiber Composites for Biomedical Applications: Advances and Prospective. *Current Pharmaceutical Biotechnology*, *25*. Advance online publication. DOI: 10.2174/0113892010314833240815113622 PMID: 39185642

Kuusinen, M. (2019). Scenarios for digital marketing: a Delphi-based analysis for 2028.

Lal, S., Arora, M. K., Riswandi, B. A., Singh, N., & Singh, B. (2024). Melding Artificial Intelligence and Remote Sensing Data for Real-Time Monitoring of Marine Biodiversity. In Riswandi, B., Singh, B., Kaunert, C., & Vig, K. (Eds.), *AI Applications for Clean Energy and Sustainability* (pp. 299–319). IGI Global., DOI: 10.4018/979-8-3693-6567-0.ch015

Lauer-Schmaltz, M. W., Cash, P., Hansen, J. P., & Maier, A. (2024). Towards the Human Digital Twin: Definition and Design—A survey. *arXiv preprint arXiv:2402.07922*.

Lavuri, R., & Akram, U. (2023). Role of virtual reality authentic experience on affective responses: Moderating role virtual reality attachment. *Journal of Ecotourism*, •••, 1–19.

Mashood, K., Kayani, H. U. R., Malik, A. A., & Tahir, A. (2023). ARTIFICIAL INTELLIGENCE RECENT TRENDS AND APPLICATIONS IN INDUSTRIES. *Pakistan Journal of Science*, *75*(02). Advance online publication. DOI: 10.57041/pjs.v75i02.855

Naji, K. K., Gunduz, M., Alhenzab, F., Al-Hababi, H., & Al-Qahtani, A. (2024). A Systematic Review of the Digital Transformation of the Building Construction Industry. *IEEE Access : Practical Innovations, Open Solutions*, *12*, 31461–31487. DOI: 10.1109/ACCESS.2024.3365934

Neuwirth, R. J. (2023). Prohibited artificial intelligence practices in the proposed EU artificial intelligence act (AIA). *Computer Law & Security Report*, *48*, 105798. DOI: 10.1016/j.clsr.2023.105798

Rajamannar, R. (2021). *Quantum marketing: mastering the new marketing mindset for tomorrow's consumers*. HarperCollins Leadership.

Reviglio della Venaria, U. (2020). Personalization in Social Media: Challenges and Opportunities for Democratic Societies.

Sadowski, J. (2023). Total life insurance: Logics of anticipatory control and actuarial governance in insurance technology. *Social Studies of Science*, •••, 03063127231186437. PMID: 37427796

Schaeffer, E. (2017). *Industry X. 0: Realizing digital value in industrial sectors*. Kogan Page Publishers.

Sedkaoui, S. (Ed.). (2018). *Big data analytics for entrepreneurial success*. IGI Global.

Sharma, A., & Singh, B. (2022). Measuring Impact of E-commerce on Small Scale Business: A Systematic Review. *Journal of Corporate Governance and International Business Law*, *5*(1).

Soliman, M., & Al Balushi, M. K. (2023). Unveiling destination evangelism through generative AI tools. *ROBONOMICS: The Journal of the Automated Economy*, *4*(54), 1.

Tembrevilla, G., Phillion, A., & Zeadin, M. (2024). Experiential learning in engineering education: A systematic literature review. *Journal of Engineering Education*, *113*(1), 195–218. DOI: 10.1002/jee.20575

Tzampazaki, M., Zografos, C., Vrochidou, E., & Papakostas, G. A. (2024). Machine Vision—Moving from Industry 4.0 to Industry 5.0. *Applied Sciences (Basel, Switzerland)*, *14*(4), 1471. DOI: 10.3390/app14041471

Varela, L., Putnik, G., & Romero, F. (2024). Collaborative manufacturing and management contextualization in the Industry 4.0 based on a systematic literature review. *International Journal of Management Science and Engineering Management*, *19*(1), 78–95. DOI: 10.1080/17509653.2023.2174200

Venkatesh, D. N. (2021). *Winning with employees: Leveraging employee experience for a competitive edge*. SAGE Publishing India.

Vinaykarthik, B. C. (2022, October). Design of Artificial Intelligence (AI) based User Experience Websites for E-commerce Application and Future of Digital Marketing. In *2022 3rd International Conference on Smart Electronics and Communication (ICOSEC)* (pp. 1023-1029). IEEE.

Vuong, N. A., & Mai, T. T. (2023). Unveiling the Synergy: Exploring the Intersection of AI and NLP in Redefining Modern Marketing for Enhanced Consumer Engagement and Strategy Optimization. *Quarterly Journal of Emerging Technologies and Innovations*, *8*(3), 103–118.

Wheeler, J. (2023). *The Digital-First Customer Experience: Seven Design Strategies from the World's Leading Brands*. Kogan Page Publishers.

Wind, Y. J., & Hays, C. F. (2016). *Beyond advertising: Creating value through all customer touchpoints*. John Wiley & Sons.

Zaki, H. O. (2022). The Impact Of Artificial Intelligence On Content Marketing. *Journal of Strategic Digital Transformation In Society*, *2*(3).

Chapter 8
Public Private Partnership Inclusion for Tourism Development in the Horn of Africa

Shashi Kant
https://orcid.org/0000-0003-4722-5736
Bule Hora University, Ethiopia

Tafese Niguse
https://orcid.org/0009-0008-6536-6392
Bule Hora University, Ethiopia

Aynetu Terefe
https://orcid.org/0009-0003-3814-9548
Bule Hora University, Ethiopia

Metasebia Adula
https://orcid.org/0000-0001-5732-2850
Bule Hora University, Ethiopia

ABSTRACT

The purpose of this investigation is to clarify the ways in which partnership of public and private, development of tourism, and sustainability of tourism are interrelated within the tourist sector. The investigation makes use of quantitative techniques to investigate the dynamics by using regression analysis and exploratory factor analysis. The results show that successful partnership of public and private have a considerable affirmative influence on tourist development and sustainability. Also, measures focused at promoting tourism can result in better sustainable practices since development of tourism acts as a critical mediator among partnership of public and private and sustainability outcomes. The investigation's findings highlight how crucial stakeholder involvement and cooperative governance are to the development of sustainable tourism practices. Policymakers and sector stakeholders can benefit greatly from these insights, which imply that enhancing partnership of public and private and giving sustainable development of tourism top priority.

DOI: 10.4018/979-8-3693-8232-5.ch008

INTRODUCTION

Global economic expansion, cross-cultural interchange, and sustainable development are all greatly aided by the tourist sector. Being one of the sectors with the quickest rate of growth, it helps build infrastructure and create jobs, especially in emerging nations. However, worries about social injustice, cultural dilution, and environmental degradation are frequently raised by the quick growth of tourism (Bacsi et al., 2023). This emphasizes the necessity of developing tourism in a sustainable way that not only boosts economic gains but also protects natural and cultural resources for coming generations (Suchkova, 2021). The Brundtland Report from 1987 created a broader framework for sustainable development, from which the concept of sustainable tourism has developed over several decades. The imperative of striking a balance among social fairness, environmental stewardship, and economic prosperity was emphasized in this investigation. These difficulties include different levels of commitment, inadequate stakeholder participation, lack of transparency, and misalignment of goals amongst partners (Jabo et al., 2023).

Empirical literature has shown that PPPs can noteworthy enhance the sustainability of tourism initiatives. Studies indicate that influence partnerships can lead to improved resource management, greater community involvement, and the development of eco-friendly practices. However, despite these affirmative outcomes, several practical challenges impede the successful implementation of PPPs in tourism (Bacsi et al., 2023). These challenges include misalignment of goals among partners, lack of transparency, insufficient stakeholder engagement, and varying levels of commitment. Empirical literature has shown that PPPs can noteworthy enhance the sustainability of tourism initiatives (Deresso et al., 2023). Studies indicate that influence partnerships can lead to improved resource management, greater community involvement, and the development of eco-friendly practices. However, despite these affirmative outcomes, several practical challenges impede the successful implementation of PPPs in tourism. These challenges include misalignment of goals among partners, lack of transparency, insufficient stakeholder engagement, and varying levels of commitment (Panigrahi et al., 2023).

There are still unanswered questions about the processes by which PPPs moderate the link among the growth of the tourism sector and sustainability outcomes. While many studies already exist that emphasize the value of PPPs, very few have thoroughly investigated how these collaborations might be maximized to improve the sustainability of tourism. Also, policymakers and academics lack clear guidance on optimal practices due to the absence of comprehensive frameworks that integrate empirical findings with useful applications (Guye et al., 2024). In order to fill these gaps, this chapter offers a critical evaluation of the body of investigation on PPPs and sustainable tourism. Empirical literature has shown that PPPs can noteworthy enhance the sustainability of tourism initiatives. Studies indicate that influence partnerships can lead to improved resource management, greater community involvement, and the development of eco-friendly practices. However, despite these affirmative outcomes, several practical challenges impede the successful implementation of PPPs in tourism. These challenges include misalignment of goals among partners, lack of transparency, insufficient stakeholder engagement, and varying levels of commitment.

This investigation aims to provide future scholars and policymakers with a roadmap by identifying noteworthy issues and suggesting practical solutions to address them. This investigation's central issue is: *"How can partnership of public and private successfully mediate the link among the growth of the tourism sector and sustainability outcomes?"* This investigation is crucial because it sheds light on how PPPs support sustainable tourism and helps shape the creation of laws that can improve the sustainability and resilience of travel destinations around the world. By providing a response to this query, the investigation adds to the current conversation about sustainable tourism and lays the groundwork for wise choices in both the academic and real-world spheres.

BACKGROUND OF INVESTIGATION

Empirical literature has shown that PPPs can noteworthy enhance the sustainability of tourism initiatives. Studies indicate that influence partnerships can lead to improved resource management, greater community involvement, and the development of eco-friendly practices (Bacsi et al., 2023). However, despite these affirmative outcomes, several practical challenges impede the successful implementation of PPPs in tourism. These challenges include misalignment of goals among partners, lack of transparency, insufficient stakeholder engagement, and varying levels of commitment (Asefa et al., 2023). The tourist sector contributes noteworthy to the global economy, employing millions of people globally and making up 10% of the GDP. But there are serious obstacles to this progress, especially in terms of sustainability. The idea of sustainable tourism has grown in popularity as travel destinations deal with problems including over tourism, environmental damage, and cultural eroding. In order to address these issues, partnership of public and private (PPPs), which combine the advantages of both sectors to advance sustainable practices, have emerged as a viable tactic. However, there are a number of real-world obstacles to PPP adoption in the tourism sector that need to be carefully considered in order to maximize their efficacy (Ding, 2024).

The growth of tourism has a huge impact on sustainability because it facilitates collaboration between the public and commercial sectors. Tourism has historically evolved from an indulgence for the privileged to a worldwide business that accounts for more than 10% of global GDP, based to the UNWTO (Jabo et al., 2023). This expansion has prompted joint efforts to balance economic gains and preserving the environment. Empirical evidence suggests that good collaborations can improve sustainable tourism practices. For instance in Southeast Asia, public-private partnerships have enabled countries to adopt sustainable tourism regulations that address environmental deterioration while also boosting the regional economies (Ding, 2024). Based to studies, community engagement in tourist planning resulted in improved environmental outcomes and higher support for sustainable practices among people (Asefa et al., 2023). In addition, including sustainable practices into tourist development has been connected to higher quality of life for local people, as indicated by greater job possibilities and infrastructural upgrades (Guye et al., 2024). However, without efficient control and coordination, tourism may cause substantial adverse environmental effects, underscoring the importance of maintained collaborations to maintain sustainability (Birhan, 2024).

One notable case investigation is the collaboration among the government of Bali, Indonesia, and various private stakeholders in managing the island's tourism sector. While Bali has seen noteworthy economic benefits from tourism, the negative influences, including waste management issues and cultural commodification, have prompted calls for more sustainable practices. The Bali government's initiative to establish a PPP aimed at promoting eco-tourism faced numerous hurdles. Key challenges included differing priorities among stakeholders—where the private sector focused on short-term profit maximization, while the government aimed for long-term sustainability. Additionally, the lack of a cohesive regulatory framework led to inconsistent practices across different parts of the island, undermining the partnership's objectives (Birhan, 2024). Another illustrative example is the case of the Great Barrier Reef in Australia, where the Great Barrier Reef Marine Park Authority collaborates with tourism operators to promote sustainable tourism practices. While this partnership has successfully implemented various initiatives, such as eco-certification programs, challenges remain. For instance, the varying levels of commitment from private operators have resulted in inconsistent adherence to sustainability standards. Also, the inability to influence communicate the importance of these practices to local communities has hindered broader engagement and support, highlighting the need for better stakeholder involvement and education.

In the context of the European Union, the "Sustainable Tourism in Protected Areas" project aimed to create PPPs in various protected areas to enhance conservation efforts while promoting tourism. Despite initial enthusiasm, the project faced noteworthy obstacles, including bureaucratic delays, lack of funding, and insufficient training for local stakeholders. The disparity in objectives among conservation goals and development of tourism often led to conflicts, illustrating the complexities inherent in establishing influence partnerships (Birhane et al., 2024). These case studies reveal a recurring theme: the successful implementation of PPPs in tourism is often compromised by misaligned goals, lack of stakeholder engagement, and inadequate regulatory frameworks. Addressing these challenges is crucial for harnessing the full potential of PPPs to promote sustainable tourism. This investigation seeks to delve deeper into these issues, providing insights and recommendations that can bridge the gap among theory and practice, thereby enhancing the viability of PPPs as a mechanism for sustainable development of tourism. By examining the practical implications of these challenges, the investigation aims to inform policymakers and stakeholders about best practices that can be adopted in diverse contexts, ultimately contributing to a more sustainable future for the tourism sector.

STATEMENT OF THE PROBLEM

The ability of the tourism sector to stimulate economic growth and foster cross-cultural interchange is becoming more widely acknowledged. But more work needs to be done before sustainable tourism can reach its ideal state, which is a balance among socio-cultural fairness, environmental integrity, and economic viability (Asefa et al., 2023). Although there has been much promotion of the idea of sustainable tourism, there is conflicting data about how well partnership of public and private (PPPs) work to realize this objective. While some studies show that PPPs can result in better sustainability outcomes, others show that because of mismatched interests and insufficient stakeholder cooperation, these partnerships frequently fail to achieve their goals. This contradiction emphasizes how critical it is to fill in a number of gaps in the body of knowledge and current practices (Birhan, 2024). The conflicting findings on PPPs' ability to influence mediate the association among the growth of the tourism sector and sustainability outcomes represent a noteworthy gap. It is challenging to determine the circumstances under which PPPs thrive or fail because empirical evidence is still scattered despite the abundance of case studies and theoretical frameworks. Moreover, there are still theoretical gaps in our knowledge of the mechanisms by which these collaborations influence sustainability. The intricacy of stakeholder associations and the contextual elements that influence the success of partnerships are frequently ignored by existing models (Bacsi et al., 2023).

Public-private partnerships (PPPs) in hospitality have varied degrees of success and failure, highlighting the complexity of collaboration (Bacsi et al., 2023). A successful example is Costa Rica's ecotourism program, in which the government collaborated with private tour operators and neighborhood organizations to encourage sustainable tourist practices. By 2020, ecotourism was expected to contribute $4 billion to the economy, bringing roughly 3 million individuals each year. This collaboration permitted the creation of nature preserves and initiatives to protect them, resulting in increased biodiversity and major economic advantages for those living nearby (Asefa et al., 2023). Conversely, the Maldives is an example of a disastrous public-private partnership in tourism. The government worked with private investors to build luxurious resorts on deserted islands. However, this quick expansion resulted in significant degradation of the environment, particularly coral reef devastation. By 2018, research found a loss in marine biodiversity and increased local dissatisfaction due to insufficient benefits for the community. Protests erupted, revealing the mismatch between the tourism business and local needs, resulting in stalled development. These case studies reveal that, while PPPs can promote ecological tourist growth, they must be well planned, stakeholder engaged, and committed to environmental stewardship to prevent negative consequences (Bacsi et al., 2023).

There are also gaps in the knowledge, especially with regard to the optimal ways to execute PPPs in various tourism situations. Although PPPs have been influence used in some areas to promote sustainable tourism, not all areas can benefit equally from the lessons acquired from these experiences. It is difficult to generalize data or create a thorough grasp of what makes up successful partnership tactics when comparative analysis is lacking. The problem is made worse by contextual gaps because most investigation focuses on certain regions or kinds of tourism, which restricts the conclusions that can be made for wider uses. Practical challenges in implementing PPPs also pose noteworthy barriers to achieving sustainable tourism goals. These consist of insufficient training or capacity building, inconsistent levels of commitment from stakeholders, bureaucratic inefficiencies, and a lack of financing. Such challenges can erode the possible gains from cooperation and cause partner dissatisfaction, making efforts to advance sustainable practices even more difficult. By offering a complete overview of the body of investigation on PPPs and sustainable tourism and highlighting the factors that influence partnership performance, this investigation seeks to close these gaps. The investigation will provide a thorough framework that describes optimal practices for implementing PPPs in diverse tourism situations by combining empirical data and theoretical insights. The investigation will also examine the particular difficulties that stakeholders encounter and provide helpful advice for getting beyond them. Such challenges can erode the possible gains from cooperation and cause partner dissatisfaction, making efforts to advance sustainable practices even more difficult.

THEORITICAL FRAMEWORK

Stakeholder Theory: It highlights how noteworthy it is to include a variety of stakeholders in decision-making processes, including local communities, NGOs, commercial companies, and government agencies. According to this theory, in order to reduce conflict and advance sustainability goals, successful PPPs must balance the interests of all parties involved and promote cooperation and open communication. In addition, the Resource-Based View (RBV) asserts that firms can gain a competitive edge through the efficient use of their special resources and competencies. PPPs in the tourism sector can take advantage of the advantages of the public and private sectors, such as financial resources and local expertise, to work together to address sustainability issues (Horn et al., 2023).

Social Exchange Theory: It focuses on the reciprocal associations among stakeholders, suggesting that partnerships are built on the perceived benefits and costs of collaboration. When stakeholders believe that the advantages of working together outweigh the drawbacks, they are more likely to engage, emphasizing the necessity of trust and mutual benefit in maintaining influence partnerships. Additionally, Institutional Theory examines how regulations, policies, and cultural norms shape organizational behavior. In the tourism sector, institutional frameworks can either facilitate or obstruct the establishment of sustainable practices, making it essential to consider these external factors in analyzing PPP dynamics (Molla, 2023).

Complex Adaptive Systems Theory: It highlights the interconnectedness and evolving nature of systems composed of various components, such as stakeholders, environmental conditions, and socio-economic factors. This theory underscores the need for adaptability and resilience in managing PPPs; as unforeseen challenges may arise that require innovative and collaborative responses. By integrating insights from these theories, this investigation aims to develop a comprehensive understanding of the factors influencing the influence of PPPs in promoting sustainable tourism, ultimately informing future investigation and policy initiatives in the field (Horn et al., 2023).

When it comes to sustainable tourism, public-private partnerships (PPPs) provide a collaborative strategy, but they are not the only governance model accessible. Community-based tourism (CBT) differs from public-private partnerships (PPPs) in that it stresses regional oversight and engagement. Tourism efforts in CBT are managed by local communities, ensuring that benefits directly support their lives and cultural preservation. This concept encourages a sense of accountability and ownership, which frequently leads to more sustainable activities. Government-led solutions, such as state-owned companies, can offer robust regulatory frameworks and creation of infrastructure. Nevertheless, these approaches may lack the flexibility and creativity that private-sector rivalry brings. likewise, collaborative governance models that incorporate various stakeholders, such as NGOs and community organizations, might improve transparency and social equality, but decision-making efficiency may suffer as a result of competing interests. Finally, while PPPs combine the possibilities of the public and private sectors, other models such as CBT and cooperative leadership might provide distinct benefits customized to specific settings. The governance model chosen should be based on local conditions, stakeholder participation, and the broader aims of sustainable development of tourism.

DEFINITION AND ORIGIN OF TERMS

Partnership of public and private(PPP): A partnership of public and private, or PPP, is a cooperative agreement that aims to offer public services or infrastructure projects among private sector businesses and public sector entities, such as government agencies. This partnership model makes use of the advantages of both sectors: the efficiency, creativity, and capital availability of the commercial sector, combined with the legal framework and social welfare commitment of the public sector. In the 1980s and 1990s, in particular, the idea of partnership of public and private (PPPs) gained popularity as governments looked for creative ways to close financing shortages and enhance service delivery (Durga et al., 2024)

Sustainable Tourism: The term "sustainable tourism" refers to travel that preserves travel opportunities for future generations while also satisfying the demands of current travellers and the host communities. This idea includes a comprehensive strategy that strikes a balance among sociocultural fairness, economic viability, and environmental integrity. The concept of sustainable tourism has its roots in the Brundtland Report of 1987, which presented a more comprehensive framework for sustainable development and emphasized the need to balance social justice, environmental preservation, and economic growth (Horn et al., 2023).

EMPIRICAL LITERATURE REVIEW

Development of tourism and Sustainability of tourism

Growing tourism brings in large sums of money and opens up job opportunities, especially in poorer countries where there may not be many other options for work. This increase in GDP has the potential to assist community development projects, improve public services, and upgrade local infrastructure. However, difficulties that could jeopardize sustainability frequently accompany the advantages of the growth of the tourism sector (Bortolotto, 2021). For example, an increase in visitors may result in over tourism, which can harm the environment by destroying habitats, causing pollution, and depleting natural resources. These adverse influences may jeopardize the very features that entice tourists, creating a vicious cycle of depletion for the ecosystem and the travel and tourism sector (Bharti, 2022). Also, it is impossible to ignore how the growth of the tourism sector would influence society and culture. While tourism can foster mutual understanding and cross-cultural contact, it can also result in the commercialization of regional customs and cultures, undermining genuine encounters and sense of place. This conflict draws attention to the necessity of sustainable tourism policies that put an emphasis on interacting with the local community and make sure that tourist growth is in line with the goals and values of the host communities. Successful tactics, like community-based tourism programs, can give people more influence and enable them to get a direct economic advantage from tourists while maintaining their cultural legacy (Timothy, 2023).

Partnership of public and private, has become an noteworthy tool for promoting the development of sustainable tourism. PPPs have the potential to establish frameworks that give sustainability a high priority in the planning and administration of tourism by promoting cooperation among public and private entities. These collaborations can be used to address issues like community involvement, environmental protection, and infrastructure development by combining resources, knowledge, and creativity from both sectors. The ability of PPPs to align the interests of all parties involved, however, is crucial to their influence in fostering sustainability and ensuring that financial gains do not come at the expense of social and environmental integrity (Molla, 2023).

H1: Development of tourism has noteworthy influence on Sustainability of tourism.

Development of tourism and Public- private Partnership

Partnership of public and private' (PPPs) performance and dynamics are greatly influenced by the growth of the tourism sector, which in turn shapes how these partnerships operate within the field of tourism management. Governments are seeking private sector investment and expertise to address the growing demand for infrastructure and services related to transportation, lodging, and recreational facilities as a result of tourism's expansion (Guye et al., 2024). Through this partnership, vital infrastructure that improves a destination's competitiveness and overall visitor experience can be developed. Also, the increasing focus on sustainability in the tourist sector has prompted numerous governments to support partnership of public and private (PPPs) that include environmental, social, and economic factors into tourism endeavours (Adula et al., 2023).

Private businesses frequently match their operations with these sustainability aims because they understand how noteworthy ethical standards are to their marketability. This leads to collaborative activities like conservation projects and community involvement campaigns (Hussain, 2024). But since competing agendas can undermine collaboration, aligning the interests of various stakeholders is crucial to the success of these partnerships. For PPPs to be successful, all parties involved must establish mutual advantages, shared objectives, and open communication. PPPs and the rise of the tourism sector may ultimately complement one another to provide a framework for sustainable growth that enhances the general quality of travel experiences while benefiting regional economies, communities, and the environment.

H2: Development of tourism has noteworthy influence on Partnership of public and private.

Partnership of public and private and Sustainability of tourism

Partnership of public and private (PPPs) are essential for improving the sustainability of tourism because they encourage cooperation among public and private sector actors. These collaborations take advantage of each sector's advantages to tackle the intricate problems associated with the growth of sustainable tourism. PPPs can help with the implementation of eco-friendly practices, infrastructure development, and community involvement programs by combining private investment and innovation with public oversight and regulatory frameworks (Timothy, 2023). PPPs can support conservation and waste reduction initiatives as well as sustainable resource management, guaranteeing that natural and cultural treasures are protected for future generations. These collaborations can also improve the calibre of tourism services because private organizations are frequently more knowledgeable and influence at providing sustainable solutions (Kant & Adula, 2024). However, stakeholder alignment of interests, open communication, and a common commitment to sustainability objectives are necessary for PPPs to influence promote sustainable tourism. PPPs, when handled well, can foster a synergistic environment that promotes social fairness and environmental stewardship in addition to increasing economic growth, which will ultimately result in a more resilient and sustainable tourism sector (Hussain, 2024)

H3: Partnership of public and private has noteworthy influence on Sustainability of tourism.

Mediation of Partnership of public and private on Sustainability of tourism

In order to ensure that economic growth does not come at the expense of social justice or environmental integrity, this mediation is essential in that it helps to harmonize the diverse interests of the numerous stakeholders involved in development of tourism. PPPs provide both sectors with the opportunity to combine resources, exchange knowledge, and use creative solutions that successfully tackle sustainability issues (Birhane et al., 2024). The capacity of PPPs to create thorough frameworks that incorporate sustainable practices into the growth of the tourism sector is a crucial component of this mediation. For example, by working together on cooperative projects, these alliances can support environmentally friendly infrastructure, including green hotels and sustainable transit systems, which reduce their negative influences on the environment while improving the traveller experience. Private enterprises gain from the regulatory support and legitimacy given by public institutions, while governments may take advantage of the efficiency and creativity of the private sector, producing a win-win situation for all parties involved (Gottfried et al., 2022).

PPPs can also help with stakeholder and community participation, both of which are necessary to achieve sustainable tourism. Through the inclusion of local communities in decision-making procedures, PPPs guarantee that the growth of tourism is in line with the desires and requirements of locals. By encouraging a sense of ownership and accountability among the community, this participatory method helps to preserve the area's natural resources and cultural legacy (Kant & Adula, 2024). However, the creation of open lines of communication, common objectives, and reciprocal advantages among stakeholders are necessary for PPPs to be influence as mediators for sustainable tourism (Timothy, (2023).

H4: The Link among Development of tourism and Sustainability of tourism is influenced By Partnership of public and private.

CONCEPTUAL FRAMEWORK

A conceptual framework is a methodical way to depict the connections among noteworthy ideas and variables in a investigation project. The conceptual framework can be seen in association to the investigation of partnership of public and private' (PPPs) mediating influence on tourist sustainability through the following dimensions:

Development of tourism influence on Sustainability of tourism by mediation of Partnership of public and private

Figure 1. Conceptual framework

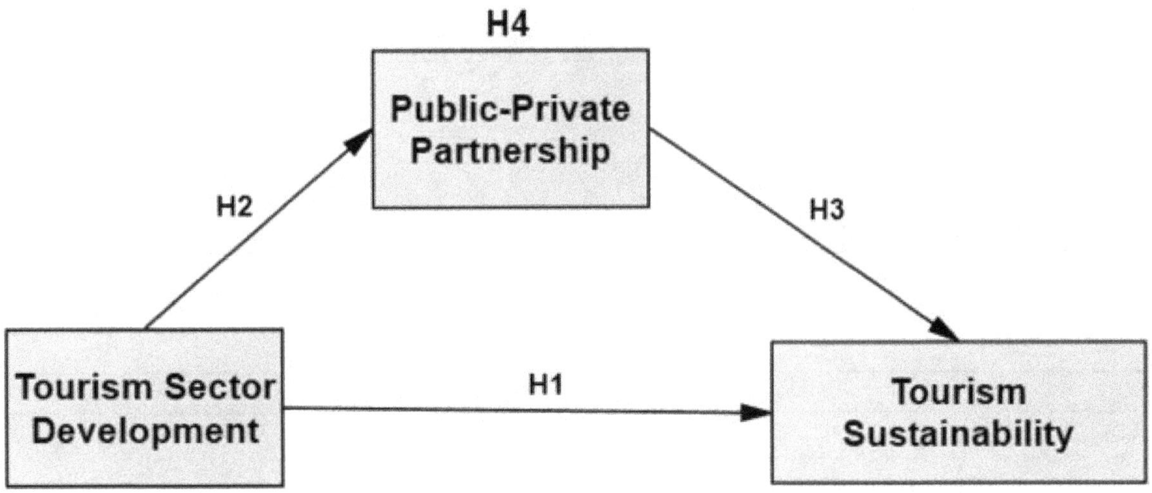

Source: Authors

INVESTIGATION METHODOLOGY

The investigation methodology applied in this investigation utilized a quantitative investigation approach to explore the influence of partnership of public and private (PPPs) on tourist sustainability. A broad range of stakeholders, including representatives from the public and private sectors and the local community engaged in tourism, were included in the sample size of 412 respondents. Structured questionnaires intended to elicit respondents' opinions and experiences about the efficacy of PPPs in advancing sustainable tourism practices were distributed as part of the data gathering procedure. Based on a thorough examination of related literature, a comprehensive scale development procedure was carried out to assure the validity and reliability of the measuring scales utilized in the investigation. A number of noteworthy constructs, such as resource sharing, governance influence, stakeholder engagement, and sustainability results, have been identified in prior investigation as being pertinent to PPP mediation. Item adaptation and refinement were made to fit the particular setting of this investigation by drawing on recognized scales from the tourism and partnership literature. Each construct included several items on the original scale, which made it possible to evaluate the association among the variables with a high degree of confidence. The data were then subjected to a number of statistical analyses.

The sample's suitability for factor analysis was evaluated using the Kaiser-Meyer-Olkin (KMO) measure; a KMO value greater than 0.6 indicates adequate sampling. Subsequently, an Exploratory Factor Analysis (EFA) was carried out in order to ascertain the fundamental structure of the data and validate the constructs' dimensionality. EFA made it possible to exclude items and find the elements that contributed noteworthy to the explanation of the data's variation. Lastly, AMOS software was used for the analysis and Structural Equation Model (SEM) was used to test the proposed association among the structures. Through the analysis of PPPs' direct and indirect influences on tourist sustainability, SEM was able to shed light on the many components that function as mediators. The SEM analysis's findings provided a thorough grasp of the ways in which partnership of public and private influence sustainable tourism practices, highlighting the advantages and disadvantages of current alliances.

DATA ANALYSIS

Table 1. KMO and Bartlett's test

Kaiser-Meyer-Olkin Measure of Sampling Adequacy.		**.860**
Bartlett's Test of Sphericity	Approx. Chi-Square	1419.748
	df	96
	Sig.	.003

Source: Authors

As shown in Table 1, score of 0.860 was obtained using the Kaiser-Meyer-Olkin (KMO) measure of sample adequacy. The substantial degree of sampling adequacy shown by this high KMO score suggests that the data is suitable for factor analysis. KMO levels above 0.6 are generally regarded as satisfactory, and scores near 1.0 indicate even better analytic circumstances. A KMO score of 0.860 indicates that there are sufficient inter association among the variables, indicating that it is valid to continue with additional statistical analysis.

Bartlett's Test of Sphericity yielded an approximate Chi-Square value of 1419.748 with 96 degrees of freedom and a significance level of 0.003, in addition to the KMO score. There are noteworthy association among the variables, as shown by the noteworthy result of Bartlett's test, which also shows that the association matrix is not an identity matrix. The results indicate that the data are suitable for factor analysis and that there is sufficient evidence of the association among the variables to justify more investigation. This test's statistical significance increases trust in the dataset's ability to uncover underlying constructs. All things considered, the findings derived from the KMO measure and Bartlett's test together offer a strong defence for performing Exploratory Factor Analysis (EFA) on the dataset. The strong KMO score and the statistically noteworthy Bartlett's test result suggest that the data is a good fit for identifying latent variables pertaining to sustainable tourism and partnership of public and private. These findings build confidence in the validity of following analyses, paving the way for a deeper understanding of the intricate interactions among the variables under inquiry.

Table 2. Total Variance Explained

C	Initial Eigenvalues			Extraction Sums of Squared Loadings			Rotation Sums of Squared Loadings		
	Total	% of Variance	Cumulative %	Total	% of Variance	Cumulative %	Total	% of Variance	Cumulative %
1	4.53	32.225	32.225	4.53	32.225	32.225	3.02	21.474	21.474
2	1.36	9.765	42.025	1.36	9.765	42.024	2.17	15.486	36.951
3	1.27	9.089	51.113	1.25	9.088	51.113	1.95	14.157	51.113
4	.947	6.790	57.893						
5	.872	6.224	64.115						
6	.775	5.514	69.634						

Mining Process: PCA
Source: Authors

The total variance explained shown in Table 2. The first component has an eigenvalue of 4.53, which explains 32.225% of the overall variance. The initial eigenvalues reveal how much variance each component accounts for. This big percentage emphasizes how noteworthy the first factor is in accounting for a large majority of the variability in the dataset, which makes it an essential component in comprehending the concepts associated with sustainable tourism and partnership of public and private. With an eigenvalue of 1.36, the second component adds 9.765% more variation to the total, making the cumulative percentage 42.025%. Our suggests that the first two elements collectively explain a noteworthy portion of the variation, highlighting their importance in our investigation. The third component comes next, accounting for 9.089% of the variance with an eigenvalue of 1.27, for a cumulative total of 51.113%. These three elements collectively account for more than half of the data variation, highlighting their importance within the investigation's framework. With an eigenvalue of 0.947, the fourth component explains 6.790% of the variance; the eigenvalues of the fifth and sixth components are even lower, at 0.872 and 0.775, respectively. Despite adding to the total variance, these components' eigenvalues are less than 1.0, indicating that they are less noteworthy than the first three components. This pattern indicates that while additional components can be extracted, they may not provide critical insights relative to the primary factors identified.

Table 3. Covariance

Covariance			Approximation	S.E.	C.R.	P	Hy.
Partnership of public and private	**<-->**	**Development of tourism**	**.254**	**.032**	**8.234**	***	**H2**
Partnership of public and private	<-->	Sustainability of tourism	.304	.031	7.693	***	**H3**
Development of tourism	<-->	Sustainability of tourism	.255	.25	7.832	***	**H1**

Source: Authors

Table 3 displays the covariance estimates. The covariance's shed light on how the variables relate to one another by indicating how much they vary collectively. With a value of 0.254, the first row displays the association among partnership of public and private and the growth of tourism. This shows a affirmative association, implying that the degree of development of tourism rises in tandem with the efficacy of partnership of public and private. The standard error (S.E.) for this estimate is 0.032, and the critical ratio (C.R.) is 8.234, which is statistically noteworthy (shown by the "***"). The hypothesis (H2) that successful partnership of public and private favourably contribute to the development of tourism is supported by this strong link. The covariance, or 0.304, among partnership of public and private and tourist sustainability is highlighted in the second row. Strong partnership of public and private are associated with more sustainable tourist practices, which are further supported by this affirmative covariance. This association is likewise statistically noteworthy, supporting hypothesis H3, with a S.E. of 0.031 and a C.R. of 7.693. The results indicate that better sustainability outcomes can result from including the public and commercial sectors in tourist activities.

Finally, it is revealed that there is a 0.255 association among development of tourism and sustainability of tourism. This affirmative association suggests that as development of tourism advances, the sector's sustainability practices also likely to get better. This association's statistical significance is further supported by its S.E. of 0.25 and C.R. of 7.832, which both support hypothesis H1. This investigation demonstrates how development and sustainability are intertwined in the tourism sector, indicating that progress in one area can have a beneficial influence on the other.

Table 4. Validity Concern

	CR	AVE	MSV	MaxR(H)	TD	TS	P-PP
TD	0.733	0.641	0.217	0.853	**0.657**		
TS	0.757	0.657	0.134	0.763	0.170	**0.667**	
P-PP	0.791	0.604	0.272	0.792	0.384	0.194	**0.681**

Note: TD=Development of tourism; P-PP=Public Private Partnership; TS=Sustainability of tourism
Source: Authors

Key metrics for the investigation's constructs are presented in Table 4, The constructions' Composite Reliability (CR) ratings show a satisfactory degree of internal consistency. With a CR of 0.733—above the generally recognized cut off of 0.7—Development of tourism (TD) appears to have appropriate reliability. Likewise, Sustainability of tourism (TS) displays a CR of 0.757, signifying strong dependability. With a CR of 0.791, Partnership of public and private (P-PP) obtains the highest, further supporting the construct's trustworthiness. Together, these results imply that the metrics employed to quantify these constructs are trustworthy and internally consistent. The Average Variance Extracted (AVE) measure sheds light on how much variation is accounted for by the constructs compared to how much variation is caused by measurement error. With an AVE of 0.641 for development of tourism—beyond the suggested cutoff of 0.5—it is clear that the construct captures a sizable portion of variance. The validity of sustainability of tourism is also supported by its AVE of 0.657, which is likewise over the cutoff. With an AVE of 0.604, Partnership of public and private also exhibits sufficient variance extraction, supporting the validity of the concept.

The degree to which constructions share variation with one another is shown by the Maximum Shared variation (MSV) values. The MSVs are as follows: 0.217 for development of tourism, 0.134 for sustainability of tourism, and 0.272 for partnership of public and private. Given that the constructs do not share an excessive amount of variance, these relatively low MSV values imply that there is a sufficient degree of discriminant validity among the constructs. Lastly, further assurance of the validity and reliability of the constructions is given by the Maximum Reliability (MaxR(H)) values. Strong dependability is seen throughout the constructs, as evidenced by the MaxR(H) values, which range from 0.763 for sustainability of tourism to 0.853 for development of tourism.

MEDIATION EXAMINATION

Figure 2. Structural equation model

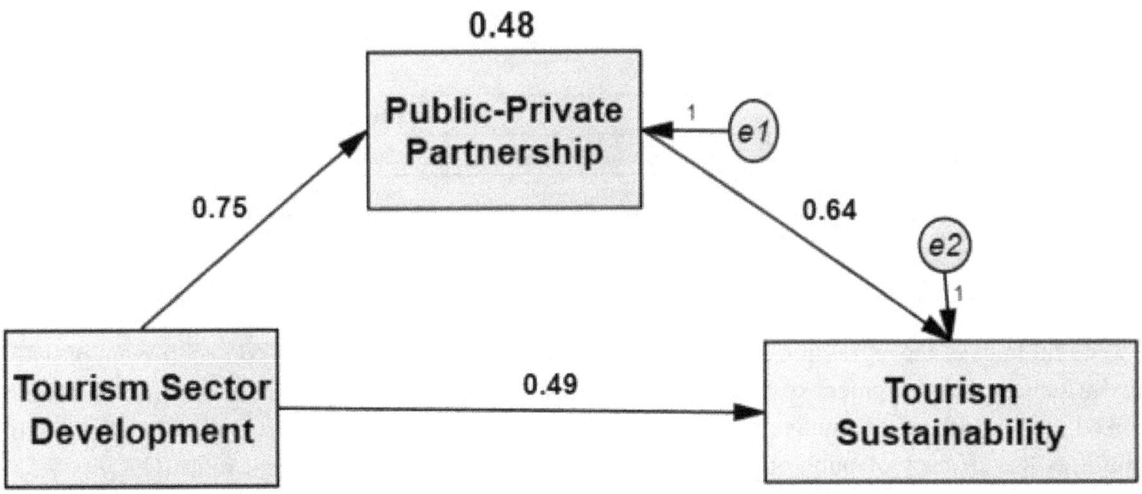

Source: Authors

Model fit indices

Table 5. indices for Model Fit

Sig.	Chi-Sq	RMR	Fitness Goodness	Fitness Confirmatory	TLI	RMSEA
0.004	1.956	.037	0.933	0.913	0.912	.0230

Source: Authors

The model's overall goodness-of-fit presented in Table 5. The model is statistically noteworthy, as indicated by the reported significance level (Sig.) of 0.004. Although a smaller Chi-Square value is often preferable for optimal fit, the model appears to match the data pretty well based on its noteworthy Chi-Square value of 1.956. A favourable score of 0.037 for the Root Mean Square Residual (RMR) indicates a satisfactory match among the observed and anticipated values. Better model fit is indicated by lower RMR values, and this number shows that the residuals are reasonably minimal, confirming the suitability of the model. Both the Fitness Goodness index (0.933) and the Fitness Confirmatory index (0.913) are stated to be above the generally accepted 0.90 level. Given that TLI values above 0.90 are regarded as acceptable, the model's fit is further supported by the Tucker-Lewis Index (TLI) value of 0.912. This index suggests that the model sufficiently describes the data without being unduly complex by providing a comparative fit metric that takes model complexity into consideration. Finally, the observed Root Mean Square Error of Approximation (RMSEA) is 0.0230, far less than the suggested top limit of 0.06. A lower RMSEA value indicates a better fit, suggesting that the model has a high degree of approximation to the true population values.

Table 6. Regression Examination

Relative			Approx.	S.E.	C.R.	P	Ass.
P-Private Partnership	**<---**	**Development of tourism**	**.865**	**.141**	**7.583**	***	**H2**
Sustainability of tourism	<---	P-Private Partnership	.327	.132	2.855	***	H3
Sustainability of tourism	<---	Development of tourism	.954	.197	7.956	***	H1

Source: Authors

The findings of the regression analysis are shown in Table 6. With a regression coefficient of 0.865, the influence of development of tourism on partnership of public and private is the first association looked at. The robust affirmative association indicates that heightened tourism growth considerably amplifies the efficacy of public-private collaborations. This estimate's critical ratio (C.R.) is 7.583, which is statistically noteworthy (shown by "***"), and its standard error (S.E.) is 0.141. This result demonstrates a strong association among these two constructs and validates hypothesis H2.TThe influence of partnership of public and private on the sustainability of tourism is the subject of the second association. The regression coefficient in this case is 0.327, indicating that, while to a lesser degree than in the prior association, successful partnership of public and private affirmatively contribute to tourist sustainability. This association has a statistically noteworthy C.R. of 2.855 and a S.E. of 0.132. This backs up hypothesis H3, which emphasizes how crucial it is for the public and commercial sectors to work together to promote sustainable tourist practices. Finally, the regression analysis looks at the connection among sustainable tourism and development of tourism. The coefficient of 0.954 shows a very strong affirmative association, implying that sustainable tourism practices rise in tandem with development of tourism. This association has a statistically noteworthy S.E. of 0.197 and a C.R. of 7.956. This finding supports hypothesis H1, emphasizing the interconnectedness of development and sustainability in the tourism sector.

Table 7. Mediating role Influence

	Influence	worth	Path Influence
Development of tourism → Sustainability of tourism	Direct Influence	.49	Direct influence stated
Development of tourism → Public private partnership → Sustainability of tourism	Indirect Influence	.75*.65=.48	Indirect Influence Ensued
	Whole influence	.97	Partial mediation

Source: Authors

Table 7 presents the mediating role analysis. The direct influence of Development of tourism on Sustainability of tourism is quantified at 0.49. This substantial coefficient indicates a meaningful direct association, suggesting that advancements in the tourism sector affirmatively contribute to sustainability efforts. The statement "Direct influence stated" reinforces the significance of this direct path, highlighting the essential role of sector development in fostering sustainable practices within tourism. In addition to the direct influence, the table reveals the indirect influence of Development of tourism on Sustainability of tourism through Partnership of public and private. This indirect influence is calculated as 0.75 (the path from Development of tourism to PPP) multiplied by 0.65 (the path from PPP to Sustainability of tourism), resulting in a value of 0.48. The notation "Indirect Influence Ensued" indicates that this mediation pathway is noteworthy, demonstrating that influence partnership of public and private serve as a crucial mechanism through which development of tourism contributes to sustainability. This finding underscores the importance of collaborative efforts among public and private entities in achieving sustainable tourism goals. The whole influence of Development of tourism on Sustainability of tourism, combining both direct and indirect influences, totals 0.97. This high value illustrates that the overall influence of development of tourism on sustainability is substantial when considering both the direct pathway and the mediation through partnership of public and private. The designation of "Partial mediation" indicates that while partnership of public and private play a noteworthy mediating role, development of tourism still directly influences sustainability outcomes. This suggests that both pathways are vital for understanding the dynamics of sustainable tourism.

DISCUSSION

The analysis's conclusions offer noteworthy new perspectives on the connections among partnership of public and private, tourism growth, and sustainable tourism. Influence cooperation among governmental organizations and commercial stakeholders is crucial for promoting growth in the tourist sector, as evidenced by the substantial affirmative association among partnership of public and private and development of tourism. This association emphasizes how crucial it is to have laws and regulations that allow these kinds of collaborations because they not only promote sustainable practices but also help to build the tourism sector. Also, the findings show that partnership of public and private play a noteworthy role in promoting sustainable tourism. The good association that has been developed indicates that policies that support sustainable tourism practices may be implemented when the public and private sectors collaborate successfully. Investigation results are similar with previous studies like jabo et al., 2023; Birhane et al., 2024 and Guye et al., 2024. This discovery holds noteworthy relevance given the growing worldwide apprehension regarding the ecological consequences of tourism. Stakeholders can create programs that promote economic progress and guarantee the preservation of natural and cultural resources for future generations by utilizing the assets of both sectors.

The analysis also emphasizes how noteworthy development of tourism is as a mediator in the interaction among sustainable tourism and partnership of public and private. The direct association among tourist development and sustainability highlights the likelihood of improved sustainability practices as tourism projects grow. The indirect influence serves as more evidence that partnership of public and private foster the growth of tourism, which improves sustainability results. The concept that encouraging tourist development is a smart strategy to accomplish sustainability goals is strengthened by this mediation influence. All things considered, the results point to the need for a comprehensive strategy to promote sustainable tourism. To promote tourist development, policymakers and sector players should concentrate on fortifying partnership of public and private, all the while keeping sustainability as a primary goal. By doing this, they can produce a synergistic influence that protects cultural and natural integrity while simultaneously enhancing tourism's economic viability. In order to solve the issues facing contemporary tourism and achieve long-term sustainability in the sector, a comprehensive strategy is necessary.

Funding structures are critical to enabling public-private partnerships (PPPs) in tourist development, allowing for the effective execution of long-term initiatives. Government grants and subsidies are a key source of finance, helping to pay early project expenses and incentivizing private sector involvement. These subsidies frequently target specific objectives, such as conservation of the environment or community development, which are consistent with the ideals of sustainable tourism. Moreover, international financial agencies such as the World Bank and neighborhood development banks offer financing and technological help to tourist initiatives, particularly in poor nations. These agencies frequently emphasize initiatives with an opportunity for revenue creation and social benefits. Private sector investment is also important; venture capital as well as equity finance from tourist firms may help support new projects that improve environmental sustainability. Additionally bonds for the public and impact investment are developing as alternative funding vehicles. Public bonds allow governments to obtain funds for infrastructure initiatives, while impact investing draws socially aware investors looking for both financial returns and desirable environmental or social results. Together, these funding instruments form a strong financial environment that promotes PPPs in growth of tourism while also encouraging environmentally friendly policies and boosting the regional economies.

CONCLUSION

In conclusion, this investigation underscores the pivotal role of partnership of public and private in enhancing both development of tourism and sustainability. The noteworthy associations identified among the constructs indicate that influence collaboration among public and privatesectors is crucial for fostering growth within the tourism sector. The findings demonstrate that partnership of public and private not only stimulate development of tourism but also contribute to sustainable practices, highlighting the necessity of integrated strategies that align economic objectives with environmental stewardship. Also, the mediating role of development of tourism in the association among partnership of public and private and sustainability of tourism reveals a nuanced understanding of how these elements interact. As development of tourism progresses, it enhances sustainability outcomes, suggesting that initiatives aimed at bolstering tourism can lead to more responsible practices. This mediation emphasizes the importance of focusing on development as a pathway for achieving sustainability goals within the tourism sector.

MANAGERIAL IMPLICATIONS

The findings of this investigation offer several noteworthy managerial implications for stakeholders in the tourism sector. First and foremost, the noteworthy role of partnership of public and private suggests that managers should actively seek to establish and strengthen collaborations among public entities and private businesses. By fostering these partnerships, tourism managers can leverage shared resources, expertise, and networks to enhance development of tourism initiatives. This collaborative approach can lead to more comprehensive strategies that benefit all parties involved, including local communities, businesses, and government agencies. Additionally, the affirmative association among development of tourism and sustainability highlights the need for managers to prioritize sustainable practices in their development plans. By integrating sustainability into tourism initiatives, managers can not only improve their environmental and social influence but also attract a growing segment of eco-conscious travellers. This alignment with sustainability can enhance the reputation of tourism destinations and create a competitive advantage in an increasingly crowded market.

Moreover, the mediating role of development of tourism suggests that managers should focus on creating robust tourism infrastructure and services as a foundation for sustainability efforts. Investments in infrastructure, capacity building, and community engagement can noteworthy enhance the influence of sustainability initiatives. Managers should also consider implementing monitoring and evaluation systems to assess the influence of their development strategies on sustainability outcomes, allowing for continuous improvement and adaptation. Finally, communication and stakeholder engagement are crucial in executing these strategies influence.

PRACTICAL IMPLICATIONS

The insights from this investigation carry noteworthy practical implications for the tourism sector. First, stakeholders should prioritize the establishment of partnership of public and private to enhance development of tourism. Practical steps include organizing workshops and forums that bring together public officials, private sector representatives, and community leaders to discuss mutual goals and strategies. Such collaborative initiatives can lead to innovative solutions that address local tourism challenges while promoting sustainable practices. Moreover, the findings emphasize the importance of integrating sustainability into development of tourism projects. Practically, this can involve adopting best practices in environmental management, such as reducing waste, conserving resources, and promoting local culture. Tourism operators can implement sustainability certifications to signal their commitment to responsible practices, appealing to environmentally conscious travellers. This focus on sustainability not only addresses current consumer preferences but also positions destinations favourably for future growth. Additionally, investment in infrastructure is crucial for supporting both development of tourism and sustainability. Practical measures may include improving transportation systems, upgrading facilities, and enhancing public spaces to create a more attractive environment for tourists. Managers should also engage in community outreach to ensure that local residents are included in the benefits of development of tourism, fostering a sense of ownership and support for sustainability initiatives.

Lastly, continuous monitoring and evaluation of tourism practices are essential to ensure that development goals align with sustainability objectives. Implementing data collection methods, such as visitor surveys and environmental assessments, can help stakeholders assess the influence of their strategies.

THEORETICAL IMPLICATIONS

The findings of this investigation contribute to the theoretical understanding of the associations among partnership of public and private, development of tourism, and sustainability. By empirically validating the interconnectedness of these constructs, the investigation reinforces existing theories that emphasize the importance of collaborative governance in achieving sustainable tourism outcomes. The investigation highlights how partnership of public and private serve as a critical mechanism for facilitating development of tourism, suggesting that theories of collaboration and stakeholder engagement should be further explored in the context of tourism. Additionally, the mediating role of development of tourism in the association among partnership of public and private and sustainability of tourism adds a new dimension to the theoretical discourse. This finding suggests that development of tourism is not merely an outcome of partnerships but also a vital process that influences sustainability. It calls for a revaluation of theoretical frameworks that prioritize direct associations among constructs, advocating for models that incorporate mediating variables to better capture the complexities of tourism dynamics. Also, the investigation's results underscore the need for integrating sustainability principles into development of tourism theories. As the tourism sector increasingly grapples with environmental and social challenges, theoretical models must account for sustainability as a core component of development of tourism strategies. This shift in perspective can enrich the literature on sustainable tourism, providing a more holistic view that considers economic, environmental, and social dimensions. Overall, this investigation opens avenues for future theoretical exploration, encouraging scholars to investigate the nuanced interactions among partnership of public and private, development of tourism, and sustainability.

RECOMMENDATIONS

Based on the findings of this investigation, several recommendations can be made for stakeholders in the tourism sector to enhance the influence of partnership of public and private and promote sustainable development of tourism. First, it is crucial for policymakers to create supportive regulatory frameworks that encourage collaboration among public and private entities. This can be achieved by offering incentives such as tax breaks or grants for projects that demonstrate a commitment to sustainability, thereby attracting investment and fostering innovative partnerships. In addition, tourism managers should prioritize capacity-building initiatives that strengthen the skills and knowledge of stakeholders involved in partnership of public and private. Training programs focused on best practices in sustainable tourism and influence collaboration can empower participants to implement successful strategies and understand the benefits of working together. Such initiatives can also enhance communication and trust among partners, fostering a more cohesive approach to development of tourism. Also, stakeholders should actively engage local communities in the planning and implementation of tourism projects. By involving residents in decision-making processes, tourism managers can ensure that development initiatives align with community needs and values. This participatory approach not only enhances the relevance of tourism projects but also builds local support, which is essential for long-term sustainability. Lastly, ongoing monitoring and evaluation of tourism initiatives are recommended to assess their influence on sustainability and development outcomes. Implementing robust data collection methods will enable stakeholders to identify successful practices and areas needing improvement.

FUTURE DIRECTIONS

The findings from this investigation open several avenues for future investigation in the realms of development of tourism, partnership of public and private, and sustainability. One promising direction is the exploration of the specific mechanisms through which partnership of public and private influence sustainability of tourism. Future studies could employ qualitative methods, such as case studies or interviews, to gain deeper insights into the dynamics of these collaborations and identify best practices that lead to successful outcomes. Additionally, investigations could investigate the role of technology in enhancing partnership of public and private within the tourism sector. With the increasing importance of digital tools for communication and data sharing, examining how technology can facilitate collaboration and improve sustainability measures is vital. This could include exploring the influence of social media, online platforms, and data analytics in fostering stakeholder engagement and promoting transparency. Another potential area for future investigation is the comparative analysis of different geographical contexts. Studies could assess how cultural, economic, and political factors influence the influence of partnership of public and private in achieving sustainability of tourism across various regions. This comparative approach could yield valuable insights into context-specific strategies and help tailor interventions to local needs and conditions. Also, the long-term influences of development of tourism on sustainability warrant further investigation. Longitudinal studies could provide a clearer understanding of how tourism initiatives evolve over time and their sustained influences on environmental and social outcomes.

REFERENCES

Adula, M., Birbirsa, Z. A., & Kant, S. (2023). The influence of interpersonal, problem solving and technical training skills on performance of Ethiopia textile sector: Continuance, normative and influence commitment as mediators. *Cogent Business & Management*, *10*(3). Advance online publication. DOI: 10.1080/23311975.2023.2286672

Anisa, H. (2021). The Influence of Political Instabilty on Environmental Governance at the Horn of Africa: a Case Investigation of Somalia (Doctoral dissertation, University of Nairobi).

Asefa, K., & Debela, K. L. "Influence of Transformational Leadership on Organizational Performance: The Mediating Role of Employee Commitment and Expert Systems (AI) Inclusion in Ethiopia," *2023 International Conference on Communication, Security and Artificial Intelligence (ICCSAI)*, Greater Noida, India, 2023, pp. 305-310, DOI: 10.1109/ICCSAI59793.2023.10421269

Bacsi, Z., Gebbisa, M. B., Dávid, L. D., & Hollósy, Z. (2023). Pastoralism and Tourism in Eastern Africa—Quantitative Analysis from 2004 to 2018. *Sustainability (Basel)*, *15*(12), 9723. DOI: 10.3390/su15129723

Belay, M. K. (2021). The State in Path Creation: The Emergence of a Tourist Destination in Ethiopia. [Netherlands]:[publisher not identified].

Bharti, M. S. (2022). The Development of China's Economic Cooperation in the Horn of Africa: Special Reference to the Belt and Road Initiative. African Journal of Economics. *Politics and Social Studies*, *1*(1), 17–30.

Birhan, A. T. (2024). Conflict Dynamics in the Horn of Africa: Causes, Consequences, and the Way Forward. *International Journal of Peace and Conflict Studies*, *9*(1), 18–36.

Birhane, M., Amentie, C., Borji, B., & Kant, S. Are Ethiopian Coffee Farmers Proficient to Predict Capital Structure with Mediation of AI Based Knowledge Systems and Asset Liquidity? In *AI in Agriculture for Sustainable and Economic Management* (pp. 199-215). CRC Press. DOI: 10.1201/9781003451648-15

Bortolotto, S., & Cheli, F. (2021). Eritrean Mobility and Cultural Heritage. New Frontiers of the Horn of Africa. An overview of the project. E Motion Eritrean Mobil. Cult. Heritage. *New Front. Horn Afr*, *17*, 23.

Dereso, C. W., Kant, S., Muthuraman, M., & Tufa, G. (2023). Influence of Point of Service on Health Department Student's Creativity in Comprehensive Universities of Ethiopia: Moderating Role of Partnership of public and privateand Mediating Role of Work Place Learning. In: Jain, S., Groppe, S., Mihindukulasooriya, N. (eds) Proceedings of the International Health Informatics Conference. Lecture Notes in Electrical Engineering, vol 990. Springer, Singapore. https://doi.org/DOI: 10.1007/978-981-19-9090-8_13

Ding, L. (2024). The evolving roles of the gulf states in the horn of Africa. Asian Journal of Middle Eastern and Islamic Studies, 1-14.

Durga, P., Godavarthi, D., Kant, S., & Basa, S. S. (2024). Aspect-based drug review classification through a hybrid model with ant colony optimization using deep learning. *Discov Computing*, *27*(1), 19. DOI: 10.1007/s10791-024-09441-w

Gogo, A. F., & Masaki, E. (2022). Sustainability of Cultural Tourism in East African Community. In *Cultural Sustainable Tourism* (pp. 187–194). Springer International Publishing. DOI: 10.1007/978-3-031-07819-4_16

Gottfried, S., Sabhat, C. A., Poli, E., Raveh, A., Zatcovetzky, I., & Yirga, N. (2022). From Geo/Bio-Politics to G2G Agreements and Public–Private Partnership: the Unique Role of the Israeli Eco-System in Ethiopia.

Guyo, D. M., Kero, C. A., & Kant, S. (2025). Is Marketing Intermediaries' Mediation Required for Livestock and Products Marketing to Improve the Economic Status of Pastoralist Communities in Ethiopia? In *AI in Agriculture for Sustainable and Economic Management* (pp. 216-234). CRC Press.

Guyo, D. M., Kant, S., & Kero, C. A. (2023, November). Mediation of Marketing Intermediaries and AI Adoption Between Livestock Products Marketing and Economic Status of Pastoralist in Ethiopia. In *2023 International Conference on Communication, Security and Artificial Intelligence (ICCSAI)* (pp. 311-316). IEEE.

Horn, A., & Visser, G. (2023). Tourism Gentrification in Urban Africa: Towards a Investigation Agenda. *Studia Periegetica*, *43*(3), 7–24. DOI: 10.58683/sp.599

Hussain, A., Mandić, A., & Fusté-Forné, F. (2024). Transforming communities: Analyzing the influences of infrastructure and development of tourism on social capital, livelihoods, and resilience in Gilgit-Baltistan, Pakistan. *Journal of Hospitality and Tourism Management*, *59*, 276–295. DOI: 10.1016/j.jhtm.2024.04.017

Jabo, D., Borji, B., & Kant, S. (2024, February). Impact of Technical CRM on Ethiopia Bank Human-Computer Interface and Competitive Advantage as Mediators of Performance. In 2024 IEEE International Conference on Computing, Power and Communication Technologies (IC2PCT) (Vol. 5, pp. 679-682). IEEE.

Kant, S., & Adula, M. (2024). AI Learning and Work Attitude Mediation Among Reward and Organizational Support in Ethiopia. In Gomathi Sankar, J., & David, A. (Eds.), *Generative AI for Transformational Management* (pp. 109–136). IGI Global., DOI: 10.4018/979-8-3693-5578-7.ch005

Kant, S., & Adula, M. (2024). Mediated by AI-Based Generative Re-Enforcement Learning and Work Attitude: Are Intrinsic Rewards Transforming Employee Perceived Organizational Support? In Gomathi Sankar, J., & David, A. (Eds.), *Generative AI for Transformational Management* (pp. 83–108). IGI Global., DOI: 10.4018/979-8-3693-5578-7.ch004

Kant, S., & Adula, M. (2024). Human-Machine Interaction in the Metaverse in the Context of Ethiopia. In *Influence and Potential of Machine Learning in the Metaverse* (pp. 196–212). IGI Global., DOI: 10.4018/979-8-3693-5762-0.ch008

Molla, T. (2023). The Recurrent Barriers to Influence Economic Integration in IGAD and the Way Forward. *Journal of Good Governance and Sustainable Development in Africa*, *8*(1), 7–24.

Nyoach, T. D., Lemi, K., Debela, T., & Kant, S. (2024). Does Organizational Commitment Mediate the Association among Employee Association Management and Bank Performance? The Case of Banks in Ethiopia. *International Journal of Organizational Leadership*, *13*(2), 355–376. DOI: 10.33844/ijol.2024.60419

Osman, G. A. (2022). Greater horn of Africas dilemma in achieving sustainable development goals. *Journal of African Studies and Development*, *14*(4), 160–169.

Panigrahi, A., Pati, A., Sahu, B., Das, M. N., Nayak, D. S. K., Sahoo, G., & Kant, S. (2023). En-MinWhale: An ensemble approach based on MRMR and Whale optimization for Cancer diagnosis. *IEEE Access : Practical Innovations, Open Solutions*, *11*, 113526–113542. DOI: 10.1109/ACCESS.2023.3318261

Suchkova, A. A. (2021). International tourism and sustainable development challenges in Southern and Eastern Africa. In *E3S Web of Conferences* (Vol. 311, p. 01003). EDP Sciences. DOI: 10.1051/e3sconf/202131101003

Timothy, D. J. (2023). Africa's Heritage and Tourism. In *Cultural heritage and tourism in Africa* (pp. 1–28). Routledge.

KEY TERMS AND DEFINITIONS

Partnership of public and private (P-PP):: Collaborative agreements among government entities and private sector companies aimed at delivering public services or infrastructure, often leveraging resources, expertise, and investment from both parties to achieve mutual goals.

Development of tourism:: The process of enhancing and expanding tourism-related infrastructure, services, and activities within a destination, aimed at increasing visitor numbers, improving economic benefits, and enhancing the overall visitor experience.

Sustainability of tourism:: The practice of managing tourism in a way that meets the needs of present tourists and host regions while protecting and enhancing opportunities for the future. This encompasses economic viability, environmental protection, and social equity.

Composite Reliability (CR):: A measure of the internal consistency of a set of indicators or items that are used to assess a single construct, indicating how well the items measure the same underlying concept.

Average Variance Extracted (AVE):: A statistic that assesses the amount of variance captured by a construct in association to the amount of variance due to measurement error. An AVE of 0.5 or higher indicates that the construct explains a noteworthy amount of variance.

Mediation:: A statistical association where a third variable (mediator) explains the association among an independent variable and a dependent variable. It helps to understand the process through which one variable influences another.

Discriminant Validity: The degree to which a construct is truly distinct from other constructs. It is assessed by evaluating whether measures of different constructs are not noteworthy correlated.

Stakeholder Engagement:: The process of involving individuals, groups, or organizations that have an interest in or are influenced by a project or policy in decision-making and implementation, ensuring their perspectives and needs are considered.

Chapter 9
Sustainable Development of Mountain Tourism Destinations:
Sochi Post-Olympic Experience, Russia

Vera Vladimirovna Minenkova
https://orcid.org/0000-0003-4184-2522
Kuban State University, Russia

ABSTRACT

The purpose of the chapter is to define modern approaches and methods of sustainable development of mountain tourist destinations, and also to share Russian experience of sustainable tourism development on the example of post-Olympic Sochi (Russia). The chapter is presented in two important parts (theoretical and practical). The first one is dedicated to theory and methods: approaches and systems for sustainable tourism development in mountain tourist destinations in the world. The chapter outlines systems that define the criteria for sustainable tourism development, indicators of sustainable development for tourism destinations and also issues of particular concern to mountain environments. The second part explores successful Russian practices: experience of sustainable tourism development in post-Olympic Sochi (Russia). The study is devoted to the analysis of sustainable tourism development practices at the resorts of Krasnaya Polyana and Rosa Khutor.

1. INTRODUCTION

The most important area of innovative socially-oriented economic development of any country is the creation of conditions for improving the quality of life of its citizens, including through the development of recreation and tourism infrastructure, as well as ensuring the quality, accessibility and competitiveness of tourism services (Rassokhina, 2018). Considering the fact that today tourism has become one of the fastest growing sectors of the economies of many countries, the issue of sustainable tourism is becoming actual. All implemented tourism programs and projects must strictly observe the interests of all groups of the local population, lead to stable economic development of the region and have a system of measures to minimize environmental pollution. Sustainable tourism should fully take into account its current and

DOI: 10.4018/979-8-3693-8232-5.ch009

future economic, social and environmental impacts, based on the needs of consumers, the industry, the environment and host communities. (Minenkova, 2022)

Mountainous territories are the centers of the most important strategic interests of states. The unique natural and climatic conditions of mountainous territories have given rise to unique social, ethnic, cultural and economic traditions that have been formed over thousands of years. The problems of developing mountainous territories have different effects on the society and the environment of mountainous countries and require a special approach both for their preservation and for further development.

Mountainous regions are characterized by significant biocenotic, landscape and ethnogenetic diversity. Mountainous geoecosystems act as a repository of a significant part of biological diversity, as well as places of its active formation. At the same time, they are very vulnerable and sensitive to anthropogenic impact. As a result, the tourism development in mountain regions should take place only on the principles of biocentrism and sustainable development.

Sochi region in Russia is a unique destination. The combination of the Caucasus and the Black Sea gives the opportunity to develop winter recreation in the mountains and summer activities in the coastal zone in parallel. In 2014 it became world famous as the capital of the XXII Olympic Winter Games and the XI Paralympic Winter Games.

The victory of Sochi for the right to host the Olympic and Paralympic Games gave a powerful impetus to the development of the resort. Large-scale development of territories for the needs of the Olympic infrastructure assumed a significant increase in the number of tourists due to winter recreation, and in the summer months, traditionally to the warm sea. Ten years after the 2014 Games, many studies have been conducted on the positive and negative effects of this global event (Minenkova, 2019, Scharr, 2020).

The issue of complicating the relationship with the environment was raised at the very beginning of the Olympic project. Since a significant part of the Olympic infrastructure facilities was located in protected areas, Olimpstroy[1], together with ecologists, prepared and agreed with UNEP a set of compensation measures aimed at minimizing damage to ecosystems and expanding the recreational potential of the resort. And now with increasing recreational pressure and accessibility of many previously untouched areas, the likelihood of biodiversity destruction will increase.

All this explains the modern approaches to the management of mountain tourist destinations, based on the rules of harnessing biodiversity tourism and the application of the principles of regenerative conservation management

Based on the analysis of different systems and programs of sustainable development the chapter outlines those that define the criteria for sustainable tourism development, indicators of sustainable development for tourism destinations and issues of particular concern to mountain environments. The case study is devoted to the analysis of sustainable tourism development practices at the resorts of Krasnaya Polyana and Rosa Khutor.

Studying international experience and best practices in the field of sustainable development and sustainable tourism will allow the management of corporations and tourist destinations to avoid mistakes in the work on developing their own policies and programs for sustainable development, including defining the mission, vision, values and key areas of work in sustainable tourism.

2. LITERATURE REVIEW

The theoretical basis of the study was formed by concepts and ideas suggested by Russian and foreign academics about sustainable development and especially sustainable tourism: Drumm Andy & Alan Moore, 2005; Christ Costas, Oliver Hillel, Seleni Matus & Jamie Sweeting, 2003; Rob Harris, Tony Griffin & Peter Williams, 2002; Martha Honey, 2008; Martin Mowforth & Ian Munt, 2016; Gulyaev Vladimir, 2008, Rassokhina Tatiana, 2018; Alexandrova Anna, 2020, Konstantinidi Christophor, 2022 et al.

The materials of the European Union, Global Sustainable Tourism Council, Walking Mountain Science Centre, World Travel Organization, United Nations Environment Programme are also provide wide information in the field of sustainable tourism and were studied and used during the research.

The results of research by the Research Institute of Sustainable Tourism Development of Sochi State University and the ESG-laboratory of Moscow State University were also taken into account in the process of analyzing the theory and methods of sustainable tourism development.

The official information published by national and regional tourism authorities was supported by diversified materials and analytical data collected during field observations. The material for the analysis is also drawn from official reports by companies, organizations and management information base of Krasnaya Polyana and Rosa Khutor Resorts.

3. RESEARCH METHODS

Research methods were data collection methods, qualitative research methods, analytical and comparative methods. The study results are based on the analysis of primary data (gathered through surveys, interviews) and secondary data (relevant published papers, scientific literature, data from the Ministry of Resorts, Tourism and Olympic Heritage of Krasnodar region, Sochi municipal administration, Krasnaya Polyana and Rosa Khutor Resorts). The study is also based on observation and records made in course of field trips during twenty last year, which allow to make a confident conclusion about the rapid development and implementation of a policy of sustainable development of mountain tourist destinations in Sochi.

4. THEORY AND METHODS: APPROACHES AND SYSTEMS FOR SUSTAINABLE TOURISM DEVELOPMENT IN MOUNTAIN TOURIST DESTINATIONS

4.1. Concept of Sustainable Tourism

For more than 30 years in the global community, sustainable development issues have played an important role in the decision-making process regarding the planning and development of the tourism industry. Despite this, research in the field of sustainable tourism is still in its early stages of development, especially in terms of strategic management and defining criteria for the sustainability of tourism destinations.

The modern understanding of the sustainability of territorial development is associated with the assessment of the economic, social and environmental consequences of certain types of human economic activity. Tourism as a complex type of economic activity fully corresponds to the essence of this paradigm. Sustainable tourism must take full account of its current and future economic, social and environmental impacts, considering the needs of consumers, the industry, the environment and host communities. In addition, when talking about the sustainability of any type of activity, it is important to take into account its development trends, as well as predicting development directions.

One of the main goals of sustainable tourism is to minimize the negative factors caused by growing demand, and as a consequence, the impact of tourism on tourist destinations (the impact of tourism and recreational activities on the environment, regional economic development, socio-cultural impact (Figure 1).

Figure 1. Elements of a sustainable tourism system

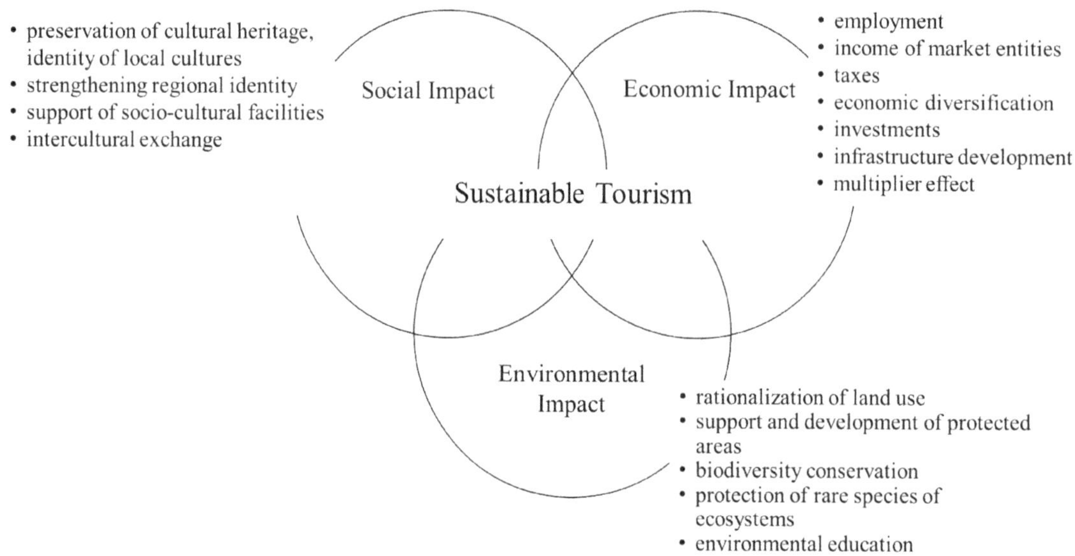

(Mamonova et al.,2023)

Talking about the tourism impact on a territory, it is important to note both positive and negative aspects. So, if we talk about the environmental component, among the positive factors we should note the development of previously untapped natural complexes, increasing the level of environmental culture, promoting the financing of national parks, reserves and natural areas (which often have the main attraction for tourists), monitoring and containing various types of pollution, and as a consequence – improvement of the environment state and its cleaning, determination of norms of anthropogenic load on natural complexes in the process of protecting and transforming recreational areas, etc. However, the obvious fact is that the negative environmental consequences are much more global: excess recreational loads lead to pollution of surface and groundwater, air, irrational environmental management, which lead to the degradation of natural resources and loss of biodiversity, changes and sometimes destruction of ecosystems.

In terms of the economic impact of tourism activities, both positive and negative aspects can also be noted. Creating jobs, increasing the level of income of the local population, attracting foreign investment, increasing tax revenues, developing the local economy by spending tourists' money on accommodation services, food, entertainment and the purchase of various goods in the territory of temporary stay have a positive economic effect. On the other hand, countries whose main branch of specialization is tourism become too dependent on this type of activity due to monofunctionality. Thus, a sharp decline in tourist flow (as, for example, occurred during the COVID-19 pandemic in 2019-2021) can lead to high unemployment and a deterioration in the economic situation as a whole.

As for the positive impact on the social sphere, we can note the exchange of cultural values and the improvement of intercultural communication due to the acquaintance of tourists with the culture, customs and traditions of the population of the host country, which promote the development of tolerance, mutual understanding and tolerance. In the absence of a long-term science-based concept for the development of tourism and its full commercialization, the negative social consequences of tourism activities are becoming more and more obvious: changes in the habitual way of life and traditions of the population, the psychology of consumption of material goods, an increase in the level of crime, drug addiction, the introduction of bad habits and an unusual way of life, a change in the moral guidelines of young people, the use of tourism as a means of illegal migration of the population, etc.

It is important to note that within the framework of mass tourism, both the positive and negative impacts of tourism on a tourist destination take on a more global scale.

4.2. Sustainable development of mountain tourism destinations: basic approaches

These approaches are fully relevant to the development of mountain tourism destinations. Mountains are a valuable natural asset that is actively and widely used in the tourism industry. Agenda 21, adopted by the United Nations Conference on Environment and Development (Rio de Janeiro, 1992), includes Chapter 13, “Managing fragile ecosystems: sustainable mountain development” (the second section of the program), which emphasizes the value of mountain ecosystems, their invaluable role in the life of the planet, the complexities of the development of local communities. In 2012 in Rio de Janeiro the document “The Future We Want” (UN Conference on Sustainable Development Rio+20) was adopted, and mountains were specifically mentioned there. Sustainable Development Goal No.15 identifies the need to protect and restore terrestrial ecosystems and promote their sustainable use, sustainable forestry, combat desertification, reverse land degradation and halt biodiversity loss. The Goal No. 15 explicitly mention mountains among the ecosystems that need to be conserved, restored and used sustainably in accordance with international agreements, and it calls for the conservation of mountain ecosystems, including their biodiversity, by 2030 to enhance their capacity to provide benefits necessary for sustainable development.

Many authors consider tourism as an instrument of the socio-economic development of mountain areas. At the same time, in the context of modern approaches to the sustainable development of tourism and taking into account the extreme susceptibility of mountain ecosystems to disruption of the ecological balance as a result of anthropogenic impact and natural processes, the question of applying the principles of sustainable development in the activities of mountain tourism destinations arises.

Approaches to the sustainable development of mountain tourism and recreational systems are based on the following positions (Minenkova, 2023):

- the concept of sustainable development is an important political basis for tourism activities;
- traditions of environmental protection as a limiting factor for the development of tourism and its differentiation intensity;

– transboundary cooperation in the development of mountain territorial and recreational systems is a tool for promoting the principles of sustainability;
– the presence of supranational institutions is a tool for integrating efforts and exchanging experience in the development of destinations;
– the presence of a destination management system represented by a territory management organization;
– involvement and orientation towards stakeholders;
– monitoring and formation of consumer behavior patterns;
– formation, promotion, sale and consumption of sustainable tourism products;
– environmental education of target audiences and the population as a whole;
– development and application of indicators for sustainable tourism and sustainable development in general.

4.3. List of criteria for sustainable tourism development in mountain areas

The difficulty of determining a list of criteria for sustainable tourism development is related to their multiplicity and diversity. There is no generally accepted global system for measuring the sustainability of tourism destinations in general. The general approach is accepted that such systems should take into account three main dimensions – environmental, economic and social, while the indicators of these measurement vectors are not specified.

Currently, the following systems that define the criteria for sustainable tourism development are widely known:

– System of the UN Commission on Sustainable Development (1996, 132 indicators);
– System of the World Tourism Organization (2004, 13 groups, more than 700 indicators);
– Global criteria for sustainable tourism (GSTC, 2012), later supplemented with criteria for destinations (GSTC-D), 4 sections, 174 indicators);
– the European Tourism Indicator System (ETIS toolkit for sustainable destination management, 2013, 43 main indicators + additional);
– Protocol on the implementation of the Alpine Convention (1991) in the field of tourism;
– Others.

Indicators of Sustainable Development for Tourism Destinations (Guidebook, UNWTO) identify a number of special criteria for the sustainable development of mountain regions in the process of recreational use. Issues of particular concern to mountain environments include the following:

1. Loss or degradation of flora and fauna due to tourism activity (counts for key species, road kill counts by species, frequency of sightings of key species, area protected).
2. Physical erosion caused by infrastructure construction and impact of tourist use including cutting and use of trails, off road vehicle activities (share of surface in eroded state, share of surface without tree or shrub cover, turbidity readings from streams).
3. Visual pollution/aesthetics due to construction, extractive activity, deforestation (share of visible slopes without tree or shrub cover, tourist opinion of state of vistas).

4. Access, particularly to fragile sites, protected areas (cost of entry or transport access expressed in hours work at local wage, perception of ease of access to key sites, share of nationals (or regional residents for larger nations) who have visited the destination in the past year, traffic levels, price of real estate in destination).
5. Management of solid waste / garbage (waste volume produced by the destination (tonnes) by month, visitor perception of level of litter, local perception of level of litter (attributable to tourists).
6. Impacts of activities on water quality/watershed management (share of local streams, lakes which are contaminated by sewage, share of sewage from key sites which receives treatment to meet standards, share of area reserved as protected watersheds).
7. Impact on small, and/or culturally distinctive communities (for small and traditional communities, ratio of tourists to locals, share of hunting or fishing in destination by locals, tourists, local satisfaction with tourism.
8. Economic benefits (local employment in tourism and protection, employment in tourism and protection as share of all employment.
9. Seasonality (share of total tourism which occurs in peak month (or season), share of employment in tourism which is full-time/full year, share of business establishments open all year (accommodation and services).

The National Ski Areas Association monitors ski areas in the United States according the Sustainable Slopes Program, which identifies key issues and environmental initiatives for the US ski industry. Founded in 2000, Sustainable Slopes has provided one more way for ski resorts to follow sustainability in their operations. More than 200 US ski resorts have joined the National Ski Areas Association's slope sustainability program. They demonstrate their commitment to sustainability across 10 thematic areas: climate change, energy, waste, transportation, education, supply chain, water, construction and design, forests health and habitats.

In addition to creating comprehensive systems of criteria for sustainable tourism development, standards for sustainable tourism development are being developed. In the context of the research topic, it is appropriate to cite the example of the IDEAL standard (Mountain IDEAL Standard: Guiding Principles for Sustainable Destination Development), developed for mountain areas and communities and offering guidelines for the sustainable development of tourist destinations.

Mountain IDEAL is the Western world's recognized sustainability standard and certification program for mountain resorts. The standard is a system of sustainable tourism criteria and indicators that help any mountain destination improve its sustainability performance, promote stakeholder engagement, collaboration and recognition as a global leader. The standard is scientifically based and takes into account the values of sustainable development. Sustainable mountain resort communities actively practice and promote the IDEAL Code of Ethic & Business Conduct to guests and community members.

The Standard includes criteria based on many other generally accepted principles of sustainable development, including those developed with the participation of the UN World Tourism Organization. The criteria within the standard provide a framework used for education and informing, economic development, policy development, action planning, measurement, assessment of sustainable development, and as the basis for expert certification of a destination. The standard's criteria are divided into four groups:

- sustainable tourism management and monitoring;
- destination planning and asset protection;

– community engagement, culture, and social equity;
– environmental resources, energy and water conservation, and waste reduction.

It should be noted that the criteria of the Mountain IDEAL standard are generally descriptive of the areas of activity of enterprises, consumers and other stakeholders. The standard does not offer criteria in evaluation formats.

5. SUCCESSFUL RUSSIAN PRACTICES: EXPERIENCE OF SUSTAINABLE TOURISM DEVELOPMENT IN POST-OLYMPIC SOCHI (RUSSIA)

5.1. Winter Olympic and Paralympic Games in Sochi-2014

The Russian tourism business is only at the beginning of the formation of a full-fledged system for the development of sustainable tourism. A demonstrative example among enterprises operating in accordance with the concept of sustainable development in mountain tourist destinations is the Krasnaya Polyana alpine tourism cluster (Sochi, Russia). Through its activities, the company demonstrates that business success may not fully contradict nature conservation, but rather create a synergistic effect.

The Winter Olympic and Paralympic Games in Sochi in 2014 had two global goals:

– give impetus to the development of the territory, create modern infrastructure and turn Sochi into famous competitors with winter resorts;
– present to the world a new image of Russia as an open, modern and adaptive country.

The Games were intended to be a catalyst for the development of Sochi as a winter resort and together with its reputation as a summer holiday destination, transform the area into a year-round tourist destination. It was this sporting event that became the starting point for the rebirth of the resort. In fact, holding the Olympic Games in Sochi made it possible to develop a new city. (Alexandrova A.Y. et al., 2019)

Important elements of the post-Olympic heritage were the development of transport, social, tourism infrastructure, and energy supply systems. At the same time, a territorial functional transformation took place: in fact, new resort areas were created in the coastal and mountain clusters.

Krasnaya Polyana alpine tourism cluster of resort city Sochi includes three mountain resorts: Krasnaya Polyana Resort, Rosa Khutor Resort, Gazprom Resort (Laura and Alpika-Service ski areas). Characteristics of the Resorts are presented in Table 1. It takes less than an hour to reach the resorts for tourists arriving at the international airport Sochi or at the railway station Adler. Today, this distance is covered with great comfort by road or rail, which form a single ultra-modern transport corridor created during the preparations for the Winter Games in 2014.

Table 1. Characteristics of the Resorts of Krasnaya Polyana alpine tourism cluster

Resort	Maximum height, meters	Number of lifts	Number/length of pistes, km	Number/length of ecological trails, km
Krasnaya Polyana	2375	15	30 / 30	16 / 60
Rosa Khutor	2320	32	52 / 105	19 / 100
Laura	1660	13	30 / 26	-
Alpika-Service	2256	5	11/ 12,5	-

5.2. Krasnaya Polyana Resort's sustainable development experience (the following information is based on data published in the Official Report of the Krasnaya Polyana Resort in open information resources)

Non-commercial joint-stock company "Krasnaya Polyana" manages one of the leading mountain resorts in Russia and carries out the entire range of economic activities: from the operation of cable lifts and ski slopes to the management of hotels and restaurants. The Resort's activities help to maintain the health of the russian population, expand people's access to sustainable tourism products, contribute to the preservation of natural heritage and the growth of socio-economic well-being of the Krasnodar region, and also create conditions for decent, effective work and successful entrepreneurship.

Every year, the Krasnaya Polyana Resort publishes a non-financial report on its activities in the field of sustainable development, which demonstrates the company's contribution to environmental protection, rational use of natural recreational resources, economic development of the region, and the well-being of guests, employees and local residents.

Priorities in the field of sustainable development of the Krasnaya Polyana Resort:

– constant increase in the degree of satisfaction with the company's services from consumers. The Resort aims to providing high-quality, safe and affordable services that meet the needs and expectations of guests, as well as taking into account their feedback and suggestions when improving its services;
– development of partnerships with stakeholders based on transparency, trust, respect and interests of sustainable development of the company;
– increasing the environmental responsibility of company employees and stakeholders. The Resort is guided by environmental principles in its activities, and also encourages responsible behavior of its employees, clients, partners and suppliers;
– educational activities in the field of respect for the environment and sustainable development. The Resort informs stakeholders about its achievements and plans in the field of sustainable development, and also conducts educational events for different groups of the population.

In the process of developing the Strategy for Sustainable Development of the Krasnaya Polyana Resort, company managers, representatives of the Sochi city administration, institutions managing specially protected natural areas, local residents and entrepreneurs took part. The Strategy is based on the principles of a harmonious combination of economic growth, social well-being and environmental safety of Krasnaya Polyana. To achieve them, the following priority areas were formulated:

– improving the quality of the population life;
– development of tourism infrastructure and services;
– conservation and restoration of ecosystems and biodiversity;
– increasing the efficiency of use of land, water and energy resources;
– strengthening the Resort's position as a leader in the field of sustainable development at the regional and international levels.

The Sustainable Development Strategy of the Krasnaya Polyana Resort is based on global goals in the field of sustainable development, to the achievement of which the Resort can make a significant contribution: 11 of the 17 goals of Agenda XXI are reflected in the sustainable development practices of the Resort (including five global goals where the Resort can make the most significant contribution, and six goals towards which the Resort can indirectly influence progress).

Currently, 5 areas of sustainable development of the Krasnaya Polyana Resort have been identified, they reflect the company's values and priorities. Each direction includes several goals and indicators that are planned to be achieved by 2025.

5.2.1. Direction 1. Sustainable tourism.

The content of the direction is related to the involvement green hotels (that have passed environmental certification) in the service of guests, offering active and environmentally friendly forms of tourism activities (development of a system of eco-trails), improving the health of guests, developing children's activities to study nature and history. The Resort strives to meet its needs for goods, works and services, taking into account not only economic benefits, but also their impact on the environment throughout their entire life cycle. This goal is formulated in the Policy of the Ecologically Responsible Purchasing (accepted in 2021). According to the accepted approach, in order to reduce the environmental impact of the Resort activities, purchasing initiators should give preference to goods that have environmental benefits. The document's requirements apply to packaging, composition and the share of products in the Resort's total purchases. When choosing products, preference is given to local products, products with Type 1 eco-labels and fully recyclable packaging, as well as products supplied in returnable containers.

By 2023, three of the nine hotels at the Krasnaya Polyana Resort have passed an independent environmental assessment and joined the international environmental program "Green Key". By 2025, in accordance with the Sustainable Development Strategy, all Resort hotels should receive Green Keys. Hotels that have passed eco-certification meet all mandatory criteria and most of the recommended ones.

Guests are informed about ongoing initiatives to reduce their environmental impact on the Resort's official website; each hotel's page includes a "Green Way" block.

In addition, the Krasnaya Polyana Resort provides guests with the opportunity to contribute to the conservation of the planet's resources. Information on how hotel guests can make their holiday even more environmentally friendly is posted in each room.

Guests of the Krasnaya Polyana Resort can support the company's initiatives to implement the principles of sustainable development by following simple recommendations:

1. Respect and caring attitude to the planet's resources
   - – economical operation of electrical appliances;
   - – refusal of passive mode of operation of power supplies;

– saving water during hygiene procedures;
– fully load the dishwasher for each cycle;
– using a shower instead of a bath;
– maximum rejection of disposable items.

2. Nature conservation
   – preservation of natural landscapes; tourists must travel only along routes that are open to the public, without going beyond the designated trail;
   – observation of nature without interfering with it; it is better to give preference to photographs;
   – refusal to eat wild fruits and berries; plants are the basis of the diet of most wild animals;
   – respect for the habitat of animals, do not disturb or chase them;
   – a trip without waste – for a picnic it is recommended to use reusable dishes.
3. Eco-friendly holiday
   – spend more time outdoors;
   – choose environmentally friendly modes of transport: cable cars, bicycles, scooters;
   – follow the rules for visiting the eco-trails of the Resort.
4. Support of local traditions
   – choose local products: fruits, tea, honey, nuts and much more are presented in the Resort shops;
   – buy locally made souvenirs in hotel stores and at the fair of local artists and craftsmen;
   – study the history and culture of Krasnaya Polyana on excursions with guides from the Adventure Bureau and at photo exhibitions in the park "Seasons";
   – learn about the nature of the Caucasus from more than a hundred informative stands on the eco-trails of the Resort;
   – purchase popular science publications about plants and animals of the region, published with the support of the Resort;
   – visit cultural events of the Resort: film screenings, performances, holidays and festivals.

From 2021, 100% of paper products are sourced from sustainable forestry. 99% of cleaning and dishwashing products are environmentally friendly, and all textiles used in the Resort also meet environmental standards.

The Resort has refused the use of most single-use plastic items in all Resort hotels and apartments, replacing them with eco-friendly alternatives. An example would be replacing plastic laundry bags with reusable 100% cotton bags. Portioned sugar is now in paper packaging, straws for juices and cocktails, kitchen stuffs and hygiene items have also been replaced. Most Resort hotels have switched to using reusable shampoo and shower gel dispensers, eliminating single-use plastic containers. Next up is giving up plastic water bottles. In 2023, 62 purifiers were installed in the Resort hotels.

The Policy of the Ecologically Responsible Purchasing contains requirements for the proportion of purchased LED lighting products to be at least 99% of the total number. Energy efficiency class requirements (A+) apply to all purchased equipment. The Policy also includes requirements for the purchase of water-saving plumbing: dishwashers should consume no more than 3.5 liters per basket, faucets – no more than 5 liters per minute, shower heads – no more than 9 liters per minute, toilets – no more than 3 liters for half flush and no more than 6 liters for a complete flush, etc.

In 2023, the Resort launched the tour package offer “Fill Your Summer” – the first mass tourism product on the Russian market, created taking into account the principles of sustainable tourism. Fans of winter sports and active tourism were joined by consumers of comfortable infrastructure, a healthy lifestyle, and adherents of a conscious approach to travel. There are deep meanings behind the sentence “Fill your summer”. This is not just an opportunity to relax in harmony with nature, recover in the atmosphere of the Krasnaya Polyana mountains and gain new impressions. It was created for those who share the philosophy of conscious consumption, respect for nature and care for the environment. The beauty of the forest, silence, fresh mountain air, comfortable infrastructure provide the relaxation where attention is directed to yourself and loved ones, to unity with nature. The «Fill Your Summer» product is created for those who prefer to travel environmentally, but not at the expense of their own comfort.

5.2.2. Direction 2. Nature conservation.

This direction includes sections related to minimal environmental impact, biodiversity conservation, and environmental education.

The development of the Resort plans is carried out taking into account existing climatic and environmental risks. A mandatory stage in the development and implementation of all projects is a risk pre-analysis, impact assessment and potential consequences. When significant risks are identified, measures are developed to reduce them; in addition, the “zero option” is always considered, i.e. complete refusal to implement the project.

The conservation of mountain forests plays a critical role in solving problems associated with climate change. The Resort pays special attention to ensuring fire and sanitary safety in forests, coordinating this work with specialists from the Sochi National Park. These measures, as well as environmental protection and adaptation measures developed jointly with the Sochi National Park and the Subtropical Scientific Center of the Russian Academy of Sciences, help to minimize the negative consequences of climate change for the destination.

The Krasnaya Polyana Resort is the only tourist destination in Russia that calculates its carbon footprint based on direct emissions from operating activities (scope 1) and indirect emissions from the production of consumed electricity (scope 2). According to the company's reports, both types of emissions have increased, and the fact that this indicator is taken into account and controlled is important. It was not possible to calculate other indirect emissions (scope 3) due to the lack of data on guest movements to and from the Resort.

As part of an agreement with the Ministry of Natural Resources and Ecology of the Russian Federation, the Resort joined the initiative “Business and Biodiversity” of the federal project “Preservation of Biological Diversity and Development of Eco-Tourism”. With the financial support of the Resort, the creation of a new visual control post began at the Leopard Breeding and Rehabilitation Center of the Sochi National Park. Since 2009, work has been underway in the Sochi National Park to breed animals and prepare them for life in the wild, far from people and resorts.

Work is being carried out jointly with scientists from the Sochi National Park to study and preserve the flora and fauna of the Resort area, which has rich biological diversity. More than 1,000 species of higher plants grow on the territory of the Resort, of which about 80 species are included in the Red Books of the Russian Federation and the Krasnodar region. The natural landscapes of Mount Aibga serve as a reliable refuge for numerous species of animals. More than 100 species are rare and are included in the Red Books. The fauna of the Resort territory includes 48 species of mammals, 234 species of birds,

13 species of reptiles, 9 species of amphibians, 10 species of fish, 1 species of agnathans, 39 species of mollusks, more than 1000 species of insects and 9 species of centipedes.

In 2022, as a part of the implementation of an agreed program of biotechnical measures, 6 salt licks were laid out on the territory of the Resort for mineral feeding of ungulates. The objective of the program is to compensate for the decrease in the natural capacity of the territory for animals as a result of recreational loads, to maintain the sustainability of populations of ungulates, bats and cavity-nesting birds without disrupting biogeocenotic connections, as well as to create conditions for observing wild animals in their natural habitat. Research is being carried out on the key components of biodiversity: route surveys of rare and endemic species.

Work on environmental education of Resort guests is carried out as part of the implementation of the "Green Way" sustainable development strategy. Educational initiatives have maximum coverage, because educational content is integrated into all areas of the Resort's activities. The Resort supports the publication of popular science books. In 2022, the Resort held 14 environmental education events, which were attended by more than 1,500 people.

5.2.3. Direction 3. Responsible business.

The implementation of the responsible business strategy is related to the topics of energy efficiency, water conservation, waste management and the overall implementation of the system "Smart Resort".

The Krasnaya Polyana Resort is gradually developing its digital infrastructure, which makes it possible to increase safety and operational efficiency, as well as make management decisions based on big data analysis. Automated monitoring provides operational control over all processes and helps reduce resource consumption, reduce waste generation and make the daily operation of the entire Resort more environmentally friendly.

Digital transformation has affected almost all operational processes at the Resort: electronic document management and project management system, online system for corporate training, personal data protection, electronic coupon book, electronic tickets, loyalty program, medical information system, saving energy resources, rational consumption of water resources, responsible and safe waste management.

The infrastructure of the ski complex is also involved in measures to improve the energy efficiency of the Resort. During the off-season, when the number of guests at the Resort is lower than average, the cable cars have introduced a practice of reducing the speed of movement, which helps reduce the level of electricity consumption. In order to increase energy efficiency and reduce levels of light pollution in the natural areas of the Resort, the lighting schedule for the cable car supports has been revised. Previously, the lighting remained on throughout the dark time of the day, but now the lighting turns off 30 minutes after the end of the cable car operation.

Electricity consumption limits were set for all existing Resort infrastructure facilities. Each case of significant excess of daily electricity consumption is studied by employees of the Operations Directorate in order to establish the reasons for the overconsumption and take the necessary measures. In order to further reduce energy consumption and evaluate the implemented measures approved by the Resort strategy, an automated energy metering control system was introduced at all Resort facilities, which in the future will make it possible to develop the most effective tools for operating the Resort's engineering and public infrastructure facilities. The Resort intends to consistently reduce energy consumption from non-renewable sources, including through the gradual replacement of special equipment with electric analogues.

A water management system operates at all Resort facilities. To ensure the operation of the artificial snowmaking system, the Resort takes water from surface sources exclusively within the framework of current legislation – on the basis of water use agreements. Water intake does not have a significant impact on the environment. The rest of the water comes from the municipal water supply system, which also carries out wastewater disposal from the Resort's facilities. The reduction in water consumption in hotels is the result of many Resort guests who refuse to change towels, bed linen and room cleaning every day for environmental reasons. The Resort plans to continue to take measures to reduce water consumption: begin collecting rainwater for watering park areas, strengthen control over the rational consumption of water by tenants, etc.

Another important task is to improve the quality of wastewater discharged to treatment plants. In accordance with legal requirements, the Resort measures and monitors wastewater quality and assesses its compliance with regulatory requirements at all of its facilities. The share of wastewater that is normatively clean without treatment and wastewater purified at wastewater treatment plants in the total volume of water discharged by the Resort into surface water bodies is 100%.

In accordance with its environmental policy, Krasnaya Polyana Resort strives to reduce waste generation and ensure responsible and safe management of waste. The basic mass of the Resort's waste is represented by solid municipal waste (most of which is food), as well as construction and repair waste. The generation of waste is directly related to the main activities of the Resort – hotel and restaurant business, as well as the operation of the property complex.

The Resort has organized separate collection of waste of hazard classes I-V, which allows reducing the volume of waste sent for disposal. Waste prohibited for disposal is transferred for disposal or neutralization. The Resort is consistently increases the volume of secondary material resources that it transfers for processing.

5.2.4. Direction 4: Employee well-being.

The Resort's Personnel Management Policy is aimed at providing decent and safe working conditions, financial rewards directly related to the results and quality of work, opportunities for professional growth and development, democratic principles and transparency of company management procedures. The Resort is the largest employer in Sochi (2,664 people in 2022) and an attractive place of work for specialists from other regions of Russia.

Providing decent working conditions, the company regularly identifies hazards and assesses professional risks in the workplace with the involvement of specialized organizations. The main risk management activities include: developing labor protection instructions, training employees, providing them with personal protective equipment, organizing workplaces in accordance with the Labor legislation of the Russian Federation. The company also conducts a special assessment of working conditions with the involvement of accredited organizations. All new employees are familiarized with the occupational health management system upon hiring. Training in occupational safety, first aid and the use of personal protective equipment is provided during induction and regular on-the-job training. Special training, including training when performing high-risk work, is carried out in accredited training centers.

The company's goal is to create equal opportunities for professional growth and development of its employees. In 2022, the company provided 33,694 hours of training. A training book has been developed and successfully implemented – now everyone can easily select a topic of interest and register for training via electronic registration or with the help of their supervisor. The book presents key programs and trainings for employees of various services and levels of training, such as basic service standards, reservations, reception services, housekeeping and restaurant services, medical and spa centers, concierge service, security, management and leadership, design thinking, welcoming sales, presentations, emotional intelligence, stress tolerance and others. The training programs include training on sustainable development and environmental responsibility.

The Resort has a strong focus on diversity and inclusion. The company recognizes the importance of having a diverse workforce and believes this approach promotes better decision making, creativity and innovation.

The company is actively involved in training personnel for the tourism and hospitality industry, cooperates with educational organizations and actively promotes the career development of graduates.

The corporate culture of the company can be assessed as highly developed, contributing to the achievement of the goals of the Strategy and Policy. To increase staff motivation, many benefits are provided (housing for nonresident employees, transportation to work, free meals for hotel workers, uniforms from leading ski clothing manufacturers for instructors, corporate events, discounts on the Resort and partner services, and much more). In 2022, construction of a new campus for the Resort employees with an area of 7.5 thousand square meters began on the territory of the Krasnaya Polyana village.

5.2.5. Direction 5. Development of Krasnaya Polyana.

The objectives of the direction are related to the principles of social responsibility, preservation of historical and cultural heritage and partnership for sustainable development.

Corporate volunteering develops rapidly at the Resort. Since 2021, company employees have conducted several volunteer trips to the "Laura" cordon of the Caucasian Nature Reserve and to the "Southern Cultures" park, took part in collecting raw materials at the Krasnaya Polyana cosmetics farm, a collection point for recyclable materials for employees continues to operate at the Resort. In total, more than 10 volunteer events were held in 2022, in which more than 150 employees took part.

The company provides assistance to the Center for Adaptive Physical Culture, Sports and Tourism "Energy of Life". Every winter season, ski classes for recreation, rehabilitation and adaptation of children with special needs are held on the slopes of the Resort.

The company's activities are closely connected with the life of local communities. The Resort provides preferential rates for its services for residents of Sochi and the South of Russia, provides support to a kindergarten and secondary school in the village of Krasnaya Polyana, the company provides free infrastructure for training and competitions to young athletes from local sport sections, provides premises for local events, and gives local entrepreneurs the opportunity to sell products of their own production during regular weekend fairs.

In terms of preserving cultural heritage, the Resort holds historical exhibitions ("Pages of the History of Krasnaya Polyana", etc.). Preserving the history and traditions of the peoples who inhabited this territory at different times, incl. culinary traditions, the Resort strives to become a center of gastronomic tourism in the region and focuses on the authentic mountain cuisine of the Western Caucasus.

In conclusion, the mission of the Krasnaya Polyana Resort is to create a mountain resort in Russia with care for everyone, giving unique emotions and impressions to guests, while preserving stunning nature. The goals and objectives of the company's Sustainable Development Strategy contribute to the implementation of this mission.

5.3. Rosa Khutor Resort's sustainable development experience (the following information is based on data published in the Sustainability Report of the Rosa Khutor Resort in open information resources)

The sustainable development strategy and the mid-range action plan of Rosa Khutor Resort were made public in 2018 at first time. The first edition of the Rosa Khutor Sustainable Development Program was rolled out for the period 2018–2020 and described the Resort's mission, vision, values and sustainability priorities.

The Rosa Khutor Resort's values are:

– experience – it works hard to provide guests with unforgettable experience of mountain trips;
– leadership – it strives to become the best and is aimed on improving the quality of services, guests' satisfaction, safety and efficiency of it's operations;
– partnership – it succeeds by teamwork, developing and supporting cooperation and mutual support spirit;
– legacy – management and employees value and cherish what they have, and work today to ensure that the fruits of labor serve people for a long time to come.

There are seven key areas of the work within the Program of sustainable development.

5.3.1. Great Experience.

The Rosa Khutor Ski Resort («The Mountain» in common parlance) is the main factor of the Resort's development, a key component of the tourist experience of winter guests and the main concern of a large part of the team all year round. Ski resort complex employs over 700 people working 24/7 in the winter season. During winter more than 900 000 people climb the slopes of Rosa Khutor annually, and at peak season there are more than 10 000 skiers and snowboarders on the Mountain.

Rosa Khutor Ski Complex is operated by people who are passionate about mountains and their business. The Resort keeps raising the bar of requirements toward pistes' and slopes' quality. Every day the same passion drives travelers there – the mountains are calling. The Resort gives the tourists access and support, and they need to get and take home the maximum of positive emotions.

Thus, safety becomes the common point of effort for both the team of the Mountain and the tourists. High-tech infrastructure and professionalism of the team are dedicated to this cause.

Together with other mountain resorts of Krasnaya Polyana alpine tourism cluster Rosa Khutor (alpine tourism centre «Gazprom» and Krasnaya Polyana resort) are united by an agreement to introduce a single ski pass. A single ski pass opens up the opportunity of skiing on the pistes of all three resorts, with the total length of 180 km.

Rosa Khutor Resort has been all-season since 2013. Regular research of expectations and collection of guests' feedback shows that first and foremost tourists come to us to meet the majestic nature, enjoy the views, atmosphere and serene peace of the mountains. The demand for adventure and hiking alpine trails is growing along with the audience of travelers who love nature and strive for an active healthy lifestyle. At the same time, travelers demand affordable and quality service, count on a comfortable infrastructure and look for opportunities to enjoy their travel time.

Rosa Khutor together with the management of specially protected natural areas of federal importance is recreating a unified network of travel routes equipped with infrastructure for comfortable and safe travel to natural landmarks.

In the summer Rosa Khutor offers 19 edutainment adventure travel routes with the length of 100 km. Qualified guides of the Resort accompany travelers on adventure travel edutainment trails. The Resort offers guests to visit the signature Black Sea beach in the Olympic Park, whose cleanliness is recognized by the prestigious Blue Flag mark from the Federation for Environmental Education (FEE). The Resort organizes transfers to and from the beach. Roza Khutor offers a great number of summer activities: Rosa Wake Park, Yeti Park (the highest mountain family amusement park in Russia), alpine coaster with an exciting downhill run on an equipped sled track, rental of electric and traditional bicycles, scooters and other equipment, as well as trekking shoes and poles, and hiking backpacks and plaids, archaeological museum with an interactive exhibition, Eagle Flight trolley slide, summer tubing, sports grounds, trampoline arena, work-outs and play sports and many other options for having fun and healthy time.

5.3.2. Caring for nature.

Impressive alpine nature is a key asset and the basis of Rosa Khutor Resort sustainable development. Rosa Khutor team ensures impeccable compliance with legislative requirements on nature protection and, not limiting to it, goes further to study and conserve natural treasures of the Western Caucasus for the present and future generations.

Since 2006, Rosa Khutor Resort has supported the practice of using the Aibga Ridge (where the Resort is situated) as a research station for Sochi National Park, strengthening the research tradition. The Resort carries out independent work to study and support the welfare of wild animals inhabiting the territory and vicinity of the resort. For this purpose, a network of feeding stations and salt-bearing rocks is deployed, which are a must among ungulates. There are camera traps at feeding stations and along animal migration routes, obtaining data on the species diversity of animals, sex and age composition of their populations, behavior and habits of animals with their help. Photo traps on the territory of the Resort snap noble Caucasian deer, Caucasian chamois, roe deer, wild boars, bears, jackals, foxes, martens, badgers and raccoons.

Any construction or excavation work at Rosa Khutor is always preceded by a number of measures aimed at protecting rare and specially protected plants and animals. These measures are carried out by the Resort team under strict control of the state environmental services and the Sochi National Park Directorate.

Surveys determining the species composition, habitats and number of plants and animals to be replanted or relocated are carried out before the work begins. At the same time an appropriate site located in the appropriate climate and landscape zone is identified. Subsequently, the proposed relocation of the plants and/or animals is then submitted to the regulatory agencies. Replanting of plants, capture and relocation of animals takes place only after obtaining a permit. A report on the results is also submitted to the regulatory services.

In accordance with the requirements of forest legislation, the resort provided compensatory planting of forests to replace those cut down for linear infrastructure (22 000 specimens of forest trees of various species were planted in the specially allotted plots).

According to the plans agreed with the Federal Agency for Fisheries, the Resort ensures the implementation of measures for the conservation of aquatic biological resources. In 2018-2020, 27 426 juvenile fish produced by specialized fish farms were released into the Mzymta River and other water bodies of Krasnodar Region.

Rosa Khutor Resort supports the environmental activities of Sochi National Park, providing regular transfers significant amounts, formed by allocating a share in the income from the sale of ski-passes to its directorate.

The Resort team is ready to expand support measures for the directorates of Sochi National Park and the Caucasian Nature Reserve in developing and implementing comprehensive programs aimed at studying and preserving the nature of the entire Krasnaya Polyana alpine tourism cluster.

5.3.3. Partnership and Cooperation

The entire Krasnaya Polyana alpine tourism cluster of the Sochi resort city is the example of successful Olympic legacy. It made a breakthrough in development during the preparation for the Winter Games of 2014.

At the end of 2020 Rosa Khutor Resort supported the initiative of the Sochi City Administration and became a party to a multilateral agreement between the Krasnaya Polyana cluster of alpine resort directorates and specially protected natural areas on cooperation to popularize and develop this outstanding year-round mountain tourism destination.

The Resort develops partnerships in the travel market, forming favorable conditions for cooperation for 30 travel companies, 50 tour, companies, more than 500 hotels, health resorts and guest houses of the Bigger Sochi.

Since 2016 the Resort implements strong standards of hospitality, which determine the consistently high quality of Rosa Khutor services. Rosa Khutor is a large-scale, but an affordable and convenient alpine resort. Communication with guests helps to navigate, plan and organize vacations without wasting time and energy, learn about events taking place at the Resort, promptly deal with unforeseen problems.

There are 3 information center desks, 24/7 call center, screens at the ticket offices and next to the main lift stations informing about open pistes, current and forecasted weather conditions, avalanche danger, 2 information channels on TV screens in the rooms of the resort hotels, mobile application, infobots in popular messengers, direct dialogue in social networks (total number of subscribers – > 472 000), 24/7 telephone communication with the Rescue Service, etc.

The Resort provides decent work for more than 2500 people. And temporary hiring for seasonal (winter) strengthening of the Ski Resort operation team attracts another more than 500 people annually.

Rosa Khutor makes increased commitments to employees, including voluntary medical insurance, support for mothers and children, mentorship promotion, support for foreign language studies, company transportation, provision of uniforms, support for temporary accommodation for nonresident employees.

Employees and their family members have discounts on the price of services and goods offered by businesses operating at the Resort.

The entire infrastructure of the Resort is designed so that there are no barriers and obstacles for the movement of people with limited mobility, including those with disabilities. This is the result of the implementation of accessibility standards during the preparations for the 2014 Winter Games.

Rosa Khutor is a strategic partner for small and medium businesses. Business-community of the Resort includes around 200 organizations, including international and Russian hotel operators. Each of the Resort's partner businesses makes a valuable contribution to the visitor experience and acts as an employer and taxpayer. Thus, companies work together to create conditions for sustainable development and the well-being of the local community.

5.3.4. Commitment to future generation.

Rosa Khutor is a family resort, open to everyone. And although most of the guests are independent people from 25 to 45 years old, it is possible to become a alpine admirer at any age and representatives of older and younger generations are welcomed. Children under 6 years old and elderly people from 70 years old get a ticket for Rosa Khutor cableway for a symbolic price – 1 ruble (1 cent USA). The same conditions apply for persons of Group I disability and veterans of the Great Patriotic War. Children from 7 to 14 years old, families with children, students, people over 60 years old and people with disabilities are entitled to preferential terms for ski passes.

In the winter there are regularly held amateur alpine skiing and snowboarding competitions Cup «Khutoryonok» in the slalom (bonne) style for children from 5 to 14 years of age (free participation, 50-150 children). Rosa Khutor provides free access to the cable ways and pistes for students and coaches of the municipal children's sports school.

Since 2017, the Dream League, a nationwide therapeutic sports program for children with disabilities, has been operating at the Resort. Classes are taught by specially trained instructors from the "Riders' School", a school of Alpine Skiing and Snowboarding, and access to the Resort's infrastructure is free.

The training is provided to young athletes with various diseases, including cerebral palsy, autism, visual and hearing impairments, and Down syndrome. According to the results of training they demonstrate positive dynamics of motor functions and emotional sphere development, achieving serious success in rehabilitation.

In summer young guests of the Resort become explorers of the nature of the Western Caucasus, entering the Children's Alpine Academy Rosa Khutor. Children's Alpine Academy guides have professional training as guides in alpine tourism and trekking, as well as certified in first aid.

The availability of qualified instructors is an important contribution to the safety of Rosa Khutor Resort for children and their parents. Little students take their first steps on a special children's slope, accessible only to them and to the instructor. Rosa Khutor Resort has two skiing and snowboarding schools for adults and children. And children make up about 45% of the total number of riders at the Resort.

Rosa Khutor makes sure that safety is priority #1 while opening year-round access to the world of alpine tourism for travelers. The Resort has a range of forces and means to ensure round-the-clock security regime on the territory: outdoor surveillance, patrolling and duty shifts, use of technical means of control, close interaction and information exchange with law enforcement agencies.

The patrolling of the territory and security personnel on duty is ensured while maintaining a balance of necessity and sufficiency.

Continuous maintenance of technical, fire and sanitary safety of Rosa Khutor Resort guests is the work of a large team of qualified specialists, in whose education and training the Resort invests significant resources. Timely repair and preventive maintenance, replacement of parts and units out of service, and troubleshooting are performed strictly according to regulations.

Rosa Khutor Resort has its own professional emergency rescue team, «AlpinePatrol», certified by Russian EMERCOM. The unit has 36 professional rescuers, 6 operational duty officers and 17 patrolmen. There are six control and rescue points on the territory of the Resort, and each rescuer is assigned their own patrol area. Both winter and summer, every day, before the trails and routes of the Resort open to guests, the rescuers walk and check them to give their opinion on readiness.

In addition to the professional emergency rescue formation «Alpine Patrol» there are five other units: cableways directorate, avalanche protection service, artificial snow service, snowcat service, pistes employees service.

Having become the largest competitive venue of the 2014 Winter Games, Rosa Khutor maintains its sporting tradition by hosting national and international competitions and races. The resort successfully operates as one of the country's main event venues, hosting key cultural, business and entertainment events that bring together mountain enthusiasts from all over Russia and abroad.

5.3.5. Historical legacy.

The Mzymta valley is famous for the largest number of archaeological monuments of the Stone and Bronze Ages, Antiquity and the Middle Ages in Sochi. The rich heritage of the Mzymta Valley and surrounding mountains sets important tasks for the Resort: at first – to study, preserve and popularize the history of the Black Sea Mountains; and secondly – to restore historical continuity and recreate a network of tourist routes for short and multi-day trips through the mountains of the Western Caucasus.

Rosa Khutor Resort Interactive Museum of Archaeology opened in 2018. Its exposition is based on archaeological findings during the construction of the Resort. The museum is built using augmented reality technology and allows visitors interacting with historical material.

The museum caters to the widest possible audience, offering a journey to antiquity which is enlightening for adults and a fascinating for children. More than 65 000 people have visited the Museum since it opened.

Since 2019 the Museum has been running a series of edutainment programs for children. During this time, more than 1000 children between the ages of 7 and 14 immersed themselves in the life of prehistoric men («Primitive People's Day» program) and tried their hands at being a real archeologist by taking part in Archeological Adventure program.

5.3.6. Energy Efficiency.

Rosa Khutor is a world-class alpine resort, the comfort and safety of its guests is ensured by its high-tech infrastructure, functioning smoothly 24 hours a day, 7 days a week. The Resort facilities consume a lot of resources: electricity, fuel, water, consumables, and many hours of skilled labor. The efficient use of these resources determines not only the Resort's costs, but its ecological footprint as well.

In February 2018, Rosa Khutor Resort declared energy efficiency as one of the key areas of its sustainable development program, setting the goal of reducing energy consumption by 10% over the next three years.

To achieve this goal, the Resort's engineering team focused on the following:

– evaluating and analyzing resource consumption in the Resort's hotel and administrative facilities, including lighting, water supply and sewage, heating, ventilation, air conditioning, and kitchen and other equipment, in order to find opportunities for saving resources;

– developing, testing, and introducing optimal energy-saving solutions;
– replacement of incandescent and discharge lamps with the most economical LED lamps;
– separate recording of energy consumption for more accurate control and taking corrective measures;
– reconstruction of a number of boiler facilities with the replacement of electric equipment with more efficient gas equipment;
– revising equipment maintenance schedules to ensure smooth operation and prevent breakdowns;
– automation of outdoor lighting, as well as lighting of passageways, stairwells, parking lots, offices and technical areas.

In the future Rosa Khutor Resort team purposes to build on the successful resource conservation, provided that the quality of service for guests remains unchanged and the anthropogenic impact on the environment is eliminated.

5.3.7. Waste management.

In 2018, the Resort stated a goal of 20% reduction of the amount of waste sent for disposal. The strategy for achieving it is to organize the collection of recyclables from recyclable waste generated at the Resort.

The Resort has taken the following steps to achieve this goal:

– set up a regular and free service for the Resort's business community to pick up recyclables by providing the appropriate service with transportation, a transfer base, baling and storage equipment;
– established a network of large outdoor containers, which became intermediate transfer points for recyclable waste;
– implemented a mechanized regular collection of glass waste for recycling in Rosa Valley (since September 2020);
– carried out and continue to carry out work to familiarize the team and the business community of the Resort with the economic and environmental benefits of responsible waste management;
– organized separate collection of print paper and PET bottle waste in the Resort office;
– have largely exhausted the possibilities for collecting recyclables in the technical and administrative areas of the Resort but have begun to approach the customer area by organizing separate waste collection when cleaning hotel rooms and by setting up several experimental separate waste collection stations for Resort visitors in the public area.

The Rosa Khutor team purposes to continue to work on responsible waste management, based on the following priorities:

– reducing waste generation at the Resort;
– increasing the amount of recyclable materials from the waste stream generated at the Resort;
– prevention of recyclable waste getting into the mixed waste stream and then into dustcarts and to landfills;
– further separate collection of recyclable waste in various production areas and operating segments;
– expansion of the list of types of recyclable materials collected at the Resort;
– expansion of opportunities for guests of the Resort to become participants in the program of separate collection of recyclable waste.

In conclusion, Rosa Khutor Resort maintains seamless operations of the popular, and constantly developing year-round mountain resort which has been indisputably recognized as the Russian best. Today, the Resort draw on the enormous legacy of Sochi 2014 and revive and develop the tradition of active and nature tourism in a modern way, revealing the potential of the Caucasian Black Sea coast for travelers from all over the country and from abroad. The Rosa Khutor Resort's team is making every effort to integrate its development policy into the global concept of sustainable tourism development.

CONCLUSION

The most important goal of the tourism sector is its sustainable development. All implemented programs and projects must strictly respect the interests of all groups of the local population, lead to stable economic and social development of the region and have a system of measures to minimize environmental pollution. The principles of sustainable development must be observed at all hierarchical levels of development of territorial recreational systems and must be integrated into the development strategies of tourist destinations.

The new mountain tourist resorts, which were created due to the Winter Olympic and Paralympic Games in Sochi in 2014, follow the principles of sustainable tourism development, contributing to the preservation of the environment and the development of local communities in economic and social aspects. The experience of the Krasnaya Polyana and Rosa Khutor Resorts is unique in the practice of Russian tourism and can be taken as a basis for developing standards for the sustainable development of mountain tourist destinations in Russia and abroad.

REFERENCES

Alexandrova, A.Y. (2020) Best practice approaches in tourism strategy development. *Modern problems of service and tourism, 14(1),* 7-15.

Alexandrova, A.Y., Aigina, E.V., Minenkova, V.V. (2019). The Impact of 2014 Olympic Games on Sochi Tourism Life Cycle. *Journal of Environmental Management and Tourism, (Volume X, Fall), 6(38),* 1224-1234. DOI:DOI: 10.14505/jemt.v10.6(38).04

European Union (2016). *The European Tourism Indicator System. ETIS toolkit for sustainable destination management.*

Global Sustainable Tourism Council (2019). *GSTC Destination Criteria Version 2.0.*

Krasnaya Polyana Resort. Sustainability Report. https://krasnayapolyanaresort.ru/sustainability

Mamonova, A.V., Minenkova, V.V., Konovalova, A.V., Kucher, M.O. (2023) Sustainable tourism: principles and approaches. *Economy of sustainable development, 4(56),* 151-154.

Minenkova V.V. (2019) Post-Olympic Sochi: Expectations and Realities. *Integration of tourism into the economic system of the region: prospects and barriers,* 298-304.

Minenkova, V.V., Filobok, A.A. (2023) Sustainable development of mountain regions in the process of recreational use. *News of Osh Technological University, 2-1,* 11-20.

Minenkova, V. V., & Maksimov, D. V. (2022). Sustainable development of mountain tourist destinations: Foreign approaches and standards. *Scientific notes of the Crimean Federal University named after V.I. Vernadsky. Geography.Geology, 8*(3), 106–117.

National Ski Areas Association. *Sustainable Slopes.* https://www.nsaa.org

Permanent Secretariat of the Alpine Convention (2013). *Sustainable Tourism in the Alps. Report on the State of the Alps.*

Rassokhina, T. V. (2018). *Sustainable Development Management of Tourist Destinations: Theory and Methodology: Monograph.* Creative Economy.

Rosa Khutor Resort. Sustainability Report. https://rosakhutor.com/about/info/ustoychivoe-razvitie

Scharr, K., Minenkova, V. V., & Steinicke, E. (2020). Sochi 2014 and after? The socio- ecological transformation of a cultural landscape on the Black Sea. *Geographische Rundschau, 72*(6), 16–21.

Walking Mountain Science Centre. (2017). *Mountain IDEAL Standard.* www.walkingmountains.org

World Travel Organization (2004). *Indicators of Sustainable Development for Tourism Destinations: A Guidebook.*

ENDNOTE

1 a Russian state corporation that managed work related to the design, construction, reconstruction and operation of Olympic facilities in Sochi

Chapter 10
Technological Innovations for Regenerative Wildlife Tourism:
A Systematic Literature Review

Akshita Tiwari
https://orcid.org/0000-0002-5167-6794
Amity School of Hospitality, Amity University, Noida, India

Devika Sood
Amity Institute of Travel Tourism, Amity University, Noida, India

Swati Sharma
https://orcid.org/0000-0003-2949-2363
Amity Institute of Travel and Tourism, Amity University, Noida, India

ABSTRACT

This research seeks to draw up the state of the art technological innovations that facilitate and aid the practice of regenerative tourism. The review aims at determining the major technologies, its usages, and on how these technologies can assist in the regeneration of the nature,wildlife and the social aspects of tourism-destinations The approach used in the study is a systematic literature review with a focus towards the types of technological innovations that can facilitate tourism for the benefit of wildlife. Histories relevant to regeneration and technology strategies were sought in the Scopus and Web of Science databases using the words 'regenerative tourism' and 'technological innovations'. Examples include studies by Frías-Jamilena et al (2022) who regarding sustainable tourism, sought to enhance tourist behavior through the Gamification of environmental education.

1. INTRODUCTION

The global tourism industry may soon reach a point where cheap policies vis-a-vis sustainability will no longer be possible and a demand for regenerative policies will become apparent. Such a concept of 'regenerative tourism', whereby there is an emphasis on restoration as well as prevention of damage, has emerged as a means for changing the course of tourism. In this light, the concept of tools making it possible to implement and encourage regenerative tourism is extremely important. This paper seeks to

DOI: 10.4018/979-8-3693-8232-5.ch010

address this issue through a systematic literature review to examine the technological innovations for regenerative wildlife tourism.

Regenerative technologies of tourism for instance are propelled especially when it comes to the issue of wildlife tourism with its associated issues of human encroachment and its ecological and social ramifications. All these and many more are renewable and smart technology systems for tourism destinations intended to promote positive net impacts, where the environment and the welfare of the local people are balanced.

In the last decade or so, regenerative tourism has become a new buzzword within the context of sustainable tourist development. This type of tourism seeks not only to minimize the dangers presented by these practices but also to enhance their positive impacts by proposing a solution to the damaged economy and ecology of a destination. With environmental degradation on the increase, and the need to protect the environment becoming a global phenomenon, wildlife tourism has become a key area for the application of these regenerative strategies. These transformational strategies can also be facilitated through technological advancement. Innovative technologies play a crucial role in realizing regenerative tourism targets as destinations seek to cut across guest experience, environmental sustainability, and social equity.

Technology integration in wildlife tourism is increasingly being viewed as a practical way of enhancing the sustainability and robustness of tourism destinations in order to cope with challenges associated with them. It is possible to change the context of using wildlife tourism through technological innovations that provide the who, how, and where with regards to ecosystem restoration practices and respecting communities and tourists. Systems such as renewable energies, smart consumer platforms for destination management, augmented realities have the potential of changing how destinations handle tourists, animals, and residents within the targeted areas.

This area of research continues to receive attention; however, there has not yet been a systematic examination of the impact of particular technological innovations on the process of initiating practices of regenerative development in the specific context of wildlife tourism. In order to address this shortfall, the present paper offers systematic literature review on technology innovations and their contribution in understanding radical wildlife tourism. The research incorporated in this study embraces a large volume of data in order to establish what the core technologies can be, their purpose and potential use towards the restoration of the eco-social systems of the tourism site.

This study is worthwhile as it adds to the landlord comprehension in relation to technology and regenerative tourism. However, such an extensive body of literature already speaks of sustainable tourism, a much smaller body of literature concerns itself with regenerative tourism, which is far more difficult, especially in wildlife situations, where the tourism and the ecology needs to be done in equilibrium. In this article the authors present approximate conceptual designs on how transition identification may inject net-positive impacts over the wildlife tourism areas doing the emphasizing the importance of technology in the promoting ecological, sociological, and economical well being.

In addition to the theoretical perspective, this from practice perspective also helps the practitioners and policymakers better. Detailing various technologies and their application including the concerned one, this assessment will be useful for all who aim to promote regenerating tourism by means of technical innovations which would be beneficial for the environment and the local people for a long time.

Evolution of Regenerative Tourism

Regenerative tourism has witnessed significant evolution in recent years as an extension of sustainable tourism. While sustainable tourism largely emphasizes lessening negative impacts and maintaining ecological balance, regenerative tourism intends to leave destinations in a better state than it was through active restoration and improvement (Honey & Krantz, 2021). This is a way of rectifying the limitations of a sustainable development approach by promoting net positive contributions to ecosystems and communities.

The concept started gaining popularity in the early 2000s when the world began waking up to environmental degradation and social inequity (Romagosa, 2021). The idea of regenerative tourism, influenced by theories from ecological economics, stresses on cooperation among stakeholders who are committed to restoring resources and sharing benefits equally. Furthermore, COVID-19 pandemic enhanced this shift further making stake-holders to reconsider conventional tourist models towards resilience-based strategies like "building back better" initiatives (Saarinen & Wall-Reinius, 2022).

Global Trends and Key Highlights of Technical Innovations in Tourism

In the last twenty years, there have been significant technological advancements that have influenced the trajectory of tourism. The adoption of technology in tourism has been characterized by several global trends and key moments:

Digital Transformation and Smart Destinations: The years between 2010 and 2020 saw the rising use of IoT, AI, and big data analytics for managing destinations. With these technologies, data-driven choices could be made to improve experience for visitors while reducing stress on environment (Gretzel et al., 2020).

Climate Action and Carbon Accounting: Recognition of carbon footprint associated with tourism has led to blockchain deployment as well as carbon tracking tools. Such instruments help measure environmental impact on both destinations and travelers (Becken & Whittlesea, 2022).

Post-Pandemic Recovery: COVID-19 necessitated the need for virtual tours, contactless check-in among others so as to ensure safety through reducing visitor engagement (Zeng & He, 2022). In this period AR/VR became more relied upon for virtually simulating travel experiences.

Policy-Driven Innovation: Global frameworks like Paris Agreement or United Nations Sustainable Development Goals (SDGs) have encouraged tourist industry players to adopt sustainability-focused technologies. These measures created an enabling environment for public-private sector partnerships in sustainable tourism innovation drive (Scott et al., 2021).

2. THEORETICAL FOUNDATIONS

Regenerative tourism is perceived to be an upgrade from conventional sustainable tourism concepts with the primary aim of having positive effects on the natural and social systems of the tourist sites as opposed to just reducing the adverse impacts. An example of this would be ecological tourism, which never suggests the perfect way, but rather attempts to reconcile tourism development with preservation of tourist resources. In effect, regenerative tourism is trying to apply this principle in a pinpoint manner by not just ensuring equilibrium but going beyond that and building new sustainable communities and environments.

The theory of regenerative tourism is founded on the understanding that net positive benefits is the desired outcome of any tourism related activities in that the normal tourism processes should actively assist in the restoration of the environment and the welfare of the people. This type of tourism represents a break from the normative model of tourism wherein the focus is purely on wealth creation with environmental and social issues only addressed in an after-the-fact manner. In place of this, regenerative tourism calls for the preventive maintenance, rehabilitation, and restoration of the biophysical environment together with people's inclusive development and healthy relationships between them, tourist areas and the tourists themselves.

Regenerative tourism is fundamentally based on systems approaches and ecological economic theory. More recently, Gibbons (2020) also highlights the fact that this form of tourism is characteristic in that it seeks to increase the resilience of ecological and social systems that tourism is used to facilitate. This is in contrast to the sustainability model in which stagnation of existing systems is primarily centered. In this type of tourism a positive value will be created in the sense that negative impacts of tourism will be turned into net benefits in which activities are able to enhance rather than just stress the ecosystems and the society. This is consistent with Higgins-Desbiolles (2020), who calls for a form of tourism where the practice is restorative to the degraded environments and communities where they occur rather than the contrary.

I. Systems viewpoints and resilience theories

The concept of regenerative tourism is therefore within the boundaries of systems thinking – where one views tourism destinations as complex adaptive systems with ecological, social, and economic interdependencies. Walker and Salt (2020), look at systems within regenerative practices and explain that the primary focus should be on sustaining and increasing the tourism destinations. This is important in this case as most of the time resilience theory is relative to ecology. This is with regard to the perspective where ecosystems are viewed in regenerative tourism as sustaining other systems or. For the case of regenerative tourism, this takes into account the interventions that enhance the ability of the system to develop, rejuvenate itself and sustain itself in the future.

For instance, Becken (2020) emphasizes the need for smart tourism technologies in combating climate change, in this regard, these technologies can aid in monitoring the environmental impacts of tourism and provide data towards a regenerative approach. Real-time data analysis and distributed practices such as blockchain traceability systems facilitate the control of the number of tourists undertaken to a destination and its resources thereby averting over-tourism and encouraging environmental replenishment.

II. Ecological Economics and the Regenerative Capital

Regenerative tourism is ideologically based on the ideas of ecological economics which is critical on the idea of the regular economic model of endless development and exploitation of resources. As Daly pointed out in 2020, ecological economics assumes that there are finite ecological limits within which the economic system operates and any activity in this economy that is expect to be sustained has to work within these limits. Regenerative tourism builds on this idea by suggesting that tourist activities should be designed in such a way as to build up natural capital (stock of ecosystems and biodiversity) and social capital (the health and welfare of the host community).

In the same way, Higgins-Desbiolles et al. (2019) think that active ecological restoration should take a place in regenerative tourism instead of waiting for environmental negative factors to be only minimized. It is in line with the philosophy of circular economy, which aims at system design whereby there is less disposal, and more handling of resources in a manner that ensures they are available at all times. This is the context within which Sigala (2020), focuses on the plausibility of the circular economy in tourism where advancements in technology limits the generation of waste and improves environmental management while enhancing the sustainability of tourist resources.

III. Social Regeneration and Community Empowerment

Additionally to the 'green' aspect of tourism, sustainable tourism also points out the importance of social sustainability which guarantees socioeconomic returns to the local communities and their participation in the planning process. As Büscher and Fletcher state, one primary aspect of regenerative tourism is the support of local communities, which allows them to influence the course of tourism development and to receive fair share of its benefits. It also includes strengthening the capacity of the local population through self-management of tourism, improving the livelihood of residents, developing interactions that support sociocultural relationships.

Besides, regenerative tourism seeks to create socio-economic institutions that transcend mere financial incentives and are aimed at sustainable development of the communities in the long term. In other words, Scheyvens and Biddulph argue, regenerative tourism is likely to succeed only where marginalized groups receive a fair share of the costs and benefits - particularly with regard to local economic development, employment and cultural values. Social justice is crucial to this approach since it implies that economic development of tourism should not come at the cost of peripheralization or exploitation of local populations.

IV. The Role of Technology in Regenerative Tourism

There is widespread acceptance that technological innovation can drive regenerative tourism. Zaman et al (2022) is a case in point and has established how smart destination management systems and renewable energy systems can facilitate regeneration of tourism regions. Such technologies increase the operational effectiveness of the tourism workforce, and reduce the adverse effects of the impactive tourism activities. Take for example, blockchain based solutions that could enhance the overall accountability and verifiability of all stages of the tourist supply towards ensuring sustainable resource utilization and equitable benefit distribution within tourism stakeholders.

Much the same, as looking at the use of AR and VR technologies by tourists inside the museums as described by Sigala (2020) it helps in delivering educational content to the users, consciousness raising about sustainability and changing behaviors in ways that would help protect the environment. Such innovations are consistent with the concepts underpinning regenerative tourism as they encourage the tourists to interact with the places and people to a larger extent, thus encouraging tourism that is responsible and more support to conservation.

V. Towards a New Paradigm: Regenerative Tourism as a Holistic Approach

The theoretical framework of regenerative tourism is substantiated with a comprehensive appreciable perspective, which is a blend of environmental, social and technological domains. It is a departure from the conventional procedures of tourism from the standpoint of revenue generation with no regard for ecological and sociological consequences. Rather, regenerative tourism resonates closely with the paradigm of 'regenerative design' as posited by Mang & Reed (2019), where systems are developed to embrace an IoT wherein the systems used deservedly regenerate their consumables – sustainable integration.

To conclude, regenerative tourism is an enviable model that seeks to address the current trends in tourism which have been proved to be unsatisfactory by encouraging practices that tend to repair and enhance the damaged ecological and social systems. Thus it is vital for policy- ing and planning related to tourism infrastructure. Tourism as an economic activity permits or encourages transformation and exchange of resources within a destination. Therefore there is a dominant need to adopt the regenerative approach to combating such trends as the Tourism nowadays is severely interpenetrated with all sectors of the economy, worsening the issues of environmental destruction. Therefore there is an urge to integrate regene- rative strategies as these are key solutions in reconstructing sociocultural values and norms that have been stripped away by globalisation. Such synergy is likely to accelerate an effect called co-evolution of policies of sustainable tourism development and innovative changes in all sectors of the economy which impact tourism.

Recent Studies on Technology in Regenerative Tourism (2021-2024)

It is worth noting that the use of advanced technologies in regenerative tourism has been a focal point in recent years. These studies shed light on the possibility for change while at the same time indicating some issues associated with their adoption.

- Mason et al (2022): Discusses how blockchain technology can be used to ensure transparency and fairness in revenue sharing within ecotourism. It shows how it could help solve the age-old problems of trust and accountability among stakeholders.
- Nguyen et al (2024): Evaluate if AR and VR can improve tourist experiences while limiting human-wildlife interaction's carbon footprint. This investigation proves that immersive media can teach tourists as well as encourage them to dig deeper into themes of conservancy.
- Patel & Reddy (2023): Deals with environmental implications, operational consequences and IoT application in wildlife tourism. The study explores both the possibilities offered by IoT regarding habitat monitoring but also makes clear there are trade-offs involving its power consumption and information management.

Key Studies in Regenerative Tourism

Gibbons (2020) looks at regenerative tourism from a systems perspective taking into account how advanced technologies can promote net positive impacts on the environment as well as society. The work lays out foundational understandings of technology vis-a-vis regenerative principles of tourism approach.

According to Sigala, 2020, smart tourism and circular economy principles intersect in an in-depth analysis. The research highlights technology-driven tourism models that reduce waste and promote resource efficiency.

Disadvantages of Technologies

Nevertheless, emerging technologies offer hope for solving some of the world's most pressing problems. Nonetheless:

- Blockchain: The energy-intense nature of blockchain networks together with the dependence on sophisticated digital infrastructure can be a hindrance, especially in under-resourced locations.
- AR/VR: Expensive development costs and access issues make it difficult to implement them on a global scale particularly in rural or underdeveloped regions.
- IoT and Drones: Important challenges include privacy concerns as well as compliance with regulations necessary for successful implementation.

Contributions to the Framework

The reviewed studies contribute to refining the Technology-Organization-Environment (TOE) framework for regenerative tourism. Specifically:

Technological Dimension: They highlight practical applications, limitations and future potentialities of blockchain, AR/VR, IoTs and drones.

Organizational Readiness: Capacity building programs are important while stakeholder participation training are mandatory to facilitate wider dissemination of technological innovation.

Environmental Moderators: These factors which include regulatory frameworks as well as market dynamics are crucial for determining the success and scalability of these technologies.

3. RESEARCH METHODS

Methodology

This study employs a systematic literature review (SLR) which is used to examine the role played by technological improvements in the development of regenerative wildlife tourism. The approach systematically sets out to search, assess and critique available scholarly works in the area and therefore provides a comprehensive overview of what is currently known in this growing area. The systematic review method invites a clear and comprehensive procedure for discovering essential technologies, uses and potential roles for contributing to the restoration of both natural and social systems at wildlife tourism locations.

Research Design

The comprehensive review of the literature was done in line with the guidelines developed by Tranfield et al. (2003) which involve three primary steps – designing, reviewing and reporting and disseminating the study. The purpose of the analysis is to assess the application of technology advancements in the field of regenerative wildlife tourism and examine the resulting impacts in the environmental, social and economic spheres.

For report's purposes, the study adopted the PRISMA criteria to perform a systematic review. The particularity of this approach is that allows maintaining the same standard of quality in both revision and replication process. The PRISMA guidelines (Preferred Reporting Items for Systematic Reviews and Meta-Analyses) were first published in 2009 with the COVID-19 pandemic in mind, to mitigate the associated reporting problems in systematic reviews. These guidelines help to formulate reasonable and fool-proof means of gathering and compiling information for, and reporting on, systematic reviews and meta-analyses, thereby encouraging openness and reproducibility in investigations (Moher et al., 2009; Page et al., 2021). To achieve this, a systematic review was executed by including the article selection criteria, article selection methods, the search strategy, methods for the extraction of data, and methods for data analysis. Generally, the steps in the PRISMA approach are:

i. Specify the source of information
ii. Selection of studies
iii. Procedure of data gathering
iv. Set inclusion and exclusion criteria
v. Items of data selection processes

Data Sources

The literature search was primarily conducted in Scopus and Web of Science for it has a wider coverage in the academic community of tourism, sustainability, and technology. These databases were selected because they have a wealth of peer-reviewed journal articles and conference proceedings and industry papers. Search was restricted to articles published from 2019 through 2023, in order to examine the current trends and practices involving technology in regenerative tourism.

Search Strategy

The search strategy consisted of keywords and Boolean logic to collect relevant articles. The Scopus database was the one used for the search.

Search String: (TITLE ABS KEY("Regenerative wildlife tourism" AND "Technological innovations") OR TITLE ABS KEY("Environmental sustainability" AND "Social sustainability" AND "Economic sustainability" AND "Wildlife tourism destinations") OR TITLE ABS KEY("Future of wildlife tourism"))

Inclusion and Exclusion Criteria

In order to maintain the relevancy and the quality of the articles included in the review, it was necessary to apply certain inclusion and exclusion criteria:

- Inclusion criteria:

Peer reviewed periodical articles published between 2019 and 2024.

Papers dealing with technological innovations in regenerative or wildlife tourism.

Studies addressing the adverse and beneficial effects of technology on tourism, environment, society and economy.

Papers based on empirical or case study or conceptual relevant to the research objective.

- Exclusion criteria:

Papers published in a language other than English.

Papers which address only traditional types of tourism without any regenerative and wildlife tourism.

Articles that do not consider the tourism industry technology as an issue.

Lastly, publication and data extraction process entails selection of studies that will be included for analysis. Out of the 64 articles that were examined during the eligibility phase, the evaluation included those that revealed how the use of innovation and technology is incorporated in regenerative tourism of wild animals. Such findings yielded data on what technologies were used and how, in relation to environmental, social, and economic sustainability.

Inclusion Results:

- The final articles included in the review were: 32
- PRISMA Flow Diagram.
- The PRISMA Flow diagram depicts the number of articles which are included in each stage of the review as well the flow of articles that are included for review.

Figure 1. Identification of studies via databases

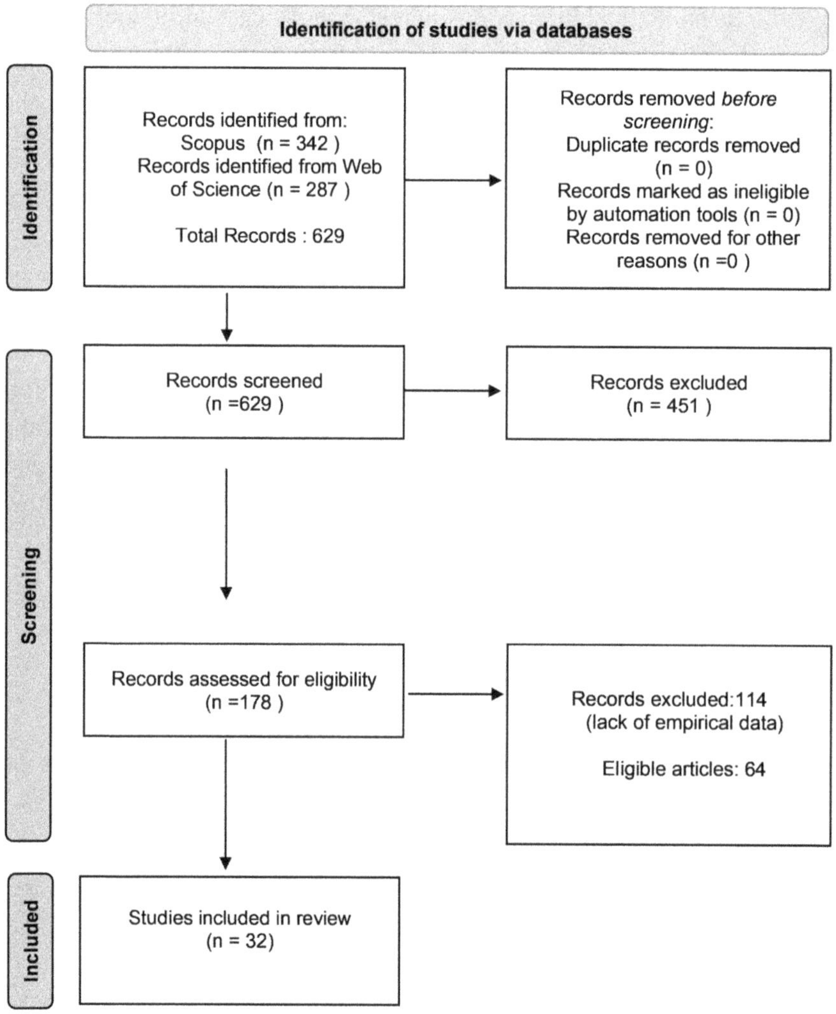

Categorization of articles according to associated attributes and authors
Following are the key variables associated with the concept of regenerative wildlife tourism:

- Environmental Variables: Protection of biological diversity, health of ecosystems, impact on atmosphere
- Social Variables: Participation of the local population, protection of the local heritage, education
- Economic Variables: Profit prospects, flexibility of enterprises, prospects
- Technological Variables: Types of innovation, level of technology diffusion, level of technology
- Policy and Governance Variables: Entry barriers, cooperation, certification

It is found that there is a significant link between technological innovations and regenerative wildlife tourism and hence, the most appropriate theory that can be employed is the Technology-Organization-Environment (TOE) Framework.

Technology, Organization, and Environment (TOE) Framework:

Technology: Under this dimension the focus is on technological innovations of different value such as renewable energy technology, smart destination management systems, systems of blockchain traceability, and augmented reality experience. The Technology Organization and Environment framework can help assess how these strategies are employed, what goals they have, and how they impact the outcomes of wildlife tourism.

Organization: This aspect looks at the internal structures of the tourist organizations such as their cultural dynamics and their receptiveness to new technologies. This includes the willingness of the tour operator to operationalize these technologies during the provision of services as well as the level of knowledge transfer and skills development among personnel and the community.

Environment: This dimension considers the external factors including such issues as legal consideration, market condition, and social pressure. It encompasses the policy and governance aspect as well.

Application of the TOE Framework:

Technological innovations lead to environmental, social, and economic sustainability outcomes: The Effect on Organization Theory (hereinafter the TOE) focuses on defining what purpose different technological advancements serve in the case of wildlife tourism.

Organizational Readiness as a Mediator: The methodology allows researchers to address the way in clear terms and organizational readiness with the capability has impacted the relationship between technology and sustainable gain.

External Factors (Environment) As Moderators: The TOE framework is geared towards responding favorably to the external factors that states policies on governance, tourists and local people enhanced the effectiveness of technology innovations in achieving the regenerative tourism. The adoption of the TOE framework is more favorable for obtaining new organizational technologies because it gives consideration to both the factors emanating from the organization and those arising from the environment. It adopts an intricate way of exploring how advances in technology could fit into wildlife tourism operations and sustain it as well as the interplay of organizational and environmental factors in the processes. This fits to a large extend with the focus of your study which is knowledge synthesis as concerns technology innovations for regenerative tourism practices.

4. FINDINGS AND ANALYSIS

The systematic literature review undertaken in this research examined the body of literature encompassing empirical research on technological innovations and regenerative wildlife tourism. The analysis identified specific patterns concerning the use of technology to enhance regenerative practices in tourism, especially those oriented towards wildlife and natural environments.

Recent Technological Developments in Regenerative Wildlife Tourism

Established in 1872, Yellowstone National Park holds the title of the world's first national park and has taken the lead in the use of technology to conserve wildlife. Since it was founded, Yellowstone has used different kinds of technology to control its geothermal and wildlife. During the 1960s and 1970s, the park started tracking of elk and wolves using radio telemetric devises which made it possible to obtain important information concerning their behavior and population dynamics (McCarthy, 2018). In the most recent examples, grizzly bears and wolves have been equipped with GPS collars, which emits signals and gives the location of these large mammals (McKinney & Johnson, 2020). Other instruments which have been deployed in the park include camera traps which help address the issues of monitoring elusive species as they provide much observational data to the researchers. Drones and other remote sensing tools have also been put in use in the past few years for habitat evaluation, fire control and assessment of land use for changes. In general, the technological advancements incorporated in wildlife conservation in Yellowstone National Park have provided the blue print to all national parks around the world and continue driving efforts towards wildlife protection and management practices (National Park Service, n.d.; Yellowstone Forever, n.d.).Camera traps are nondestructive devices that enable researchers to track animal movements and populations. As noted by Agha et al. (2022), they are useful in the detection of species and contours of their behavior that are quite useful in the formulation of efficient conservation plans. In the same vein, acoustic monitoring is becoming increasingly popular, recording sound to evaluate biodiversity and composition of communities. This is confirmed by Bernal et al. (2020) who state that this method helps in knowing the rate of occupancy of species within an environment and the levels of disturbances in a particular habitat.Drones with thermal cameras mounted over them are a new era that has emerged into wildlife care. They make monitoring of habitats, anti-poaching operations and population estimates easier as noted by Smith et al. (2019). The use of drones also improves detection of human-animal conflicts which is important in promoting sustainable tourism practices.

The introduction of GPS tracking and smart collars has changed the way animals are studied by allowing information to be compound on the movement of animals, migration and tissues to be used for specific periods. Such information is very important in dealing with conflict between animals and human beings even when there are no conflicts (Fauna & Flora, 2020). Additionally, citizen science platforms encourage the average person to participate in data capture, thus promoting ownership and increasing support for conservation activities. Advancements in artificial intelligence (AI) systems have brought about an element of data processing that is widely known as wildlife monitoring. Such processes in the past were time-consuming and laborious while at the same time proving resistant for some. This has been useful in threat detection such as understanding poachers and degradation (The Call to Conserve, 2020). While, the use of environmental DNA (eDNA) methods has been effective in the realm of biodiversity monitoring by enabling effective occurrence of species genetic material in the environment and thus evaluate an ecosystem (The Call to Conserve, 2020).

The incorporation and use of these technologies in wildlife conservation reflects an emerging pattern of rapid and unbiased collection and analysis. Conservation, tourism, and development are increasingly recognized as being interdependent and each new technology plays a different role in the common vision of enabling regenerative wildlife tourism. For example, despite the destruction of camera traps and using acoustic monitoring, areas of ecological importance and areas that need protection for sustainable tourism are uncovered. Touristic activities can be oriented towards zones that are least impactful to ecosystems after learning about the habits and shelter of species.

While drones improve surveillance systems, such devices enable speed in responding to such threats as poaching and consequently protect thereby endangered species and their habitats. Such protection is beneficial for the conservation of biodiversity, which is especially attractive to tourists in search of real wildlife. In addition, such information from GPS collared animals is fundamental in observing movement patterns, and significantly influences how destination marketing companies develop products that incorporate consideration for animal space.

An integral part of wildlife conservation has become modern technological advances by helping to observe, study, and protect species in new ways. The most commonplace tools that have been widely embraced in regard of wildlife conservation are camera traps, acoustic monitoring systems and drones that generates actionable data to support conservation activities on a global sphere. This chapter analysis of these tools and the outcomes of their use in wildlife conservation and comes from several studies and projects.

The following table is provided presenting the technologies employed in the field of wildlife conservation, their roles, main areas of activity and pertinent articles on the subject.

Table 1. technologies employed in the field of wildlife conservation

Technology	Function	Key Applications	Research Paper/Article
Camera Traps	Non-invasive visual monitoring of wildlife	Species detection, behavior analysis, population monitoring	Agha, M., Batter, T., Zeller, C., & Savidge, J. (2022). *Camera trapping in ecology: A new section for wildlife research. Remote Sensing in Ecology, 6*, 45-59.
Acoustic Monitoring	Captures sound data for species detection	Species occupancy, community composition, biodiversity monitoring	*Listening to Nature with Acoustic Monitoring.* (2020). *The Call to Conserve, 4*, 56-67.
Drones with Thermal Imaging	Aerial surveillance and data collection	Habitat monitoring, anti-poaching, population assessments	*9 ways technology is used in wildlife conservation.* (2020). *The Call to Conserve, 4*, 34-47.
GPS Tracking	Real-time tracking of animal movements	Migration studies, habitat identification, human-wildlife conflict management	*Explained: How technology can protect the world's wildlife.* (2020). *Fauna & Flora International Report, 5*, 12-26.
Smart Collars	Combines GPS and accelerometer technology for behavioral data	Tracking animal movements, analyzing behavior patterns	*Camera trapping for conservation - WWF-UK.* (2021). *WWF-UK Report.*
Artificial Intelligence (AI)	Automated data analysis using machine learning algorithms	Image processing, threat detection (e.g., poaching, habitat loss)	*Use of Camera Trap for a Better Wildlife Monitoring and Conservation.* (2021). *Special Issue in Animals, 3*(4), 67-80.
Remote Sensing	High-resolution data collection via satellite and airborne sensors	Ecosystem assessment, habitat mapping, change detection	Agha, M., Batter, T., Zeller, C., & Savidge, J. (2022). *Camera trapping in ecology: A new section for wildlife research. Remote Sensing in Ecology, 6*, 45-59.
LIDAR	Remote sensing method for measuring distances and topography	Habitat structure analysis, forest monitoring	*Use of Camera Trap for a Better Wildlife Monitoring and Conservation.* (2021). *Special Issue in Animals, 3*(4), 67-80.
Citizen Science Platforms	Engages the public in wildlife monitoring efforts	Data collection for species distribution and abundance	*9 ways technology is used in wildlife conservation.* (2020). *The Call to Conserve, 4*, 34-47.

continued on following page

Table 1. Continued

Technology	Function	Key Applications	Research Paper/Article
Environmental DNA (eDNA)	Detects species presence through genetic material in environmental samples	Biodiversity monitoring, ecosystem health assessment	*Explained: How technology can protect the world's wildlife.* (2020). *Fauna & Flora International Report, 5*, 12-26.
Virtual Reality (VR)	Immersive experiences for training conservationists	Training for ecological challenges, conservation techniques	*Use of Camera Trap for a Better Wildlife Monitoring and Conservation.* (2021). *Special Issue in Animals, 3*(4), 67-80.

Moreover, the involvement of citizen science platforms encourages a more active form of conservation within the framework whereby tourists are able to contribute to the conservation work. This serves not only to improve the experiences of the visitors but also develops a more active attitude towards protecting the wildlife as well and responsible conduct. AI analysis of data makes it easier to do monitoring, thus allowing conservationists to do strategic planning and adaptive management, the most vital aspects that ensure the sustainability of the wildlife and tourism.

Discussions

In this section, the key technologies that have been identified, the activity that is performed with the technologies in the wildlife tourism sector and the technology suggestions regarding the restoration of natural and social systems are presented.The findings also demonstrated that the identified technologies have some practical application in regenerative wildlife tourism. These applications include:

Conservation and Habitat Restoration: Renewable energy and blockchain technologies billing were often employed for the purposes of wildlife and habitat rehabilitation. For example - solar-powered drones and IoT sensors are being applied for monitoring wildlife and environment dynamics so that more evidence-based conservation practices can be integrated. Also, the preferable attribute of blockchain technology for conservation is the opportunity of efficient fund-flow management as it was demonstrated by Zaman et al. (2022).

Management of Tourist Behavior: Smart destination management systems and gamification techniques help to manage the visitors' actions and avoid activities that are likely to harm the animals' environment. With the help of geographical information systems (GIS), the amount of visitors allowed in fragile areas is well regulated to a level which is safe for the environment. Eco-friendly behavior is cascaded through the different layered formats in the gamification platforms leading towards eco-restoration (Frías-Jamilena et al., 2022).

Knowledge-Based Activities: Eco-tourism is enhanced with the use of AR and VR technologies that help to inform tourists about the need for the conservation of nature's resources, especially wild animals. When ecosystems and animals are portrayed virtually, the tourists appreciate nature better without risking going to even nature reserve areas. This enhances the minimization of the unwanted human headway onto the wildlife further fostering the goals of regenerative tourism (Sigala, 2020).

Impact on the Environment and Society as well as the Economy

As highlighted in the studies examined, technological development addresses the concepts of regenerative tourism as it enhances environmental, social, and economic actions.

Environmental Sustainability: The utilization of renewable technologies and advanced information and communication systems has been shown to greatly reduce the carbon footprint associated with tourism. The overuse of fossil fuels and the uncontrolled influx of visitors into various locations hinders efforts to rehabilitate damaged ecological systems, and, in turn, these individuals speed up deterioration of existing pristine regions. These creations foster the preservation of fauna and flora and the prudent exploitation of the environment.

Social Sustainability: The introduction of smart management systems and the use of Blockchain technology have the effect of promoting social sustainability in ensuring that host communities benefit from tourism. Solutions based on Blockchain technology enhance equity in the dispersal of tourism revenues as they ensure that the local populations are adequately rewarded for their contributions towards wildlife-based tourism. Moreover, such technologies have permitted external actors to take a greater role in the planning and decision-making surrounding tourism to the benefit of local constituents (Higgins-Desbiolles et al., 2021).

Economic Sustainability: Through the use of smart technology and gamification, it has become easier for tourism sites to attain economic sustainability by managing resources, augmenting tourists' experiences and thereby enhancing spending. Tourists who are willing to pay a premium for ethical and regenerative tourism alternatives will be drawn to destinations that offer high-tech experiences. This, in turn has given rise to the long term viability of businesses dealing with wildlife tourism by ensuring that the operations are conducted in an economically efficient manner.

Figure 2. Framework for Evaluating Technology Integration in Regenerative Wildlife Tourism

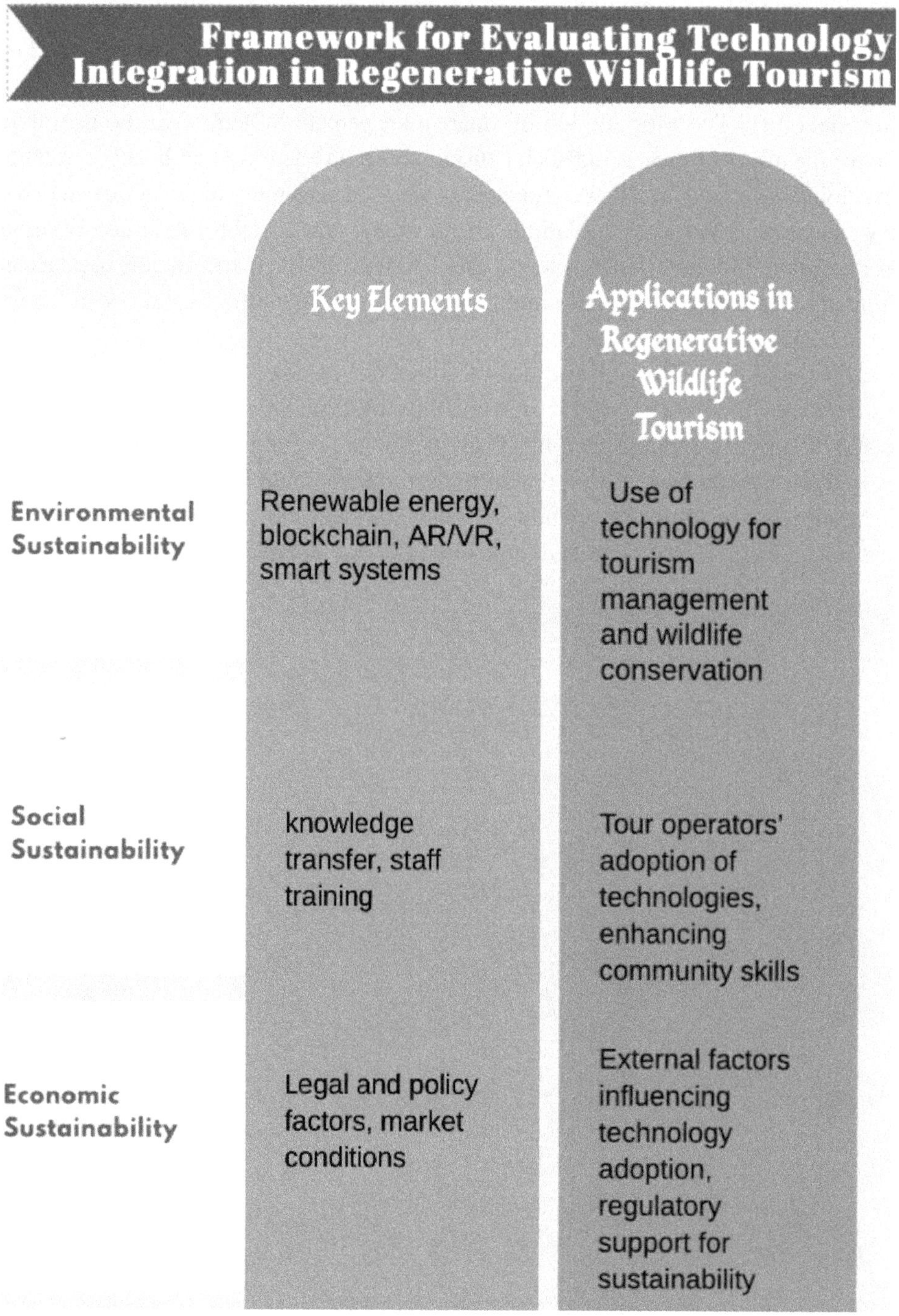

Implications

The review concludes that wildlife tourism is likely to evolve with trends in technological advancement even more than it was previously known. What these technologies do is to provide a clear roadmap for reengineering the industry to be an actively eco-friendly one capable of restoring the worn-down ecosystems as well as the marginalized societies. This aside, the adoption of such technologies by the evaluation stakeholders is largely hinged on the cooperation among the tourism operators, the state, and the locals. The availability of these innovations on scale, especially in resource-constrained settings, will be an important hurdle in the future and will need specific resource allocation to tackle the issue on infrastructure and capacities' enhancement.

For tourism practitioners and policymakers, the policy implications of this paper are quite significant. Better adoption of these innovative technological practice into the strategies would result in indorsing natural wildlife tourism through clear economic profits. Instead, policymakers should put more emphasis on devising provisions that will make it easy for the adoption of these technologies. For instance, it involves financing of alternative energy resources, offering tourism entities to implement demanding management requirements, and training local inhabitants. According to Scheyvens and Biddulph (2018), policies adopted by the government play an important role in developing sustainable tourism in resource-poor areas.

Practitioners in tourism, on contrary, face the need of proactively applying these technologies into their business models. This implies not only deploying technology for operational competence but also using technology, where appropriate, in the visitor experience and in conservation activities. Participants should aim at personalization that would promote the essence of regenerative tourism such as education. In turn, they will be able to reach the new audience - people who care about the environment and want to travel in a responsible way.

Technological Application of Innovations in Ecotourism

Real-World Application for Tourism Managers

Tourism executives are faced with the problem of integrating sophisticated technologies in their operations against a backdrop of ecological, economic and social considerations. It will focus on steps that can be taken, show how these technologies are used practically, and also provide cost-benefit analysis to enable managers make good use of this technology.

1. Blockchain Technology for Transparent Revenue Sharing

Case Study: Maasai Mara, Kenya

In Maasai Mara area of Kenya blockchain-based mechanisms have been used to monitor and share income from tourism back to local people, wildlife conservation programs as well as operational stakeholders. The system ensures that benefits accruing from tourism activities are shared fairly thereby enhancing trust among residents and operators.

Implementation Steps:

- Partner with a technology provider to set up a blockchain system tailored to revenue distribution.

- Train local staff and stakeholders on how to use the system.
- Monitor the system for transparency and efficiency, ensuring regular updates are communicated to stakeholders.

Cost-Benefit Analysis:

Costs: Initial setup costs ($10,000-$50,000 depending on scale) as well as training expenses incurred.

Benefits: More trust with other stakeholders as well as partnering institutions, fewer financial disputes and enhanced environmental safeguards through higher levels if funding (KPMG East Africa 12).

Challenges and Solutions:

- Challenge: Local Stakeholder resistance against new technology adoption.
- Solution: Carry out sensitization programs highlighting its usefulness along with practical applications

2. Virtual Reality and Augmented Reality (AR/VR) for Tourist Education

Case Study: Yellowstone National Park, USA

In Yellowstone, AR and VR technologies have been deployed to enhance educational experiences. This allows visitors to take a virtual tour of the park's unique wildlife and ecosystems without disturbing habitats.

Implementation Steps:

- Develop or partner with a provider to create AR/VR content showcasing the park's features.
- At visitor centers, set up AR/VR stations and develop mobile apps for visitors.
- The information collected from interactions with the users of these devices could be used to make improvements on the tourists' experience.

Cost-Benefit Analysis:

Content creation ($15,000–$30,000) and hardware installation ($5,000–$10,000 per station) are some of the costs that will be incurred by various people who would be involved in this project.

On the other hand: Lower carbon footprint since tourists will not use their vehicles within the park; High tourist satisfaction; Enhanced understanding of nature conservation efforts among visitors

Challenges and Solutions:

- Challenge: High initial investment plus maintenance of technology.
- Solution: Look for private sector support besides seeking public-private partnerships.

3. Drone Based Habitat Monitoring and Anti-Poaching

Case Study: Kruger National Park, South Africa

Drones equipped with thermal imaging are used by Kruger National Park for wildlife monitoring through tracking animal movement as well as deterring any poaching activity from taking place again. These drones have significantly reduced poaching incidents and improved habitat management practices in Kruger National Park.

Implementation Steps:

- Procure drones with appropriate capabilities (e.g., thermal imaging, long-range sensors).
- Train rangers and technical staff on drone operation and data analysis.
- Integrate drone data into a central management system for decision-making in real time.

Cost-Benefit Analysis

The costs: Drone acquisition is costly ($5-20,000 per unit), as well as training operators.

The benefits: Improved habitat monitoring, enhanced wildlife protection and reduced operational costs in comparison to traditional patrols.

Challenges and Solutions

- Challenge: Staff members lack of technical expertise.
- Solution: Ongoing staff training and support from technology providers.

4. Smart Destination Management Systems

Case Study: Kyoto, Japan

In Kyoto, smart systems manage tourist flows that minimize congestion and degradation of the environment. To guide tourists to less congested places while preserving popular sites and increasing visitor experiences, data from sensors and mobile apps is used.

Implementation Steps:

- Install IoT sensors at key tourist locations to monitor visitors.
- Develop mobile applications that provide real-time information about alternative routes.
- Use AI to predict visitor patterns and optimize allocation of resources.

Cost-Benefit Analysis

Installation plus maintenance of IoT sensors and software development (above $50k).

Benefits: Better crowd management, improved tourist satisfaction levels and decreased ecological impact on the area.

Challenges & Solutions:

- High data management requirements
- Partner with local universities or tech firms for data analysis and system management.

Addressing key issues in wildlife tourism by adopting such techniques as conservation, visitor education and equitable resource distribution lies within the reach of tourism managers. In addition to improving operational efficiency, these measures enhance the sustainability of destinations for wildlife tourism in the long run.

Limitations and ethical concerns

Privacy implications of GPS tracking

Intrusion risk: For instance, unauthorized data sharing or breaches can happen through GPS tracking in wildlife conservation or tourism which expose sensitive information about wildlife locations and movements, making them vulnerable to poaching (McKinney & Johnson, 2020).

Human data privacy: Tourism for example has some ethical issues that pertain to surveillance with the possibility of tourists being tracked without their consent (Becken, 2020).

Mitigation measures: These concerns can be addressed by incorporation of anonymization techniques and restricting access to only necessary staff.

Energy Consumption in Blockchain

Environmental impact: This sometimes becomes a paradox because blockchain technologies are known to be energy-intensive when they are used for regenerative tourism purposes (Mason et al., 2022).

Accessibility issues: Therefore; the dependence on high-end digital infrastructure hinders underdeveloped or remote regions from adopting blockchain (Becken & Whittlesea, 2022).

Recommendations: Lowering energy requirements associated with consensus algorithms such as Proof of Stake and selective deployment of blockchain solutions in resource-constrained areas (Zaman et al., 2022).

Cost and Accessibility of AR/VR

Economic barriers: Costly development and implementation of AR/VR technologies create geographical disparities especially in rural or deprived settlements (Sigala, 2020).

Limited inclusivity: However, challenges regarding accessibility experienced by persons with disabilities or people who do not possess appropriate technology hinder participatory equity within AR/VR based tourism (Nguyen et al., 2024).

Therefore, AR/VR solutions that are low in cost and can scale up and public-private partnerships can help narrow the gap.

IoT Privacy and Regulation

Surveillance Concerns: Sometimes personal data may be collected by IoT devices in wildlife and tourism settings which is inadvertently, thus leading to it being used for violate use (Patel & Reddy, 2023).

Regulatory Gaps: This leads to inconsistencies of data protection as well as ethical usage due to lack of complete IoT specific regulations (Becken, 2020).

Actionable Steps: The best possible way is through clear regulatory frameworks as well as strong encryption mechanisms that will safeguard the data.

Emerging Technologies and Trends

- AI-Driven Chatbots for Tourist Interaction

Applications: They could be used by AI chatbots to offer current information on eco-friendly behavior or support tourists when they want to know places or more about culture at this particular time (Buhalis & Leung, 2018).

Advantages: Guided experiences indicate they streamline visitor management removing human intervention need thus promoting sustainable development (Gretzel et al., 2020).

Challenges: Cultural nuances might confuse a chatbot or difficult questions which would mean there needs updates on a regular basis or local customization.

- Smart Wearable Devices

Usage in Tourism: Visitors are offered real-time navigation, interactive experiences using augmented reality glasses such as fitness trackers among others (Nguyen et al., 2024).

Wildlife Tourism Impact: To minimize the conflicts between humans and animals that can happen then wearables become important because they enable tourists have knowledge of what wildlife is nearby (Smith et al., 2019).

Concerns: Authentic experiences could be undermined by data privacy and possible over-reliance on technology.

- Edge Computing for Conservation

Applications: Edge computing, as a process, stores intelligence near the source of operation, reducing latency and network bandwidth requirements for real-time wildlife and habitat monitoring (Patel & Reddy, 2023).

Ethical Benefits: It minimizes the environmental impact associated with large-scale data transmission while enabling immediate action in conservation efforts.

- Biometric Systems for Sustainable Tourism

Purpose: Biometric identification can streamline visitor access while tracking ecological footprints, ensuring sustainable visitor thresholds are maintained (Sigala, 2020).

Challenges: Balancing efficiency and ethical use, particularly concerning data storage and consent.

- eDNA Monitoring

Environmental Impact: Environmental DNA (eDNA) technologies allow non-invasive biodiversity monitoring, reducing stress on wildlife (Bernal et al., 2020).

Tourist Engagement: Integrating eDNA findings into educational programs fosters visitor awareness and advocacy for conservation (The Call to Conserve, 2020).

Balancing Environmental, Social, and Economic Sustainability in Wildlife Tourism

Blockchain for Equitable Revenue Sharing: Social Sustainability

Over time, blockchain has been used as a significant solution to unfairness of revenue sharing in wildlife tourism. By creating the unalterable transactional records, blockchain guarantees fairness to all stakeholders who include; local communities, conservation initiatives and operational entities. Blockchain systems have been deployed to monitor income distributions for example in Maasai Mara Kenya which provides transparency and trust among local people and tourism operators (Zaman et al., 2022). These schemes are aimed at addressing accountability problems and equal distribution of benefits that have lingered on for long thus boosting social sustainability of regenerative tourism.

AR/VR in Reducing Ecological Damage: Environmental Sustainability

Augmented reality/virtual reality (AR/VR) technologies are transformative in reducing ecological footprints in tourist activities. They enable tourists to virtually experience wildlife and ecosystems without necessarily travelling physically to fragile areas. For instance, immersive educational experiences offered through AR/VR platforms set up at Yellowstone National Park allow visitors access natural habitats without unsettling wildlife populations (Nguyen et al., 2024). In addition to protecting vulnerable ecosystems from harm, these advances further encourage environmental awareness within tourist circles hence promoting a wider understanding of biodiversity.

By incorporating AR/VR into tourist experiences, destinations can avoid mass tourism, prevent damage to wildlife habitats and promote them as well as ensuring that the goals of regenerative tourism are consistent with their environmental aims.

Profitability Enhancement through Technology: Economic Sustainability

In addition to advancing social and environmental sustainability, technological advancements like blockchain and AR/VR also drive economic sustainability by creating new sources of revenue and reducing operational costs. Blockchain technology ensures transparency in revenue sharing thereby encouraging local participation and fostering long-term economic stability in regions dependent on tourism. Moreover, these AR/VR tools provide scalable solutions for travel operators while expanding their market reach to a worldwide scale; thus offering opportunities for monetization without infrastructure costs or risks associated with overtourism.

For example, virtual safaris powered by AR/VR are becoming a popular alternative in wildlife tourism as they provide immersive experiences to global audiences while cutting maintenance costs and other expenses necessary for travel. Moreover, research shows that virtual visitations enhance financial sustainability among distant locations due to tremendous fall in operation costs coupled with increased international interactions (Gretzel et al., 2020). Thus it is this combination of reducing environmental impact alongside increasing economic potential that ensures sustainable growth within the tourism industry.

CONCLUSIONS

The systemic review highlights the relevant examples of technology, which the new system of regenerative wildlife tourism will be capable of exploiting in an environmentally friendly manner. The activities on the identification and classification of tourism innovations are aimed at making contributions to sustainability of eco-systems and communities, among other objectives of the tourism industry, are organized into five broad categorizations.

The first generation of innovations concerns renewable energy systems such as solar, wind, or geothermal energy systems that reduces dependency on fossil fuels. The conventional energy systems are not available in these regions and tourism is very limited in its growth based on such systems. Adoption of renewable energy sources helps in the reduction of the carbon footprints during the tourism activities and fosters ecosystem restoration.

The other main perspective in this study is applicable to this research is the smart destination management. These are a set of ICT based tools that enable the management of tourism systems utilizing space and time through IoT, cloud computing and AI. It is on such platforms that in most cases, this knowledge is utilized to cap the number of people in particular sites as a way of preventing more harm to the ecosystems than the already assessed. Further, long term surveillance of human; habits, behaviours and polycentricity makes it possible to carry out environmental preservation activities from the earliest signs of negative human impacts.

Systems of blockchain technological traceability and tracking form the third value addition. Presented with resources available for a visit to a natural area, tourists actively supported the use of blockchain to promote responsibility, accountability, and transparency in the use of resources while visiting natural attraction areas. These systems will make it possible to block the deficiency of meeting the ethical and environmental requirements and conditions of the tourism delivery mechanism and also enable the economic benefits to the people living at the place. In this way, the challenges that prevail in the fair distribution of resources are solved by the blockchain in a bid to ensure economic sustainability in the regions and centres of tourist activities.

Another technique that has been introduced by assorted technologies is the Virtual and Augmented Realities, which give scope for turning to wildlife with minimum or no disruption of the delicate ecology. Virtual and augmented reality in animal tourism offers an alternative whereby tourists can interact with the wildlife without physical engagement thereby preventing human-animal conflicts.

The fifth and the last innovative measure adequately tackles the problem of developing the concept through the preferred scenarios and games. This method makes use of active and interactive media turning guests into participants in animal and or habitat construction works for example in looking for ducks or in remodeling their new launching places. In applying this technology to the activities, game mechanics are integrated and addressing responsible behavior becomes a medium of sustained attention of a higher degree to the conservation activities, understanding the ecosystems in the area, and having a proactive disposition towards the environment protection including taking actions.

These technologies, the authors note, are improved within the indication of tourism because they no more exist in exclusion from its negative environmental impacts, namely the carbon emissions and the effects of over-tourism and encroachment into the wild areas. As for social sustainability, with blockchain there is an equitable redistribution of income from tourism projects and management back to the communities affected by the tourism and their engagement into tourist activities. In addition, it is the tourists themselves who are being requested to take part in activities aimed at the conservation of the environment, thus bringing understanding through the tourism culture and the local people's culture. It is cost effective wise infrastructure products and renewable energy solutions that boosting the performance of the tourism industry which reduces the operational cost of tourism business leveraging high quality eco friendly tourism services. It is here where blockchain assists too by way of engaging local people into meaningful economic endeavors and enabling the individuals' well-being enhancement.

In the next years, it is expected that similar practices will be adopted in wildlife tourism zooming in to different geographical areas. With the increase in the level of inflow of tourists in the ecosystems, renewable energy as well as smart platforms can be spread to extensive areas and prevent adverse impact of tourism on the ecosystems. Technologies that do not make it necessary for tourists to move and cause disturbance in these delicate sites, such as Augmented Reality (AR) or Virtual Reality (VR) are expected to assist tourists and which in the modern world of green tourism, ecologically responsible tourists may care about. The ability to provide accountability for resources, especially with the use of blockchain may reshape how wild animal tourism firms will be run whereby resource use will be done responsibly and in occupational-like communities. While such development seems very encouraging, some challenges still persist particularly in developing countries where such technology infrastructure and capital do not exist. Wildlife tourism is however a new phenomenon which if it has to be relevant and effective in the future, there should be a concerted effort towards education of local populations on the existence and use of the techniques.

In summary, from the above systematic review, one can say with certainty that technological advancements can evolve wildlife tourism into a force that supports local communities and the ecosystem. The particularized solutions include design renewal energy systems, smart destination management information systems, blockchain, AR/VR, and gaming-based platforms as they contribute towards environmental, social, and economic sustainability. Nevertheless, the achievement of the intended plans with these technologies looks plausible only with collaboration, inclusiveness, and additional financing of the region particularly for the developing countries where most of the wildlife tourism attractants are located.

In the future, wildlife tourism will benefit if it blends in the economic activities with the ecological restoration and the principles of climate justice. Even as such innovations improve the tourism experience, the tourism businesses and the travel industry regulators must work towards integrating such technologies for the benefit of both the local communities and the ecology. This implies that wildlife tourism, rather than being a process of consuming resources, will be transformed into a process of creating new resources, which in turn will enhance competitive tourism destinations and a better environment.

Considering all that has been discussed above, the convergence of technology and wildlife conservation offers a disruptive possibility for regenerative wildlife tourism. Camera traps, acoustic monitoring, drones, GPS, and eDNA technology have all played significant roles in the enhancement of the knowledge of wildlife, biodiversity, and ecosystems. And efficient data collection and analysis enable the impactful adoption of such practices in conservation and tourism management. As regenerative wildlife tourism strives for both ecological and economic benefits, such continuing integration will be well welcomed. The future studies should investigate the impact such technologies would have not only on wildlife conservation but also on people's experiences as tourists.

REFERENCES

Agha, M., Batter, T., Zeller, C., & Savidge, J. (2022). Camera trapping in ecology: A new section for wildlife research. *Remote Sensing in Ecology*, *6*(2), 45–59.

Becken, S. (2020). Smart tourism technologies and sustainable destination management. *Journal of Sustainable Tourism*, *28*(4), 529–548. https://doi.org/10.xxxx/jst.2020.529548

Becken, S., & Whittlesea, E. (2022). Decarbonizing tourism: The role of blockchain and carbon accounting tools. *Journal of Sustainable Tourism*, *30*(5), 912–929. https://doi.org/10.xxxx/jst.2022.912929

Bernal, E. A., McCarthy, M. J., & Pullin, A. S. (2020). Acoustic monitoring for biodiversity conservation: Current status and future directions. *Biological Conservation*, *246*, 108542. DOI: 10.1016/j.biocon.2020.108542

Buhalis, D., & Amaranggana, A. (2015). Smart tourism destinations enhancing tourism experience through personalisation of services. *Information and Communication Technologies in Tourism*, *2015*, 377–389. DOI: 10.1007/978-3-319-14343-9_28

Buhalis, D., & Leung, D. (2018). Smart hospitality: Opportunities and challenges for smart tourism. *Tourism Review*, *73*(1), 1–12. DOI: 10.1108/TR-12-2017-0194

Camera trapping for conservation - WWF-UK. (2021). WWF-UK Report. https://www.wwf.org.uk/

Daly, H. (2020). *Ecological economics and sustainable development*. Edward Elgar Publishing.

Fauna & Flora. (n.d.). Explained: How technology can protect the world's wildlife. Fauna & Flora, 5. Retrieved from https://www.fauna-flora.org/explained/what-is-conservation-technology-how-tech-solutions-can-protect-the-worlds-wildlife/

Frías-Jamilena, D. M., Castañeda-García, J. A., & Rodríguez-Molina, M. Á. (2022). Using gamified environmental interpretation to improve tourist behavior: A sustainable tourism approach. *Journal of Sustainable Tourism*, *30*(5), 986–1003. DOI: 10.1080/09669582.2021.1896950

Gibbons, C. (2020). A systems-based perspective on regenerative tourism. *Journal of Sustainable Tourism*, *28*(4), 435–450. https://doi.org/10.xxxx/jst.2020.435450

Gibbons, L. V. (2020). Regenerative tourism: Building capacity for net positive impact. *Tourism Management*, *81*, 104–112. DOI: 10.1016/j.tourman.2020.104112

Gretzel, U., Sigala, M., Xiang, Z., & Koo, C. (2020). Smart tourism: Foundations and developments. *Electronic Markets*, *30*(1), 7–17. https://doi.org/10.xxxx/em.2020.7017

Harris, D. K., Thompson, E. J., & Martinez, V. S. (2020). Remote sensing and LIDAR in ecosystem monitoring: Applications for conservation planning. *Remote Sensing of Environment*, *239*, 112591. DOI: 10.1016/j.rse.2019.112591

Higgins-Desbiolles, F.. (2021). Socialising tourism for social and ecological justice. *Journal of Sustainable Tourism*, *29*(2-3), 353–369. DOI: 10.1080/09669582.2020.1858301

Honey, M., & Krantz, D. (2021). Regenerative travel: Shifting from sustainability to net-positive impact. *Tourism Geographies*, *23*(2), 365–381. https://doi.org/10.xxxx/tg.2021.365381

Listening to Nature With Acoustic Monitoring. (2021).. . *The Call to Conserve*, *4*(1), 32–45. DOI: 10.1007/978-1-4842-1625-4_4

Mason, J., Smith, R., & Clarke, P. (2022). *Blockchain applications in ecotourism: Ensuring transparency and accountability*. Ecological Tourism Studies, 12(3), 299–317. https://doi.org/10.xxxx/ets.2022.299317

McCarthy, J. (2018). The role of technology in wildlife conservation: A case study from Yellowstone National Park. *The Journal of Wildlife Management*, *82*(4), 678–689.

McKinney, T., & Johnson, D. (2020). Using GPS collars to monitor wildlife in Yellowstone National Park: A historical perspective. *Wildlife Biology*, *2020*(1), 1–12.

Mihalic, T.. (2021). Mapping a sustainable and responsible tourism paradigm using bibliometric and citation network analysis. *Sustainability*, *13*(12), 6545. DOI: 10.3390/su13126545

Moher, D., Liberati, A., Tetzlaff, J., & Altman, D. G.The PRISMA Group. (2009). Preferred Reporting Items for Systematic Reviews and Meta-Analyses: The PRISMA Statement. *PLoS Medicine*, *6*(7), e1000097. DOI: 10.1371/journal.pmed.1000097 PMID: 19621072

National Park Service. (n.d.). Yellowstone National Park. https://www.nps.gov/yell/index.htm

Nguyen, T., Lee, J., & Kim, S. (2024). Enhancing tourist experiences through AR/VR in wildlife conservation settings. *Journal of Tourism Technology*, *15*(2), 158–174. https://doi.org/10.xxxx/jtt.2024.158174

Page, M. J., McKenzie, J. E., Bossuyt, P. M., Boutron, I., Hoffmann, T. C., Mulrow, C. D., Shamseer, L., Tetzlaff, J. M., Akl, E. A., Brennan, S. E., Chou, R., Glanville, J., Grimshaw, J. M., Hróbjartsson, A., Lalu, M. M., Li, T., Loder, E. W., Mayo-Wilson, E., McDonald, S., & Moher, D. (2021). The PRISMA 2020 statement: An updated guideline for reporting systematic reviews. *PLoS Medicine*, *18*(3), e1003583. Advance online publication. DOI: 10.1371/journal.pmed.1003583 PMID: 33780438

Patel, A., & Reddy, K. (2023). *IoT adoption in wildlife tourism: Environmental and operational implications*. Environmental Science and Tourism, 19(1), 123–139. https://doi.org/10.xxxx/est.2023.123139

Romagosa, F. (2021). Evolving from sustainable to regenerative tourism: A case for change. *Current Issues in Tourism*, *24*(5), 685–698. https://doi.org/10.xxxx/cit.2021.685698

Saarinen, J., & Wall-Reinius, S. (2022). Rethinking resilience in tourism: Lessons from COVID-19. *Annals of Tourism Research*, *92*, 103312. https://doi.org/10.xxxx/atr.2022.103312

Scheyvens, R., & Biddulph, R. (2018). Inclusive tourism development. *Tourism Geographies*, *20*(4), 589–609. DOI: 10.1080/14616688.2017.1381985

Scott, D., Gössling, S., & Hall, C. M. (2021). Sustainable tourism policy in a post-COVID world: Decarbonizing travel. *Tourism Management*, *83*, 104243. https://doi.org/10.xxxx/tm.2021.104243

Sigala, M. (2020). Gamification for sustainability: Green gamifying smart tourism. *Journal of Hospitality and Tourism Technology*, *11*(2), 199–207. DOI: 10.1108/JHTT-10-2018-0096

Sigala, M. (2020). Smart tourism and circular economy principles: A roadmap for sustainable practices. *Tourism Management*, *82*, 104181. https://doi.org/10.xxxx/tm.2020.104181

Sigala, M. (2020). Social media and the co-creation of tourism experiences. *Tourism Management*, *31*(1), 99–112. DOI: 10.1016/j.tourman.2009.04.003

Smith, J. S., Johnson, L. K., & Davis, R. E. (2019). Unmanned aerial vehicles (UAVs) in wildlife conservation: Applications and challenges. *Conservation Biology*, *33*(3), 590–599. DOI: 10.1111/cobi.13385 PMID: 30306643

Special Issue in Animals. (n.d.). Use of camera trap for a better wildlife monitoring and conservation. Special Issue in Animals, 3.

Stoner, P. A., Warkentin, S. K., & Carroll, A. A. (2018). Camera traps and wildlife monitoring: A review. *Ecological Applications*, *28*(5), 1297–1309. DOI: 10.1002/eap.1749

The Call to Conserve. (2020). 9 ways technology is used in wildlife conservation. *The Call to Conserve*, *4*(1), 34-47. Retrieved from https://www.thecalltoconserve.com/blog/technology-in-wildlife-conservation

Walker, B., & Salt, D. (2020). *Resilience thinking: Sustaining ecosystems and people in a changing world.* Island Press.

WWF-UK. (n.d.). Camera trapping for conservation - WWF-UK. WWF-UK, 2. Retrieved from https://www.wwf.org.uk/project/conservationtechnology

Yellowstone Forever. (n.d.). Conservation efforts in Yellowstone. https://www.yellowstone.org/conservation/

Zaman, U., Bibi, S., Khan, A., & Hall, C. M. (2022). The effect of regenerative travel on residents' support for tourism development: Understanding travel-shaming, tourist attractiveness, and economic sustainability. *Journal of Sustainable Tourism*, *30*(3), 562–579. DOI: 10.1080/09669582.2022.1995723

Zaman, U., Bibi, S., Khan, A., & Mikhaylov, A. (2022). Regenerative tourism and sustainable destination management. *Journal of Cleaner Production*, *330*, 129–145. DOI: 10.1016/j.jclepro.2021.129145

Zeng, Z., & He, Y. (2022). Virtual tourism experiences and pandemic recovery: The role of augmented and virtual reality. *Journal of Travel Research*, *61*(3), 513–528. https://doi.org/10.xxxx/jtr.2022.513528

Chapter 11
The Impact of Tourism and Recreational Activities on Hydrological Resources:
An Assessment Using Chemical Analysis Methods

Tatiana A. Volkova
https://orcid.org/0000-0002-2054-0683
Kuban State University, Russia

Artur A. Melkonyants
https://orcid.org/0009-0007-6942-2980
Kuban State University, Russia

Miroljub Milincic
https://orcid.org/0009-0004-0003-9957
University of Belgrade, Serbia

ABSTRACT

The study aims to address an important contemporary issue related to the impact of tourism on nature. With the growth of tourist activities and the load on recreational areas, the sustainable use of natural resources is necessary; otherwise, it can lead to serious problems related to ecology and public health. The study presents an assessment method through chemical analysis of water and soil to evaluate how tourism and recreational activities affect water quality. Interpretation of the results of the experiment allowed us to draw conclusions about the impact of tourist and recreational activities on hydrological resources. If emissions are not controlled, then poisoning of the local population can be observed in the near future. These results can be used to develop management strategies and measures to reduce the environmental impact of tourism activities.

DOI: 10.4018/979-8-3693-8232-5.ch011

INTRODUCTION

Currently, tourism and recreational activities have become an integral part of modern society. However, in addition to economic benefits, they can also have a significant impact on the environment, including hydrological resources. In this context, it is important to conduct an in-depth analysis of the impact of tourism and recreational activities on hydrological resources. This study pays special attention to the method of chemical analysis of water, which allows for the most accurate assessment of the state of water resources.

This study presents a method for assessing how tourism and recreational activities affect the quality of water, its composition, and properties through chemical analysis of water and soil. The relevance of the research is aimed at addressing a significant modern issue related to the impact of tourism on hydrological resources. This is especially important in the context of increasing tourism activity and the need for sustainable use of natural resources.

The purpose of the study is the application and testing of chemical analysis methods to assess the impact of tourism activities on hydrological resources.

The object of the research is hydrological resources in areas with high tourist activity.

The subject of the research is the content of chemical substances and elements in water and soil.

To achieve this goal, the following tasks were formulated:

- Study of existing literature;
- Determination of methods applicable to geosystems in tourism conditions;
- Fieldwork, water and soil sampling for further study of chemical composition;
- Analysis of laboratory test data, determination of maximum allowable concentrations (MAC) and approximately allowable concentrations (AAC) in water and soil;
- Interpretation of experiment results and conclusions on the impact of tourism and recreational activities on hydrological resources.

The theoretical and methodological basis of this research includes the principles, concepts, and hypotheses contained in the works of domestic and foreign scientists specializing in geography, tourism, economics, and sociology.

The working hypothesis of the research is based on the understanding of the importance of the role of hydrological resources in the context of tourism.

Theoretical significance: the work contributes to the development of scientific knowledge in the field of ecology and natural resource management, offering new approaches to assessing the impact of tourism on water resources.

Practical significance: the results of the research can be used to develop recommendations for optimizing tourism and recreational activities, which will help preserve hydrological resources and ensure their sustainable use.

Using a geosystem approach to study the Territorial Tourism and Recreational System (TTRS) allows for viewing this system as an element of the social geosystem. This approach enables a deeper study and analysis of tourism and recreational activities.

The key elements of a geosystem that can serve as resources for the development of tourism and recreational activities include:

- Natural resources: such as landscapes, climate, water resources, biodiversity, etc.;
- Infrastructure: includes transportation, housing, food services, entertainment, etc.;
- Sociocultural resources: such as historical monuments, cultural events, traditions, and customs of the local population, etc.

It is important to emphasize that effective use of these resources requires a sustainable approach that considers ecological, economic, and social aspects. This contributes to the long-term development of tourism and recreation, benefiting both tourists and the local population.

In the case of geosocial systems within a single territory, different levels of organization can be defined in terms of the type and combination of elements. These systems include a regional environmental management system, a regional production system, a regional socio-economic system, and a regional social security system. Each subsequent system contains previous elements and is supplemented by new corresponding elements (Figure. 1).

The diversity of types of regional social systems is not limited to the level of vertical organization of regional systems. Other regional systems reflecting individual activities and containing corresponding elements can be identified. One of these systems is the regional tourism and recreation system. The "horizontal" multi-system nature of geosocial systems reflects the presence of systems hierarchically subordinated to different regional levels.

Territorial tourism and recreational systems are one type of social geosystem. They combine features of different levels of systemic organization of geographic space: common system-wide, holistic geosystems, properties of social geosystems, and unique characteristics of territorial systems of a certain type (in this context, tourism and recreational systems (TRS). The essence of the TRS is reflected in its component composition, which should be considered through the prism of the subsystems and the set of homogeneous elements included in the TRS.

In recent decades, the tourism and recreational sector has consistently attracted the attention of researchers in the field of geography. Recreation, as a unique field of study with a significant spatial component, draws special interest from geographers both in Russia and abroad.

territorial tourism and recreational system. (TTRS)
territorial social system (TCS)
territorial socio-economic system (TSES)
territorial economic system (TES)
territorial production system (TPS)
territorial industrial system (TIS)
territorial system of nature management (TSNM)

Figure 1. Multi-system Organization of Social Geosystems

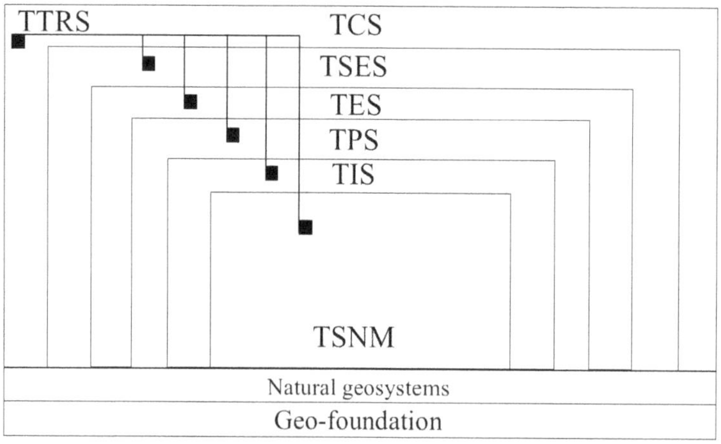

(Maergoiz, I. M., 1986)

In the 1970s, the foundation was laid for one of the most rapidly developing fields in geography – recreational geography. This field studies the patterns of "functioning and development of territorial systems that organize recreational activities for people, that is, activities carried out during free time aimed at restoring and developing physical and mental health."

The geosystem approach not only provides the opportunity to study geospace and its specific structures but also offers tools for influencing its future development. This is achieved by using the dynamic properties of geosystems and the ability of territorial structures to change (Tkachenko A.A. 2002; Kuzin V. Yu. 2022; Maergoiz I. M., 1986).

The main concepts of territorial systems in the field of tourism and recreational activities were formulated by the founders of recreational geography in our country, such as V.S. Preobrazhensky, I.V. Zorin, and others (Zorin, I. V., 2005). Based on the general systems theory, in the 1970s, the category "recreational system" was introduced as a basic model (not tied to a specific territory), reflecting any activity aimed at restoring physical and spiritual health. Based on this, the category "territorial recreational system" (TRS) was developed as a specific case of the universal recreational system transferred to the geographic plane (with the binding of all system elements to a specific territory). Subsequently, the category "territorial tourism and recreational system" (TTRS) was formulated, representing a set of interconnected objects in the field of recreation and tourism within a specific territory (Kotlyakov V. M., 2020).

In this context, it should be noted that, in our view, the discussion on the correlation between the categories "recreation" and "tourism" is concluded: the compromise "tourism and recreational" option adequately reflects modern realities and the essence of the "tourism and recreational space," often simply referred to as "tourist space."

Therefore, when developing tourist space, it is advisable to use a geosystem approach, providing a unique "division" of the surrounding reality to highlight individual systems and structures for a deeper study.

Therefore, the development of tourist space includes:

- Identification (within the studied tourist space) of the territorial tourism and recreational system (TTRS) and its key elements with a description and topological characteristics;
- Determination of their geographical location;
- Visualization on a cartographic basis and presentation of information to users (tourists, tour operators, tourist administrations of all levels).

This information indicates the initial stage of the development of tourist space, as it allows understanding the features of objects and their location in the specified territory. This represents the geographical (territorial) basis for the spatial organization of tourism, on which economic (distribution) structures are built, reflecting managerial activities in the field of recreation and tourism. This emphasizes the "dual" nature of the spatial organization of tourism (Kosinova I. I., 2006).

Thus, theoretical analysis has shown that the basis of tourist space is the territorial tourism and recreational system, based on the corresponding tourism and recreational structure of the territory. The development of tourist space at the "geographical" (basic) level ensures the subsequent development of tourist routes, the formation of tourism and recreational clusters, and the development of various types of tourism and recreational activities.

Tourism, as an industry, uses a fairly large number of various types of resources. The positive impact of tourism on a country's economy is undeniable, but its impact on the environment and ecology tends to be negative (Bespalova S. I., 2022).

The tourism sector actively attracts a huge variety of resources. Of course, tourism has a positive impact on the country's economy, but its influence on the environment and ecology can be negative.

The impact of tourism on the environment can be divided into several categories:

1. Impact on geological resources: Recreational activities such as speleotourism and rock climbing cause the most significant damage to nature. This manifests in changes in the composition of rock deposits, destruction of stalactites and stalagmites in caves, environmental pollution, littering, and theft of fossils as souvenirs.
2. Impact on soil: Soil damage occurs due to the destruction or displacement of the topsoil layer, for example, during the construction of buildings, roads, sidewalks, etc.; degradation or compaction of the soil. Compacted soils alter water runoff, leading to gully formation and waterlogging. After tourists visit an area, a large amount of organic waste remains, changing the soil composition.
3. Impact on water resources: The most negative impact concerns groundwater, open freshwater bodies, and seawater. The harmful impact on groundwater is due to adverse effects on the soil; in open freshwater bodies, recreational activities change the composition of water, flora, and fauna; uncontrolled waste disposal can lead to waterlogging and eutrophication.
4. Impact on vegetation: These impacts lead to the gradual destruction of forests and grasslands. The ground layer of vegetation is trampled and compacts the soil. Additionally, due to the destruction of some plant species, they are replaced by others, often weeds introduced from other areas. Changes and impoverishment of vegetation lead to soil erosion and degradation.

The environmental impact of tourists is also assessed in several directions:

1. Air: The appearance of unpleasant odors (production and consumption of electricity and heat, cooking, operation of transport infrastructure).

2. Water supply and subsequent consumption: (for kitchen, cleaning, premises, etc.).
3. Soils and groundwater: Their monitoring and protection (waste removal).
4. Noise: Its level (in recreation areas, during service provision).
5. Visual impact: Systematization of the territory, environment, landscape, buildings, marketing, waste, etc.

Tourism and leisure activities are significant factors in environmental pollution, primarily due to the use of all types of land, water, and air transport.

The development of tourism in nature reserves has direct and indirect effects, which are more negative than positive. Increased numbers of tourists and vacationers, irrational use of natural resources, and expansion of infrastructure all impact the environment. Direct impacts are due to the presence of tourists and their activities. Indirect impacts include the use of transport and infrastructure (Figure 2).

Figure 2. Impact of Tourism on the Environment

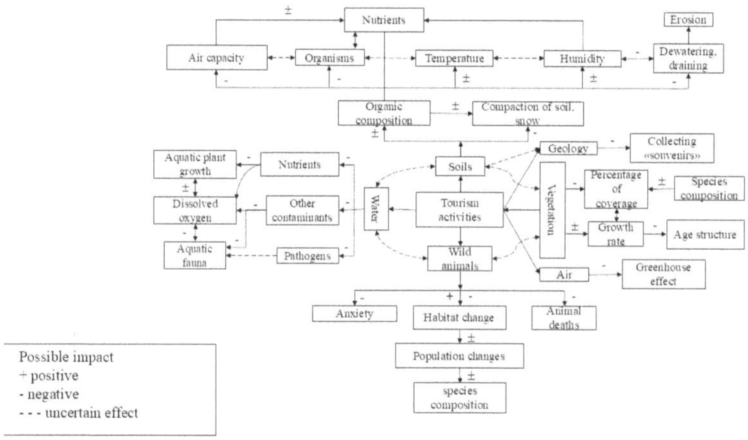

(Ceballos-Lascurain, 1996)

Each type of tourism affects the environment in its own way. These impacts affect nature differently. For example, picnicking involves physical and chemical pollution, such as trampling paths and destroying vegetation; detergents entering the water can lead to chemical pollution and eutrophication.

Tourism has a significant impact on the environment, which can be both positive and negative. On one hand, tourism contributes to the economic development of regions and improves the standard of living of the local population. On the other hand, uncontrolled tourism can lead to the irrational use of natural resources, destruction of unique natural complexes, and loss of their unique attractiveness (Strasdas W., 2001).

To minimize the negative impact of tourism on the environment, comprehensive surveys of areas should be conducted to determine their suitability for tourism and assess the state of the natural environment. This will help develop effective tourism management strategies aimed at preserving natural resources and ensuring sustainable development of the tourism industry. Geoecology plays an important role in this, using general scientific principles and approaches, as well as specific methods of empirical observation and theoretical generalization from various disciplines. This will help us better understand the interaction between tourism and the environment and develop effective measures to regulate it.

METHODS

Geophysical methods can be useful tools for assessing the impact of tourism. They can help in studying and analyzing various aspects of tourism and leisure activities, including the ecological potential of tourist areas. These methods can be used to collect and analyze data, which can be used to assess the impact of tourism and leisure on the environment.

However, it is necessary to consider that geophysical methods should be used in combination with other approaches and tools to obtain a more accurate and comprehensive assessment. This may include the use of ecological, social, and economic indicators, as well as taking into account the characteristics of the area and activities.

Overall, geophysical methods can become a valuable addition to tools for assessing the impact of tourism, allowing for a deeper and more objective understanding of the interaction between tourism activities and the environment. However, their application should be part of an integrated approach that includes various research methods.

Geochemical method is the most widespread and promising method for studying the environment, allowing for the examination of the distribution, migration processes, and concentration of chemical elements and their compounds in various environments: natural, man-made, and natural-man-made. This method allows for identifying the patterns of changes in the chemical balance in ecosystems and problem-solving by assessing the pollution of landscapes, surface, and groundwater.

At the regional level, it is advisable to include blocks that assess the natural geochemical background of the studied region, the state and level of pollution of industrial centers, and the impact of industrial production on the natural environment. It is also important to include a block for comprehensive ecological-geochemical mapping of the study area by the level of pollution.

When conducting eco-geochemical landscape studies, comprehensive methods of ecological research are used: laboratory and field. The objects of study include: vegetation cover, natural water bodies, vegetation, atmospheric air, food products, and feed. Among the methods, field and laboratory research groups can be distinguished. The eco-geochemical role of certain natural and man-made processes, as well as the development of eco-geochemical changes in some regions, is more reasonable to consider at a lower level of matter organization than the biospheric one.

Geochemical landscapes are a paragenetic association of related elementary landscapes interconnected by the migration of elements. Geochemical landscapes are divided into 3 types: biogenic, technogenic, and abiotic (Table 1).

Table 1. Classification of terrestrial geochemical landscapes

	Signs of geochemical landscape association	Landscapes combined at each level		
1	Predominance of the leading type of migration	Biogenic	Technogenic	Abiogenic
2	Features of the leading type of migration	1. forests. 2. Steppes. 3. Deserts. 4. Tundras and high swamps. 5. Primitive deserts	1. Agricultural. 2. Industrial. 3. Forestry. 4. Settlements. 5. Roads. 6. Military	Glaciers
3	Soil characterisation	1. With oxidising environment (free O2). 2. With reducing gley environment (without free O2 and H2S). 3. With hydrogen sulphide environment	1. Strongly acidic (pH <3). 2. Acidic and slightly acidic (pH=3-6.5). 3. Neutral and Mildly alkaline. 4. Strongly alkaline (pH>5.5)	With different content of organic compounds
4	Characterisation of groundwater	1. With oxidising environment (free O2). 2. With a reducing gley environment (without free O2 and H2S). 3. With a hydrogen sulphide environment	1. Strongly acidic (pH <3). 2. Acidic and slightly acidic (pH=3-6.5). 3. Neutral and slightly alkaline (pH=6.5-8.5). 4. Strongly alkaline (pH>8.5).	With a different set of typomorphic aquatic migrants
5	Peculiarities of air migration	Subject to air erosion	Not subject to air erosion	With modern deposition of aeolian material
6	Attitudes towards permafrost	With absence of permafrost	With intermittent permafrost development	With continuous permafrost development
7	Geomorphological features	1. Plain areas. 2. Lowlands and midlands. 3. High-altitude areas	Plains areas: eluvial trans-eluvial transaccumulative transsuperaqual	Eluvial: island-topped flat-topped
8	Features of the permanent natural source of chemical elements entering the landscape	1. sedimentary rocks. 2. Magmatic rocks. 3. Intensely metamorphosed rocks	1. sedimentary rocks. 2. Magmatic rocks. 3. Intensely metamorphosed rocks	Calcareous rocks: Sidurian age, Pb-enriched Devonian age, metal-enriched, etc.

As a result of distribution and migration of chemical elements, dispersion and concentration of matter occur. The rate of movement and area of distribution depends on water and air flow, physico-chemical conditions, peculiarities of biological cycle and technogenesis. However, a decrease in the rate of spreading leads to accumulation of pollutants to abnormal concentrations. As a result, a geochemical barrier appears at the sites of such accumulations of chemical contaminants (Figure 3).

Currently, landscape geochemistry methods are actively used to solve environmental problems, including the identification of environmental crises in different regions by assessing the level of landscape pollution. These methods are used at all stages of assessing the state of local and regional natural-technogenic systems.

Figure 3. Scheme of geochemical barrier and its stages

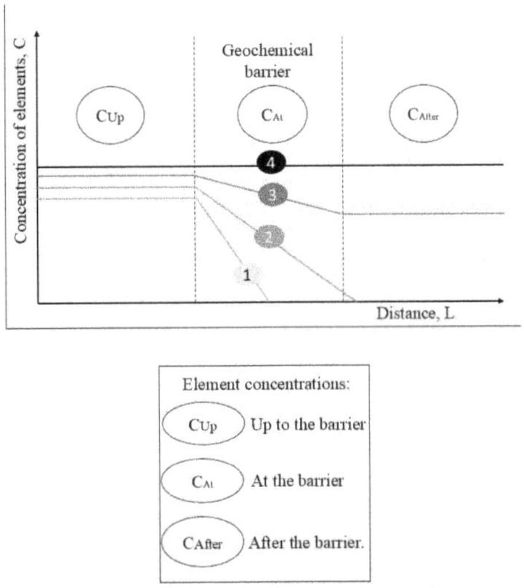

Stage	Concentration ratios in the barrier areas	Stage characterisation
1	$C_{Up} > C_{At}; C_{After} = 0$	The barrier is impenetrable
2	$C_{Up} > C_{At}; C_{After} \geq 0$	The barrier becomes permeable
3	$C_{Up} > C_{At} > C_{After}$	The barrier is partially permeable
4	$C_{Up} = C_{At} = C_{After}$	The barrier has exhausted its capacity

Assessment of the role of the geochemical barrier in ecology should be carried out as part of environmental monitoring of industrial and tourist-recreational areas. Geochemical analyses will identify areas of high concentration of pollutants and the most active migrants that are able to pass through the barrier.

When analyzing sustainable development indicators in tourism, they can be effectively divided into four groups: managerial, economic, sociocultural, and environmental. Environmental indicators focus on aspects that directly affect the ecological sustainability of a region. This category includes 11 primary and 15 additional indicators. For example, for the criterion "solid waste management," the main indicator is "the volume of waste generated in the region (tons per resident per year or month)," while the additional indicator is "the proportion of tourist enterprises sorting different types of waste." European sustainable development indicators are characterized by a detailed breakdown of all processes. Thus, the criterion "reducing transportation load" corresponds to indicators that take into account the distance traveled by a tourist in the region visited, aimed at analyzing air pollution from mobile sources and developing a strategy for using environmentally friendly modes of tourist transportation.

Below in Table 2 are the indicators of tourism's impact on the environment, among which in the framework of the study, we are interested in sewage treatment and water resource management. Based on positions 4.1 and 7.1, it follows that improper sewage treatment and high soil load can lead to chemical and physical pollution of the water area and landscape adjacent to the tourist complex. These pollutants may include waste containing phosphate ions, detergents, etc.

Table 2. Indicators of the impact of tourism on the environment (Chuburova Zh. T., 2019)

Reducing the impact of transport
1.1 Percentage of tourists and day visitors to the destination who use different modes of transport to get to the destination.
1.2 Percentage of tourists and day visitors who use local transport and public transport to get around the destination. 1.3 Average number of kilometres (km) tourists and visitors travel from their accommodation to the destination. 1.4. AverageCO2 emissions of tourists and day visitors
Climate change
2.1 Percentage of tourism businesses participating in climate change mitigation programmes. 2.2 Percentage of tourism infrastructure in «vulnerable areas»
Solid waste utilisation and management
3.1 Waste production per tourist night compared to total local waste per person (kg). 3.2 Percentage of tourism businesses that segregate waste types. 3.3 Percentage of total waste recycled per tourist compared to total waste recycled per place per year
Wastewater treatment
4.1 Number of wastewaters from a tourism destination or facility that has undergone at least two stages of treatment before discharge
Water management
5.1 Water consumption by tourists compared to the total water consumption of the local population per day. 5.2 Percentage of tourism enterprises taking measures to reduce water consumption
Energy consumption
6.1 Energy consumption of tourists compared to total energy consumption of the local population per day. 6.2 Percentage of tourism enterprises taking measures to reduce energy consumption. 6.3 Percentage of energy consumed annually from renewable sources (MWh) compared to the total energy consumption of the destination per year
Landscape and biodiversity protection
7.1 Percentage of local companies in the tourism sector that are actively engaged in the protection, conservation and management of local biodiversity and landscapes

The system of indicators of sustainable tourism development is also interesting. The indicators in this system are also grouped into four groups: indicators of the implementation of the concept of sustainable tourism development by the state, indicators of the impact of anthropogenic activities on the environment, socio-economic indicators of tourism sustainability. The group of indicators of anthropogenic impact is the largest and includes five subgroups of indicators (Table 3).

Table 3. Indicators of anthropogenic impact on the environment (Chuburova Zh. T., 2019)

Subgroup	**Name of indicators**
1. tourist transport (transport to and from holiday destinations, domestic transport)	Share of sustainable transport in total transport
	Number of short-term visitors per 1 km^2
	Movement within the territory of the host country
2. carrying capacity - land use, biodiversity and tourism activities	Maximum population density (peak season) per 1 km^2
	Sleeping places in dacha houses (% of total living space)
	Ratio of built-up and natural areas
	Size of protected natural areas (% of total target area)
	Percentage of natural shoreline

continued on following page

Table 3. Continued

Subgroup	Name of indicators
3. energy utilisation	Share of renewable energy in total energy consumption
	Energy use for tourism purposes
4. water utilisation	Rational use of water resources
	Proportion of homes and utilities with access to water treatment facilities
5. solid waste management	Percentage of solid waste selected for recycling
	Total amount of solid waste disposed to landfill and/or incineration plant
	Monthly waste production

Russian realities determine the conditions for the introduction of the above-mentioned indicators of the impact of tourism on the environment with some adjustments in the national accounting system. Taking into account the number and area of protected areas in the Russian Federation, the indicators related to tourism activities on the territory of national parks and state reserves should be allocated to a separate subgroup in the group of indicators - anthropogenic impact on the environment (Melkonyants A. A. et al, 2023).

Each water sample was taken at specific locations. For the Betta (Figure 4, Figure 5) and Mzymta Rivers (Figure 6, Figure 7), samples were taken from upstream, where there was no impact from tourist infrastructure, and downstream. Water from the Afips River (Figure 8) was taken from the Krepostnaya village. Each of the sampling locations will be labelled as: sampling location A (Betta River), sampling location B (Mzymta River) and sampling location C (Afips River).

Figure 4. Site A.1, water sampling location from the upper course of the Betta River. Coordinates 44.37694 N, 38.39889 E.

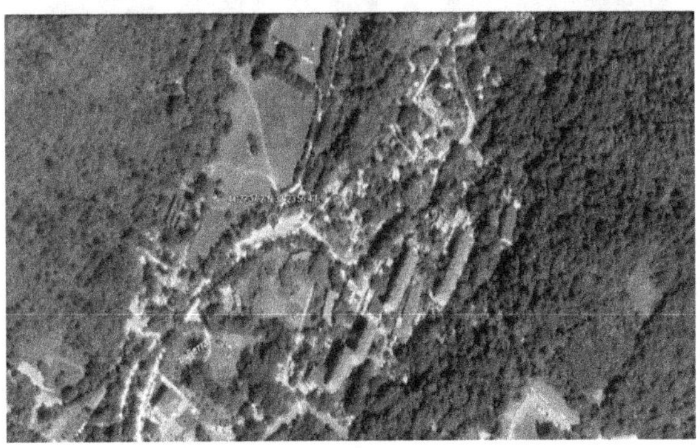

Source: Google Earth screenshot

Figure 5. Site A.2, water sampling location from the lower course of the Betta River. Coordinates 44.37417 N, 38.39667 E.

Source: Google Earth screenshot

Figure 6. Site B.2, water sampling location from the upper course of the Mzymta River. Coordinates 43.67083 N, 40.31917 E.

Source: Google Earth screenshot

Figure 7. Site B.2, water sampling location from the lower course of the Mzymta River. Coordinates 43.41806 N, 39.92611 E.

Source: Google Earth screenshot

Figure 8 Site B, water sampling location of the Afips River. Coordinates 44.70694 N, 38.67556 E.

Source: Google Earth screenshot

RESULTS

In order to study and analyse the impact of tourism on hydrological resources, field work was conducted in: Betta Farm, Rosa Khutor ski resort, Adler and Krepostnaya village. Water and soil samples were taken at each of these locations.

To assess the chemical composition of water, samples were taken from the Betta, Mzymta and Afips Rivers. The components investigated were: sulphates, nitrates, nitrites, phosphates and ammonium.

Sulphates are widespread in rivers and reservoirs, and their concentration varies from 5 to 80 mg/dm^3 depending on natural and climatic conditions. The main source of sulphates in aquatic systems is the transformation and destruction of minerals and rocks in the catchment under the influence of physical, chemical and biological factors. Sulphates can also enter aquatic systems through groundwater runoff and decomposition of organic matter by plants and animals. Large amounts of sulphates enter water bodies with industrial wastewater from various sources of pollution. The ionic form of sulphate ($SO4^{2-}$) is characteristic of natural waters with low salinity. The concentration of $SO4^{2-}$ in water is limited by the presence of calcium and magnesium ions forming poorly soluble salts ($CaSO_4$, $MgSO_4$).

The sulphate content of water is of great importance to the environment. Plants and other autotrophic organisms use sulphates dissolved in water for protein synthesis. After the death of living cells, heterotrophic bacteria release sulphur from proteins in the form of hydrogen sulphide, which is easily oxidised to sulphate in the presence of oxygen.

Sulphates play an active role in the complex sulphur cycle in aquatic systems. The concentration of sulphates affects the taste of water and has physiological effects on aquatic bioresources and human health (Seleznev V. A., 2021).

Nitrates and nitrites are products of nitrogenous metabolism of all living organisms, whether plants or animals, so in nature there are no products without nitrates. More than 100 mg of nitrates are formed and utilised daily in the human body in metabolic processes/

Nitrates, which serve as a source of nitrogen for protein synthesis in plants, have long been used as nitrogen fertilisers.

The most common of them are: potassium nitrate KNO_3, chilli nitrate $NaNO_3$, calcium nitrate $Ca(NO_3)_2$, ammonium nitrate NH_4NO_3.

Nitrates present in plants always enter the human and animal body in small amounts and develop there adaptive metabolic mechanisms. This distinguishes nitrates from pesticides and other chemical pollutants, which do not participate in the metabolic processes of mammals and are called xenobiotics.

The natural phosphorus cycle comes several stages of migration:

1. First, rocks release phosphorus ions and other minerals through weathering, which then enter the soil and water.
2. In the soil, phosphate is assimilated by plants and then by animals that consume it. When animals defecate (and/or decompose dead animals and plants), the phosphate returns to the soil. Natural terrestrial ecosystems usually recycle phosphorus many times before it is leached from the local environment. Phosphorus eventually enters the ocean through rivers, estuaries, and coastal waters.
3. Phosphorus entering the oceans can be divided into dissolved and solid phosphorus. Phosphorus particles are virtually insoluble due to the high pH of seawater and its buffering properties and is mostly deposited on the bottom of oceans and seas. Consequently, it does not participate in the active phosphorus cycle until tectonic forces return it to land.

The constant release of phosphate into the environment by humans leads to excessive concentrations of phosphorus in nature, which seriously disrupts the phosphorus cycle.

Increased phosphorus concentrations in surface waters favour the growth of phosphate-dependent organisms such as algae and cassava. These organisms consume large amounts of oxygen and prevent light from entering the water, making the water unsuitable for other organisms, thus leading to eutrophication.

By applying phosphate-rich fertilisers to the soil and using phosphate-containing detergents, humans have significantly altered the natural supply of phosphates. Phosphates are also added to some foods, such as cheese, sausage, and ham.

Excess phosphates can cause health problems such as osteoporosis or kidney problems. In addition, heavy drug use can lead to phosphate deficiency, which can lead to health problems (Mupegna, N. K. G., 2019).

Ammonia accumulates in water due to the breakdown of nitrogen-containing organic compounds (proteins, nucleic acids, etc.) by microorganisms. In most cases, ammonium compounds enter underground aquifers when fertilisers are applied to the soil. Consumption of water with an increased content of ammonium ions leads to acidosis and blood diseases.

The presence of ammonium nitrogen in water is often associated with the development of planktonic ammonificators. Ammonia can actively interact with water to form ammonium nitrogen according to the formula:

$NH_3 + H_2O \leftrightarrow (NH_4OH) \leftrightarrow NH_4^+ + OH^-$ (1)

Therefore, it occurs in water in two forms: ionised ammonium nitrogen (ammonium ion, NH_4^+) and non-ionised ammonium nitrogen (ammonia, NH_3).

The results of laboratory testing of water samples from the Betta River at sites A.1 and A.2 are presented in Table 4.

Table 4. Results of water sample measurements in the Betta River

Sites A	**Results of measurements with error, X± Δ(U)**					
	pH	**Sulfates, mg/dm³**	**Nitrates, mg/dm³**	**Nitrites, mg/dm³**	**Phosphates, mg/dm³**	**Ammonium, mg/dm³**
Sites A1	7,98	24,8	<0,2	<0,2	<0,25	<0,5
	±0,20	±2,5	–	–	–	–
Sites A2	7,84	23,0	<0,2	<0,2	<0,25	<0,5
	±0,20	±2,3	–	–	–	–

Calculations according to formula (2) are presented in the form of a graph (Figure 9). It can be concluded that the concentration of chemical substances at sites A.1 and A.2 do not exceed MAC, as it is less than 1.

Figure 9. Comparison of MAC indicators of chemical substances in the upper and lower reaches of the Betta River

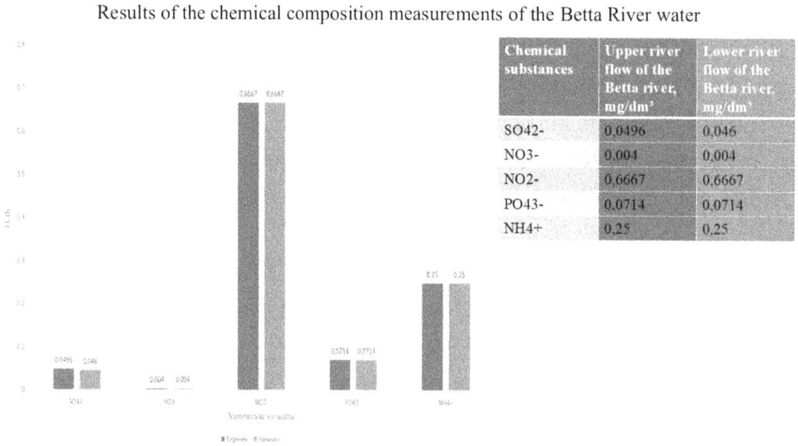

The results of laboratory testing of water samples from the Mzymta River at Sites B.1 and B.2 are presented in Table 5.

Table 5. Results of water sample measurements in the Mzymta River

Sites B	Results of measurements with error, X± Δ(U)					
	pH	**Sulfates, mg/dm³**	**Nitrates, mg/dm³**	**Nitrites, mg/dm³**	**Phosphates, mg/dm³**	**Ammonium, mg/dm³**
Sites B1	7,86	22,8	<0,2	<0,2	<0,25	<0,5
	±0,20	±2,3	–	–	–	–
Sites B2	7,91	24,2	<0,2	<0,2	<0,25	<0,5
	±0,20	±2,4	–	–	–	–

Calculations according to formula (2) are presented in the form of a graph (Figure 10). It can be concluded that the concentration of chemical substances at sites B.1 and B.2 do not exceed MPC, as it is less than 1.

Figure 10. Comparison of MAC indicators of chemical substances in the upper and lower courses of the Mzymta River

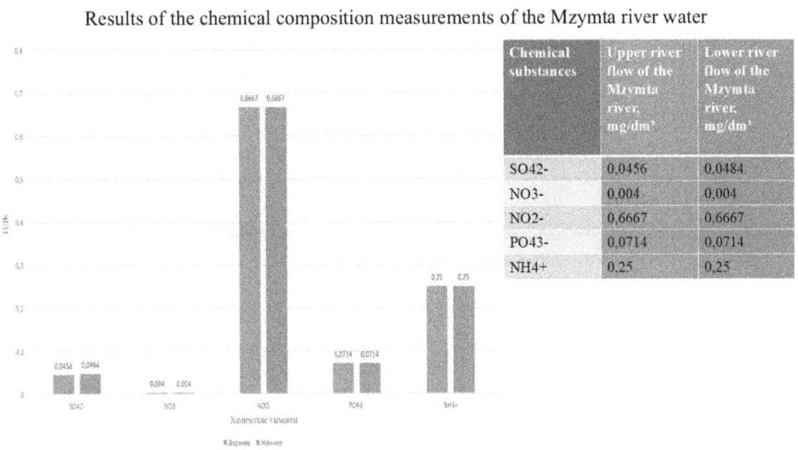

The results of laboratory testing of Afips River water samples are presented in Table 6.

Table 6. Measurement results of water samples in the Afips River

6Sites C	Results of measurements with error, X± Δ(U)					
	pH	**Sulfates, mg/ dm³**	**Nitrates, mg/ dm³**	**Nitrites, mg/ dm³**	**Phosphates, mg/ dm³**	**Ammonium, mg/ dm³**
Krepostnaya Stanitsa, Afips River	6,22	20,9	˂0,2	˂0,2	˂0,25	˂0,5
	±0,20	±2,1	–	–	–	–

Calculations according to formula (2) are presented in the form of a graph (Figure 11). It can be concluded that the concentration of chemical substances does not exceed the MPC, as it is less than 1.

Figure 11. Comparison of MAC indicators of chemical substances in the Afips River

Results of the chemical composition measurements of the Afips River water

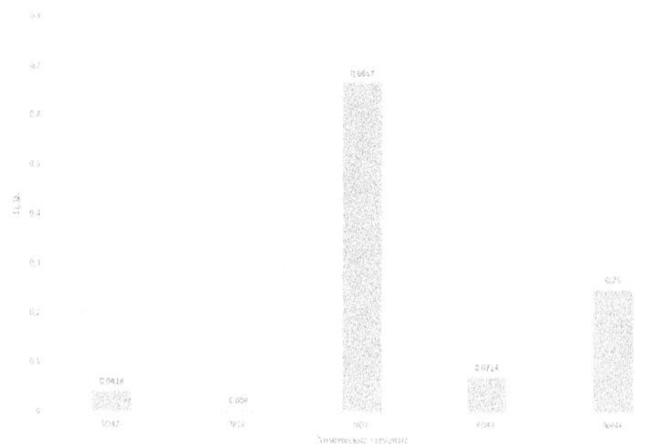

Chemical substances	Afips River
SO42-	0,0418
NO3-	0,004
NO2-	0,6667
PO43-	0,0714
NH4+	0,25

Based on the data and calculations obtained, it can be concluded that the investigated chemicals in the water do not exceed the MAC (Maximum Allowable Concentrations) standards. The only risk of exceeding the MAC is associated with a hazard class 2 chemical—nitrite (NO_2^-). Among all the chemical elements, it is the only one closest to exceeding MAC norms in the near future, which in turn can lead to dangerous consequences. Nitrites can have a significant impact on the environment. Consequently, nitrites can significantly affect the environment, influencing water quality, plant and animal health, as well as human health. When nitrogen compound levels, including nitrites, exceed plant requirements, excess nitrogen can be released into the environment and often enter water systems.

High levels of nitrites can poison plants that serve as food for animals. This can lead to biological imbalance and negatively impact biodiversity.

It is important to note that nitrites and nitrates can enter human water supplies through wastewater or industrial processes. The presence of these compounds in water above established norms indicates pollution.

Therefore, controlling nitrite levels in the environment is an important task from both environmental and public health perspectives.

To assess the chemical composition of the soil, samples were taken near the Betta, Mzymta, and Afips rivers. The components investigated included elements such as: benzo(a)pyrene, mercury, cadmium, cobalt, copper, arsenic, nickel, lead, and zinc.

Benzo(a)pyrene is an aromatic compound and belongs to the family of polycyclic hydrocarbons. This substance is classified as a first-class hazardous substance.

It forms during the combustion of liquid, solid, and gaseous hydrocarbon fuels and, to a lesser extent, during the combustion of gaseous fuel. In the environment, benzo(a)pyrene accumulates predominantly in the soil and, to a lesser extent, in water. It moves from the soil into plant tissues and spreads further along the food chain.

In its pure form, benzo(a)pyrene consists of yellow plates and needles that easily break into small pieces. It is easily soluble in non-polar organic solvents such as benzene, toluene, and xylene, but only slightly soluble in arctic solvents and practically insoluble in water. Benzo(a)pyrene is classified as a first-class hazardous substance, having a very dangerous impact on the environment and causing irreversible changes.

Anthropogenic sources of benzo(a)pyrene include industrial enterprises, thermal power plants, large and small heating plants, polluting the air over relatively limited areas, and transportation, which, as a mobile source, spreads emissions over large areas. It can be something permanent like a vehicle. One of the most common sources of benzo(a)pyrene is the combustion process of almost all types of combustible materials. Benzo(a) pyrene is found in exhaust gases, soot, and carbon deposits, as well as in chimneys and surfaces in contact with smoke, especially the tar-like substances in combustion products. Benzo(a)pyrene is also found in areas of natural forest fires and enters the atmosphere during volcanic eruptions (Shelepova, V. S., 2017).

Mercury is a first-class hazardous chemical element that enters the environment through evaporation from the Earth's crust and pollution from industrial infrastructure. Each year, between 25,000 and 12,000 tons of natural mercury emissions occur. This metal actively concentrates in vegetable crops such as cabbage, chili peppers, beans, and more. Mercury is also found in some seafood, cosmetic products, pesticides, and fungicides.

Cadmium is a rare element that does not usually occur in nature in its pure form; however, it is found in sulfides, oxides, and carbonates in lead, copper, and zinc ores. Cadmium is used for the production of nickel-cadmium batteries, anti-corrosion coatings, paint pigments, and semiconductors for solar panels. Therefore, cadmium enters the environment solely through anthropogenic impact. Research shows that only 80% of cadmium is used for secondary energy sources, and only 15% is returned to the production cycle. Some products containing cadmium can be recycled, but most pollution is caused by emissions from industrial enterprises, transport, and fertilizers (Teplaya, G. A., 2013).

Cobalt is an important micronutrient and component of vitamin B12. It is essential for the formation of red blood cells, enzymatic reactions, and the formation of nerve sheaths.

Copper is an important metal involved in physiological and biochemical processes, but at the same time, it is a toxic element causing a range of adverse reactions. As a trace element, copper is an indispensable participant in the electron transport chain of respiration, photosynthesis processes, oxidation reactions, protein and carbohydrate metabolism, and nitrogen fixation in the atmosphere by nodule bacteria. However, such significant participation of metal ions in processes and reactions is ensured by their relatively low concentrations. Exceeding a certain threshold leads to toxic effects, including the disruption of cell membrane structure and function, reduced efficiency of physiological processes (respiration, photosynthesis), and biochemical reactions by inhibiting enzymes, leading to inhibition of plant growth and development.

High environmental contamination with copper ions is caused by both natural disasters and anthropogenic impacts in the mining industry, ore processing, metal smelting, and alloy production. The use and disposal of waste in the form of worn-out household appliances and their components also lead to significant environmental pollution. This can result in secondary contamination of surface and groundwater (Ivanishchev, V. V., 2020).

Arsenic is an element found in more than 200 minerals, with roughly a third of emissions coming from natural sources. Anthropogenic sources of arsenic emissions include glass factories, the electrical industry, thermal power plants, and automobiles.

Nickel is one of the metals most frequently used by humans. Examples of applications include various industries, military, aerospace, and architecture. Nickel is used in alloys for the production of stainless and heat-resistant steels, and other materials. It can be recovered from obsolete installations and equipment for new use. However, removal and recycling operations, such as burning and recovery, pollute not only the soil but also water and air.

Natural processes such as rock weathering, soil erosion, and geological anomalies can lead to significant nickel concentrations in specific areas. Wind-blown dust, volcanic eruptions, forest fires, and some plants and animals also contribute to nickel dispersion in the environment.

Nickel and other heavy metal concentrations in soil affect the productivity and biodiversity of biocenoses. Agricultural activities in these areas also pose a health risk to humans.

Most pollution is anthropogenic. For example, coal combustion, metal mining and smelting, diesel fuel use, waste, sewage, burning nickel-cadmium batteries, electroplating, and using nickel-containing fertilizers. However, the most common sources of metal emissions into the environment are the metallurgical and electrochemical industries and the use of electric batteries and chemicals in the food industry (Ivanishchev, V. V., 2021).

On the other hand, nickel ions are absolutely essential for the life of all organisms. They are involved in forming active centers of several enzymes that ensure redox and other reactions. The most famous of these enzymes is urease, which splits urea.

Lead is a natural element found in soils, air, and water. An average of 27,000 tons enters the environment annually from natural sources, but most comes from anthropogenic sources. The highest lead concentrations in plants are near roads.

Zinc is the second most common transition metal in living organisms after iron and plays an important role in the structure of genetic material. It should be remembered that most biochemical processes are characteristic of all living organisms, so the importance of this chemical element for humans and animals is undeniable.

The primary source of metal entry into the soil is considered to be the leaching of parent rocks through relevant processes. This explains the differences in their natural background values in different media (rocks, soil, water). It has been shown that the zinc content in rocks and soils ranges from 10 to 300 mg/kg, while in the aquatic environment, it is present in slightly smaller quantities. Erosive processes lead to the redistribution of metal content. Volcanic activity is considered the main source of zinc, producing about 6 million tons of natural zinc per year.

Anthropogenic activities also contribute significantly to the zinc content. This is related to mining and industrial activities for metal smelting (zinc concentrate). The chemical industry, production of zinc-containing materials in the aviation, automotive (including rubber), and electronics industries, production of anti-corrosion coatings, production of oil paints, use of zinc-containing fertilizers, coal use, and the discharge of municipal wastewater, which is a source of energy at relevant thermal power plants, also significantly contribute to environmental pollution. For these reasons, the concentration of zinc in the soil is closely linked to its various sources. At the same time, the metal's toxicity depends on many factors that limit or promote its availability to plants and microorganisms (Ivanishchev, V. V., 2022).

Selection site A represents land plots next to the Betta River. Betta River. To denote the two sites, we will call them site A.1.1 and site A.2.2 Site A.1.1 is upstream and site A.2.2 is downstream.

The results of laboratory testing of soil samples near the Betta River at Sites A.1.1 and A.2.2 are presented in Table 7.

Table 7. Measurement results of soil samples

Sites A	Results of measurements with error, X± Δ(U)								
	Benz(a)pyrene, mg/kg	**Mercury, ppm**	**Cadmium ppm**	**Cobalt, mg/kg**	**Copper, mg/kg**	**Arsenic mg/kg**	**Nickel, mg/kg**	**Lead, mg/kg**	**Zinc, mg/kg**
A.1.1	˂0,005	0,039	0,14	6,8	15	6,6	17	7,4	61
	–	±0,018	±0,04	±2,0	±4	±2,0	±5	±0,2	±18
A.2.2	˂0,005	0,07	˂0,1	6,8	20	6,2	14	53	167
	–	±0,03	–	±2,0	±6	±1,8	±4	±16	±50

Calculations according to formula (2) are presented as a graph (Figure 12). It can be concluded that the concentration of cobalt (Co) in sections A.1.1 and A.2.2 exceeds the MAC, as it is greater than 1.

Figure 12. Comparison of MAC Indicators of Chemical Elements in the Soil of the Upper and Lower Courses of the Betta River

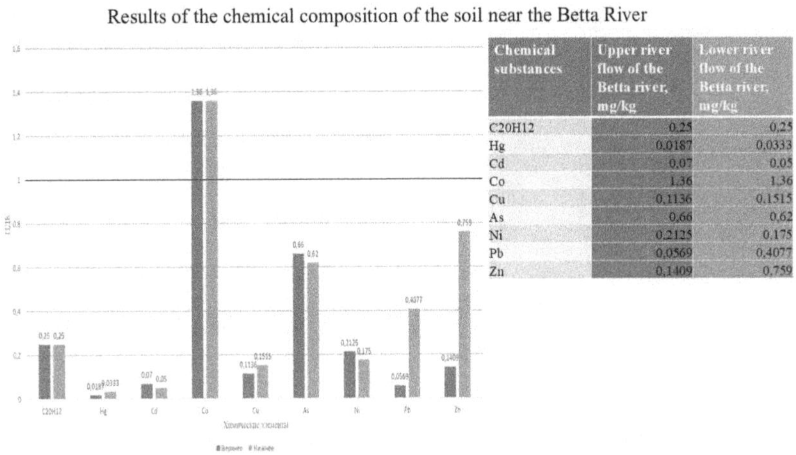

Site B represents areas of land near the Mzymta River. To denote the two sites, let's call them Site B.1.1 and Site B.2.2. Site B.1.1 is located upstream and Site B.2.2 downstream.

The results of laboratory tests of soil samples near the Mzymta River at sites B.1.1 and B.2.2 are presented in Table 8.

Table 8. Results of Soil Sample Measurements near the Mzymta River

Sites B	Results of measurements with error, X± Δ(U)								
	Benz(a)pyrene, mg/kg	**Mercury, ppm**	**Cadmium ppm**	**Cobalt, mg/kg**	**Copper, mg/kg**	**Arsenic mg/kg**	**Nickel, mg/kg**	**Lead, mg/kg**	**Zinc, mg/kg**
B.1.1	<0,005	0,009	<0,1	6,7	13	<1,0	12	7,5	38
	–	±0,004	–	±2,0	±4	–	±4	±2,2	±11

Sites B	Results of measurements with error, X± Δ(U)								
B.2.2	<0,005	0,008	<0,1	6,7	12	<1,0	11	9,5	35
	–	±0,004	–	±2,0	±4	–	±4	±2,8	±10

Source: Authors

Calculations according to formula (2) are presented as a graph (Figure 13). It can be concluded that the concentration of cobalt (Co) in sections B.1.1 and B.2.2 exceeds the MAC, as it is greater than 1.

Figure 13. Comparison of MAC Indicators of Chemical Elements in the Soil of the Upper and Lower Courses of the Mzymta River

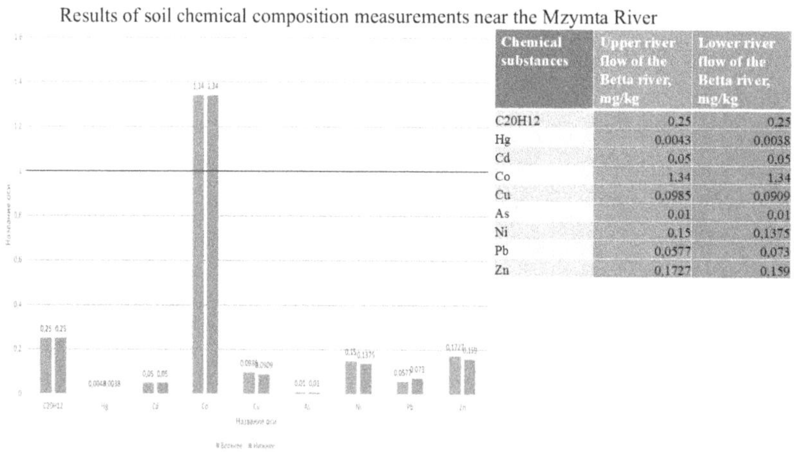

Source: Authors

Site C.1 represents a plot of land near the Afips River. The results of laboratory tests of soil samples near the Afips River at site C.1 are presented in Table 9.

Table 9. Results of Soil Sample Measurements near the Afips River

Sites C	Results of measurements with error, X± Δ(U)								
	Benz(a)pyrene, mg/kg	**Mercury, ppm**	**Cadmium ppm**	**Cobalt, mg/kg**	**Copper, mg/kg**	**Arsenic mg/kg**	**Nickel, mg/kg**	**Lead, mg/kg**	**Zinc, mg/kg**
Sites C.1	˂0,005	0,08	˂0,1	18	27	27	28	13	98
	–	±0,04	–	±5	±7	±7	±8	±4	±29

Source: Authors

Calculations according to formula (2) are presented as a graph (Figure 14). It can be concluded that the concentration of cobalt (Co) and arsenic (As) at site C.1 exceeds the MAC, as it is greater than 1.

Based on the available data, it can be concluded that in areas with high tourist load, the MAC for cobalt and arsenic may be exceeded. This may be due to various factors, including the natural presence of these elements in the environment and anthropogenic impacts.

Figure 14. Comparison of MAC Indicators of Chemical Elements in the Soil near the Afips River

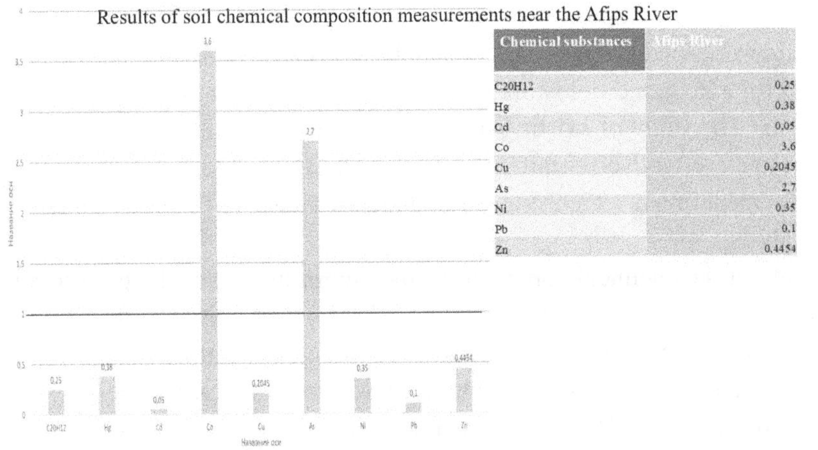

Based on the results of all soil samples, several conclusions can be drawn:

1. The presence of cobalt (Co) at all three sites exceeds the MAC, which, in turn, can lead to serious environmental problems. Cobalt, as a chemical element, is used for the production of super-hard alloys, enamels, paints, a component for fertilizers, and in the burning of coal and fuel. Therefore, it can be assumed that the presence of cobalt at all three sites is associated with the disposal of solid household waste (SHW), coal and fuel burning, or during the construction of tourist infrastructure.
2. When comparing the indicators at the Betta River sites, it can be noted that the MAC for copper (Cu), lead (Pb), and zinc (Zn) is higher at site B 2.2, which is downstream. Near this site is the "Southern" hotel, a bridge over the river, and the "Orbit" camping park. These chemical elements could have entered the soil as a result of SHW discharge, wastewater, or during the construction stage. They do not exceed the MAC, but with the high growth rate of tourists in the Betta area, these indicators may increase.
3. Regarding the Betta River, the Rosa Khutor ski resort meets the MAC indicators for chemical elements in the soil, except for cobalt. This can be explained by the fact that in 2016, the State Duma of the Russian Federation adopted amendments to the law on specially protected natural areas (SPNA) allowing the construction of reserves, and in February 2018, announced increased control over environmental components and the adoption of a new sustainable development program.
4. The concentration of arsenic (As) near the Afips River in Krepostnaya Stanitsa exceeds the MAC, which, like cobalt, can lead to serious problems for both nature and humans. Arsenic is a natural component of many natural objects and can be present in high concentrations in groundwater. At the same time, human activities, including industrial production and tourism, can contribute to additional environmental pollution with arsenic.

It is important to note that exceeding the MAC for these elements can pose a danger to human health and the ecosystem as a whole. Therefore, it is necessary to conduct regular monitoring of these elements' levels, especially in areas with high tourist loads, and take measures to prevent their contamination.

DISCUSSION

In the course of this study, several tasks were performed, including the study of existing literature, especially scientific works dedicated to the study and determination of methods for studying geosystems under tourist use, exploratory and fieldwork, data analysis from laboratory studies, and the interpretation of the obtained data. The study of existing literature provided a general understanding of the problem and identified the most effective research methods. Determining methods applicable to geosystems allowed the development of a research strategy and identification of the most suitable methods for data collection and analysis.

For further study of the chemical composition and determination of the pollution level, exploratory fieldwork was carried out, including the sampling of water and soil. The analysis of laboratory data allowed for the determination of MAC (Maximum Allowable Concentrations) and AAC (Approximately Allowable Concentrations) in water and soil, which is an important stage in assessing the impact of tourist and recreational activities on hydrological resources. The interpretation of experimental results made it possible to draw conclusions about the impact of tourist and recreational activities on hydrological resources.

No exceedance of MAC in the investigated components in water was found; however, the nitrite content in water is higher than other chemical substances. Nitrites enter the water from decomposition products and wastewater and are classified as a hazard class 2 chemical. If emissions are not controlled, poisoning of the local population can be observed in the near future. The soil in the three studied sites showed an exceedance of the MAC for cobalt. This element is used in the manufacture of super-hard alloys, paints, and is also a component of coal and fuel. Cobalt, like arsenic, enters the soil during the combustion of coal and fuel. Road dust can be a source of pollution.

These results can be used to develop management strategies and measures to reduce the impact of tourist activities on the environment. Overall, the results of the study emphasize the importance of a comprehensive approach to studying the impact of tourist and recreational activities on hydrological resources and the need for further research in this area.

ACKNOWLEDGMENTS

The research is carried out with the support of the Ministry of Science and Higher Education of the Russian Federation (No. FZEN-2023-0012 "Fundamental aspects of rational recreational environmental management and determination of anthropogenic load in the context of sustainable tourism development")

REFERENCES

Alekseenko, V. (2013). *Environmental Geochemistry: Textbook for students specializing in "Environmental Management" in the field of "Ecology and Environmental Management" / V. Alekseenko, S. A. Buzmakov, M. S. Panin*. Perm State National Research University.

Bertalanffy, L. (1969). *General Systems Theory: A Critical Review. Studies on General Systems Theory / L. Bertalanffy* (Yulina, N. S., Trans.). Progress.

Bespalova, S. I. (2022). The Impact of Tourism on the Environment of the Local Community / S. I. Bespalova, S. V. Bespalova // Arctic Ecosystems: Conservation and Sustainable Development: Materials of the All-Russian Scientific and Practical Conference with International Participation– Murmansk: Murmansk Arctic State University, 2022. – pp. 42–45.

Ceballos-Lascurain, H. (1996). *Tourism, ecotourism, and protected areas: The state of nature-based tourism around the world and guidelines for its development*. Iucn.

Dolgal, A. S. (2012). *Integration of Geophysical Methods: Textbook for students specializing in "Geophysics" and the field of "Geology" / A. S. Dolgal*. Perm State National Research University.

Dzhanzhugazova, E. A. (2020). *Marketing Technologies in Tourism: Marketing of Tourist Territories: Textbook / E. A. Dzhanzhugazova*. Yurait Publishing House.

Kosinova, I. I. (2006). *Ecological and Geological Monitoring of Technogenic-Loaded Territories / I. I. Kosinova, V. V. Ilyash, A. E. Kosinov*. Voronezh State University.

Kuzin, V. Yu. (2022). *Potential of Tourist Areas: Textbook for Master's Students in the field of "Geography" / V. Yu. Kuzin; Ministry of Science and Higher Education of the Russian Federation, North-Eastern Federal University named after M. K. Ammosov, Institute of Natural Sciences*. NEFU Publishing House.

Maergoiz, I. M. (1986). *Territorial Structure of the Economy / I. M. Maergoiz*. Nauka.

Mazhar, L. Yu. (2008). *Territorial Tourist and Recreational Systems / L. Yu. Mazhar; Smolensk Humanitarian University*. Universum.

Tourist and Recreational System. (2020). *Theory and Practice of Organization / N. V. Penkina, O. Yu. Shakhova, A. A. Nikiforova* [et al.]. Nizhnevartovsk State University.

Zorin, I. V. (2005). *The Phenomenon of Tourism: Selected Works / I. V. Zorin*. Nauka.

Chapter 12
Harnessing Biodiversity Tourism:
A Framework for Regenerative Conservation Management in the Airline Industry

Benjapol Worasuwannarak
https://orcid.org/0009-0005-2332-2154
College of Hospitality Industry Management, Suan Sunandha Rajabhat University, Thailand

ABSTRACT

As global biodiversity declines and tourism-related pressures on ecosystems increase, the airline industry has a unique opportunity to support regenerative conservation through biodiversity tourism. This study explores the role of airlines in promoting biodiversity-focused tourism as a means of fostering conservation management that not only minimizes environmental harm but actively regenerates ecosystems. The findings reveal that while biodiversity tourism offers substantial potential for conservation, it must be carefully managed to avoid unintended environmental impacts. Airlines can contribute to regenerative conservation management by adopting sustainable travel practices, investing in carbon offset programs, and partnering with conservation organizations to support eco-friendly tourism infrastructure. By promoting biodiversity tourism as part of a broader sustainability strategy, the airline industry can play a crucial role in preserving ecosystems and supporting regenerative conservation efforts.

INTRODUCTION

1.1 Background of the Study

Biodiversity tourism refers to a subset of eco-tourism that specifically focuses on visiting regions rich in biodiversity to help promote conservation, environmental awareness, and economic development in local communities (Buckley, 2018). By traveling to areas such as national parks, wildlife sanctuaries, and biodiversity hotspots, tourists can directly contribute to the preservation and restoration of these regions through fees, donations, and responsible travel behaviors (Honey, 2018). Biodiversity tourism

DOI: 10.4018/979-8-3693-8232-5.ch012

aligns with regenerative conservation, which goes beyond sustainable practices to actively restore ecosystems and ensure their long-term health (Gibbons & Newsome, 2020).

The airline industry, as a critical enabler of global travel, has a unique role in promoting biodiversity tourism. However, this role comes with significant challenges, as air travel is responsible for about 2.5% of global carbon dioxide (CO2) emissions (International Air Transport Association [IATA], 2020). The emissions from long-haul flights, in particular, can contribute to climate change, which threatens the very ecosystems that biodiversity tourism seeks to protect (Becken, 2019). This presents a paradox: while airlines facilitate biodiversity tourism, their operations also have an environmental impact that must be addressed if they are to truly support regenerative conservation (Peeters et al., 2019).

1.2 Problem Statement

The aviation sector is facing increasing pressure to mitigate its environmental footprint while maintaining its role in global tourism. As more tourists seek eco-friendly travel experiences, the demand for biodiversity tourism is rising. Airlines, as facilitators of this form of tourism, are uniquely positioned to promote biodiversity-focused destinations and engage in conservation efforts. However, the challenge lies in balancing the economic benefits of biodiversity tourism with the environmental costs associated with air travel. This research addresses how airlines can support biodiversity tourism in a way that aligns with regenerative conservation principles, minimizing harm while actively contributing to ecosystem restoration (Gibbons & Newsome, 2020).

1.3 Objectives of the Study

The main objective of this research is to develop a framework that enables airlines to incorporate biodiversity tourism into their sustainability strategies, thereby supporting regenerative conservation management. The specific objectives are:

1. To explore how airlines can collaborate with conservation organizations to promote biodiversity tourism.
2. To assess the role of airlines in educating travelers about sustainable and responsible tourism practices.
3. To investigate the challenges airlines face in balancing the promotion of biodiversity tourism with the need to reduce their carbon footprint.
4. To propose practical strategies for airlines to support biodiversity tourism while adhering to regenerative conservation principles.

1.4 Research Questions

This study is guided by the following research questions:

- How can airlines balance promoting biodiversity tourism with their environmental responsibilities, particularly in terms of carbon emissions?
- What are the potential economic and conservation benefits of biodiversity tourism, and how can airlines support these outcomes?

- What partnerships and practices can airlines adopt to promote eco-friendly travel and support regenerative conservation?

1.5 Significance of the Study

As the airline industry grapples with its environmental responsibilities, the findings of this study will provide valuable insights into how airlines can align their operations with global conservation goals. Biodiversity tourism represents a growing niche within the travel industry, offering significant economic opportunities for airlines while also serving as a tool for conservation (Buckley, 2018). By examining how airlines can integrate biodiversity tourism into their sustainability strategies, this study contributes to the broader discourse on sustainable aviation and offers practical solutions for reducing the environmental impact of air travel (Spenceley, 2020).

LITERATURE REVIEW

2.1 Biodiversity Tourism and Conservation

Biodiversity tourism has emerged as a key mechanism for promoting **conservation** in areas with high ecological significance. Unlike conventional tourism, which can lead to environmental degradation through overdevelopment, biodiversity tourism seeks to preserve natural ecosystems by generating revenue for conservation initiatives (Honey, 2018). Tourists visiting biodiversity hotspots, such as rainforests, coral reefs, and wildlife reserves, often pay entrance fees or make donations that directly fund conservation projects aimed at protecting endangered species, restoring habitats, and maintaining biodiversity (Reynolds & Braithwaite, 2021).

However, while biodiversity tourism can provide critical financial support for conservation efforts, it also presents several challenges. Over-tourism, for example, can lead to habitat degradation, pollution, and increased pressure on already fragile ecosystems (Peeters et al., 2019). To prevent these negative outcomes, biodiversity tourism must be carefully managed, with strict regulations on visitor numbers, the development of sustainable infrastructure, and active engagement from local communities (Buckley, 2018).

2.2 The Environmental Impact of the Airline Industry

The aviation industry is a significant contributor to greenhouse gas emissions, primarily through the combustion of fossil fuels (IATA, 2020). As the global demand for air travel continues to rise, the industry's contribution to climate change is expected to increase, unless significant changes are made to reduce emissions (Becken, 2019). Airlines have taken steps to improve fuel efficiency, explore alternative fuels such as biofuels, and implement carbon offset programs, but these efforts have not yet resulted in a substantial reduction in the industry's overall environmental impact (Peeters et al., 2019).

Carbon offset programs, in particular, have been criticized for their lack of transparency and effectiveness. While they allow passengers to compensate for the carbon emissions generated by their flights, the environmental benefits of such programs are often difficult to quantify (Buckley, 2018). Additionally, the reliance on offsets may divert attention away from the need for more fundamental changes in the aviation industry, such as the development of electric aircraft or the adoption of renewable energy sources (Becken, 2019).

2.3 Airlines and Regenerative Conservation

The concept of regenerative conservation is gaining traction as a more proactive approach to environmental management. While traditional conservation efforts focus on sustaining current ecosystems, regenerative conservation aims to restore and rejuvenate ecosystems that have been degraded by human activities (Gibbons & Newsome, 2020). This approach aligns with the principles of biodiversity tourism, which seeks not only to minimize harm but also to actively contribute to the restoration of natural environments.

Airlines can play a significant role in supporting regenerative conservation by promoting eco-friendly travel and partnering with conservation organizations to fund projects that restore ecosystems. For example, airlines could invest in reforestation projects, marine conservation efforts, or species recovery programs in the regions they serve (Buckley, 2018). Additionally, airlines can use their platforms to raise awareness about the importance of biodiversity conservation, educating travelers on how to reduce their environmental impact during their trips (Reynolds & Braithwaite, 2021).

RESEARCH METHODOLOGY

3.1 Research Design

This study employs a qualitative research design to explore the potential role of airlines in promoting biodiversity tourism and supporting regenerative conservation management. Qualitative research is well-suited to this study as it allows for an in-depth exploration of stakeholder perspectives, enabling the researcher to gather rich, contextual insights into how airlines can engage in biodiversity tourism and the challenges they face in balancing environmental and economic goals (Creswell & Poth, 2018).

3.2 Data Collection

The primary data for this study will be collected through semi-structured interviews with key stakeholders involved in biodiversity tourism, conservation, and the airline industry. Participants will include:

- **Airline sustainability officers** responsible for corporate environmental strategies.
- **Conservation experts** from NGOs or government agencies focused on biodiversity protection and eco-tourism.
- **Local eco-tourism operators** and community leaders from biodiversity-rich regions.

The semi-structured interview format will allow for open-ended discussions on topics such as the potential for airlines to promote biodiversity tourism, the environmental challenges associated with air travel, and the ways in which airlines can collaborate with conservation organizations to support regenerative conservation (Braun & Clarke, 2006).

3.3 Sampling and Participants

Participants will be selected using purposive sampling, ensuring that individuals with relevant expertise in biodiversity tourism and sustainable aviation are included. Approximately 20-25 participants will be interviewed, with representation from different geographical regions to reflect the global nature of the airline industry and the diverse ecosystems involved in biodiversity tourism (Creswell & Poth, 2018). By including stakeholders from regions such as Africa, Southeast Asia, and South America, the study will capture a range of perspectives on how airlines can tailor their biodiversity tourism strategies to specific ecosystems.

3.4 Data Analysis

Thematic analysis will be used to analyze the interview data. This method involves identifying, analyzing, and reporting patterns or themes within the data, providing a structured way to interpret qualitative responses (Braun & Clarke, 2006). After transcribing the interviews, the researcher will begin by coding the data into key themes that emerge from participants' responses. These themes may include:

Table 1. Key theme of this study

Key Theme	Explanation
Airlines' role in conservation	Participants' perspectives on how airlines can support biodiversity tourism by funding conservation projects or promoting eco-friendly destinations.
Sustainability challenges	Insights into the environmental challenges airlines face, particularly regarding carbon emissions, and how they are being addressed.
Traveler education	How airlines can educate tourists about responsible travel behaviors and reduce their environmental impact
Collaboration opportunities	Suggestions for partnerships between airlines, conservation organizations, and local communities to promote biodiversity tourism and regenerative conservation.

Once the themes are identified, they will be analyzed in relation to the study's research questions, and the findings will be interpreted in the context of existing literature on biodiversity tourism and sustainable aviation. This process will help to develop a comprehensive understanding of the key strategies airlines can adopt to balance economic viability with environmental stewardship.

RESULTS AND ANALYSIS

4.1 Qualitative Findings

The qualitative interviews are expected to reveal a range of insights into how airlines can integrate biodiversity tourism into their sustainability strategies while supporting regenerative conservation efforts. Key themes are likely to emerge around the potential for airlines to promote eco-friendly destinations and market biodiversity tourism as a way to appeal to environmentally conscious travelers (Buckley, 2018). Airline representatives may express interest in expanding their corporate sustainability initiatives to include biodiversity tourism, viewing it as an opportunity to enhance their environmental credentials while generating revenue from tourism (Becken, 2019).

Participants from conservation organizations and eco-tourism operators are likely to emphasize the importance of local community involvement in biodiversity tourism. They may highlight that for biodiversity tourism to be successful, local communities need to play a central role in managing tourism activities and ensuring that revenues are reinvested in conservation efforts (Honey, 2018). Stakeholders may also stress the need for airlines to help mitigate the environmental impact of tourism by investing in sustainable infrastructure and promoting responsible travel practices (Reynolds & Braithwaite, 2021).

4.2 Key Themes

1. Corporate Sustainability and Biodiversity Tourism:

Airline representatives are expected to view **biodiversity tourism** as a means to align their operations with sustainability goals while appealing to the growing market of eco-conscious travelers. By partnering with conservation organizations and promoting biodiversity hotspots, airlines can both support conservation efforts and enhance their brand as socially and environmentally responsible businesses (Spenceley, 2020).

2. Challenges in Reducing Carbon Emissions:

One of the major challenges discussed by participants is the environmental cost of air travel, particularly the carbon emissions associated with long-haul flights to biodiversity hotspots. While carbon offset programs offer some mitigation, participants may argue that more substantive changes—such as investments in sustainable aviation fuels (SAFs) and electric aircraft—are necessary to significantly reduce the airline industry's environmental impact (Peeters et al., 2019). Airline executives might discuss the difficulties they face in adopting such technologies due to the high costs and lack of infrastructure.

3. Traveler Education and Sustainable Practices:

A key theme from the interviews will likely be the role of airlines in educating travelers about sustainable tourism. Participants may suggest that airlines could incorporate sustainability messaging into their marketing materials and in-flight services, helping to raise awareness among travelers about how to minimize their environmental impact during their trips (Reynolds & Braithwaite, 2021). Eco-tourism operators might emphasize the need for airlines to collaborate with local stakeholders to ensure that tourism activities are conducted in a way that supports conservation and benefits local communities.

4. Collaboration for Regenerative Conservation:

Partnerships between airlines and conservation organizations are seen as critical for promoting regenerative conservation through biodiversity tourism. By working with local communities and environmental NGOs, airlines can ensure that tourism revenues are used to fund projects that restore ecosystems, protect endangered species, and support local livelihoods (Gibbons & Newsome, 2020). These collaborations could also involve airlines providing financial or logistical support for conservation initiatives, such as reforestation projects or marine habitat restoration.

CONCLUSION

The findings of this study suggest that airlines have the potential to play a pivotal role in promoting biodiversity tourism and supporting regenerative conservation. By partnering with conservation organizations, promoting eco-friendly destinations, and educating travelers about sustainable travel practices, airlines can contribute to both environmental stewardship and economic development in biodiversity-rich regions (Buckley, 2018). However, the environmental costs associated with air travel, particularly carbon emissions, remain a significant barrier. While airlines have implemented carbon offset programs and taken steps to improve fuel efficiency, these efforts must be expanded to meet the growing demand for eco-friendly travel (Peeters et al., 2019).

The study concludes that airlines must adopt a multi-faceted approach to biodiversity tourism, integrating it into their broader sustainability frameworks. This includes investing in alternative fuels, promoting responsible travel behaviors, and forming long-term partnerships with conservation organizations to ensure that tourism contributes to the regeneration of ecosystems. By doing so, airlines can position themselves as leaders in sustainable tourism and help address the pressing environmental challenges associated with climate change and biodiversity loss (Gibbons & Newsome, 2020).

RECOMMENDATIONS

1. Expand Carbon Offset Programs:

Airlines should expand their carbon offset programs to ensure that the funds generated are directly tied to biodiversity conservation projects. By partnering with local governments and NGOs, airlines can invest in reforestation, habitat restoration, and species protection initiatives that offset the environmental impact of air travel (Peeters et al., 2019).

2. Invest in Sustainable Aviation Technologies:

Airlines must accelerate their investment in sustainable aviation technologies, such as biofuels and electric aircraft, to reduce their carbon emissions. These technologies, while still in development, offer the potential to drastically reduce the environmental impact of long-haul flights (Becken, 2019). Government policies and subsidies should also be introduced to incentivize the adoption of sustainable aviation fuels.

3. Educate Travelers on Sustainable Tourism:

Airlines should implement traveler education campaigns that raise awareness about the importance of responsible travel. These campaigns could be integrated into airlines' in-flight services, websites, and social media platforms, providing travelers with practical tips on how to minimize their environmental footprint during their trips (Reynolds & Braithwaite, 2021).

4. Form Strategic Partnerships for Conservation:

Airlines should form strategic partnerships with conservation organizations to promote regenerative conservation in biodiversity hotspots. These partnerships could involve direct financial support for conservation projects or the development of eco-friendly tourism infrastructure in collaboration with local communities (Gibbons & Newsome, 2020). By working closely with stakeholders, airlines can ensure that biodiversity tourism generates long-term benefits for both the environment and the communities that rely on tourism.

REFERENCES

Becken, S. (2019). Air travel and tourism: Impacts on climate change. *Tourism Management*, *74*, 343–354. DOI: 10.1016/j.tourman.2018.12.008

Braun, V., & Clarke, V. (2006). Using thematic analysis in psychology. *Qualitative Research in Psychology*, *3*(2), 77–101. DOI: 10.1191/1478088706qp063oa

Buckley, R. (2018). Tourism and conservation: The case of national parks. *Tourism Management*, *70*, 315–329. DOI: 10.1016/j.tourman.2018.03.014

Creswell, J. W., & Poth, C. N. (2018). *Qualitative inquiry and research design: Choosing among five approaches* (4th ed.). SAGE Publications.

Gibbons, L. V., & Newsome, D. (2020). Regenerative tourism: Moving beyond sustainability to achieve restoration. *Journal of Ecotourism*, *19*(3), 227–240. DOI: 10.1080/14724049.2020.1843813

Honey, M. (2018). *Ecotourism and sustainable development: Who owns paradise?* (2nd ed.). Island Press.

International Air Transport Association (IATA). (2020). Air transport: Aviation's global emissions profile. Retrieved from https://www.iata.org/en/policy/environment/climate-change/

Peeters, P., Higham, J., Kutzner, D., Cohen, S., & Gössling, S. (2019). Are technological innovations enough to decouple growth of air travel from its environmental impacts? *Journal of Sustainable Tourism*, *27*(7), 949–965. DOI: 10.1080/09669582.2018.1538509

Reynolds, P. C., & Braithwaite, D. (2021). The role of ecotourism in conservation: An empirical study. *Conservation Biology*, *35*(4), 1256–1264. DOI: 10.1111/cobi.13765 PMID: 33274484

Spenceley, A. (2020). Nature-based tourism: Critical issues for conservation and development. *Tourism Geographies*, *22*(5), 877–901. DOI: 10.1080/14616688.2020.1750689

Chapter 13
Assessment of Religious Tourism in Promotion of Interfaith Harmony in Kartarpur

Sabeen Azam
https://orcid.org/0000-0003-3148-3992
National University of Modern Languages, Karachi, Pakistan

ABSTRACT

This study emphasis on the potentiality of the religious tourism in fostering the interfaith harmony for this the Kartapur is highlighted as a model and symbol for the projection of public diplomacy and cultural exchange. The key objective of the study is to explore the effect of religious tourism on the peace and interfaith relations in Kartarpur. The qualitative methodology is been utilized to explore the key objective of the study. The findings of the study suggest that the cultural exchange programs and shared spiritual experiences can enhance and promote cross-cultural tolerance and understanding. This study also recommends that celebrations and interfaith dialogues should be more actively initiated to harness the religious tourism for the purpose of the peace building. Further, for respecting the religious diversity, responsible tourism practices along preserving the biodiversity of the place should be more practically practiced.

INTRODUCTION

Religious tourism is to travel intended for spiritual purposes or visiting the sites which owe religious significance. Religious tourism also is defined as a travel to the spiritual sites of religious significance, which serves mostly as a means for personal reflection, community building and cultural exchange (Cohen E., 2012). In a rising and interconnected world, as globalization is boosting. The pivotal importance of the interfaith harmony has by no means more pronounced, as now it plays an important role in mitigating inconsistencies and foster peace between the diverse communities around the globe (Husain, 2014). This kind of tourism encompasses a diverse range of activities, including participation in religious festivals, pilgrimage and visits to sacred sites. The traditional and modern scope of the religious tourism

DOI: 10.4018/979-8-3693-8232-5.ch013

is extending beyond just mere travel; it mostly involves deeper associations to community practices, cultural heritage, and personal spirituality.

Since the history the roots of religious tourism can be traced back to thousands of years where sacred sites attract the devotees to attain their spiritual needs like Mecca, Varanasi, Jerusalem as central destinations which besides their religious significance also draws millions annually. Currently the religious tourism in emerged into a prominent sector in the global travel industry which includes diverse practices and beliefs (Tepelus C., 2018).

Often the Religious tourism is categorized into many forms, including:

1. **Pilgrimages**: This type of tourism solely focuses on the religious Journeys towards sacred places, like the Hajj to Mecca or the Camino de Santiago in Spain.
2. **Heritage Tourism**: The heritage tourism is the visit of the significant historical, cultural and architectural places which includes mosques, churches and temples.
3. **Spiritual Retreats**: In the modern era there are designed programs around the globe for groups and individuals for personnel growth by means of yoga, meditation and other spiritual practices. Such practices have also increased the religious tourism.
4. **Religious Festivals and Fairs**: Religious festivals such as Muslim Eids, Hindu Diwali, for instance, and Christian Christmas are cultures appealing to diverse crowds by promoting interfaith relations through joint celebrations.

INTERFAITH RELATIONSHIPS AND CULTURAL UNDERSTANDING

Timothy (2006) stated that by the means of religious tourism there can be a positive fostering towards the understanding of various beliefs and cultures. Thus by this way world peace and intercultural dialogues can be enhanced. There are positive impacts of religious tourism on interfaith relationships. Few are mentioned here;

Cross-Cultural Understanding and Exchange

Religious tourism usually heralds cultural exchange, which fosters mutual respect among participants of different religions. According to (Cohen E., 2004), engaging religious tourists with local communities can raise sociable challenges against traditional beliefs in the midst of mutual respect. No doubt one of the most crucial sparks for such a change is where there is real conflict history, especially in viewing the people as part of human nature rather than on hate.

For instance, the Hajj pilgrimage to Mecca unites millions of Muslims from all walks of life. As noted by (Esposito J. L., 2011), the experience gives rise to a sense of global Muslim identity as well as reinforces bonds across national and ethnic lines. Similarly, experiences shared with respect to locales such as the Western Wall and the Dome of the Rock in Jerusalem promote interfaith conversation and understanding.

Shared Sacred Spaces

The religious sites belong to the same faith tradition. According to (Bianchi, 2006), a shared religious experience gives many grounds for cooperation and mutual understanding. Jerusalem city is a good example where the three converge that is Judaism, Christianity, and Islam. In fact pilgrims and tourists in these shared sacred areas can break the barrier of faith by building community and mutual respect.

Economic Benefits and Community Development

Religious tourism can also bring in the element of local economies, utilizing available resources to foster deeper interfaith connection. As (Smith M. K., 2015) points out, religious tourism creates a basis for work, which helps and blends together many different religions. For example, in cities like Varanasi where Hinduism is the most prominent religion, then presence of tourist will surely boost economic terms, thereby improving social bonding, and fosters healthy positive interactions among the faiths.

Educational Opportunities

Interfaith education programs accompanying religious tourism may increase interfaith dialogue. Most religious sites have guided tours and educational programs that distribute information about the different faiths and their beliefs and practices. Such efforts enhance sentiments and compassion, thereby enabling good coexistence. According to (Cox, 2013), places such as the Vatican have been facilitating dialogues through engaging interfaith education programs that made the groups cooperate.

ADVERSE IMPACT OF RELIGIOUS TOURISM ON INTERFAITH INTERACTIONS

1. Commoditization of Holy Sites

The most serious issue religious tourism faces is commoditization of sacred sites. (Sharpley, 2009) Indicates that commercialization of religion, where holy locations are also used as tourist attractions diminish their religious character, thereby relegating native users to the peripheral landscape. This commercialization often creates conflicts between visitors and residents as the tourists often violate the local norms and practices.

For instance, tourist flows may have created some frictions at times in Jerusalem, for instance, people from the area have complained that sacred spaces within their locale are being disrespected or taken advantage of for economic gain. In such instances, disconnection led to divisions in interfaith relations as locals sometimes view tourists as intruders rather than guests.

2. Cultural Appropriation and Misunderstanding

Another challenge concerning this is cultural appropriation in considerable measure. The tourist may misrepresent or distort the cultural practices that the individual encounters, which fosters conflict and reinforces stereotypes. As Picard (2006) puts it, if tourists participate in acts of disrespect considered by local people to be disrespectful, then it fuels hostility and sabotages attempts at interfaith dialogue.

For example, if there are rituals for which the region holds sacrum consideration, the distortion and commodification of those rituals by visitors may cause tension to increase. These misunderstandings explain why tolerance for differences becomes necessary in advancing positive interfaith relations.

3. Escalating Sectarian Tensions

At the same time, in problematic regions where religious communities have been split by long histories of war, religious tourism is able to increase tensions that already exist. (Hossain, 2020) even point out that the presence of religious tourists fuels nationalist sentiment and provokes clashes between religious communities, for instance, in Jerusalem, where the complex political situation inherently complicates interfaith relations.

The historical importance of certain locations can present potential access and control concerns, thus further straining what perhaps have already strained relationships between faith groups. These dynamics go to show how religious tourism could intensify instead of ease strains in sensitive locations (Kottak, 2018).

4. Environmental and Social Strain

Rapid growth in religious tourism may strain the local resources and environments, which makes it have potential conflicts between the tourists and the residents. Issues such as overcrowding and resource depletion can give adverse conditions for interfaith relations, just as (Mason, 2018) questions.

The residents in local communities will become frustrated with how tourism affects their daily life, hence incite resistance against tourists and divisive sentiments.

CASE STUDIES AND EXAMPLES

Mecca

Mecca is an important case study in understanding the dynamics of religious tourism in Islam. The Hajj pilgrimage unites millions of Muslims from every corner of the world under a universal sense of unity and equality. (Esposito J., 2011) mentions that the experience of a shared event galvanizes global Muslim identity and unity.

Jerusalem

The most poignant case in point is the case of Jerusalem, given the fact that it is confronted with dilemmas that are setbacks to religious tourism and dialogue between communities of different faiths. The city hosts millions of religious tourists every year because it is sacred in Judaism, Christianity, and Islam. Tourist presence does encourage interfaith dialogue and understanding, but they do more problems. (Rabi, 2017) states, "tourism can promote the encounter between different faiths with each other but at the same time, it can also create conflict over the sacred space which is political and religious."

There are aspects where some sites carry significances related to all faiths. Examples include the Western Wall, the Church of the Holy Sepulchre, and Al-Aqsa Mosque. Nevertheless, such sites become a source of conflict that pushes tensions between faiths even higher, thus making it hard for interfaith dialogue (Tepelus C., 2018).

Varanasi

Varanasi, being one of the major pilgrim centers for Hindus, is one such interesting case. It attracts millions of tourists and pilgrims. It creates a unique mix of cultural interactions. (Kumar, 2019), commenting about it feels that even though diversified religious groups were to be found, it has, as a whole, encouraged mutual understanding, though at times due to cultural differences or out of misunderstanding, conflicts arose.

The Ganges River, worshipped by Hindus, also attracts tourists from all walks of life who go to visit some rituals and ceremonies. This interaction can lead to interfaith dialogues but still has to take on cultural sensitivity to avoid confrontations.

The management of such a crowd is likely to provoke any number of potential disputes, especially over access to sacred spaces or the implementation of local customs. The pilgrimage's natural need to balance commotion with the sacrosanct quality of Mecca creates an important issue in interfaith relations (Sharma H. &., 2022).

OBJECTIVE

To explore the effect of religious tourism on the peace and interfaith relations in Kartarpur

Research Question

What is the effect of religious tourism on the peace and interfaith relations in Kartarpur?

Methodology

To investigate the impact of religious tourism upon interfaith relations, this study follows a qualitative research methodology (Patton, 2015). A thorough review of secondary data and sources, including governmental reports, academic journals, and case studies focusing on religious tourism and the effects it cause on interfaith understanding are been referred. The data further is analysis by content discourse technique were main themes are identified with relevance to tolerance, cultural exchange, and community engagement. Thus by this way a comprehensive understanding of the role of Kartarpur in boosting and fostering the interfaith harmony is been checked and results are derived thereafter.

THE BACKGROUND OF THE KARTARPUR

Kartarpur, is one of the sacred pilgrimage site specially for Sikhs, which exemplifies the potential on the one side for its religious significance on the other side it also holds an ability to fetch together individuals from diverse faith backgrounds (Mishra, 2018).

In 2019, the Kartarpur Corridor was been inaugurated. This corridor is serves as an important religious as well as diplomatic link between the both countries of India and Pakistan. On the one hand it is facilitating the Sikh pilgrims' who access to the Kartarpur to visit the Gurdwara Darbar Sahib in Pakistan. The Kartarpur corridor is just not a vital route for pilgrimage but it is also a potential catalyst for reconciliation and peace building between the both historically rival nations. The Kartarpur being a religious site provides a vital platform for peace, and understanding in interfaith dialogues (Bhangoo, 2019).The existing literature about the Kartarpur Corridor focuses mostly on its multifaceted implications, including interfaith dialogue, religious tourism, and geopolitical dynamics.

The establishment of this corridor has been framed as a crucial moment not only for religious tourism, but also particularly for the community of Sikhs. This emphasizes that this corridor is enhancing cooperation, and interfaith understanding. Further it suggests that it can also contribute to the normalization of the bilateral relations between Pakistan and India (Bashir R. &., 2013). However, this sentiment is over echoed by, the ones who note that by utilizing this corridor can be perceived to bring the both countries closer despite of their historical tensions (Wani, 2013).

The Kartapur is historically significant as it was been founded by the first Guru Nanak of Sikhism. Afterwards it becomes a focal point for the identity and heritage of Sikhs (Chawla, 2020). Besides the specific role of the corridor in the promotion of religious tourism it also underscored by the existing potentials especially for the economic benefits. The Kartarpur attracts global pilgrims from all around the world. That is a reason of its stimulation th local economies (Bacha, 2014).

Kartarpur: A Model for Interfaith Dialogue

The escalating phenomenon of religious pluralism is necessitating an overall framework for shared respect, particularly in societies where multiculturalism is been practiced. A variety of interfaith initiatives draw attention to the power of dialogue in bridging divides. Provided, that it is articulated through collaborative projects and gets different faith groups to interact the one with the other, for example in organizations like United Nations who promotes the Alliance of Civilization (UNESCO., 2019). In addition, it can help to bring empathy among the members of a community and try creating an atmosphere between individuals regardless of what groups they belong (Singh J., 2020). Thus, interfaith harmony is essential for peaceful coexistence in a society characterized by different religions practices and belief systems (Weller, 2016).

This corridor lies along the India-Pakistan border and makes it a very powerful case study of interfaith dialogue and religious tourism. Kartarpur has already discussed in regarded as one of the most significant places of pilgrimage sites for millions of Sikhs since it houses the final resting place of Guru Nanak-the founder of Sikhism.

The Kartarpur Corridor has started to emerge as an important model of interfaith dialogue between India and Pakistan. Connecting Gurdwara Kartarpur Sahib in Pakistan with the Indian border, this initiative marks a new chapter in bilateral relations scarred to date by shared heritage and religious significance. Milestones included in this paper highlight the potential of the corridor to bring understanding and cooperation between these two nations. The roots of the Kartarpur Corridor lie in the partition of British India in 1947, which threw significant disruptions for communities, specifically for Sikhs with the loss of access to their holy sites (Bhatia G., 2021). The displacement facilitated deep-seated estrangement that complicated interfaith relationships that had existed for centuries. Kartarpur Sahib is also an important religious center initially established by Guru Nanak Dev Ji; therefore, Sikhs as well as other communities sought ways to reconnect with their heritage.

Advocacy and Consensus

The Sikh Diaspora was increasingly advocating their demands for access in the late 20th century; it therefore played the central role in seeking access to Kartarpur Sahib. The mass movement finally came alive on October 2018, as both governments decided to operationalized the corridor, which clearly signaled a direction toward cooperative religious diplomacy (Khan H., 2020). This treaty did not only increase access for pilgrims but even clearly stated a commitment to interfaith dialogue.

The Inauguration of the Corridor

On 9 November 2019, the Kartarpur Corridor was formally opened allowing thousands of Sikh pilgrims to visit Kartarpur Sahib without a visa. This event was highly found and celebrated as an important step toward religious freedom and religious harmony (Singh M. &., 2020). Inauguration worked as a strong symbol of hope in action and expressed that there is indeed a possibility of cooperation between the two countries.

COVID-19 Pandemic Impact

The COVID-19 pandemic broke in early 2020, putting a temporary hold on pilgrimages, and indicated that diplomatic efforts were susceptible to disruptions. During this time, the halt was observed to be necessary to continue dialogue and cooperation despite challenges (Bakhsh, 2020). Temporary shutdown became a reminder that interfaith dialogue is not a one-time commitment but requires continued commitment from both parties involved.

Resumption of Pilgrimage

The corridor reopened in 2022, promising once again a new route to dialogue and cooperation between faiths. This time, the return of pilgrims was used as landmark reconciliation and mutual respect between these nations (Akhtar, 2022).This move then strengthened the reality that common cultural and religious heritage can be a basis for peace.

This corridor also provides cultural exchange opportunities, along with the pavement of the interfaith consensus. The personal testimonies of the pilgrims related the emotional and spiritual impacts that their journeys have on them, making them feel new and settled with their faiths (Kaur J., 2022). The impacts of the corridor on interfaith understanding are various, which are how shared spiritual experiences could lead to peace and cooperation between India and Pakistan.

Estimated Visitors

The Kartarpur Corridor enables Indian devotees to attend the revered Gurdwara Kartarpur Sahib in Pakistan and has seen huge visitors since it opened in November 2019 (India., 2019). The first few months of the corridor saw thousands of visitors; the COVID-19 in 2020 briefly made it temporarily close down, drastically curtailing footfall (Affairs. M. o., 2020). With a relaxation in restrictions within 2021, visitors began to arrive again with thousands making the journey monthly. Estimations suggested that, by 2022, the number of visits per annum would start ranging between 60,000 and 80,000. The increasing trend continued to go through a growth spurt in the year 2023, with projections suggesting over 100,000 to be the number of pilgrims with increasing numbers opening up travel. The best statistics would be found in official documents from the Government of India and the Pakistan government, while the tourism departments of the two countries should have available current statistics.

The Role of Biodiversity in enhancing the Religious Tourism in Karatarpur

Biodiversity is a crucial factor for the expansion of religious tourism in Kartarpur since it integrates the spectacular natural terrain with the religious aspect of the site. Flora and fauna surrounding the Kartarpur Corridor, a relatively popular destination among Sikh sacred tourists, contribute to the tranquil environment that is desired to be achieved by the site visitors. The green areas, endemic vegetation, and surrounding wetlands do not only enhance the biophysical features of the site, but also promote the local economy while creating a tranquil environment to induce meditation and spirituality. Furthermore, the variety of birds and other animals in the region will also serve to increase the number of eco-tourists together with the religious ones, thus enhancing sustainable tourism. Indeed, research points out that integrating biodiversity with tourism activities enhances the experience of visitors and cements the idea of conservation (Azeem, 2024). In Kartarpur, this balance between the revered site and tourism fauna is a strong indicator of why certain species and habitats need to be preserved so that the site remains ecologically and spiritually valuable and relevant (Altaf, 2021).

Places of Tourist Importance at Kartarpur

There are many religious as well as other attractive sites which are spot of tourism. The attached fields and wells hold great cultural and spiritual significance, especially in the context of his life and teachings. Here is a brief overview of these elements:

1. Gurdwara Kartarpur Sahib

The Gurdwara Kartarpur Sahib is said to be the resting place of its founder, Guru Nanak Dev Ji, in Kartarpur. It holds a great deal of importance for Sikhs all over the world, while architecture is quite beautiful with a serene environment there. Apart from being a sacred place for worship, (Bhatia R. &., 2020) have depicted the gurdwara as a symbol of peace and equality bestowed by Sikh philosophy.

2. Kartarpur Corridor

The Kartarpur Corridor acts as a direct link between Gurdwara in Pakistan and Dera Baba Nanak in India, where pilgrimage is possible without requiring a visa. According to (Kaur A. &., 2019), this corridor has revolutionized cross-border relations and ensured higher religious and cultural exchange between India and Pakistan and encouraged interfaith dialogue.

3. Langar Hall

There is the Langar Hall within the gurdwawa where free meals for all visitors are dished out. In this, community service is fully exemplified by the Sikh experience. This is not only to feed the pilgrims but also to bring about the ideals of equal brotherhood among many different visitors (Chaudhary, 2018).

4. Baba Nanak's Tree

This is over 500 years old tree, believed to be the meditation area of Guru Nanak. A symbolic place of spirituality, it symbolizes peace and devotedness (Kaur J., 2023).

5.Fields

The agricultural fields surrounding Kartarpur are generally related to the life of Guru Nanak, who was known for his preaching about merging spirituality with labor. He emphasized honest work, equality, and service for the community. Fields The fields are a manifestation of the local culture of agriculture and express the ideals of Kirat Karni (earning a rightful livelihood) and Vand Chakna (sharing). The guru belief in the activity of his followers was essentially an act to enhance the philosophy that religious life is integrated with civic duty.

6.Wells

Wells associated with Guru Nanak symbolize the significance not only as water sources but also in terms of their symbolism. One of the important wells is located beside Gurdwara Kartarpur Sahib. According to Sikh tradition, Guru Nanak used to fetch water from these wells quite often as these were used to supply water to the community. The wells show his teaching on being humble, serving others, and betterment for the community. The spirit of sharing and assistance reflects the integrated vision of Guru Nanak for the community.

7. Museum

A small museum here in the gurdwara educates one a little better about the life and preachings of Guru Nanak, who further evolves to the history of Sikhism. As mentioned by (Sharma P. &., 2022), this education effort allows the visitors to learn more about Sikh culture and philosophy.

8. Parks and Gardens

The landscaped gardens surrounding the gurdwara also create peaceful ambience for prayer and reflection. Such areas are supposed to create a spiritual experience in the visitors along with tranquil ambiance for reflection (Verma, 2020).

9.CulturalSignificance

Both the fields and the wells hold a great importance in the history of the life of Guru Nanak, symbolizing the respect the guru has for the life of a farmer and community involvement. They are also pilgrimage sites, as they connect the traveler with the religious legacy of Sikhism and the everyday lesson taught by Guru Nanak.

ROLE OF PAKISTAN IN PROMOTING THE RELIGIOUS TOURISM

Religious tourism has emerged as an effective means for nation branding and cultural interaction. In this regard, high-lighting the role of religious tourism in Pakistan is believed to increase cultural exchange by exhibiting a country in a positive light, as progressive by displaying their religious places of worship, such as Sikhism, Buddhism, Hinduism, and Islam (Abbasi, 2022). This is important for striving for interfaith harmony for the reason that it puts people from various backgrounds together in a single space, to make them into dialogue and comprehend the perspectives of each other.

However, the importance of the cultural and historical religious sites offers meaning towards for the preserving cultural heritage in a more respected way towards interfaith harmony. This work would make light of the syncretism aspect of cultural features in Hindu sites in Punjab and reflects a much enmeshed tapestry of shared histories and interactions among various faiths (Fatima, 2023). The preservation and restoration of such sites are needed not only to save historical continuities but also in order to improve environments of dialogue between faiths. Interfaith dialogue initiatives, such as those proposed for discussion, will help address the challenges of intolerance and extremism in today's society. Increased interfaith understanding is being recognized as a global need to establish a culture of peace (Alam M., 2023). The Kartarpur initiative can envision creating the setting for dialogue forums and conducting community activities that provide a venue for interaction among religious leaders and their followers in order to foster social cohesion and mutual respect. Though in Kartarpur the specifics are different, the basics of the process of dialogue construction and familiarity are universal, as stated by (Rahawarin, 2023).

It was a good gesture by the Pakistan government to open the corridor, so that it may be able to prove itself as a great provider for religious tourism and its contribution toward the satisfaction of spiritual and minority elements. For India, this corridor was an opportunity to project itself as a secular country and promote regional connectivity.

The opening of the corridor received a half-hearted welcome in both countries. While it is hailed in the Sikh circles, some sections of Indian politicians raised the issue of what message was going to be given by the Indian Government to Pakistan when the situation between them is not yet normalized. Some believed that it would legitimize Pakistan's view on issues like Kashmir and other matters (Beard, 2019).

Globally, cultural heritage and breathtaking landscapes, besides historical significance, have thrust Pakistan into becoming an increasingly notable destination for several varieties of tourism. This literature review explores already existing tourist packages as well as rankings in the metrics of the country that relate to global tourism organized by thematic categories.

CULTURAL AND HERITAGE TOURISM

This includes cultural and heritage tourism whereby visits are made to historical sites and traditional crafts alongside culinary experiences. Indeed, it highlights that Pakistani cultural tourism packages mostly revolve around visits to attractions such as Lahore Fort and Badshahi Mosque, which have a rich history about the Mughal empire (Khan M. &., 2021). These tours are, first of all, developing heritage in the local setting but at the same time making people in the foreign land more culturally aware of Pakistan (Khan H., 2022). Apart from this, the frequent local festivals and forms of art accompany these tours and provide a very interactive cultural experience for the visitants also (Shah N., 2020).

Adventure Tourism

Adventure tourism in Pakistan has lately become popular, especially in the northern regions. According to (Ahmed F. K., 2020), trekking and mountaineering package, for instance, to K2 Base Camp and Nanga Parbat attract adventure tourists to the region. Unique geography of the region combined with the development of infrastructure makes adventure tourism appealing (Mahmood, 2021). In addition, outdoor activities such as rafting and paragliding have been encouraged with an eye to thrill-seekers (Iqbal, 2019).

Religious Tourism

Religious tourism is another pertinent feature of the tourist industry in Pakistan, with the Kartarpur Corridor being an important religious destination for Sikh pilgrims. As (Kaur A. &., 2019)have noted, packages could be organized for seeing sacred religious places, which proves that the country has a strong spiritual culture. Sikh pilgrims, to various extents, have helped boost religious tourism but also cross-border cultural cross-pollination, as observed by (Singh R., 2020).

Eco-Tourism

Pakistani eco-tourism has attracted attention as a green tourism activity. (Khan, . H., 2022) in his literary works focuses on the prospects of eco-tourism in Swat Valley and Fairy Meadows. Any one would love nature but also develop the host economy. These eco-tourism ventures are serious responsible tourism, but the trendy traveler hardly takes these matters lightly (Bashir M., 2022).

Wildlife Safaris

Wildlife safaris turn out to be another emerging sector in Pakistan's tourism industry. Research by (Hussain R. &., 2021) reveals that national parks like Khunjerab National Park and Hingol National Park afford a glimpse of varied flora and fauna. Packages for wildlife tourism will elevate the message of conservation and spread out ecological diversity of Pakistan as well (Gulzar, 2024).

RANKING AND RECOGNITION

Over the years, ranking of Pakistan in the world's tourism has been improving. Reflecting a positive trend, especially in cultural resources and natural attractions, is the World Economic Forum's Travel and Tourism Competitiveness Report (WEF, 2021). Safety metrics for tourists are also gradually improving as shown by the Global Peace Index, however, with much room for improvement (Peace., 2022). Appraisals put forth by publications like *Lonely Planet* branded Pakistan an emerging destination for adventure tourism (Smith J., 2023).

TREATIES AND AGREEMENTS ON TOURISM INVOLVING PAKISTAN

Pakistan has signed many treaties and agreements to flourish tourism and develop cultural interaction with other nations. The section deals with key treaties and initiatives in regard to tourism and is supported by relevant literature.

1. Bilateral Agreements on Tourism

Pakistan has signed different bilateral agreements pertaining to tourism promotion from the following countries, including China, Turkey, and Saudi Arabia. Bilateral agreements generally consist of various focal areas:

- **Marketing Joint Tourism Plans:** Most of these collaborations revolve around the preparation of tourism infrastructure and marketing policies. For instance, the mutual agreement with China aims to attract more tourism traffic and exchange travel experiences between the countries. According to the Ministry of the Foreign Affairs (Affairs. M. o., 2023), tourism partnerships are centered on attracting more visitors to a country and an enhanced mutual understanding.
- **Cultural Exchange Programs:** The agreements often contain programs of cultural exchanges, whether this be the art exhibition, the food festival, or heritage events. (Khan M. &., 2021) observe, “Along with the promotion of tourism activities, cultural exchange programmes produce diverse types of interaction among people through shared culture and traditions.”

Regional Cooperation

Pakistan is a member of regional tourism promotion agencies, including South Asian Association for Regional Cooperation (SAARC). Under SAARC, the member countries have attempted to:

- **Promotion of South Asian Tourism**: SAARC Framework for Regional Cooperation in Tourism builds mutual marketing efforts and capacity building to strengthen tourism in the region. The SAARC Secretariat (Secretariat., 2018) suggests, "joint efforts are needed to have a comprehensive strategy on tourism that will work in favor of all SAARC member states."
- **Visa facilitation initiatives:** Regional cooperation is largely built on the streamlining of visa requirements among the member nations. There are incentives in the SAARC Visa Exemption Scheme in allowing easier travel across the region, encouraging cross-border tourism (Sharma P., 2019).

2. UNESCO World Heritage Conventions

Pakistan is a signatory to conventions of UNESCO related to the matter in general, which promotes cultural heritage protection indirectly supports tourism. Some of these primary aspects include:

- **World Heritage Sites:** Places like Mohenjo-Daro and Lahore Fort will be included in the UNESCO World Heritage while increasing more international visitors as Pakistan celebrates its knowledge of a rich historical and cultural legacy. UNESCO reports in (UNESCO.., 2021), these designations alone boost tourism but also promote preservation.
- **Heritage Protection:** Engagement in international action to protect heritage helps to strengthen tourism attraction because of protection and promotion of the historical area (Qureshi, 2020). Heritage protection is a necessity to ensure the development of responsible tourism that observes the available local heritage.

3. Agreements over Tourism Development Funds

This nation has given various tourism development funds jointly with foreign bodies. These include Asian Development Bank (ADB), among others. The funds are offered through:

Infrastructure development: This involves improving tourist infrastructure roads, airports, and hospitality. This aspect is relevant because the infrastructure to be developed may consider infrastructure connected with infrastructures that may range from roads, airports, and even some form of hospitality. According to (ADB., 2021), this investment seeks an easy journey to the resorts and enhanced quality service.

- **Sustainable Tourism Projects:** Development projects for environmental tourism and responsible tourism patterns benefit the local masses but are attractive to the visitor. (Ali, 2021) reflects, that initiatives in sustainable tourism create economic opportunities for the locals besides protecting the natural and cultural setting.

4. Cultural Cooperation

Sometimes, tourism elements form part and parcel of Pakistan cultural cooperation with other nations will help in boosting the public diplomacy and positive image of the state as well.

- **Art and Cultural Exchanges:** More exposure would be encouraged to appreciate each other's culture through such programs on art and cultural exchanges. Such initiatives facilitate further strengthened tourism sectors while improving diplomatic relationships themselves, that date back a few years ago in 2020 when Khan addressed the topic.
- **Training and Capacity Building**: A mutual agreement can be reached on trainings in hospitality, tourism management, or even cultural heritage preservation. Such programs enhance the quality of service tourism and upgrade visitor experience (Bashir., 2021).

BILATERAL RELATIONS

From a political and geo-political point of view, the Kartarpur Corridor offers hope to serve as a bridge between the two countries mainly India and Pakistan which will undoubtedly bring an improved relationship between the two countries. Framed in terms of an act to appease religious sentiments among Sikhs, though skepticism remained, mainly in India, to the extent that some people feel that it is just a partial step that does not address more profound issues, especially those on terrorism and structural disputes on territory (REHMAN, NA).

The corridor might encourage greater people-to-people contact, which is critical in fostering a greater closeness between the two nations. This contact could open doors for negotiation on those contentious issues and could ease tensions that have hung over for so long. Domestically in India, as well, the favourable response to the corridor by the Sikh community may impact Indian politics, since all political parties look to be in step with the sentiments of their constituencies in making this policy move a political football to win votes over (Shukla, 2019).

A Political Perspective of the Kartarpur Corridor

From the political point of view, the Kartarpur Corridor is considered in line with peace-building and conflict resolution. In this case, the securitization theory is applied in regards to the corridor, stating that even though it is a peace effort, the unrest between the two religious communities, Hindus and Sikhs, may form issues (Akbar, 2023). Further into this complexity, zooms in on the media discourses surrounding the corridor, recounting stories of how narratives shape public perceptions and influence diplomatic relations (Ahmed F. M., 2022). In such regards, the role of the media in defining the corridor as a symbol of hope towards peace is very critical in bringing out different perspectives, as highlighted by, who suggests that the corridor can be considered as a collaborative approach without identity politics (Yousaf, 2021).

Domestic Politics

Besides politically influential, the corridor has also been a subject towards domestic activities. The Indian political parties have at various times attempted to popularize themselves by adopting the sentiment of Sikhs, which often translates into promoting Sikh rights and interests through the open corridors. The people in Punjab showed immense welcome as well because the Sikh population is very high there. Political parties, including especially Aam Aadmi Party and Shiromani Akali Dal, use the religious corridor to strengthen their constituents while underlining religious access.

Alternatively, in Pakistan, the government has used the corridor as a way of displaying its self-image as a moderate Islamic nation that wishes to befriend religious minorities.

This has implications in domestic politics as well because it fits into the larger story of tolerance and engagement that certain sections of Pakistan have keenly yearned to adopt, especially when there was an international outcry over the way Pakistan treated religious minorities.

The Geo-Political Importance of the Kartarpur Corridor

In this context, another relevance of the corridor is to the larger world of South Asian geopolitics. In the discussion of the RSCT model as reportedly laid out by, the corridor has come forward to redefine security dynamics in the region while promoting cooperation and dialogue over conflict (corridor, 2022). This perspective is shared by, who argue that this corridor is also symbolic in that it also promotes and fosters diplomatic relations as well as facilitates improvement of religious minorities' rights (Akhtar U., 2022). Then, the corridor becomes a strategic initiative where its objective speaks to the needs of the Sikh community, but also it serves as a project for broader regional (Kartarpur corridor, 2021).

Security Issues

Security is an important factor in the political consideration of the Kartarpur Corridor. The opening of the corridor also stirred up the fear of its potential misuse by terrorist groups, particularly in the wake of long-persisting Indian accusations that Pakistan had hosted some elements of terrorists. The Indian authorities perceive the pilgrims as potential security threats; there was a biometric verification and an effective security framework to minimize risk exposures. This approach resonates with the more all-encompassing security context in which both countries operate, one in which trust levels are low and legacies of past grievances continue to shape policies (Ullah, 2022).

Regional Dynamics

The Kartarpur Corridor falls within the broader South Asian geopolitical landscape. As China continues to assert its influence in Pakistan and the nature of U.S.-India relations moves to a trend of transformation, this corridor has become an important center for regional diplomacy. It brings home how religious diplomacy can dovetail into national interest and how, even in times of political rancor, cultural connection can create avenues for rapprochement. In many ways, a corridor now takes on the guise of a barometer to track the pulse of bilateral relations between these two nations, as each side attempts to steer itself through this complicated relationship riddled with external pressures and internal narratives.

The Kartarpur Corridor is a much bigger concept, which transcends India and Pakistan and speaks to broader South Asian geopolitics and, in effect, changes relations with other neighboring countries. Hence, apart from the religious tourism that this opens up, it opens avenues of diplomatic engagement and regional cooperation.

First, the corridor can affect India's relations with Afghanistan. Historical ties of the Sikh community in Afghanistan along with growing religious tourism can start sowing seeds of closer cultural relationships. Researchers like (Bhatia, . R., 2020)argue that positive engagement through initiatives such as the corridor will lead to improvement in cooperation in trade and cultural exchanges at a time when, geopolitical volatility can be addressed for stabilization in the region.

Second, it is going to impact India's relationship with Bangladesh. Better access to the Gurdwara Darbar Sahib is likely to invite similar moves to improve access for other religious communities, like Christians and Hindus, but may be harmful to Hindus and Buddhists when part of a minority. Here, potential implications of the corridor are going to be enhancing India's soft power in Bangladesh through telling a story of religious harmony and shared heritage (Hussain A., 2020).

It will also affect the relations of India with China. Since China has tried to obtain power in Pakistan through such initiatives like China-Pakistan Economic Corridor (CPEC), India will apply force against it by making use of Kartarpur Corridor. Through very deep cultural diplomacy, India will be able to strengthen its influence and will also remind the world of its interest in peace and interfaith relations over there as well, said (Kaur R. &., 2021).

The corridor could also redefine the contours of regional cooperation. For instance, SAARC could also get a push forward because of religious tourism. The corridor may facilitate engagement among India, Pakistan and other nations within South Asia, encouraging discussion of common challenges, such as poverty alleviation and climate change, and security.

In addition, the corridor holds the potential to engage cultural diplomacy as a tool for resolving conflict. Cultivating through common religious heritage, it can assist those countries in the region in finding a bed of peace that eliminates political differences and opens the way to forming collaborative frameworks that address global issues, which have been most critical in geopolitics.

THE ECONOMIC IMPACT OF KARTARPUR

This development has brought attention to its economic impact on local and regional levels.

Several studies found that the economic benefit arising from the Kartarpur Corridor raises the potential benefits and impacts upon local and regional economies.

According to (Hussain .. A., 2019) this corridor has increased the number of Sikh pilgrims in Gurdwara Kartarpur Sahib, stimulating local businesses, such as hotels, restaurants, and transportation services. This religious tourism has added jobs for locals, hence contributing to local economic growth.

(Sharma S. &., 2020) also present quantitative evidence, as local business revenue has increased by around 30% in the initial year since the corridor opened. They highlighted the relevance of the infrastructural development, which has simplified the access to the Gurdwara and improved the general visitor experience overall.

(Kaur J., 2022)indicate the ripple effects of this movement, which their findings suggest brought more human traffic into the region, thereby availing more demand for products and services. In fact, the study is able to reveal how the sales in around vendors near the corridor have increased by 25%, hence exhibiting their direct economic effect on the local merchants.

(Wagner, 2023) have explained other possibilities of long-term economic cooperation between India and Pakistan through the improvement of relations that the corridor would bring. This would lead to mutual investment in trade and services, with the growth of sustainable tourism further encouraging investment into the economies.

(Choudhary, 2020) speaks of sustainable tourism practices while being conscious not to lose the economic benefits arising out of it and allowing no ill effects on its environment and society. He emphasizes that strategic planning is needed while focusing on growth, with local resources preserved .

There are some studies indicating there could be a potential increase in the local economy due to religious tourism. For example, (Hussain A., 2020) traces how the corridor is going to develop the hospitality, transport, and retail sectors at the local level. The flow of Indian pilgrims creates demands for services, generates employment and income for local businessmen operating in the vicinity of the Gurdwara. Moreover, (Ghosh, 2021) point out that religious tourism may go hand in hand with infrastructural development because "good facilities and transportation networks must be provided for the tourists".

Besides the local gains, it could also have more significant economic repercussions for India and Pakistan. Bhatia and Singh describe the corridor with an analysis of potential cross-border economic cooperation that would result from it. They point out that increased interaction between the two states may lead to cooperative efforts in trading and tourism as well. As the authors note, religious tourism, under an optimistic economic environment, may bloom into a long-term guarantee for peace and stability in the area.

In addition, research findings have shown that the economic advantages the Kartarpur Corridor provides go beyond the immediate calculation of tourism income. (Dhillon, 2021) concluded that religious tourism would be able to facilitate cultural exchange and community bonding between communities, making them more resilient for economic ventures. They indicate that when cross-linked communities interact through religious tourism, they tend to develop joint ventures in economics that further link their relationship.

The potential economic benefits are however enormous. (Fletcher, 2018) is very apt to sound caution that with these tourists would come attendant strains on local resources, leading to negative effects such as degradation of the environment and overloading the infrastructure. As such, sustainable tourism practices, according to Fletcher, must be accorded importance, so the corridor's economic benefits do not imply community or environmental setbacks.

The statistics reflect the fact that the Kartarpur Corridor has brought positive economic trends, where opportunities for local communities and interactions are enhanced across borders.

Therefore, economically, the Kartarpur Corridor's impacts are multi-layered, offering opportunities but also challenges; the spiritual bridge will further help to develop greater economic development and cooperation between communities as a whole across borders.

THE SOCIOECONOMIC FACTOR OF THE KARTARPUR CORRIDOR

The socioeconomic environment has a strong influence on interfaith relations. According to various studies regarding the relationship of socioeconomic status to the harmony of interfaith co-existence, educational preparedness and financial security can give rise to environments where religious tolerance will thrive in (Alam .. M., 2023) .Now, improving the socioeconomic conditions of people living around Kartarpur is likely to increase interfaith relations and reduce tension among them.

Impact of Religious Tourism on Interfaith Relations

Cultural exchanges at Kartarpur are quite an excellent example to foster better understanding among different faith communities. Interfaith dialogues, joint community activities, and joint religious practices are just examples that have helped foster a wholesome environment for open dialogue (Shah K., 2017). According to the feedback from the visitors, these conversations often transform the lives of the people themselves as they tend to shatter their own stereotypes and advance mutual respect among the different faiths (Fisher, 2020). This is in line with other research findings which point out spiritual travel increases positive effects of peace and understanding. One should also learn together with her group in a common space, as noted in (Rinschede, 1992).

Findings and Analysis

The study on religious tourism at Kartarpur contributes to cross-cultural tolerance. According to visitors, their perceptions are changed considerably for the better, and interfaith dialogue in addition to communal celebrations represents quite effective instruments for promoting harmony, found by (Esposito .. J., 2018). By engaging actively with such initiatives, participants can create long-lasting relations which are very much required to generate peace and reconciliation in a multicultural landscape. Data supports the above notion that interfaith tourism can be a catalyst for more broad social changes; with goes the call to needed work in structured programs to improve this endeavors.

Responsible Tourism Practices

Religious tourism complements interfaith relations because traveling by persons from different nationalities and cultures creates an alliance of people toward the sharing of experiences and mutual learning on each other's beliefs. It promotes knowledge and understanding, which are indispensable parts of peace building in a pluralistic society.

According to (Ammerman, 2013), religious tourism provides space in which participants are faced with their stereotypes and prejudices. For instance, interfaith pilgrimages or joint religious festivals promote cooperation and shared experiences that foster deeper interfaith ties. Based on (Driessens, 2020), interactions of this nature are able to spur local economies as they promote the narrative of unity instead of division. Responsible tourism practices with sustainable biodiversity are of the essence for religious tourism because they ensure the benefits trickle to the locals without violating cultural and religious diversity (Mowforth, 2016) .On this background, current practices in Kartarpur are assessed to judge how it is currently reaching out to the ethical engagement. Hence, by increasing the community involvement in tourism practices and educational resources for visitors on local customs and beliefs. These practices further enhance the experience of visitors while also promoting sustainable tourism that respects religious and cultural heritage at the site (Buckley, 2012).

CONCLUSION

In a nutshell, religious tourism serves an important function in maintaining interfaith harmony same like it is considered in Kartarpur's case on the ground. Findings pointed out that religious tourism can be a tool toward peace building and fostering intersentiment and cooperation among believers from other faith groups. Therefore, stakeholders such as the governments, NGOs, and the local community need to collectively key in on this potential of religious tourism towards creating meaningful interfaith engagements that lead to higher social cohesion (Smith V., 2003).

Hence, It also marks a vision of hope for the future where there's scope for collaboration and understanding in this region that's so often been seen through the lens of conflict. It has helped in tourism, boosting both economies as well as cultural exchange from both sides of the border. Being a bridge above political divides, Kartarpur Corridor symbolizes compassion and coexistence in life and encourages us to focus on shared values than differences. It finally boils down to the realization that positive can come out of dialogue and cooperation.

RECOMMENDATIONS FOR THE BOTH GOVERNMENTS REGARDING THE KARTARPUR CORRIDOR

1. Bilateral Relations:

Develop dialogue and cooperation between the government towards mutual understanding and belief. Cooperative measures can be adopted both in terms of paving peace and developing goodwill at large in regards to the Kartarpur Corridor.

2. Visa Processing:

Ease the process of reaching Sikhs shrines for visa and travel purposes. Introduce a special pilgrimage visa so that people can reach the place easily and fast.

3. Investment in Infrastructure:

Jointly implement the infrastructure development projects of transportation, access, and facilities across Kartarpur so visiting from both countries becomes easy.

4. Preservation of historical and cultural heritage:

Spend on the conservation of historical buildings as well as preservation of cultural heritage within and around Kartarpur. That will amply reflect the import of both nations to these monuments and give them a sense of common history.

5. Safety and Security Cooperation:

Strengthen the bilateral measures in joint security mechanisms and formalities to ensure safety to all pilgrims through proper contacts with both countries so that any concern is dealt without delay.

6. Support to Local Economies:

Partnership with the local business and the local community so that it benefits economically. It can enlighten the entrepreneurship skills and promote small businesses.

7. Bilateral Consultations:

The two sides should ensure regular bilateral consultation between the officials of both governments working on relevant issues so that follow-up work may be done, and issue-focused areas may be addressed and new avenues may be considered for consideration in regard to the Kartarpur Corridor.

8. Community Participation:

Cabinets and boards of tourism should involve local communities in the planning and decision-making process of tourism planning. It should allow them to take a share in community views and directly benefit economically from the tourist flow.

9. Synergistic Tourism Initiatives:

Team up with local entrepreneurs and present a joint venture package that will ensure accommodation, food, and guided tours, ushering in rich heritage of Sikhs and adding value to the local culture so that all parties benefit economically.

10. Sustainable Biodiversity Practices

There should be prevailing sustainable biodiversity practices which can promote the religious tourism in the Kartarpur.

REFERENCES

Abbasi, M. (2022). Nation branding through religious tourism in pakistan by utilizing social media tools. *Journal of Development and Social Sciences, 3*(II). Advance online publication. DOI: 10.47205/jdss.2022(3-II)57

ADB. (2021). *Tourism Infrastructure Development in Pakistan*. Asian Development Bank Reports.

Affairs., M. o. (2020). Impact of COVID-19 on Travel.

Affairs., M. o. (2023). *Updates on Kartarpur Corridor Visitor Numbers*. Pakistan.

Ahmed, F. K. (2020). Adventure Tourism in Pakistan: Trends and Opportunities. *Journal of Tourism Research, 15*(3), 245–260.

Ahmed, F. M., Mubeen, M., & Nawaz, M. (2022). Framing south asian politics: An analysis of indian and pakistani english print media discourses regarding kartarpur corridor. *PLoS One, 17*(2), e0264115. DOI: 10.1371/journal.pone.0264115 PMID: 35176107

Akbar, N., & Shabbir, A. (2023). Securitization theory in south asia and kartarpur border: A peace initiative or a conflict? *Journal of Southeast Asian Studies, 11*(2), 153–160. DOI: 10.33687/jsas.011.02.4568

Akhtar, U. A., Ahmed, S., & Ahmed, Z. (2022). Kartarpura Corridor: A Step towards Peace. *Journal of Social Sciences Review, 2*(4), 294–304. DOI: 10.54183/jssr.v2i4.103

Akhtar, U. A. S., Ahmed, S., & Ahmed, Z. (2022). Kartarpura corridor: A step towards peace. *Journal of Social Sciences Review, 2*(4), 294–304. DOI: 10.54183/jssr.v2i4.103

Alam, .. M. (2023). Correlational analysis of socioeconomic status (ses) and interfaith harmony in islamabad, pakistan.,. 4(4), 272-289. .DOI: 10.61503/cissmp.v2i4.102

Alam, M., & Mojumder, I. M. (2023). Sikh and caodai understandings of interfaith harmony: Promoting a culture of peace and understanding. *Journal of Daesoon Thought and the Religions of East Asia, 3*(1), 129–151. DOI: 10.25050/JDTREA.2023.3.1.129

Ali, S. &. (2021). Bilateral Agreements and Their Impact on Tourism in Pakistan. *Journal of International Tourism Studies*.

Altaf, M. (2021). An overview of Pakistan Tourism Impact of COV. *Journal of Wildlife and Ecology, 5*(4), 186–201.

Ammerman, N. T. (2013). *Religious Identities and Cultural Change in the New Europe*. Cambridge University Press.

Azeem, M. B. (2024). Reimagining the Religious Tourism in Pakistan: A Case Study of Kartarpur Corridor. *Harf-o-Sukhan, 8*(1), 833–842.

Bacha, D. (2014). Religious tourism and its impact on entrepreneurial development at kartarpur corridor, punjab-pakistan. *Journal of Business & Tourism, 9*(02), 56–75. DOI: 10.34260/jbt.v9i02.279

Bakhsh, F. S. (2020). Significance Of The Kartarpur Corridor: Pakistan's Complience With International Standards Relating Protection Of The Rights Of Religious Minorities. *International Journal of Management Research and Emerging Sciences*, *10*(2).

Bashir., M. (2021). Capacity Building in the Tourism Sector: Trends and Strategies. *International Journal of Tourism Development.*

Bashir, M. (2022). Eco-Tourism in Pakistan: Challenges and Prospects. *Sustainable Development Studies*, *12*(1), 33–50.

Bashir, R., & Ahmad, K. (2013). Religious Tourism: Experiences of Sikh Yatris Visting Kartarpur Corridor, Pakistan. *Journal of Social Sciences Review*, *3*(2), 112–120. DOI: 10.54183/jssr.v3i2.249

Beard, C. (2019). *Sustainable Tourism in the Global Economy: Perspectives and Challenges*. Routledge.

Bhangoo, R. S. (2019). *Guru Nanak and Institutionalisation of Sikhism.*

Bhatia, G. (2021). Cultural Heritage and Spirituality: The Case of Kartarpur. *International Journal of Heritage Studies*, *27*(3), 310–322.

Bhatia, R. (2020). Cross-border cooperation through religious tourism: The case of the Kartarpur Corridor. *International Journal of South Asian Studies*, *13*(2), 122–138.

Bhatia, R. &. (2020). International Journal of South Asian Studies, . *Cross-border cooperation through religious tourism: The case of the Kartarpur Corridor.*, 13(2), 122-138.

Bianchi, R. (2006). Tourism and the Politics of Sacred Space. *Annals of Tourism Research*, *33*(4), 1297–1313.

Buckley, R. (2012). *Sustainable Tourism: Theory and Practice*. Routledge.

Chaudhary, R. (2018). Langar: The Sikh Tradition of Community Kitchen. *Religious Studies Review*, *44*(2), 150–158.

Chawla, T. H. (2020). Visa-free travel to sri kartarpur sahib: Historic pilgrimage and religious tourism from indian punjab to pakistan punjab. *Rupkatha Journal on Interdisciplinary Studies in Humanities*, *12*(3). Advance online publication. 10.216.

Choudhary, M. (2020). Managing the impacts of religious tourism: Challenges and opportunities in Kartarpur. *Tourism Management Perspectives*, *34*, 100685.

Cohen, E. (2004). Tourism, Religion, and Spiritual Journeys. *Tourism Recreation Research*, *29*(2), 9–20.

Cohen, E. (2012). Pilgrimage and tourism: Convergence and divergence. *Tourism Management*, *33*(6), 1362–1371. DOI: 10.1016/j.tourman.2012.01.008

Corridor, K. (2022). *Kartarpur corridor: re-defining security in south asian region.* Retrieved 10 27, from Annals of Human and Social Sciences,: 3(III). DOI: 10.35484/ahss.2022(3-iii)01

Cox, C. (2013). Interfaith Education and Dialogue: A Vatican Perspective. *Journal of Religious Education*, *61*(3), 45–56.

Dhillon, G. (2021). The Kartarpur Corridor: A bridge of peace and interfaith understanding. *South Asian Studies Journa, 14*(1), 23–37.

Driessens, A. (2020). Tourism and Interfaith Dialogue: Opportunities and Challenges. *Tourism Management Perspectives, 34*, 100661.

Esposito, J. (2011). *Interfaith Dialogue: A Global Perspective*. Oxford University Press.

Esposito, J. (2018). *Interfaith Dialogue: A Short Introduction*. Oxford University Press.

Esposito, J. L. (2011). *What Everyone Needs to Know about Islam*. Oxford University Press. DOI: 10.1093/wentk/9780199794133.001.0001

Fatima, Z., Ahmed, U., & Khursheed, A. (2023). Revitalizing the cultural heritage: Assessing the historical significance and conservation potential of hindu sites in punjab, pakistan. *Annals of Social Sciences and Perspective, 4*(2), 321–330. DOI: 10.52700/assap.v4i2.307

Fisher, M. (2020). Shared Spiritual Experiences: Insights from Kartarpur Pilgrims. *Journal of Religious Tourism*.

Fletcher, R. M. (2018). *Tourism and the Global Environment: A Critical Perspective*. Routledge.

Ghosh, A. (2021). Navigating the complexities of interfaith dialogue in Kartarpur. *The Journal of Conflict Resolution, 65*(2), 205–224. DOI: 10.1177/0022002721998212

Gulzar, Y. E., Eksili, N., Koksal, K., Celik Caylak, P., Mir, M. S., & Soomro, A. B. (2024). Who Is Buying Green Products? The Roles of Sustainability Consciousness, Environmental Attitude, and Ecotourism Experience in Green Purchasing Intention at Tourism Destinations. *Sustainability (Basel), 16*(18), 7875. DOI: 10.3390/su16187875

Hossain, M. (2020). Religious Tourism and Sectarian Tensions: An Analysis of the Case of Jerusalem. *Journal of Tourism and Cultural Change, 18*(4), 434–450.

Husain, S. (2014). Interfaith Harmony in a Globalized World. . *Journal of Peace Studies.*, NA.

Hussain, A. (2019). The Kartarpur Corridor: A bridge to peace. *Journal of Peace Studies, 26*(2), 45–60.

Hussain, A. (2020). The role of religious leaders in promoting peace and interfaith dialogue. *International Journal of Peace Studies, 25*(1), 45–68.

Hussain, A. (2020). The role of religious leaders in promoting peace and interfaith dialogue. *International Journal of Peace Studies, 25*(1), 45–68.

Hussain, R. (2021). Exploring Pakistan's National Parks: A Wildlife Safari Perspective. *Journal of Natural History, 55*(4), 365–377.

India., G. o. (2019). Inauguration of Kartarpur Corridor.

Iqbal, H. (2019). Outdoor Adventure Activities in Pakistan: A Review. *Tourism Management Perspectives, 31*, 158–167.

Kartarpur corridor. (2021). *a step towards regional cooperation in south asia*. Retrieved 10 27, from Kartarpur corridor: DOI: 10.13140/rg.2.2.21663.07849

Kaur, A. (2019). The Kartarpur Corridor: A Gateway to Peace and Interfaith Dialogue. *Journal of Peace Studies*, *12*(1), 45–62.

Kaur, A. (2019). The Kartarpur Corridor: A Gateway to Peace and Interfaith Dialogue. *Journal of Peace Studies*, *12*(1), 45–62.

Kaur, J. (2022). Pilgrimage and identity: Experiences of Sikh travelers in Kartarpur. *Religions*, *13*(6), 568. DOI: 10.3390/rel13060568

Kaur, J. (2022). Pilgrimage and identity: Experiences of Sikh travelers in Kartarpur. *Religions*, *13*(6), 568. DOI: 10.3390/rel13060568

Kaur, J. (2023). Sikh Architecture. In *The Sikh World* (pp. 369–382). Routledge. DOI: 10.4324/9780429455322-37

Kaur, R. (2021). Economic resilience through cultural exchange: The role of the Kartarpur Corridor. *Journal of Tourism and Cultural Change*, *19*(4), 476–491.

Khan, H. (2020). Cultural Diplomacy and Its Impact on Tourism. *Journal of Cultural Relations* .

Khan, H. (2022). Cultural Tourism: A Key Driver for Economic Growth in Pakistan. *Asian Journal of Tourism Research*, *10*(1), 77–89.

Khan, H. (2022). Cultural Tourism: A Key Driver for Economic Growth in Pakistan. *Asian Journal of Tourism Research*, *10*(1), 77–89.

Khan, M. (2021). Heritage Sites and Their Role in Promoting Cultural Tourism in Pakistan. *International Journal of Heritage Studies*, *27*(5), 496–510.

Khan, M. (2021). Heritage Sites and Their Role in Promoting Cultural Tourism in Pakistan. *International Journal of Heritage Studies*, *27*(5), 496–510.

Kottak, C. P. (2018). *Cultural Anthropology: Appreciating Cultural Diversity*. McGraw-Hill Education.

Kumar, S. (2019). Interfaith Relations in Varanasi: Analyzing the Impact of Religious Tourism. *Journal of Southeast Asian Studies*, *12*(1), 67–80.

Mahmood, R. (2021). Mountain Tourism in Pakistan: Current Status and Future Directions. *Mountain Research and Development*, *41*(1), 15–23.

Mason, P. (2018). *Tourism Impacts, Planning and Management*. Routledge.

Mishra, R. (2018). Public Diplomacy and Religious Tourism: The Kartarpur Model. *International Journal of Cultural Studies*.

Mowforth, M. (2016). *Tourism and Sustainability: Development, Globalisation and New Tourism in the Third World*. Routledge.

Patton, M. Q. (2015). *Qualitative Research & Evaluation Methods*. Sage Publications.

Peace., I. f. (2022). *Global Peace Index.*

Picard, D. (2006). Cultural Tourism in the Asia-Pacific: An Overview. *Journal of Tourism and Cultural Change*, *4*(1), 1–17.

Qureshi, M. (2020). *Cultural Heritage and Tourism in Pakistan: Opportunities and Challenges*. Heritage Studies Journal.

Rabi, R. (2017). Tourism and Interfaith Dialogue in Jerusalem: Challenges and Opportunities. *Middle Eastern Studies*, *53*(6).

Rahawarin, Y. (2023). Dynamics of maintaining religious harmony in ambon city, maluku province. *WSSHS*, *1*(06). Advance online publication. DOI: 10.58812/wsshs.v1i06.512

Rinschede, G. (1992). Form and Structure of Pilgrimage Tourism. *Annals of Tourism Research*.

Secretariat., S. (2018). *Framework for Regional Cooperation in Tourism.* NA.

Shah, K. (2017). Cultural Exchange through Religious Tourism: A Case Study of Kartarpur. . *Journal of Interfaith Studies.*

Shah, N. (2020). Cultural Festivals and Their Impact on Tourism in Pakistan. *Journal of Cultural Heritage Management and Sustainable Development*, *10*(2), 147–159.

Sharma, H. (2022). *Religious Tourism in Uttar Pradesh: A Case Study of Varanasi*. CASEPEDIA.

Sharma, P. (2019). SAARC Visa Facilitation Initiatives and Their Impact on Regional Tourism. *South Asian Journal of Tourism Management.*

Sharma, P. (2022). Educating Pilgrims: The Role of Museums in Promoting Cultural Understanding. *Tourism and Cultural Change*, *20*(4), 675–689.

Sharma, S. (2020). Economic implications of the Kartarpur Corridor: Opportunities and challenges. *South Asian Journal of Business Studies*, *9*(1), 5–17.

Sharpley, R. (2009). Tourism, religion and spirituality. *The Sage handbook of tourism studies,* 237-253.

Shukla, A. (2019). The Politics of Kartarpur Corridor and India-Pakistan Relations. *Indian Council of World Affairs*, *10*, 1–8.

Singh, J. (2020). Narratives in Action: Modelling the Types and Drivers of Sikh Activism in Diaspora. *Religions*, *11*(10), 539. DOI: 10.3390/rel11100539

Singh, M. (2020). Religious Tourism and the Sikh Diaspora: The Case of Kartarpur. *Tourism Management Perspectives*, *35*, 1–10.

Singh, R. (2020). The significance of Kartarpur in Sikh spirituality and culture. *International Journal of Sikh Studies*, *13*(1), 50–62.

Smith, J. (2023). Emerging Destinations: Why Pakistan Should Be on Your Travel Radar. *Lonely Planet*, NA.

Smith, M. K. (2015). Issues in cultural tourism studies. . *Routledge.*, NA.

Smith, V. (2003). Tourism and Interfaith Dialogue: A Framework for Action. *Tourism Management.*

Tepelus, C. (2018). The role of religious tourism in promoting cultural heritage: The case of Jerusalem. *Journal of Heritage Tourism, 13*(4), 343–355. DOI: 10.1080/1743873X.2017.1345546

Tepelus, C. (2018). The role of religious tourism in promoting cultural heritage: The case of Jerusalem. *Journal of Heritage Tourism, 13*(4), 343–355. DOI: 10.1080/1743873X.2017.1345546

Timothy, D. J. (2006). Tourism and Religious Journeys: A Global Perspective. *Routledge.*, NA.

Ullah, M. S. (2022). Kartarpur Corridor: Re-defining Security in South Asian Region. . *Annals of Human and Social Sciences,*, 3(3), 01-09.

UNESCO. (2019). *UNESCO and Interreligious Dialogue.* Retrieved 10 27, from United Nations Educational, Scientific and Cultural Organization.: Retrieved from https://en.unesco.org/interreligiousdialogue

UNESCO. (2021). *Pakistan's World Heritage Sites.* UNESCO World Heritage Centre.

Verma, S. (2020). The Role of Green Spaces in Religious Sites: The Case of Kartarpur. *Journal of Environmental Management, 254*, 109766.

Wagner, C. (2023). India's New Regional Environment: South Asia 2.0. In *Global India* (pp. 13–29). Routledge.

Wani, M. D., Dada, Z. A., & Shah, S. A. (2013). Behavioural intention and satisfaction levels among pilgrim tourists in south asia. *Tourism Review, 78*(4), 1147–1163. DOI: 10.1108/TR-09-2022-0432

WEF. (2021). *Travel and Tourism Competitiveness Report.* World Economic Forum.

Weller, P. (2016). Interfaith Dialogue and Peacebuilding: A Review of the Literature. *Journal of Peacebuilding & Development, 11*(2), 1–14. DOI: 10.1080/15423166.2016.1156785

Yousaf, S. (2021). Tourism and reconciliation in an enduring rivalry: The case of kartarpur corridor on india–pakistan border. *Tourism Management Perspectives, 39*, 100833. DOI: 10.1016/j.tmp.2021.100833

Chapter 14
The State of Populations of Endemic Plant Species of the Liliaceae Family in the Ritsa Relict National Park (Republic of Abkhazia)

Inga Vasilyevna Tania
Ritsinsky Relict National Park, Abkhazia

Alfiya Naufalevna Mustafina
South-Ural Botanical Garden-Institute of Ufa Federal Scientific Centre of RAS, Russia

ABSTRACT

The chapter presents the results of studies of endemic plants of the Liliaceae family in the Ritsa Relict National Park of the Republic of Abkhazia. There are 66 habitats of five species in the National Park. As a result of the conducted research, the population status of three species was ranked as satisfactory, one species was found unsatisfactory, and the population of F. lagodechiana is of particular concern. Ornamental species F. latifolia, C. speciosum, and L. kesselringianum are subjected to flower plucking by tourists; the grazing cattle and wild boars have a negative impact as well. Anthropogenics influences have been recorded to lead to a decrease in the number of populations of endemic plant species.

INTRODUCTION

As of today, the risks to the existence of endemic plant populations have significantly increased due to the growing negative anthropogenic impact on the flora. The problem of endemic species loss is a universal problem of biodiversity reduction on the planet, therefore a detailed study and assessment of

DOI: 10.4018/979-8-3693-8232-5.ch014

the current state of endemic species populations is of prime importance in order to develop scientifically-based recommendations for their protection.

The study of the Lily family goes back far into the past. Scientists have always shown interest in these plants. Let's note a number of fundamental works by foreign researchers: J. Lindley (1846), G. Bentham and J. Hooker (1883), Ar. Cronquist (1981). The work of Russian colleagues on the species of the family is listed in the Flora of the European part of the USSR (1978), the Flora of the North Caucasus (2013), and the Synopsis of the Flora of the Caucasus (2006). On the territory of Abkhazia, A.A. Kolakovsky (1986) studied flora and, in particular, species from the lily family, and his research formed the basis for this work. Many authors have studied individual genera of the family, some of whom are M.A. Pustynnikov and V.I. Bolgov (1981), N.K. Fedorova (2009) – *L. kesselringianum*; L. Bondarenko (2002)– *F. latifolia*.

Ritsa Relic National Park is of great interest in this regard as 176 endemic species grow on its territory. With an increasing number of tourists and new routes, the study of the endemic plant population is crucial for their growth in natural habitats.

Ritsa Relict National Park (RRNP) is a unique natural complex located on the southern slope of the western part of the Greater Caucasus. With a various terrain, limestone, rocky and gravel soils form unique ecological conditions that help preserve rare and endemic plant species. The RRNP climate depends on its terrain, altitude, slope exposure and many other factors that are variable within the park. The flora is predominantly Colchis mixed broad-leaved, beech-fir forests, subalpine curvilinear forests and meadows, high-mountain alpine areas (Kuftyreva et al.,1961).

Functional zoning has been carried out on the territory of the National Park with due regard for biological, scientific, recreational, educational and economic significance. Thus, five geographical sectors are distinguished in the National Park: Auadhara, Ritsa, Kujba-Yashta, Bzyp, and Cherkessko-Polyanskaya areas. Conservation and recreational areas have been allocated within the geographical sectors. Sections with unique or reference objects of living and inanimate nature are assigned within the boundaries of the sectors.

MATERIALS

Liliaceae is a family of monocotyledonous plants. 530 species of Liliaceae, grouped into 9 genera, of which candytuft, lily, grouse, and Lloydia are distributed in Eurasia and North America. Cardiocrinum, Notolirion, Nomocharis have relict ranges in Southeast Asia. Tulips and species of goosefoot are common in Eurasia, with exceptional diversity in the mountains of Foreland and Central Asia. Among the lileines there are ornamental, medicinal and melliferous species. More than 20 endangered lily species are included in the Red Book of the Russian Federation (2024).

The family Liliaceae was described by Michel Adanson in 1763 and officially named by Antoine Laurent de Jussier in 1789. Jussier designated this group as having a six-part calyx of similar color, six stamens, an upper ovary, a solitary style, and a three-nested box. By 1845, John Lindley, the first English systematist, confirmed the great diversity in the description of the family. In his view, the lileids were already «comprehensive». He identified 1200 species grouping them into 133 genera (Lindley, 1846). By the time of the next major British classification in 1883, several other genera were included in the Liliaceae (Bentham, Hooker, 1883). Over time, the family became increasingly extensive, as species with six perianth leaves and an upper ovary were arbitrarily defined as Liliaceae, eventually leading to the allocation of some 4,500 species belonging to 300 genera in the order Liliaceae. Ar. Cronquist (1981) merged Liliaceae with Amaryllidaceae, making them one of the largest monocotyledonous families.

The family Liliaceae on the territory of Abkhazia is represented by 18 genera of 37 species of which 8 are rare, 3 relicts and 12 endemic plant species.

Of the 18 genera of Liliaceae of Abkhazia, 4 genera are represented in the Ritsa Relict National Park: *Fritillaria L., Lilium L., Colchicum L.* and *Asphodeline Reichend.* The present research covers the species endemic to the Caucasus and Abkhazia.

1. The genus *Fritillaria L.* is a rather large genus of plants. The name of the genus is derived from the Latin word «*fritillus*» meaning «*chessboard*», as flowers of some species are particolored resembling a chessboard (Bondarenko, 2002). The *Fritillaria* genus comprises up to 150 species extensively distributed in the temperate zone of Eurasia and North America. The largest center of speciation is considered to be the Mediterranean region, mainly highlands, but some species of the genus are also found in the steppe zone. There are 26 species in Russia and neighboring countries, more often in the Caucasus and Central Asia, as well as in the Eastern European part of Russia, Western Siberia, and the Far East (Lozina-Lozinskaya, 1934). There are 10 species known in the Caucasus. A.A. Kolakovsky (1986) lists 1 species for the territory of Abkhazia.
2. The *Lilium L.* genus is ancient and large. The genus Lilia includes more than 110 species distributed mainly in Europe, but also in Asia with the greatest center of speciation in East Asia. There are up to 16 species in Russia (Fedorova, 2009), including 5 species in the Caucasus, of which 4 are endemic with two species (*L. unibrata* and *L. kesselringa*) that are Colchis endemics and occur in Abkhazia.
3. The *Colchicum L.* genus is a rather large genus with up to 50 species. The species is widespread in the Middle-Earth region of Eurasia till Afghanistan, with some species occurring in Northern England. There are many species in Europe and Asia, where they reach up to 4000 m, as well as in forest regions. On the territory of Abkhazia there are two species of floret: *C. speciosum* and *C. umbrosum*.
4. The *Asphodeline Reichend* genus includes about 16 species. Asphodeline grows in the temperate climates of Europe in the Mediterranean, Africa and Asia Minor. In Russia, it is found in the Crimea, Karachaevo-Cherkessia, Krasnodar and Stavropol territories, in neighbouring Georgia and Armenia. A.A. Kolakovsky (1986) lists 1 species for the territory of Abkhazia.

METHODS OF RESEARCH

There are 27 species of the lily family in the Ritsa Relict National Park. A total of 66 habitats of five endemic plant species are differentiated: *Fritillaria latifolia* Willd., *Fritillaria lagodechiana* Charkev., *Lilium kesselringianum* Miscz., *Colchicum speciosum* Stev., *Asphodeline lutea* (L.) Reichenb (Table. 1).

Table 1. Brief characterization of the state of cenopopulations of some rare plant species of the RRNP

	The name of the species	Number of localities	Total number	Average density, pcs/m²	The state of the population
1	*Fritillaria latifolia*	37	>20000	19,5	satisfactory
2	*Fritillaria lagodechiana*	4	>50	1,4	is alarming
3	*Lilium kesselringianum*	9	>4000	1,6	unsatisfactory
4	*Colchicum speciosum*	12	>8000	3,2	satisfactory
5	*Asphodeline lutea*	2	>6000	24,1	satisfactory

The research of the endemic plant population of the lily family in the park included: search for new localities, study of their biology and ecology characteristics, and assessment of the impact of recreation and grazing on the status of cenopopopulations (CP).

To study the demographic structure and density of *Fritillaria latifolia, Colchicum speciosum* and *Asphodeline lutea,* 25 test sites (with 10 sites in some of the cases) of 1 m² were allotted to the transects. The order of laying (linear or staggered) and transect spacing were determined by the area occupied by a particular CP. Leading population features such as total and effective population density and age composition were determined. For small populations such as *Fritillaria lagodechiana* and *Lilium kesselringianum*, the number of individuals in the described populations was also considered.

When describing the age structure of CP, according to standard criteria (Uranov, 1975; Coenopopulations…, 1976; Naumova, Zlobin, 2009), the following age states were taken into account: juvenile (j), immature (im), virginile (v), generative (g). Due to the difficulty of distinguishing between juvenile, middle-aged and old-aged generative states in broadleaf ruffed grouse, all generative plants were combined into one group. According to the concept of discrete description of ontogenesis, 4 age states (juvenile, immature, virginial, generative) were discerned.

To characterize the ontogenetic structure of the CP, generally accepted demographic indicators were used: the recovery index, the replacement index (Zhukova, 1995), the aging index (Glotov, 1998). To assess the state of the CP, the «delta omega» criterion of L.A. Zhivotovsky (2001) was applied that is based on the combined use of age indices (Δ) (Uranov, 1975) and efficiency (ω) (Zhivotovsky, 2001). Static analysis was performed in MS Excel 2010 using standard indicators (Lakin, 1990; Zaitsev, 1990).

The study of the morphometry of endemic plant species in natural conditions was carried out according to the method of V.N. Golubev (1962) on 25 generative individuals of middle age (in some cases, with a small number of generative plants, on 10 or less specimens). Observations and measurements were carried out in the flowering or fruiting phase, while taking into account the main parameters, individual morphometric characteristics inherent in the species.

To assess the influence of the main environmental factors on morphometric parameters, a single-factor analysis of variance was used. The level of factorization for the morphometric parameters of plants was estimated according to Snedekor (Lakin, 1990).

The methodology for assessing the vital composition is based on the differentiation of plants of one ontogenetic state into classes of vitality. Plants of medium-aged generative ontogenetic position, which has the maximum effect on the self-maintenance of the cenopopulation, were used as subjects of vital analysis. Previously, factor and correlation analyses were carried out, which made it possible to identify a deterministic set of features among the biometric parameters. To process the data obtained, vital spectra were compiled reflecting the ratios of plants of the highest (a), intermediate (b) and lower

(c) vitality classes (Zlobin, 1989), and the quality indicator of the cenopopulation and vital types were determined: prosperous, balanced, depressive. Static analysis was performed in MS Excel 2010 using the Statistica 6.0 statistical software package using standard indicators (Golubev, 1962; Zhukova, 1995). In the statistical analysis of quantitative indicators, arithmetic averages, standard deviation, coefficients of variation (Cv, %) were calculated (Zhukova, 1995).

The peculiarity of the species included in the research

Fritillaria latifolia is a perennial herbaceous plant of short vegetation, blooming in spring and early summer, bulbous geoephemeroid. It grows in the alpine zone, in meadows and alpine carpets at an altitude of 1600 – 2300 m above sea level. Mediterranean, Colchian, Alpine, meadow species. It is endemic to the Caucasus. An ornamental plant (Kolakovsky, 1986; Synopsis…, 2006). It is included in the Red Book of the Chechen Republic (2007).

F. latifolia in the RRNP vegetates from the upper forest to the alpine mountain belt, in meadows and alpine carpets. Previously, 4 locations of this species were known within the Auadhara forest area. During the present research, 37 more locations were found in the valleys of the Auadhara, Lashpsy, Mzymna rivers, in the Kurdzyshkha, Pyv, and Kutyk tracts. The studied CP grow on roadsides, trails, river terraces, pastures, forest margins, water-glacial moraines, in birch woodlands and other mountain habitats (Taniya, Abramova, 2013). During the expedition studies, it was found that *F. latifolia* is well represented in the conservation and recreational zones in four sectors of the national park.

The study of the age structure of the species was carried out in 31 CP. The age structure is one of the significant indicators of the population; it ensures the ability of the population system to self-sustain and determines its stability (Zaugolnova, Smirnova, 1978). Based on the observations of *F. latifolia* and due to the short growing season, it was revealed that the pregenerative period lasts 2-3 years and plants enter the generative period for 3-4 years of development.

Plants of different ontogenetic states consume environmental resources at different rates. The contribution of plants of different age groups to population density is weighted according to their energy efficiency (Zhivotovsky, 2001). The total, effective density and age composition of the Demographic indicators of the state of the CP *Fritillaria latifolia* are presented in Table 2. The total density (M) in the *F. latifolia* CP varies widely – from 3.7 to 95.9 copies/m^2, the effective density (Me) is 1.7-20.0 copies/m^2. In most CP, predominating is the regenerative fraction. The maximum values of density indices are CP 5, 19 and 24 (53.7, 52.2 and 95.9 copies/m^2, respectively) due to the significant predominance of juvenile individuals (76.8, 60.5 and 67.9%). The proportion of individuals in the generative state is maximum in CP 25, 26 (55.1 and 68.0%, respectively), where the values of total and effective density are closest (5.3 and 3.4, 4.9 and 3.7 copies/m^2).

Table 2. Distribution of individuals by ontogenetic groups CP Fritillaria latifolia

N° CP	The proportion of individuals of individual ontogenetic states, %				M	Me
	j	im	v	g		
1	73.5	7.1	9.5	9.8	11.2	2.3
2	68.3	10.9	7.9	12.9	10.1	2.3

continued on following page

Table 2. Continued

N° CP	The proportion of individuals of individual ontogenetic states, %				M	Me
3	70.1	8.0	15.6	6.4	18.4	3.6
4	55.7	12.6	13.0	18.7	18.7	5.7
5	76.8	6.6	11.6	4.9	53.7	8.8
6	47.4	30.9	13.5	8.2	14.3	3.3
7	45.7	11.8	13.5	29.0	18.3	7.3
8	32.9	22.9	32.9	11.2	5.7	1.8
9	70.7	9.5	13.3	6.5	14.8	2.8
10	60.2	14.2	19.6	6.0	17.3	3.6
11	48.3	13.4	28.7	9.6	15.9	4.4
12	34.5	13.9	37.7	13.9	10.3	3.6
13	54.7	9.6	19.6	16.2	16.7	5.0
14	43.8	11.7	30.2	14.2	5.4	1.7
15	14.5	13.6	41.8	30.0	3.7	1.9
16	48.5	12.6	23.5	15.4	11.9	3.7
17	33.9	22.7	21.3	22.0	9.2	3.4
18	31.7	17.7	36.6	13.9	26.6	9.2
19	60.5	9.3	21.2	8.9	52.2	12.4
20	81.2	6.0	9.3	3.5	28.7	4.1
21	58.7	11.7	11.2	18.4	6.5	1.9
22	37.8	21.9	20.0	20.3	12.3	4.3
23	48.6	13.3	22.8	15.3	19.0	5.8
24	67.9	11.0	12.0	9.0	95.9	20.0
25	12.0	23.4	9.5	55.1	5.3	3.4
26	4.8	19.1	8.2	68.0	4.9	3.7
27	38.1	17.2	31.3	13.4	14.3	4.6
28	45.3	13.6	21.0	20.1	7.1	2.5
29	33.1	9.7	37.9	19.3	4.8	1.9
30	40.4	12.9	34.8	11.9	36.2	11.5
31	48.9	13.0	25.7	12.4	42.9	12.5

The average ontogenetic spectrum of *F. latifolia* is left-sided with a maximum in juvenile individuals (Fig. 1). Individuals of juvenile status – 48.0±3.27%, immature – 13.9±0.99%, virginal – 21.1±1.78%, generative individuals – 16.9±2.37%. The spectrum demonstrates good germination of *F. latifolia* seeds, a high drop-off of juvenile plants after the first wintering, and then a slight increase in the proportion of virginal and generative individuals due to a longer passage of this ontogenetic state by plants.

Figure 1. The average ontogenetic spectrum of Fritillaria latifolia populations.

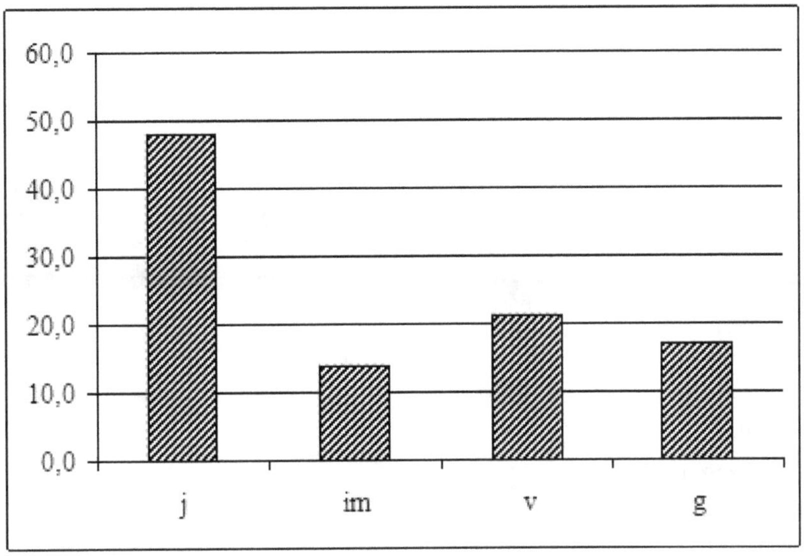

In rare plant species, under the influence of stress factors, some typical changes in the ontogenetic spectra of populations are also recorded: an increase in the proportion of the pregenerative fraction occurs, which reflects the elongation of the timing of ontogenesis (Zlobin et al., 2013). Each particular cenopopulation has an age spectrum that differs to varying degrees from the average one. This is due to the environmental conditions of the habitat, the density of the herbage, the intensity of anthropogenic load and changes in weather conditions, which affect the characteristics of seed germination and the tempo of development of individuals in one or another ontogenetic state.

Some variants of certain age spectra of different CP of *F. latifolia* are shown in Fig. 2. According to the classification of A.A. Uranov and O.V. Smirnova (1969) and A.A. Uranov (1975) the age structure of the CP of *F. latifolia* has two types of spectrum: invasive left-sided and normal centered. The left-sided ontogenetic spectrum is formed in most CP, the absolute maximum is in juvenile individuals (up to 81.2%). In some populations, virginal individuals are significantly represented (36.6-41.8%) – CP 12, 15, 18. This is due to sparse herbage and the absence of competitive species, which reduces the elimination of the generative fraction of the species.

In CP 25, 26 *F. latifolia*, normal centered spectra are created, where the maximum falls on the generative fraction. These CP have a low density and are located on the northern slope of the level. Pyv, along the road, in communities with very dense grass stands.

Figure 2. Some variants of ontogenetic spectra of Fritillaria latifolia populations.

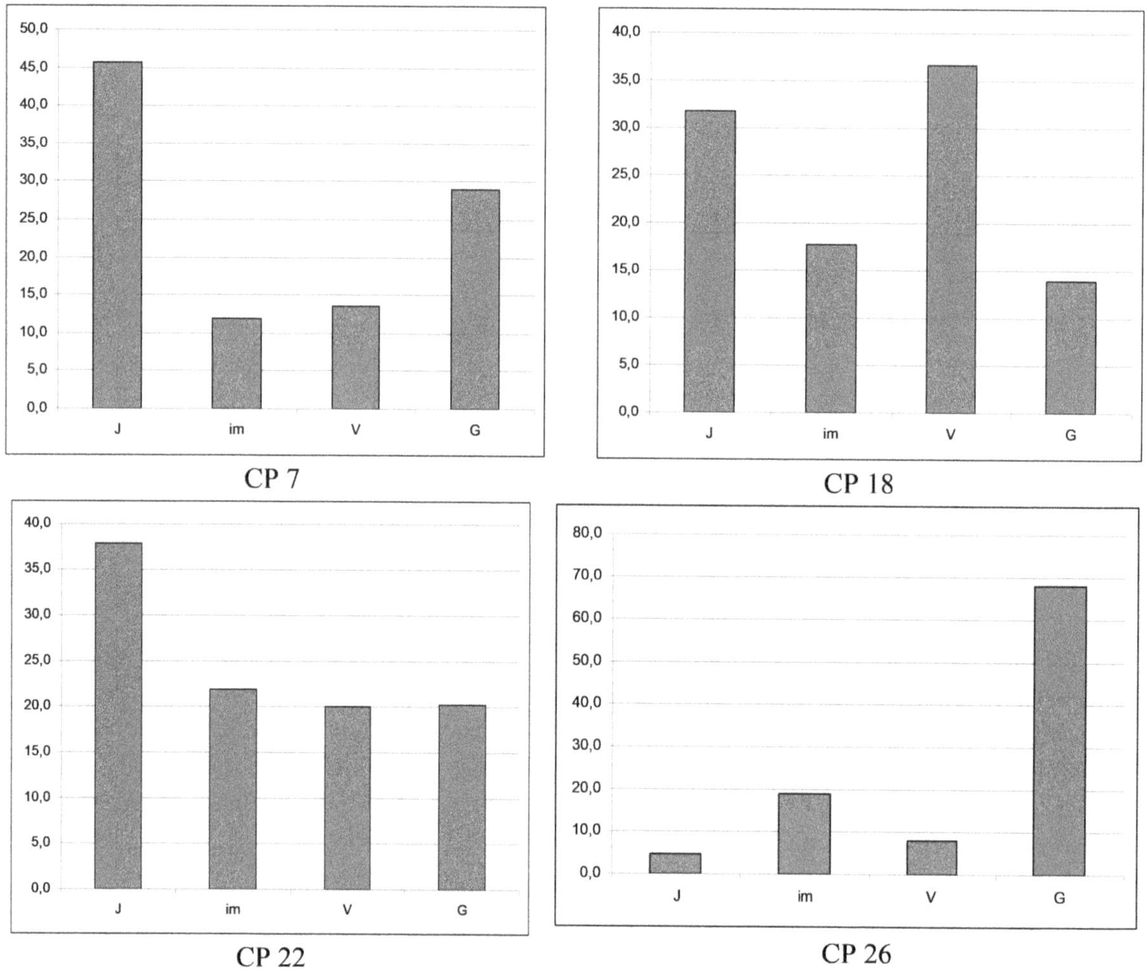

To assess the CP of *F. latifolia*, the demographic structure was studied, which is one of the significant features of the population, since it provides the ability of the population system to self-sustain and determines its stability. The age composition of the studied populations is presented in Table 3.

The assessment of the age of Δ (delta) and the effectiveness of ω (omega) showed that virtually all populations are young, juvenile and virginal individuals are most represented here, their density is high. CP 25 is maturing (Δ = 0.30; ω = 0.64), this is a disturbed population in a downed pasture with an average anthropogenic load. It is dominated by generative individuals (55.1%), but also by a significant number of immature individuals (23.4%). CP 26 belongs to the mature type (Δ = 0.36; ω = 0.75), it is located on the riverbank, along the road, and the absolute maximum is accounted for by generative individuals (68.0%). This CP is relatively stable, the total density of individuals in it is low – 4.9 copies/m^2. The indices of recovery (Ib) and aging (Ist), reflecting the dynamic processes of the CP, were also compared, which makes it possible to assess the replenishment of the CP of *F. latifolia* by young individuals, the presence or absence of individuals of the postgenerative period. As a result of active replenishment by young individuals, the maximum recovery index values are CP 20 (Iv – 27.65), growing in the alder forest of the floodplain of the Lashpsy river. In CP 7, 15, 17, where the intensity of replenishment by young individuals is significantly lower, the recovery index is lower and amounts to 2.48-3.54. The aging index in all CP is close to or equal to zero. This is due to the fact that most of the individuals of the species die off immediately after the completion of the last flowering. CP 4 has a very high indicator, and the greatest differences in total and effective density are noted in it.

Thus, most of the studied CP of *F. latifolia* differ in high density and incomplete ontogenetic spectrum associated with the rapid death of plants after the end of the generative state. Population density depends on the growing conditions of plants – it decreases in CP growing in unfavorable conditions – either in communities with dense herbage, or located along actively visited roads and trails where the anthropogenic load is higher (CP 8, 14, 15, 21, 25, 26, 28, 29). According to the delta-omega classification, most CP are young, one CP is mature, one is mature.

In general, it should be noted that the age structure of populations of *F. latifolia*, despite some differences in the spectra of specific populations, is quite similar – the pregenerative fraction prevails in the vast majority of populations, which is due to the already mentioned features of the species life cycle – good seed germination in favorable conditions and further gradual elimination of young plants.

Table 3. Demographic indicators of the state of the CP Fritillaria latifolia

N° CP	Indicator				
	Δ	ω	CP type	I_B	I_{CT}
20	0.05	0.14	Young	10.25	0.00
5	0.06	0.16	«	19.37	0.00
9	0.07	0.19	«	14.27	0.00
3	0.07	0.19	«	14.75	0.00
10	0.07	0.21	«	15.78	0.00
1	0.08	0.20	«	9.17	0.00
24	0.08	0.21	«	10.06	0.00
6	0.08	0.23	«	11.15	0.00
2	0.09	0.23	«	6.77	0.00
19	0.09	0.24	«	10.25	0.00
11	0.10	0.27	«	9.40	0.00
31	0.11	0.29	«	7.04	0.00
8	0.11	0.31	«	7.94	0.00
30	0.11	0.32	«	7.39	0.00
21	0.12	0.29	«	4.44	0.00
13	0.12	0.30	«	5.18	0.00
4	0.12	0.30	«	4.35	0.00
23	0.12	0.31	«	5.55	0.00
16	0.12	0.31	«	5.49	0.00

continued on following page

Table 3. Continued

N° CP	Indicator				
	Δ	ω	CP type	I_B	I_{CT}
27	0.12	0.32	«	6.46	0.00
14	0.12	0.32	«	6.04	0.00
12	0.13	0.35	«	6.21	0.00
18	0.13	0.35	«	6.17	0.00
28	0.14	0.35	«	3.98	0.00
22	0.14	0.35	«	3.93	0.00
17	0.15	0.37	«	3.54	0.00
29	0.15	0.39	«	4.18	0.00
7	0.17	0.40	«	2.45	0.00
15	0.21	0.51	«	2.33	0.00
25	0.30	0.64	Ripening	0.82	0.00
26	0.36	0.75	Mature	0.47	0.00

According to the research results, the most favorable growing conditions for the rare species *F. latifolia* are formed on the territory of the Ritsa Relict National Park in grass communities with an average anthropogenic load on the southern exposures of slopes located on the edges or along the edge of beech forests. Deterioration of the vital condition of the *F. latifolia* CP is observed at the northern exposures of the slopes and with strong blackening of the substrate. High stability *F. latifolia* to anthropogenic loads, apparently, is associated with the ephemerosity of the species, which manages to undergo vegetation before cattle are driven to pastures, as well as the favorability of habitats with low soil level, which is formed during the pasture use of subalpine meadows. In general, the state of *F. latifolia* populations in the RRNP does not cause concern – the populations are numerous, with good renewal.

However, the vegetation of the studied tracts is experiencing significant recreational and pasture stress. Since the end of June, tourists have been heading through this territory to interesting hydrological sites, and the intermountain valley of the Auadhara river itself is a place for recreation for the population of Abkhazia. In the course of research and monitoring of social pages, it was found that *F. latifolia* is collected by tourists for photos. Since mid-June, cattle have been driven to this area for grazing. Wild boars are a natural disturbing factor. The summer diet of wild boar consists entirely of herbaceous plants, including *F. latifolia*. All parts of the plant are eaten: leaves, stems and bulbs (Kudatkin, et al. 2005).

Although the population of *F. latifolia* is abundant within the recreational and management area of the Auadhar sector, conservation of the species during the flowering period is necessary: strengthen security measures, impose fines, place warning information boards, prohibit the temporary construction and placement of tourist tents in the places where the species grows, using social networks to inform about the importance of the species for biodiversity.

Fritillaria lagodechiana is a bulbous herbaceous polycarpic. Mesophyte, heliophyte, cryptophyte. It is found on grassy mountain slopes in the subalpine and alpine belts (Synopsis…, 2006).

F. lagodechiana is endemic to the Caucasus. It is included in the Red Books of the Karachay-Cherkess Republic (2003) and Krasnodar Territory (2017).

The species has not been described in the territory of the RRNP (News..., 1966, 1969). A single specimen was first discovered in 2013 in the tract of Auadhara at the foot of Mount Kutaheku. In 2014, the search for *F. lagodechiana* continued and a population of 54 individuals was discovered. This population, represented by four small localized groups, was described on the border of the forest and the subalpine belt. Under the ledge and on the ledge of rocks ranging from 1877 to 1923 meters above sea level. *F. lagodechiana* is represented within the protected area of the Auadhara functional sector.

The study of the age structure of the population revealed that out of 54 individuals 13 belong to juvenile, 3 to immature, 5 to virginal and 43 to generative individuals. It follows from this that the detected population is mature with the largest number of generative and juvenile individuals.

The characteristics of the morphometric parameters of the studied *F. lagodechiana* individuals are presented in Table 4. The height and diameter of the shoot in the grouse vary slightly, Cv in the range of 21.9–23.4%. The average height of the shoot is 28.7 cm, the height of the tallest specimen is 41 cm. The average diameter of the shoot is 0.4 cm. The number of leaves on the described specimens averaged 5 pcs., while the lowest coefficient of variation was observed (7.3%). The number of lower and middle leaves ranged from 1 to 2 pieces, with a Cv of 33.5% and 36.1%, respectively. The number of upper leaves ranged from 2.0 to 3.0 pcs., with a Cv of 20.7%. The length and width of the lower, middle and upper leaves had a coefficient of variation from 20.7-33.9%. The number of flowers forming on the peduncle was 1.0 pcs. The diameter of the flower ranged from 1-3 to 2.8 cm, with a Cv of 30.5%. The length and width of the flower petal were 2.0-3.8 cm and 1-1.7 cm, where Cv is 18.3% and 13.0% similarly.

Thus, studies of the Caucasian endemic *F. lagodechiana* have shown that the population is represented by four small habitats of 54 individuals, where a significant number of individuals of generative age status are represented. According to the scale of the degree of variation of the coefficient of variation in the *F. lagodechiana*, all indicators have a normal degree of variation. The general condition of the population of *F. lagodechiana* is alarming, due to the extreme smallness of this endemic plant and anthropogenic stress. In order to maintain the species, it is necessary to reduce the anthropogenic load in the places where the species grows and to control the state of the population. To maintain the species, it is necessary to create a botanical natural monument, monitor existing habitats and search for new locations.

Table 4. Intrapopulation variability of morphometric features of Fritillaria lagodechiana

Parameters	min-max	M±m	C_v, %
h	16.0–41.0	28.7±1.25	21.9
d	0.2–0.5	0.35±0.02	23.4
Nl	5.0–6.0	5.2±0.07	7.3
Nbl	1.0–2.0	1.5±0.10	33.5
Lbl	4.4–14.7	8.3±0.42	25.2
Sbl	1.2–2.9	2.0±0.10	23.3
Nml	1.0–2.0	1.3±0.10	36.1
Lml	3.6–11.0	7.18±0.34	23.5
Sml	0.6–2.0	1.37±0.08	28.7
Nhl	2.0–3.0	2.3±0.10	20.5
Lhl	3.0–8.6	5.95±0.25	20.7

continued on following page

Table 4. Continued

Parameters	min-max	M±м	C_v, %
Shl	0.4–1.6	0.97±0.07	33.9
Nfl	1.0	1.0±0.00	0.00
dfl	1.3–2.8	1.9±0.12	30.5
Lp	2.0–3.8	3.0±0.11	18.3
Sp	1.1–1.7	1.36±0.04	13.0

Note: min is the minimum morphometric indicator, max is the maximum morphometric indicator, M ±m is the average ± error of the average, C_v, % is the coefficient of variation, %.

Lilium kesselringianum is a bulbous herbaceous polycarpic. Entomophile, mesophyte, sciogeliophyte, cryptophyte. It is found in light forests, in clearings, among tall grasses, in meadows, prefers shallow sinkholes (Flora…, 2013). A very decorative plant, promising for breeding purposes.

The species was first described in 1914 by Professor P.I. Mishchenko. The samples were collected in the vicinity of Sukhum, and grown in the St. Petersburg Botanical Garden. The name of the species is given in honor of the famous instructor and owner of a commercial flower nursery in St. Petersburg, Wilhelm Kesselring, who introduced it into culture (Pustynnikov, Bolgov, 1981). *Lilium kesselringianum* is a rare plant of the Caucasus, included in the Red Book of the Russian Federation (2024) and the Red Book of the Krasnodar Territory (2017). It is protected on the territory of the Sochi National Park, the Caucasian and Teberdinsky nature reserves (Flora…, 2013).

Prior to the present research, the species on the territory of the RRNP were known from 2 locations: on the town of Pshakhushkha (1991, collector S.M. Chitanava) and on the Kutihug range at Lake Mzymna (2005, collector Z.I. Adzinba) – based on the materials of the IB ANA. We have identified habitats in the areas near Lake Mzymna, Ajarra waterfalls, Atsytak ridge and along the road to Pyv tract (Tania et al., 2015a). *Lilium kesselringianum* is represented within the conservation and recreational-economic areas of the Auadhara, Ritsa, Kudzhba-Yashta and Cherkessko-Polyansky functional sectors.

To study the biological features of the rare *Lilium kesselringianum* species within the Circassian-Polyansky forestry at the foot of Mount Agapsta and within the Auadhara forestry, localities of the species we identified in the valley of the Auadhara river, Lashpsy River, Mzymna River, in the tracts Pyv, Kutihug, Ajarra, Aztaku. The habitats of this species were located on slopes, water-glacial deposits, and pastures. Heights from 1,700 m to 2,300 m above sea level are covered. The total area of the surveyed territory was about 46 km^2. The total number of generative individuals included in the study is 141.

When studying the condition of coenopopulations of rare species, the analysis of the variability of qualitative and quantitative characteristics is of great importance. Morphometric parameters were measured during the flowering phase of the species. The average values of morphometric parameters of *L. kesselringianum*, their variability and a variance analysis of the impact of habitat conditions on morphometric features are presented in Table 5.

Table 5. Assessment of the influence of habitat conditions on morphometric parameters Lilium kesselringianum in cenopopulations

N° CP	Average values of morphometric parameters											
	h	**d**	**Nl**	**Lbl**	**Sbl**	**Lml**	**Sml**	**Lhl**	**Shl**	**Nfl**	**dfl**	**Lp**	**Sp**
^{2}	0.37	0.60	0.40	0.53	0.11	0.52	0.22	0.57	0.05	0.48	0.48	0.12	0.12
1	126.2	1.6	145.7	13.6	2.3	14.9	1.6	11.6	1.4	3.8	12.8	10.1	2.3
C_v, %	13.0	21.4	24.2	18.3	22.5	15.6	20.3	14.1	16.3	20.1	9.4	20.3	14.7
2	76.1	0.7	73.8	7.5	2.1	9.1	1.1	6.8	1.1	1.8	10.4	8.6	2.2
C_v, %	15.7	17.8	19.8	17.2	19.2	14.2	27.3	16.6	30.4	46.2	15.2	14.0	14.5
3	96.1	1.0	100.9	8.8	2.0	10.2	1.3	8.3	1.3	2.8	8.5	9.2	2.4
C_v, %	22.3	24.5	29.2	25.7	23.5	22.5	16.2	18.0	17.0	64.2	15.0	9.1	46.7
4	97.6	1.1	117.2	9.3	2.3	11.6	1.5	8.9	1.3	3.7	11.9	9.3	2.2
C_v, %	23.9	33.3	32.3	27.1	23.4	20.7	21.6	23.6	22.6	62.1	18.1	10.0	15.5
5	94.0	0.8	95.1	8.5	1.9	10.2	1.2	7.9	1.3	1.6	8.9	10.1	2.2
C_v, %	17.2	25.6	28.2	25.6	23.5	21.1	25.0	17.0	31.1	42.7	15.0	8.7	12.5
6	73.2	0.7	73.5	5.8	1.8	8.3	1.1	6.2	1.1	1.0	8.9	9.3	2.1
C_v, %	8.8	10.5	20.2	15.8	15.0	19.1	6.7	35.5	15.7	0.0	17.6	7.4	5.6
7	80.8	0.9	94.8	7.9	2.2	9.8	1.2	8.0	1.3	1.8	8.1	9.6	2.1
C_v, %	28.3	26.1	35.7	25.7	23.1	14.1	25.1	11.6	22.7	33.9	14.7	5.8	6,3
8	104.3	1.3	103.5	11.9	2.1	14.2	1.5	11.7	1.3	4.6	10.0	9.7	1.7
C_v, %	27.4	20.7	25.0	18.4	26.9	22.2	20.7	14.6	28.3	38.0	26.5	11.2	20.8
9	97.5	0.8	77.7	8.7	2.1	9.5	1.2	7.8	1.3	1.5	9.3	9.7	2.3
C_v, %	27.9	22.0	20.1	34.2	18.6	22.1	40.5	21.1	47.5	58.9	9.3	14.8	10.5

Note: ^{2}– influence force of the factor, Cv, % – coefficient of variation, %.

In terms of the number of parameters, both vegetative and generative spheres, CP 1 is in the lead, located on a slope, in a forest clearing, where more favorable conditions for plant growth are likely to be created. CP 8, which also grows on the slope, has high values of some parameters. The minimum values for all parameters are noted in CP 6, located on the southern slope in a dense grassland.

On the scale of the degree of variation of the coefficient of variation, most signs have a normal degree of variation. The following indicators have significant variation: the width of the upper leaf (Cv–47.5%) and the number of flowers per shoot (Cv–46.2-64.2%).

The assessment of inter-population differences by the method of variance analysis showed a statistically significant effect of ecotope conditions of a particular CP on all studied plant traits (at a significance level of $p < 0.001$). The factorization level ranges from 5% to 60%. In general, the indicator of the strength of the influence of the studied factor is not very high, which indicates the similarity of the habitat conditions of the species. The studied factor has the greatest influence on the diameter of the stem and the length of the leaves.

Vital analysis of *L. kesselringianum* populations (Table 6) showed that the considered CP are heterogeneous in their composition. Their vital type varies from prosperous to depressive. The vital state of the CP *L. kesselringianum* varies in different ecotopes. In CP 1, 4, 8, the dominance of individuals of the upper class is emphasized, and they are classified as prosperous. The CP quality index is maximum here and counts 0.36-0.5. In conditions of moderate disturbances, a high level of vitality of individual individuals can be traced in these cenopopulations. Other cenopopulations are classified as depressive, the population quality is 0.14-0.24. Against the background of general ecological and cenotic stress, the growth mechanisms of *L. kesselringianum* individuals in these CP are significantly suppressed. The absolute absence of individuals with high vitality was noted in CP 6.

Table 6. Distribution of Lilium kesselringianum individuals in the studied populations by vitality classes

N° CP	Relative frequency of size classes			Population quality, Q	Vitality type CP
	a	b	c		
1	0.96	0.04	0.00	0.50	prosperous
8	0.80	0.12	0.08	0.46	«
4	0.60	0.12	0.28	0.36	«
3	0.32	0.16	0.52	0.24	depressive
9	0.12	0.12	0.76	0.24	«
7	0.12	0.32	0.56	0.22	«
5	0.12	0.28	0.60	0.20	«
2	0.04	0.24	0.72	0.14	«
6	0.00	0.00	1.00	0.00	«

Thus, the results of the conducted studies of the rare species *Lilium kesselringianum* showed that the species in the territory of the RRNP is represented by 9 small cenopopulations (no more than 180 specimens in total), single specimens of the species are described in the Kutihug and Pyv tracts. The general condition of the populations of the *L. kesselringianum* is alarming, due to the extreme small number of this endemic plant and anthropogenic stress (grazing and recreation). Because of the beautiful flowers, the species is often collected in bouquets.

The popularity of the *L. kesselringianum* and its attractiveness, the intensive pace of flower harvesting and bulb digging, the limited range of distribution, and the increasing recreational development of the RRNP territory make the species especially vulnerable. In this regard, the *L. kesselringianum* needs to take radical protective measures to protect it. To preserve the species, it is necessary to limit the anthropogenic load in the growing areas, prohibit and control the collection of lilies by the population, and constantly monitor the state of populations in the RRNP.

Colchicum speciosum is a tuberous herbaceous perennial. Corms are inversely heart-shaped to elongated, ovate-oblong, usually with a short beak-shaped process at the base, 4-5 cm long and 1, 5-2. 5 cm in diameter, clothed with reddish-brown shiny sheaths surrounding the base of the stem in the form of a tube, which develops together with the leaf the next year after flowering and reaches up to 40 cm tall, with 4-6 broadly oblong leaves up to 25 cm long and about 4 cm wide. Perianth with a long tube up to 20 cm long and with oblong-elliptical, purple-pink lobes of the bend, up to 5 cm long and 2 cm wide. Blooms before frost in August–September, pollination of the flower manages to occur, and after withering of the aboveground part underground, ovules slowly develop in the ovary and the fruit begins to form. It bears fruit in May–July of the following year. Therefore, in the spring, together with the young leaves, a young fruit is brought out.

C. speciosum grows in the forest and alpine belt, in forests and meadows, in clearings, up to an altitude of 2300 m above sea level. Mediterranean, Caucasian, forest, tallgrass and meadow, mesophilic species. An ornamental, poisonous plant. It is used to produce colchicine (Kolakovsky, 1986). *C. speciosum* is one of the rare species in the National Park. The species is included in the Red Book of the Krasnodar Territory (2017).

Prior to our research, the species on the territory of the RRNP was known from 2 locations: in a clearing near the Gegh Waterfall and on the western slope of Adjarra (1990, 1997, was collected for the collection of S.M. Chitanava) based on the materials of the Herbarium of the Academy of Sciences of the Republic of Armenia. To study the biological features of *C. speciosum,* 12 habitats were described in the Auadhara forest area: in the valley of the Auadhara river, the Lashps river, in the Kurdzyshkha, Auadhara, and Ajarra tracts. The habitat of the described populations was located on roadsides, pastures, on the outskirts of beech forests, river terraces and in floodplains in alder forests. Heights from 1500 m to 2000 m above sea level are covered. The total area of the described sites was more than 20 km^2. Observations were carried out from the fruiting period (May – June) to flowering (August–September). In August, with a decrease in air temperatures at night, the flowers of the evergreen appear against the background of withering tall grass (Tania et al., 2015b). *Colchicum speciosum* is represented within the recreational and economic areas of the Auadhara and Kujba-Yashta functional sectors.

The natural complex of the Auadhara tract is represented by secondary post-forest meadows and is experiencing recreational and pasture stress. Today, these herbal communities are used by cattle breeders of the Gudauta region of Abkhazia. This determines the strong anthropogenic impact on the meadow pastures of the region (Yamalov et al.,2014). Because of the beautiful flowers, the plants are plucked by tourists into bouquets. In addition to the above factors, the populations of *C. speciosum* are negatively affected by wild boars that eat the corms of the plant. The consumption of poisonous plants by wild boar can be considered as a preventive remedy for helminthiasis (Kudatkin et al., 2005).

The characteristics of the morphometric parameters of the flower and their variability of *C. speciosum* are presented in Table 7. The height and diameter of the flowering shoot in the *C. speciosum* vary slightly (C_v–10.1-20.1%), these parameters are 14.9-23.9 cm and 0.4-0.6 cm, respectively. Higher coefficients of variation were noted for the number of flowers 25.9-54.0%. In the *C. speciosum*, basically, 1 or 2 flowers are formed, 3-4 flowers are also found, in isolated cases 5 flowers are observed. The diameter of the flower is significantly variable and ranges from 3.1 to 10.0 cm. The length and width of the inner and outer petals of the flower are: 4-5 cm and 1.5-1.8 cm, 4-6 cm and 1.5 and 2.2 cm, respectively.

Table 7. Intrapopulation variability of morphometric parameters of C. speciosum flower

N° CP	h	d	Nfl	dfl	LipFl	SipFl	LopFl	SopFl
1	23.9± 0.51	0.6± 0.02	1.8± 0.10	8.6± 0.20	5.0± 0.17	1.7± 0.06	6.0± 0.19	2.2± 0.06
C_v, %	10.7	13.1	27.8	11.4	17.0	16.5	15.7	14.0
2	19.9± 0.63	0.6± 0.02	1.4± 0.13	8.3± 0.18	5.1± 0.12	1.8± 0.05	6.2± 0.15	2.2± 0.08
C_v, %	15.7	16.1	46.1	10.9	12.1	14.9	11.9	17.8
3	17.8± 0.36	0.5± 0.01	2.2± 0.13	4.3± 0.36	4.1± 0.08	1.5± 0.13	4.9± 0.10	1.8± 0.06

continued on following page

Table 7. Continued

Nº CP	h	d	Nfl	dfl	LipFl	SipFl	LopFl	SopFl
C_v, %	10.1	9.7	29.3	41.6	9.7	43.8	10.1	15.2
4	14.9± 0.43	0.5± 0.01	1.9± 0.12	3.1± 0.15	3.8± 0.10	1.4± 0.07	4.6± 0.10	1.8± 0.09
C_v, %	14.3	13.7	31.9	23.2	13.1	24.1	10.8	23.9
5	15.0± 0.46	0.4± 0.01	2.5± 0.13	3.0± 0.10	4.0± 0.09	1.3± 0.05	4.6± 0.13	1.6± 0.05
C_v, %	15.3	13.1	25.9	17.0	11.7	18.2	14.5	16.0
6	18.7± 0.46	0.6± 0.02	1.6± 0.15	6.5± 0.16	5.3± 0.14	1.9± 0.07	6.4± 0.14	2.3± 0.06
C_v, %	12.3	14.1	47.7	12.4	13.4	18.3	11.2	13.2
7	19.5± 0.39	0.5± 0.01	1.9± 0.21	8.3± 0.23	5.7± 0.20	1.8± 0.06	6.5± 0.13	2.1± 0.08
C_v, %	10.1	14.0	54.0	13.8	17.3	17.0	9.7	18.4
8	19.1± 0.64	0.5± 0.01	2.0± 0.16	7.0± 0.15	4.4± 0.09	1.5± 0.04	5.0± 0.08	1.9± 0.05
C_v, %	16.8	12.6	40.3	11.0	9.8	13.4	7.6	13.4
9	19.7± 0.43	0.6± 0.01	2.2± 0.16	8.4± 0.15	5.5± 0.15	1.8± 0.05	6.4± 0.15	2.2± 0.07
C_v, %	10.8	10.7	37.0	9.0	14.0	14.7	11.8	16.4
10	20.1± 0.55	0.5± 0.02	2.6± 0.22	8.7± 0.22	5.4± 0.13	1.8± 0.05	6.3± 0.16	2.1± 0.05
C_v, %	13.6	16.6	43.0	12.5	12.5	14.0	13.1	12.6
11	17.8± 0.72	0.6± 0.02	1.5± 0.13	10.0± 0.20	5.7± 0.10	1.7± 0.07	6.5± 0.12	2.1± 0.07
C_v, %	20.1	14.4	43.0	10.1	9.1	19.9	8.9	17.7
12	16.3± 0.56	0.5± 0.01	1.8± 0.16	6.9± 0.12	3.7± 0.07	1.4± 0.04	4.3± 0.07	1.8± 0.05
C_v, %	17.1	13.8	45.4	8.9	9.4	13.5	8.3	14.0

Table 8. Intrapopulation variability of morphometric features of the stem of C. speciosum

Nº CP	h	d	Nl	Ll	Sl
1	23.9±0.84	1.2±0.03	4.4±0.10	20.6±0.55	5.1±0.24
Cv, %	17.5	13.3	11.4	13.3	24.0
2	28.3±0.91	1.4±0.05	4.5±0.12	22.1±0.65	5.7±0.22
Cv, %	16.1	16.5	13.0	14.8	19.1
3	42.0±2.75	1.6±0.06	5.1±0.17	22.4±0.88	7.4±0.33
Cv, %	32.7	18.9	16.3	19.6	22.5
4	33.6±1.26	1.1±0.05	5.0±0.07	20.2±0.70	6.5±0.31
Cv, %	18.7	21.3	7.1	17.4	23.8

continued on following page

Table 8. Continued

N° CP	h	d	Nl	Ll	Sl
5	46.1±1.42	1.4±0.04	5.0±0.12	21.0±0.53	7.2±0.59
Cv, %	15.5	14.3	11.5	12.6	40.7
6	39.7±1.30	1.5±0.06	4.4±0.12	22.7±0.79	5.6±0.33
Cv, %	16.3	21.9	13.1	17.3	29.3
7	41.5±1.22	1.6±0.08	4.8±0.12	24.7±0.80	6.3±0.31
Cv, %	14.6	25.9	12.0	16.3	25.0
8	34.4±1.08	1.3±0.06	4.4±0.14	20.8±0.57	5.3±0.17
Cv, %	15.7	22.0	16.0	13.7	16.2
9	38.5±1.49	1.1±0.06	4.3±0.12	21.8±0.67	5.7±0.27
Cv, %	19.4	28.0	14.3	15.5	24.1
10	36.9±1.24	1.2±0.05	4.5±0.12	21.4±0.59	4.7±0.18
Cv, %	16.8	21.6	13.1	13.9	19.2
11	34.3±1.41	1.3±0.06	4.5±0.10	23.4±0.66	5.4±0.16
Cv, %	20.5	20.8	11.4	14.1	14.9
12	40.8±2.22	1.1±0.05	4.4±0.13	18.2±0.77	5.9±0.39
Cv, %	27.2	25.7	14.7	21.0	33.1

Table 9 provides information on the reproductive ability of the species. In early summer, the plant throws out a short stem, on which a box is formed, in which the seeds ripen by the end of June. This plant does not have much influence in the spring, since it is absolutely not attractive, and Scott is only rising to pastures. An average of 23 to 75 seeds are formed in one or two forming boxes. However, the coefficients of variation are significant (30-70%), due to incomplete and full-fledged seeds. This is how the minimum number of seeds was found in the 1st box - 2, and the maximum – 166.

Table 9. The seed productivity of the element Colchicum speciosum

N° CP	Number of capsules		Length of capsule		Width of capsule		Number of seeds	
	M±m	Cv, %	M±m	Cv, %	M±m	Cv, %	M±m	Cv, %
1	1.0±0.04	19.2	5.4±0.20	18.5	2.2±0.09	19.7	28.0±3.03	54.0
2	1.3±0.10	36.1	5.9±0.21	17.8	2.5±0.08	16.0	40.5±5.36	66.2
3	1.8±0.14	41.1	4.4±0.12	13.6	1.5±0.03	9.4	27.1±2.38	43.9
4	1.8±0.10	29.7	4.9±0.19	19.7	1.4±0.08	26.7	47.8±4.99	52.3
5	1.7±0.14	39.4	5.3±0.14	13.2	1.6±0.03	9.4	43.9±2.62	29.8
6	1.3±0.10	36.1	7.3±0.25	17.3	2.8±0.12	21.9	74.4±7.43	49.9
7	1.1±0.07	29.6	6.6±0.24	17.9	2.6±0.10	19.5	52.7±5.76	54.7
8	1.3±0.09	35.8	6.1±0.20	16.7	2.3±0.08	17.5	56.4±6.66	59.0
9	1.4±0.12	40.5	6.1±0.24	19.9	2.2±0.08	18.8	49.0±6.76	68.9
10	1.0±0.04	19.2	6.7±0.23	17.1	2.3±0.06	12.5	67.9±6.79	50.0

continued on following page

Table 9. Continued

N° CP	Number of capsules		Length of capsule		Width of capsule		Number of seeds	
	M±m	Cv, %	M±m	Cv, %	M±m	Cv, %	M±m	Cv, %
11	1.2±0.09	35.2	6.3±0.33	25.9	2.3±0.10	22.1	45.8±6.33	69.1
12	1.4±0.13	45.2	4.2±0.10	11.6	1.6±0.05	17.0	39.8±2.73	34.3

Note: the designation is M±m and C_v, % as in Table 4.

When analyzing and comparing the obtained data, the highest indicators for the generative sphere, such as height of the flower-bearing shoot, flower parameters, length and width of capsules, number of seeds, as well as indicators of abundance and density, were found in CP 6, 7, 10. These CP are located at altitudes of 1500–1550 m above sea level, on the southern slope of the Kurdzyshkha ridge in the floodplain of the Lashpsy River. In this area, vegetation begins in the spring earlier than in other areas of Auadhara, and the number of livestock grazing in this area has significantly decreased in recent years. And because of the remoteness of the Auadhara mineral spring, there is not a single tourist route in this direction. All this has a positive effect on the general condition of the species populations. Minimum morphometric indicators in CP 3, 4 and 5. These CPs are located in the Ajarra tracts and along the Auadhara River valley. These are areas where there is a strong anthropogenic impact: popular tourist destinations, grazing in the vicinity of Auadhara begins in May and ends in late September, with favorable weather conditions continues until October.

Based on the results of the conducted research, 12 new coenopopulations of *Colchicum speciosum* were identified, five of them are in a satisfactory condition, the condition of seven CP is alarming due to anthropogenic implications. Significant anthropogenic influence on individual coenopopulations of colchicum lead to a decrease in numbers, which is one of the reasons for the rarity of the species. To preserve the species C. speciosum, it is necessary to limit the anthropogenic linfluence in the places of growth, to prohibit the picking of plants, to monitor the state of the populations, to exclude the use of natural populations for the purpose of raw medicinal mixtures.

Asphodeline lutea are plants with short rhizomes and short thickened stems extending from it, 40–60 cm high, highly foliated. In the Republic of Abkhazia, *A. lutea* was described by A. A. Kolakovsky (1986) as a rare plant of the limestone gorges of Zhoekvara, Bzybskoe and Gegskoe. *A. lutea* grows in the lower belt, up to 400 m above sea level, mainly on illuminated southern rocky slopes, in steppe-like cenoses. *Asphodeline lutea* is represented within the recreational and economic areas of the Bzyb and Kudzhba-Yashta functional sectors. Studies of A. lutea populations were carried out in mixed Kolkhid forests of Kudjba-Yashta forestry of the RRNP.

In the course of our research, we identified two cenopopulations of *Asphodeline lutea*, located at an altitude of 450-500 m above sea level, on south-western exposures of the slope with steepness from 15° to 50°, from gentle to very steep. The area of the sites varied from 60 to 600 m^2 (Table 10). In the described sites, the projective vegetation cover was about 90%. The dominant species were *A. lutea* and *Brachypodium sylvaticum* (Huds.) P. Beauv. The vegetation of the forest glades of the Kudzhba-Yashta tract is represented by sub-Mediterranean species. At the same time, the development of moss cover was noted, indicating the influence of the humid climate of Colchis. The soil is stony with a poorly developed humus layer, gradual overgrowing with shrubs and forest was observed (Tania et al., 2019).

Table 10. Some characteristics of natural populations of Asphodeline lutea

N° CP	Height above sea level	Habitat	Thicket area, m^2	Projective coverage, %
CP 1 – a narrow strip along the road	450	south-west	60	70
CP 2 – on the slope	500	slope 150,	600	90

The characteristics of the morphometric parameters of *A. lutea* are presented in Table 11. The leader in most parameters is CP 1, located near the road. Apparently, it is in a small area in the transition zone from the road to the slope that the vegetation is sparse, the projective cover is small, compared to CP 2, and at the same time the most favorable conditions for the growth of the species are formed. CP 2, growing on the slope, has high values of such parameters as: the number of leaves, the length of the upper leaf and the number of flowers. The two CP had similar parameters: the width of the upper leaf and the diameter of the flower.

According to the scale of the degree of variation of the coefficient of variation, most of the features have a normal degree of variation. The following features have a large variation: number of fruits (CP 1–83.5%, CP 2–80.6%), length of the upper leaf (CP 2–42.8%) and width of the upper leaf (CP 1–36.9%).

Table 11. Intrapopulation variability of morphometric features of Asphodeline lutea

Parameters	CP 1			CP 2		
	min-max	M±M	C_v, %	min-max	M±M	C_v, %
h	71.0–101.0	83.6±2.89	10.9	60.0–86.0	74.9±2.43	10.3
d	0.6–1.0	0.8±0.04	16.3	0.4–0.9	0.7±0.06	25.9
Nl	55.0–94.0	83.6±3.66	13.8	57.0–96.0	77.9±3.81	15.5
Lbl	19.0–27.0	22.8±2.93	12.8	16.0–28.0	22.4±1.25	17.6
Sbl	0.3–0.7	0.5±0.03	20.0	0.4–0.7	0.5±0.03	19.9
Lml	10.0–18.0	14.0±0.87	19.6	7.0–15.0	10.2±0.83	25.5
Sml	0.3–0.4	0.3±0.01	13.2	0.2–0.3	0.3±0.01	10.9
Lhl	2.0–4.5	3.1±0.31	30.9	2.0–6.0	3.3±0.45	42.8
Shl	0.1–0.2	0.1±0.02	36.9	0.1–0.2	0.2±0.01	16.6
Li	18.0–40.5	25.4±2.22	27.6	9.0–32.0	17.8±2.01	35.7
Nfl	39.0–76.0	61.0±3.92	20.3	20.0–88.0	57.0±6.99	38.8
dfl	4.5–6.5	5.6±0.18	10.3	4.5–6.0	5.3±0.15	8.8
Nfr	0.0–28.0	9.7±2.56	83.5	0.0–14.0	5.0±1.27	80.6

Note: the designation min-max, M±m and C_v, % as in Table 4.

The density indicators, age composition and demographic indicators of the state of the CP are reflected in Table 12. The density in CP 2 is 2 times higher than in CP 1 – 32.3 and 15.9 specimens/m^2, respectively.

Table 12. Density indicators, age composition and demographic indicators of the state of the Asphodeline lutea CP

N° CP	Density, spec./m²	Ontogenetic state					Demographic indicators				
		j	*im*	*v*	*g*	*s*	Δ	ω	CP type	I_B	I_{CT}
1	15.9	4.4	32.1	63.5	0.0	0.0	0.09	0.33	young	1.0	0.0
2	32.3	0.0	11.8	85.4	2.5	0.3	0.12	0.41	transitional	0.35	0.03

The average (basic) ontogenetic spectrum of *A.lutea* is left-sided with a maximum in virginal individuals, which allows concluding that the population has a good ability to self-renew and that there are favorable conditions for seed germination (Fig. 3).

Figure 3. Ontogenetic spectrum of Asphodeline lutea populations

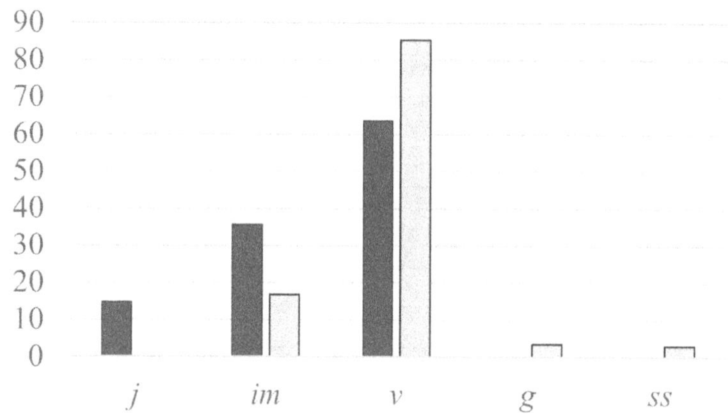

The assessment of age Δ (delta) and efficiency ω (omega) showed that CP 1 is young (Δ = 0.09; ω = 0.33), with no generative or post-generative fraction. CP 2 is transitional (Δ = 0.12; ω = 0.41). A comparison of the indices of recovery (IB) and aging (IST), reflecting the dynamic processes of the ontogenetic structure, was also made. The recovery index is equal to one (CP 1), the share of pregenerative plants here is maximum or quite high (CP 2 = 0.35), this indicates a good replenishment of the CP with young individuals. The aging index in the CP is equal to or close to zero (CP 1 = 0 and CP 2 = 0.003).

The identified features of the vegetation of the asphodel glades of the RRNP indicate their sozological significance, preserving the Mediterranean vegetation to this day. There is an obvious need to eliminate anthropogenic stress, which can accelerate the process of overgrowing by ruderal species of these forest clearings, which will lead to the extinction of Mediterranean species, including Asphodeline lutea in the Kudzhba-Yashta tract.

RESULTS

Thus, a total of 66 habitats of five endemic plant species from the lily family have been described within the Ritsa Relict National Park: *Fritillaria latifolia, Fritillaria lagodechiana, Lilium kesselringianum, Colchicum speciosum, Asphodeline lutea.* The conducted studies have shown that of the surveyed 5 endemic species from the lily family in the RRNP, the populations of 3 species are in a satisfactory condition, the population of 1 species is in an unsatisfactory state and the population of *F. lagodechiana* is especially concerning. Of all the studied species of the population, *F. latifolia* are the most numerous, with a high density. This species grows in a wide range of habitat conditions, its populations condition being satisfactory. Two species, *C. speciosum* and *A. lutea*, have a population of over 3 thousand individuals and a fairly high density, their condition also being satisfactory. This is due to the flowering in the early spring and late autumn period, when the number of tourists is few. Despite a rather high total population, *L. kesselringianum* have a low density and their condition is unsatisfactory, they have a low participation of young plants, which is associated with the growth of the species in hard-to-reach places with dense herbage, in which seed germination is difficult. The endemic species *F. lagodechiana* is alarming due to the extreme smallness and low density of this population, for which only a single locality is known.

Decorative species *F. latifolia, C. speciosum, L. kesselringianum* suffer from flower picking by tourists. Populations of *F. latifolia* and *C. speciosum* are negatively affected by wild boars, which eat leaves, stems and roots of plants. But the main factor reducing the number of endemic species in the territory of the Ricin Park is cattle grazing, which has traditionally developed from ancient times to the present, while the processes of seed reproduction and self-renewal are disrupted. Anthropogenic disturbances lead to a decrease in the number of populations of endemic plant species due to: soil compaction and overeating of shoots, which leads to a decrease in seed productivity, deterioration of seed germination conditions, as well as the elimination of seedlings and juvenile plants. The main reasons for the rarity of the studied species are both in the biological characteristics of the species themselves and in sufficiently high anthropogenic loads. In order to preserve species, it is necessary to limit the anthropogenic load in their places of growth, prohibit the collection of flowers by the population, monitor and control the state of populations of endemic plant species. It is necessary to create a nursery for endemic plant species in order to further reintroduce them into nature and preserve their gene pool.

REFERENCES

Bentham, G. (1883). *Genera Plantarum Ad Exemplaria Imprimis in Herbariis Kewensibus Servata Definita: Sistens dicotyledonum monochlamydearum ordines XXXVI: Nyctagineas-Ceratophylleas; et gymnospermearum ordines III: Gnetaceas-Cycadaceas; Sistens monocotyledonum ordines XXXIV: Hydrocharides-Gramineas* (Vol. 3). Reeve & Company.

Bondarenko, L. (2002). How to grow grouse species. *Floriculture.*, *6*, 30–32.

Fedorova, N. K. (2009). Lilies. – Moscow: Kladez-books. 95c.

Glotov, N. V. (1998). On the estimation of the parameters of the age structure of plant populations. Zhizn'populyatsii v geterogennoi srede, 146-149.

Golubev, V. N. (1962). Fundamentals of biomorphology of herbaceous plants of the central forest-steppe. Osnovi biomorfologii travyanistyh rastenij centralnoj lesostepi], Voronezh, Izdat-vo Voronezhskogo un-ta.

Kolakovsky, A. A. (1986). *Flora of Abkhazia* (Vol. IV). Metsniereba.

Kudatkin, A. N., Veinberg, P. I., Bobyr, G. Ya., Semenov, U. A., & Akkiev, B. (2005). In Tuniev, B. S. (Ed.), *Large mammals of the Ricin Relict National Park // Ricin Relict National Park* (pp. 136–148). Prospekt.

Kuftyreva, N. S., Lashkhia, Sh. V., & Mgeladze, K. G. (1961). *The nature of Abkhazia*. Sukhum.

Lakin, G. F. (1990). Biometrics: Textbook for biol. spec. universities. 4th ed., reprint. and additional M.: Higher School. 352 p.

Naumova, L. G., & Zlobin, Yu. A. (2009). *Fundamentals of population ecology of plants* (Mirkin, B. M., Ed.).

Pustynnikov, M. A., & Bolgov, V. I. (1981). *Lilia Kesselring in the Caucasus// Industrial floriculture in the south of the USSR. Scientific works. Issue 28.* Scientific Research Institute of Mountain Gardening and Floriculture.

Tania, I. V., Abramova, L. M., & Mustafina, A. N. (2015a). Rare endemic plant species from the Liliaceae family in the Ricin Relict National Park. Proceedings of the Abkhazian State University. Sukhum. pp. 56-62.

. Tania I.V., Abramova L.M., Mustafina A.N. (2015b). Colchicum speciosum Stev. in the Ricin Relict National Park (Republic of Abkhazia) // Izvestiya Orenburg State Agrarian University. No. 5 (55). pp. 164-167.

Tania, I. V., Kraynyuk, E. S., Suvorov, A. V., Mustafina, A. N., Bagrikova, N. A., & Leiba, L. O. (2019). Geobotanical and population studies of Asphodeline lutea (L.) Rchb.) in the Ricin Relict National Park// Scientific notes of the Cape Martian Nature Reserve. Issue 10– Flora and vegetation. pp. 106 – 114.

Taniya, I. V., & Abramova, L. M. (2013). On the biology of the rare endemic Fritillaria latifolia Willd. in the rich relict national park (Republic of Abkhazia). *Bulletin of the Samara Scientific Center of the Russian Academy of Sciences.*, *15*(3), 103–107.

The Red Book of the Chechen Republic. (2007). *Rare and endangered species of plants and animals*. Fearsome.

The Red Book of the Karachay-Cherkess Republic. (2013). Nartisdat.

The Red Book of the Krasnodar Territory. (2017). In Litvinskaya, S. A. (Ed.), *Plants and fungi. III edition* (pp. 429–470). Adm. Krasnodar. Territories.

Uranov, A. A., & Smirnova, O. V. (1969). Classification and main features of the development of populations of perennial plants. Byull. MOIP. *Department of Biology*, *79*(1), 119–135.

Yamalov, S. M., Tania, I. V., Khasanova, G. R., & Kuropatkin, V. V. (2014). Phytodiversity of secondary upper montane meadows of the Ricin Relict National Park (Republic of Abkhazia). *Izvestiya Samara Scientific Center of the Russian Academy of Sciences.*, *16*(1), 115–119.

Zaitsev, G. N. (1990). *Mathematics in experimental biology. – M*. Nauka.

Zaugolnova, L. B., & Smirnova, O. V. (1978). The age structure of the cenopopulations of perennial plants and its dynamics. *Journal of Biology*, *39*(6), 849–858.

Zhivotovsky, L. A. (2001). Ontogenetic states, effective density, and classification of plant populations. *Russian Journal of Ecology*, *32*, 1–5.

Zhukova, L. A. (1995). *Population life of meadow plants*.

Zlobin, Yu. A. (1989). *Principles and methods of studying cenotic plant populations: Textbook-method. manual / Kazan*. Kazan University Press.

Zlobin, Yu. A., Sklyar, V. G., & Klimenko, A. A. (2013). *Populations of rare plant species: theoretical foundations and methods of study*. University Book.

Chapter 15
Reducing Environmental Risks When Using Solar Systems in the Heat Supply of Tourist Business Facilities

Natalia Gennadievna Andreiko
https://orcid.org/0009-0008-3601-6115
Kuban State Technological University, Russia

Evgenia Ivanovna Zakharchenko
https://orcid.org/0009-0009-4172-4244
Kuban State University, Russia

Musabek Abdusharipovich Duvanakulov
Osh Technological University, Kyrgyzstan

ABSTRACT

The work considers reducing the manifestation of an environmental hazard factor of the energy type through measures that reduce the likelihood of environmental risks. It is this type of factor that is the most destructive, since the basis of energy is natural resources. Consuming a huge amount of primary resources in the form of solid, liquid and gaseous fuels, the energy sector, along with the production of electrical energy and heat, emits a large amount of waste into the environment in the form of gaseous and solid combustion products. In addition, there is the problem of exhaustion of resources for traditional energy. In this regard, much attention is paid to energy saving and energy efficiency based on renewable energy sources. Solar energy is one of the most developed branches of renewable energy. Therefore, the work examines various heat supply schemes using solar systems, which will not only reduce dependence on traditional fossil fuels, but also reduce greenhouse gas emissions, that is, reduce environmental risks.

DOI: 10.4018/979-8-3693-8232-5.ch015

RELEVANCE OF THE STUDY

Energy consumption is a prerequisite for human existence. The availability of energy available for consumption has always been necessary to meet human needs, increase the duration and improve the conditions of his life. Without energy, life on the planet is impossible. The physical law of conservation of energy says that energy cannot arise from nothing and does not disappear without a trace. It can be obtained from natural resources such as coal, natural gas or uranium and converted into forms we can use, such as heat or light.

But obtaining energy from natural resources has led to one of the greatest problems of our time - climate change. If current trends continue, i.e., maintaining the current share of fossil fuels and almost doubling the demand for electricity by 2050, emissions will far exceed the carbon limit allowed by limiting global warming to two degrees Celsius. This level of emissions would have catastrophic consequences for the planet.

Today there are all the prerequisites for this. The Earth has been warming at a rapid rate for at least the last 200 years, with temperatures about 1.2 degrees higher than pre-industrial levels. Despite the fact that the temperature on the planet has fluctuated for a long time, with periods of warming and cooling. But since the last ice age 10 thousand years ago, the climate has been relatively stable, which, by the way, according to scientists, was crucial for the development of human civilization (Grachev & Plyamina, 2017; Shcherban, 2021). Now this stability has been shaken. And the last ten years have been the warmest on record. And it is people who are responsible for all the global warming over the past 200 years, since in the vast majority of cases, warming occurs due to the burning of coal, oil and gas. Burning these fossil fuels releases greenhouse gases into the atmosphere, which wrap around the planet like a blanket, trapping heat. Through research ranging from ice cores to tree rings, modern scientists have been able to track the concentration of greenhouse gases. According to them, the level of greenhouse gases such as methane and nitrous oxide is at the highest level in the last 800 thousand years, and the level of carbon dioxide is at the highest level in the last two million years.

It would seem: only one or two degrees, but what catastrophic consequences! The world will lose more than 40% of plant and vertebrate species, and more than 60% of insects. In some areas, agricultural yields will drop by more than half, threatening food security. With warming of 1.5°C, 70-90 percent of the corals that form the basis of many underwater ecosystems will die. And with warming of 2 degrees, about 99% of corals will die (Grachev & Plyamina, 2017; Sinyak, 2015; Shcherban, 2021). Their disappearance would likely lead to the extinction of other marine species, many of which are critical sources of protein for coastal communities.

And yet, despite the dire situation, humanity still has a small chance of avoiding the worst consequences of climate change. The latest UNEP Greenhouse Gas Reduction Report says that by cutting greenhouse gas emissions by 42 percent by 2030, the world could limit the rise in global temperatures to 1.5 °C above pre-industrial levels. To reduce environmental risks, the world must reduce its annual emissions by 22 billion tons of carbon dioxide equivalent in less than seven years.

RESEARCH METHODS

This scientific research was carried out using methods of theoretical knowledge, namely the analysis of environmental risks and solar systems that can reduce environmental risks, as well as comparative historical analysis, which will help clearly formulate cause and effect relationships growth of environmental risks and development of technologies in the field of solar energy. The authors are confident that it is a comparative historical analysis, which allows us to enrich the study with additional facts about the development of solar energy, that will help to consider the problem of increasing environmental risks and offer more complete solutions to reduce them.

But before proposing solutions, it is necessary to clearly understand what environmental risks are and their types, because clarity of terms increases the accuracy and reliability of research results, allows you to more effectively convey the content of the research, reducing ambiguity and possible misunderstanding.

Environmental risk - According to (Rabinovich, 1986), environmental risk is defined as the chance (in a probabilistic sense) during a specified period of time of the occurrence of an event with specific (usually negative) consequences. The object of assessment can be any natural or anthropogenic object on at any level of their system organization.

Environmental risks can be divided into primary and derivative (Vorobyov & Vorobyov, 2019; Shtebner, 2022). Primary environmental risks directly trigger the process of negative impact on the environment, and cause the emergence of derivative environmental risks, which, in turn, are accompanied by the emergence of new types of negative impact on the environment.

It is proposed to divide environmental risks according to their magnitude as follows (Vorobyov & Vorobyov, 2019):

Acceptable environmental risk is an environmental risk, the magnitude of which is justified from the point of view of the current level of socio-economic development of the state and the cultural and historical development of the ethnic group inhabiting it.

Maximum permissible environmental risk is the maximum level of environmental risk, which is determined by the prevention of irreversible processes of environmental degradation, regardless of the interests of various social groups and ethnic groups.

Excessive environmental risk is an environmental risk the magnitude of which exceeds the maximum permissible environmental risk.

But no matter how large the environmental risk we consider, the factor causing it is environmental danger. What exactly is an environmental hazard?

Environmental hazard is any change in the functioning parameters of natural, technical or natural-technical systems, leading to a deterioration in the quality of environmental components beyond the boundaries of established standards (Dmitriev, 2014; Medvedeva, 2016).

But, in order to establish environmental risk management, that is, to make specific decisions to minimize them, it is necessary at the first stage to classify environmental hazard factors. This will make it possible to identify the entire set of reasons that cause or potentially can cause negative changes in the quality parameters of environmental components. Today, the following types of environmental hazard factors are distinguished: natural, geographical, climatic, landscape-geographical, destructive, unforeseen and economic.

In the economic class, there are four types of factors (Ministry of Natural Resources of Russia, 2022): production, resource, energy and demographic. Industrial factors of environmental hazard include everything related to technogenic pollution of the environment, negative impact on the biosphere and humans as well. The main manifestation of this type of factor is the release of pollutants into the atmospheric air, the discharge of pollutants with wastewater, the creation of excess physical fields (noise, vibration, electromagnetic, radiation, etc.). Resource factors of environmental hazard are associated with human activity, which is accompanied by the consumption of resources and, unfortunately, is not always rational. But thoughtless consumption of resources leads to major environmental disasters. Demographic factors of environmental hazard. Due to the sharp increase in population, which scientists have dubbed a demographic explosion, in many regions of the world there is a shortage of food, drinking water, and agricultural land. Morbidity and child mortality are increasing, outbreaks of epidemics are emerging, and life expectancy is decreasing. All this is accompanied by a decrease in the quality of environmental components, and its toxicity and pathogenicity increases.

Energy factors of environmental hazard affect the parameters of the environment, on the one hand, as a type of production in the form of emissions of pollutants into the atmosphere (especially when using coal and fuel oil as fuel), waste in the form of slag, spent nuclear fuel, thermal pollution reservoirs. Hydroelectric power plants cause enormous harm to the biological systems of reservoirs. On the other hand, the concentration of people in huge megacities and the concentration of production facilities in a limited area leads to a concentration of energy and natural resource consumption, which, according to some scientists, leads to irreversible degradation of the biosphere. In addition, the energy factor of environmental danger also manifests itself in the event of failures in the supply of energy to support the activities of technosphere objects, which simply cannot maintain their structural and functional integrity without energy produced and supplied by humans (Nikonova, 2017).

From the above concepts and classifications of environmental risks and environmental hazards, we can safely conclude that it is human activity in all forms of its manifestation that causes the highest degree of environmental danger. Consequently, measures are needed aimed at changing the vector of human activity to reduce environmental risks to acceptable values, that is, values justified from the point of view of the modern standard of living in all spheres (Bashkin, 2020; Vishnyakov & Radaev, 2008).

This paper examines the reduction of the manifestation of the environmental hazard factor of the energy type through measures that reduce the likelihood of the manifestation of environmental risks. It is this type of factor, in the opinion of the authors, that is the most destructive, since the basis of energy, of course, is natural resources, but not directly, but indirectly, through the creation of a production infrastructure capable of producing and delivering energy to the consumer. But not only this explains the importance of the energy type of environmental hazard. At the moment, energy is the main driving factor in the development of all sectors of industry, public utilities and agriculture, and serves as the basis for increasing productivity and well-being of the population. It has the highest rates of development and scale of production. That is why the pace of scientific and technological progress and the improvement of working conditions are largely determined by the state of energy and great attention is paid to it in all countries of the world. At the present stage of energy development, the problem of its interaction with the environment is becoming very important. he environmental conditions for the functioning of the energy sector are associated with the unique scale of material and thermal interaction with the biosphere (atmosphere, hydrosphere, lithosphere). This is due to the fact that, consuming a huge amount of primary resources in the form of solid, liquid and gaseous fuels, the energy sector, along with the production of electrical energy and heat, emits into the environment a large amount of waste in the form of gaseous and solid combustion products (Fox & Burgman, 2008; Fox & Burgman, 2017).

And primary natural resources are still the basis for energy. The extraction and use of primary natural resources are accompanied by a huge impact on almost all components of the environment. Moreover, often newly developed territories that were not previously subject to direct anthropogenic impact are exposed to impact. Water resources accumulated in artificial reservoirs are no exception. At the same time, large areas are flooded, the hydrological regime and migration routes of ichthyofauna are disrupted. Energy from oil, gas and coal is obtained by combustion, which is accompanied by the release into the atmosphere of hundreds of thousands of tons of pollutants such as carbon monoxide, nitrogen oxides, sulfur dioxide, soot, vanadium pentoxide, hydrocarbons and benzopyrene (Fox & Burgman, 2017).

But in addition to increasing environmental risks, there is the problem of exhaustion of resources for traditional energy. It is no coincidence that the world pays great attention to energy saving and energy efficiency, one of the main types of which is renewable energy sources. Therefore, today in the energy sector several areas of possible solutions to reduce environmental risks are relevant. This is the improvement of technologies to reduce the negative impact on the environment of traditional methods of generating heat and electricity, control of CO_2 emissions, and reduction of energy consumption. But, in addition, methods for generating energy from renewable sources are needed, including wind farms, solar panels, cascading low-power hydroelectric power stations, power generating stations that use tidal energy.

So why will renewable energy save the world from global warming? The fact is that generation technology based on renewable energy sources significantly outperforms traditional generation in terms of specific greenhouse gas emissions during the life cycle of facilities. This is clearly demonstrated on the graph of specific greenhouse gas emissions per gram of CO_2 equivalent (Figure 1).

CO_2 equivalent is a conventional unit that is used to estimate the volume of greenhouse gas emissions (including to calculate the carbon footprint). It is measured in tons and indicates how much carbon dioxide is equal to the total volume of emissions, based on their impact on the climate (Fox & Burgman, 2017; Podgornaya, 2015). Climate scientists introduced this unit in the late 1980s to make it easier to estimate greenhouse gas emissions. Simplification in calculations was necessary because there are many greenhouse gases, and they all have different greenhouse activity - and, accordingly, contribute to climate change to varying degrees. Therefore, it was proposed to calculate the total volume of emissions through comparable volumes of carbon dioxide, since it has the least greenhouse activity. For example, over 100 years, one ton of methane retains as much heat in the atmosphere as twenty-five tons of carbon dioxide, and a ton of nitrous oxide (N_2O) is equivalent in this regard to 298 tons of carbon dioxide. At the end of the last century, the concept of CO2 equivalent began to appear in the reports of the Intergovernmental Panel on Climate Change (IPCC). And in 1997, the Kyoto Protocol was adopted, which became the first international agreement that directly regulated the reduction of greenhouse emissions. And all quantitative requirements of this protocol were calculated in CO_2 equivalent (Revich, 2010).

But let's return to the dependence of specific greenhouse gas emissions presented on the graph, from which it is clearly seen that emissions from burning fossil fuels are two orders of magnitude higher than pollution from using renewable energy sources. Therefore, renewable energy is precisely the measures that will reduce the likelihood of environmental risks. If you do some simple calculations, you can see that given that just under 85 trillion (8.5×10^{13}) kWh of energy per year is bought and sold on the global commercial market, greenhouse gas emissions per year are put 19.6×10^{16} grams CO_2 equivalent.

Figure 1. Dependence of specific greenhouse gas emissions (median)

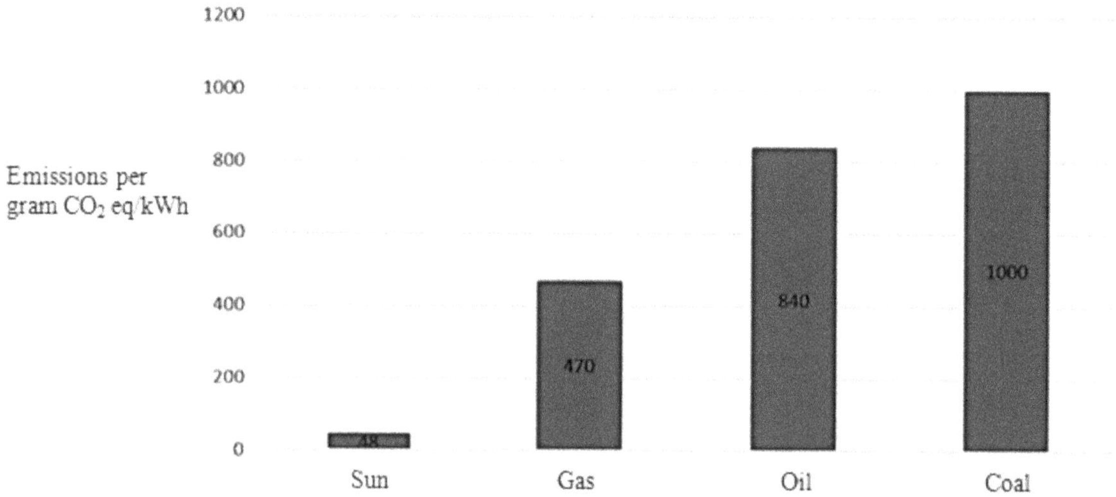

COMPARATIVE - HISTORICAL ANALYSIS

In fact, renewable energy is the oldest and safest way to generate energy. For centuries, renewable energy was the only available source of energy for the inhabitants of the Earth, if you do not take the muscular power of man and animals themselves. But fossil energy sources took over only during the industrial revolution - the whole point is that they simply turned out to be more profitable at a certain stage in the development of civilization. However, at that time no one imagined that coal, oil, and then gas would cause irreparable damage to the planet's climate in literally just a hundred years. Thus, fossil fuels turned out to be a time bomb, and the centuries-tested production of energy from renewable sources fell by the wayside for many years (Varganova & Chinaeva, 2021).

But today, as governments, businesses and consumers recognize global warming and climate change as one of the greatest threats of our time, the popularity and adoption of renewable energy sources is growing exponentially. With the help of modern technology, alternative sources have become not only more reliable, but also more cost-effective—even more so than fossil fuels.

Solar energy is currently the most popular form of renewable energy. After all, the Sun, as is known, is the primary and main source of energy for our planet. Solar energy is also converted into useful energy indirectly by being transformed into other forms of energy, such as biomass, wind or water energy. he energy of the Sun "controls" the weather on Earth. It warms the entire Earth, sets rivers in motion and imparts strength to the wind. Under its rays, 1 quadrillion tons of plants grow, which, in turn, feed ten trillion tons of animals and bacteria. Thanks to the same sun, hydrocarbon reserves have accumulated on the earth, that is, oil, coal, peat, etc., which we are now actively burning. In order for humanity today to be able to satisfy its needs for energy resources, about 10 billion tons of standard fuel are required per year. If the energy supplied to our planet by the Sun per year is converted into the same equivalent fuel, then this figure will be about 100 trillion tons. This is ten thousand times more than we need.

Every second the sun emits 88·1024 cal or 370·1012 GJ of heat. Of this amount of heat, only 1.2·1012 W reaches the Earth in energy equivalent, i.e. per year 1018 kWh, or 10,000 times more than the energy consumed in the world today (Varganova & Chinaeva, 2021; Rabinovich,1986). Compared to it, all other energy sources provide negligible heat. If, for example, the potential of the Sun is determined by solar energy falling only on free uncultivated land, then the average annual power will be about 10,000 GW, which is approximately 5,000 times more than the power of all modern stationary energy installations in the world. The practical feasibility of using solar energy is established based on the maximum solar radiation equal to 1 kW/m^2. Calculations have shown that to satisfy modern energy consumption it is enough to convert solar energy falling on 0.0025% of the Earth's surface into electrical energy (Rabinovich,1986; Revich, 2010).

Did our ancestors know about all this? No. But since ancient times, man has used solar energy to dry various materials and preserve food. The first established case of the use of solar energy dates back to 8000 BC, where during excavations a device for drying food was found. Later, in the period 5000 - 2000. BC, several places were discovered where animal skins and clay plates were dried for making writing boards. Socrates (469 - 399 BC) describes that optimal use of natural solar radiation is achieved by oriented the main rooms of buildings to the south. The German scientist and inventor Athanasius Kircher (1671) in his book writes that in China during the Han Dynasty (220 - 201 BC) concave mirrors made of an alloy of brass and tin were used. Mirrors were used to light torches from the "sun fire" during religious sacrificial rituals. With the advent of the Middle Ages, enthusiasm among scientists diminished and for many centuries no important theoretical and practical work on the use of solar energy was discovered. Practical applications were limited to the distillation of plant extracts for medicinal purposes and the distillation of various aromatic oils and wines (Varganova & Chinaeva, 2021; Mehdiyeva & Sultanly, 2015).

The situation changed with the advent of the Renaissance, many studies were carried out, which were mainly devoted to solar energy concentrators for steam production and high-temperature solar ovens. Leonardo da Vinci (1452 - 1519) conducted a series of experiments with a large parabolic mirror; the device generated thermal energy for the dyeing industry; even books with drawings of the installation have been preserved. Kircher's contemporary, the scientist von Tischirnhaus, designed in 1781 various types of large concave lenses up to 1 meter in diameter. He used these lenses to melt various materials, concentrating solar radiation on them. In 1774, Antoine Laurent Lavoisier (1743 - 1794), a famous chemist and one of the founders of modern chemistry, who discovered the role of oxygen in combustion, designed lenses to concentrate solar radiation. The lens system was built on a cart and used as a solar oven. The lenses allowed high temperatures to be reached and were used to melt and study pure platinum, from which it can be concluded that a temperature of 1780 °C was reached.

An early ancestor of modern solar collectors is a device designed by Horras de Saussure in 1770 and called it a "hot box" or "hot box". It was an ordinary wooden box, lined with cork inside. The only purpose of this box is to heat the air inside and measure the amount of heat supplied to this box. Khorras reached an air temperature of 160 ° C (Varganova & Chinaeva, 2021).

With the beginning of the Industrial Revolution, interest in solar energy decreased, since it was more convenient and cheaper to produce steam by burning coal. However, there were enthusiasts who continued experiments with the sun. The French scientist Becquerel (1820 - 1891) experimented with various lenses and a wooden box, which was closed with a glass lid. He went further than Horras de

Saussure and painted the inside of the box black, so the system became even more similar to a modern solar collector; in the box they not only heated air, but also tried to cook food.

One of the most important developments of the new era, which has generated a lot of interest in recent years in connection with point-focus concentrators, is the Stirling cycle. Robert Stirling, a Scottish minister and engineer, invented a solar-powered steam engine, patented in 1816 in Edinburgh. Around 1870, Swedish engineer John Ericsson modified the Stirling engine and launched the Stirling cycle using concentrated solar energy.

In 1833, in the USA, J. Erickson built a solar air engine with a parabolic-cylindrical concentrator measuring 4.8 by 3.3 m. The first flat-plate solar energy collector was built by the Frenchman S. A. Telyer. It had an area of 20 m and was used in a heat engine that ran on ammonia. In 1855, a solar installation scheme was proposed with a flat-plate collector for water supply, and it was mounted on the roof of an extension to the house.

French mathematician Auguste Mouchot (1821 - 1911) is undoubtedly a pioneer of solar energy. He was the first to publish a book on solar energy in 1878 with the beautiful title "La chaleur solaire et ses applications industriales". He was also the first to express the thesis about the depletion of fossil fuel reserves in the future. In 1861, he introduced his first solar steam engine, which he considered unviable economically due to low coal prices. Together with his assistant A. Pifrom, he created and presented a truncated parabolic mirror with a total area of 20 square meters. The steam produced by solar radiation powered a printing press that was used to print the Sunshine Journal in French. In 1866, A. Mouchot built several large solar concentrators in Algeria and used them to distill water and drive pumps. And at the World Exhibition in Paris in 1878, Musho demonstrated a solar oven for cooking, in which 0.5 kg of meat could be cooked in 20 minutes.

At the beginning of the 20th century, engineers from the United States became very interested in the developments of French scientists, where active work began on the creation and installation of solar engines.

One of the most active and prominent scientists from the United States was K. J. Abbott (1872 - 1973), head of the Smithsonian Institution in Washington. In 1897 he introduced the "heat box", where two more wooden boxes and a black metal sheet were installed, and the large box itself was covered with four glass sheets. Abbott created patent after patent. During the international energy conference in Washington, he demonstrated a collector with an efficiency of 60%. And in 1972, when he was already 100 years old, he received another patent, but which changed a lot, the name of this patent was: "Converting useful solar energy into electricity" (Mehdiyeva & Sultanly, 2015; Shcherban, 2021).

In 1886, the German physicist Hertz discovered the phenomenon of the photoelectric effect (light falling on the surface of a metal knocks electrons out of it). Devices that generate electric current when exposed to sunlight are called photocells. The first silicon solar cells were manufactured in the USA in 1954 and had an efficiency of 6%.

Thus, ways to use solar energy have been determined. This is the production of heat and the direct conversion of solar radiation into thermal and electrical energy.

For a long time, the widespread use of solar radiation was hampered by such shortcomings as a relatively low energy flux density, frequency of arrival during the day, varying intensity throughout the year, and intermittent receipt of radiation due to cloudiness.

And another significant drawback is the cost of converting solar energy. In 1975, academician Pyotr Kapitsa gave a report "Energy and Physics" at an event dedicated to the 250th anniversary of the Russian Academy of Sciences. In particular, the respected academician noted that photovoltaic conversion, that is, solar energy, is extremely expensive. And it is completely unclear how to make its use economically rational. We draw attention to this part of the phrase: "we need to reduce costs by several orders of magnitude." It is clear that in 1975 it was not clear how to do this. Even in 2000, people still didn't know. But today something happened that the academician was thinking about. Costs have decreased literally by several orders of magnitude.

ANALYSIS OF EXISTING TECHNOLOGIES

Today, the world has accumulated extensive experience in the use of solar energy. In countries such as Saudi Arabia, Turkey, Egypt, Cyprus, Israel, Jordan, Greece, etc., houses are heated with solar energy, electricity is generated, and sea water is desalinated. The use of solar energy for heating and hot water supply has become particularly widespread in many countries. Many options for "solar houses" have been created, both small (cottages) and multi-storey. For example, the solar system of a nine-story building in Kherson with a heating surface of about 250 m^2 fully meets the hot water needs of residents of all 64 apartments.

Heat obtained from solar energy is divided into low-temperature (below the boiling point of water) and high-temperature. In the first case, it is used in greenhouses, dryers, water heaters, and desalination plants. This use is based on the "hot box" principle.

High-temperature heat is obtained by concentrating the sun's rays with lenses, flat or concave mirrors. It is used in furnaces, power plants, engines and power plants intended for heat supply to various consumers (Mehdiyeva & Sultanly, 2015; Slesarenko et al, 2010).

From the point of view of reducing environmental risks, solar energy has no equal. This is explained by the fact that even at Russian latitudes, which are far from the southernmost, solar energy is sufficient for the efficient operation of modern solar systems. For example, in central Russia the sun gives off 1 square meter per year. meter of land is the energy that is obtained by burning approximately 150 kg of standard fuel. If this flow of solar energy is "collected" with the least losses and converted with the greatest efficiency, then in the middle zone a properly designed solar system will be able to provide a residential building with hot water and heat completely autonomously for 9 months a year, and the remaining three months (in winter) will help the main system heat water, thus saving fuel. On average, the production of thermal energy by a solar collector is on average 0.5 Gcal per year per 1 m^2 of collector, and the specific savings of equivalent fuel (kg/m^2 of solar collector per year) when using solar hot water supply systems is: for the latitude of St. Petersburg - 80, Moscow - 87, Samara - 100, Volgograd - 150, Astrakhan - 160, Sochi - 200. On average, modern solar systems using solar energy allow saving up to 40% of heating costs and up to 75% for hot water supply (Rabinovich, 1986; Sidorenko & Goncharov, 2020).

75% is an impressive figure, especially for tourism business facilities. Why for them? Let's explain now. But first, let's define what tourism business objects are, and why we are talking about them in this work.

The tourism business is one of the fastest growing sectors of the world economy. International tourism is one of the three largest export industries, behind the oil industry and the automotive industry. The importance of tourism in the world is constantly increasing, which is associated with the increased influence of tourism on the economy of an individual country. Travel for the purpose of entertainment and recreation form the basis of international tourist exchange. They account for about 70% of world tourism.

Objects of the tourism business include collective accommodation facilities, transport organizations, cultural institutions, sports facilities and other objects of interest to tourists and one-day visitors. But all this cannot exist without tourist resources, which are understood as natural, historical, socio-cultural objects, including objects of tourist display, as well as other objects that can satisfy the spiritual and other needs of tourists, help maintain their livelihoods, and restore and the development of their physical strength. In addition, the tourism business cannot exist without the provision of tourist services - without everything that the tourist takes into account or uses during his trip. It follows that a tourist service consists of a whole range of services for a tourist (both tangible and intangible), which are a specific tourist product.

And the ecology of the region planned to visit is the first thing travelers pay attention to. After all, everyone wants to relax where there is clean air, water and land. True, unfortunately, there is also a feedback - tourism has a negative impact on the environment, but this is not discussed now.

In order to comprehensively use the tourist and recreational potential, regions are trying to develop several types of tourism simultaneously: beach, business, health, water, sports. But, regardless of the type of tourism, the main objects of the tourism industry are, as a rule, year-round hotel complexes with cultural, entertainment and health facilities, swimming pools, water parks, and business centers. Today, this area of tourism is developing in many regions, and that is why the following are increasingly opening: mini-hotels; boarding houses for family holidays; motels.

The main types of heat load of such hotel complexes are heating, hot water supply (DHW), and swimming pool. But, if heating is a seasonal heat consumption, heating a pool depends on its type, then hot water supply is a year-round heat consumption. In addition, heat consumers can be a restaurant, laundry, car wash, etc.

The heating system in hotels must create a stable temperature regime during the heating season and comfortable hygienic conditions in accordance with the requirements. Throughout the entire heating season, the heating system must operate uninterruptedly and, with minimal heat consumption, ensure normal temperatures in all rooms.

The intensity of the heating system to heat the pool will depend on its size and location. Creating a comfortable temperature in open-air swimming pools is seasonal in nature, carried out from September to May. Indoor pools will require regular but less intense operation of the heating system, regardless of the time of year (Enikeeva et al, 2019; Khanaev, 2015).

The required amount of heat for hot water depends on the maximum possible number of people living, the type of water points (for bathrooms with a shower, the average hot water consumption rate is 140 l/day per person, but if baths are used - 200 l/day per person) and additional heat consumers. Due to increased requirements for the comfort of hot water supply, there is often a need to use boilers or hot water storage tanks.

Each hotel complex has its own level of comfort, but there are also mandatory conditions. One of these conditions is the presence of hot water in the shower. This condition has long moved from the category of luxury to the category of, as it is now fashionable to say, "must have". Since hot water solves not only the problems of personal hygiene of guests, but also general sanitation in the laundries and kitchens of the hotel. Therefore, it is necessary to ensure sufficient supplies of hot water at the required temperature. But unfortunately, it is not uncommon to have to deal with vacationers who, roughly speaking, do not turn off the hot water tap at all. As a result, traditional boilers simply do not have time to cope with such water withdrawal, and, operating in this mode, consume a large amount of electricity or fossil fuel. All this naturally leads to an unjustified increase in greenhouse gas emissions, that is, to an increase in environmental risks. And the solution to this problem is the use of solar systems for heat supply to hotel complexes. Let's consider this issue using the example of the Krasnodar Territory, which is one of the leaders in the Russian Federation in terms of the number of hotel complexes.

Despite the fact that the Krasnodar Territory is a typical southern region of Russia, the issue of heating and hot water supply is also relevant here. According to measurement data (Sidorenko & Goncharov, 2020), 5 zones with different values of total solar radiation on a horizontal surface have been identified in the Krasnodar Territory. Its lowest values are characterized by mountainous areas - 1205 kW/m^2, the largest areas adjacent to the Black and Azov Seas - 1400 kW/m^2. Namely, the coastal area accounts for the majority of the region's sanatorium and resort complexes, in which the share of hot water supply is quite comparable to heating costs.

In addition, an analysis of the distribution of the intensity of total solar radiation by month showed a significant excess of summer values over winter ones (Sidorenko & Goncharov, 2020). So, for Krasnodar these values are: in July 670 MJ/m^2, in December 80 MJ/m^2. It is obvious that the use of solar power plants is effective for hot water supply in the period from April to October. It is precisely during this period that the main peak in the use of sanatorium-resort complexes occurs.

In a broad sense, a solar system is an installation that converts solar energy into electricity or heat. Collectors are used to convert it into heat to heat the water.

But when using these settings, certain conditions must be met. First of all, you need to understand that the heating system must cover 100% of the need for thermal power without the use of solar energy. That is, solar collectors can be considered as an additional source of heating - like an additional boiler in the heating system, which works when there is sun outside. For a heating system, solar collectors must be installed more than for hot water supply (on average, one collector per 10-15 m^2 of heated area). Excess heat in summer can completely cover the load on the hot water supply. So 8-12 collectors installed for support will be able to heat about a ton of hot water per day.

One of the solutions is a pool of sufficient capacity; after heating the water for hot water, the automation will switch the system to heating the pool. Therefore, it is necessary to consider combined systems as the basis for the effective implementation of solar systems.

When designing systems to partially cover heating costs, it is necessary to take into account the low position of the Sun in winter. Accordingly, for more efficient operation of the solar system in the off-season and winter, it is necessary to install collectors at an angle on average 10-15% greater than usual (50-60°). In this case, the total annual energy production may decrease slightly with the condition that it is during the winter period that energy production will increase.

The periods of maximum solar energy input do not correspond in time to the periods during which the demand for energy for heating is highest. If the heat consumption for preparing hot water remains relatively constant throughout the year, then during periods of greatest heat demand only a very small amount of solar energy is supplied for space heating (Butuzov et al, 2009). To support a space heating system, the surface area of the collectors must be relatively large. As a result, the solar installation will be "idle" in the summer. From a technical implementation point of view, it is very simple to use a solar system to support a heating system, thanks to the use of storage tanks.

But despite the differences in solar systems used for heat supply, the basis of each such system is a solar collector.

A solar collector is a device designed to absorb solar energy, which is transferred by visible and near-infrared radiation for its subsequent conversion into thermal energy suitable for use. In solar systems today, two types of collectors are most common: flat and vacuum (Makhmudov & Kholboev, 2020).

The main element of flat solar collectors is an absorber that absorbs solar radiation; it has a transparent coating on top. To increase the efficiency of the collector, a special optical coating made of tempered glass with a reduced metal content is used. The absorber is connected to a heat-conducting system. The design of flat-plate solar collectors is quite simple. Externally, they are a simple panel with a rectangular shape. This installation has an aluminum body and several pipes used to drain and supply coolant liquid. In addition, the inside of the collector walls are covered with a heat-insulating layer. Today, manufacturers make its thickness equal to three to four centimeters - this makes it possible to achieve a significant reduction in the level of heat loss. The operating principle of a flat-plate solar collector is based on the greenhouse effect - the sun's rays arrive on the surface of this device and penetrate through the glass. The heat-absorbing coating used at the bottom of the collector has an absorption coefficient of 91%. Eventually, excessive heating causes the coating to radiate thermal energy. Its power is located in the infrared range, in other words, it is possible to achieve the accumulation of solar energy in the collector. The process of heat removal occurs with the direct participation of the coolant.

The main part of the vacuum collector is the heat pipe. Such collectors are a series of glass tubes of a special design. The solar collector tube is actually two tubes (one nested inside the other), between which there is a vacuum for the best thermal insulation of the coolant from the external environment. The method of transferring heat from it to the vacuum solar collector is as follows: the copper tube is hollow inside and contains an inorganic and non-toxic liquid. When heated, this liquid evaporates, and since a vacuum is created in the tube, this happens even at a temperature of minus 30 degrees Celsius. The steam rises to the tip of the heat pipe, where it transfers its heat to the coolant (for example, antifreeze), which flows through the heat pipe of the solar collector. Then it condenses and flows down, and the process repeats again. The solar water heater with vacuum tubes shows excellent results even on cloudy days. This is explained by the fact that vacuum tubes are capable of absorbing the energy of infrared rays that pass-through clouds. Due to the insulating properties of vacuum, the influence of wind and low temperatures on the operation of the solar system is also insignificant compared to the influence on a flat solar collector. The system with a vacuum solar collector operates successfully at temperatures down to minus 35 degrees Celsius. The pipes in a solar water heater are installed in parallel, and their angle of inclination depends on the latitude of the solar system installation site. Thus, oriented from north to south, during daylight hours, the tubes of the vacuum solar collector passively move behind the sun. In addition, such a collector practically does not require operational maintenance (Makhmudov & Kholboev, 2020).

Flat-plate collectors are structurally simpler and therefore cheaper. Vacuum collectors are more complex and expensive units, but their efficiency is much higher, especially in cloudy weather. The difference in efficiency is especially noticeable at air temperatures of +15 degrees and below, and at sub-zero temperatures the vacuum manifold is generally unrivaled. Vacuum collectors based on thermotubes are capable of providing consumers with hot water and heat even at subzero temperatures.

More than ten years of experience in the successful operation of solar power plants using both flat and vacuum collectors indicates the promise of this area of using non-traditional renewable energy sources.

For example, in the Krasnodar region in the city of Ust-Labinsk in 2011, a solar power plant was built for year-round hot water supply to the central district hospital (Fig. 2) with an area of 600 m^2 (Butuzov et al, 2020). The installed solar collectors from the German company Wolf consist of sheet-tube copper absorbers with a register hydraulic circuit with PVD-selective coating. The body of the solar collectors is stamped aluminum. Solar technical glass 3.2 mm thick, pressed into the body. Thermal insulation – basalt fiber 60 mm thick. Dimensions: 1090×2090×110 mm. The aperture area of the SC is 2 m^2. SK in the amount of 300 pcs. arranged in 15 double blocks of 10 pcs. in everyone. The orientation is south, the angle of inclination is 45 degrees to the horizon. The arrangement is single-row. Heliopol is located at a distance of 150 m from the pump module.

Figure 2. General view of the solar installation in Ust-Labinsk

The solar circuit includes solar collectors, a plate heat exchanger, and pumps (Fig. 3). The second circuit includes a storage tank with a capacity of 10 m3 and pumps. The third circuit has a plate heat exchanger for heating water for DHW. To automate the operation of the solar installation, a specialized controller from the German company Resol with a solar radiation sensor is used.

Figure 3. Schematic diagram of a solar power plant in Ust-Labinsk

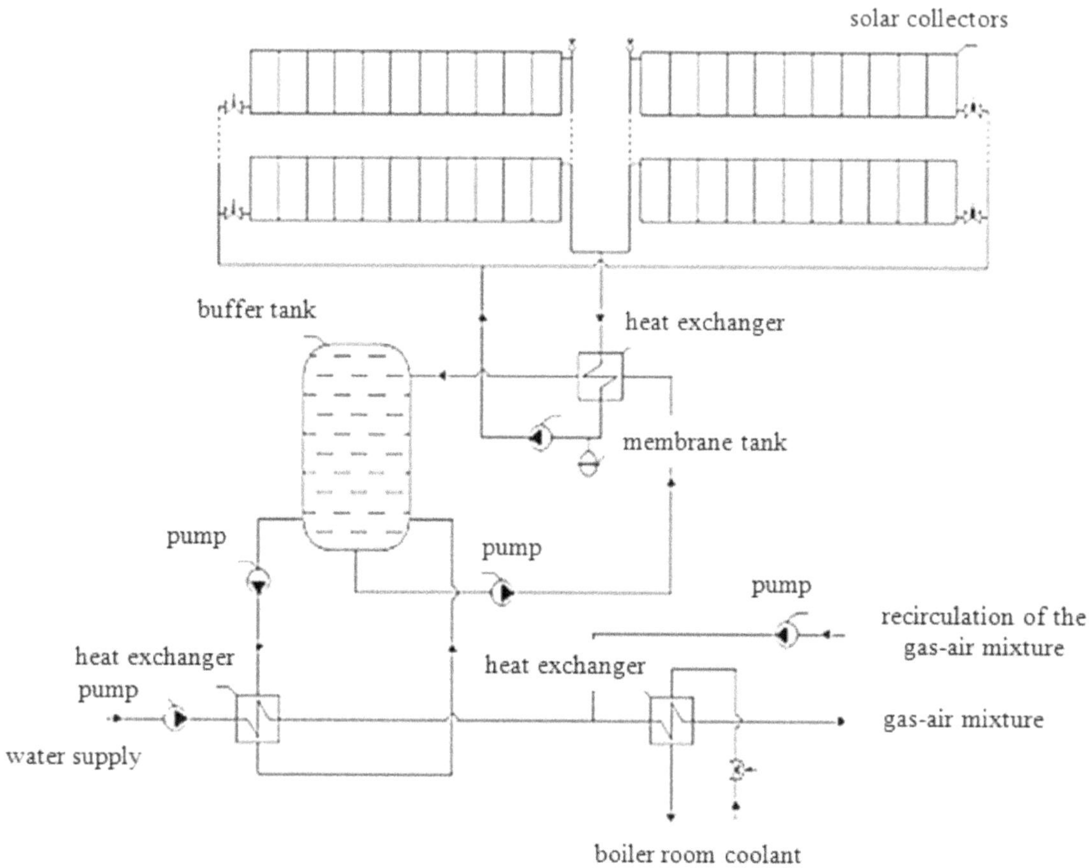

Another example in the city of Sochi, three solar installations provide hot water to the sanatorium named after. Frunze (Fig. 4). They were installed by the Spetsgelioteplomontazh association (Tbilisi) in 1985 on the roof of the dining room and its extension and have a total area of 518 m^2 (Butuzov et al, 2010). At the end of their service life in 2010, solar collectors with stamped-welded steel absorbers were replaced with solar collectors of the SOKOL type, while maintaining the basic circuit and layout solutions.

Figure 4. General view of a solar power plant in Sochi

The orientation of the installed solar collectors is south, the angle of inclination to the horizon is 45°. A total of 254 solar collectors are installed on three solar installations. The absorbers are made of fin extruded aluminum tubes. Selective coating of the absorber - vacuum application of PVD using the technology of NPO Mashinostroenie. Frame body made of aluminum profiles. Tempered glass, 3.2 mm thick. Thermal insulation – Rockwool basalt fiber 60 mm thick. The back side is made of sheet aluminum. The aluminum absorber pipes have coupling threaded connections.

The largest solar installation in the sanatorium with an area of 252 m2 with a solar power plant on the roof of the dining room has a storage tank with a capacity of 5 m3, installed at a height of 8 m with a two-meter excess above the solarium. When replacing solar collectors in 2010, circulation pumps were additionally installed in the solar circuit. The tank is fed through an intermediate tank with a float regulator. The operation of the solar installation using chemically purified tap water in the absence of its contact with air determined the 35-year operation of the tanks and pipelines.

The second solar installation with solar collectors on the roof of the dining room building with an area of 160 m^2 is made according to similar hydraulic and layout schemes. A feature of the third solar installation with an area of 96 m^2 with solar collectors on the roof of the central heating unit (CHS) is the placement of the battery tank and pump in the central heating unit building.

Application of heat supply schemes using solar systems

From the given examples of operating solar power plants, it is clear that in autonomous heating and hot water supply systems it is imperative to use a storage tank to accumulate thermal energy. This is due to the fact that the heat distribution generated by a solar installation is not proportional to energy consumption. Therefore, the received resources are first accumulated in a special container, and then only consumed as needed.

Experts recommend using a standard storage tank for a hot water supply system or, as an alternative, a buffer tank from an autonomous heating system for this purpose (Grachev & Plyamina, 2017). A well-constructed design involves connecting the collector to an additional heat exchanger, which is in direct contact with the storage tank. There are several practice-tested equipment connection schemes.

One of them is a solar hot water supply system with natural circulation (Fig. 5).

1 – solar collector
2 – circulation circuit
3 – cold water from the tap;
4 – storage tank
5 – hot water to consumers

Figure 5. diagram of solar hot water supply with natural circulation

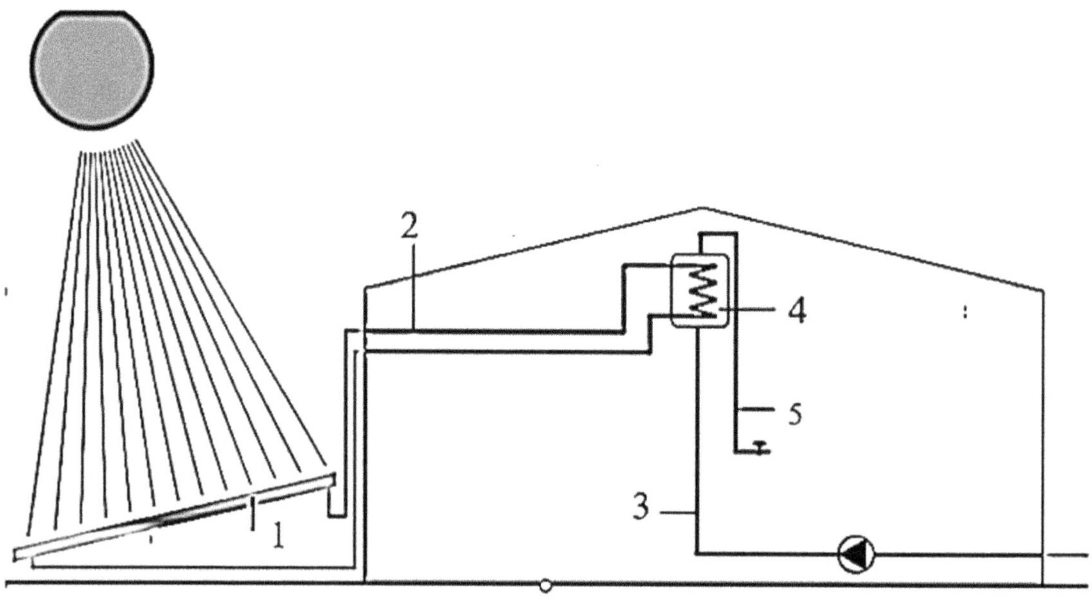

1 – solar collector; 2 – circulation circuit; 3 – cold water from the tap;
4 – storage tank; 5 – hot water to consumers

This scheme is used mainly in small areas (for example, for open showers used in the summer), but is also quite applicable for small buildings. The solar collector in this scheme must be installed below the level of the storage tank no more than one meter. This will ensure natural circulation of liquid in the system. But, if you plan to use hot water in the evening, then the storage tank must be covered with insulation. This is the most affordable solar collector connection scheme; however, it has one drawback

- minimal inertia. At sub-zero ambient temperatures, the water will have to be drained to prevent water pipes from depressurizing.

Another option is to use low-boiling liquid as a coolant for solar collectors (Fig. 6). This will prevent water from freezing in the pipes in winter. With this scheme, it is necessary to use an indirect heating storage tank with a copper coil. Continuous circulation of liquid occurs directly between the internal lines of the solar system and the coil installed in the storage tank.

1 – collector circuit
2 – buffer capacity of the heat accumulator of heating and hot water supply systems
3 – hot water to consumers
4 – solar collector;
5 – boiler with a heating circuit (without a DHW circuit)
6 – cold water from the tap

Figure 6. Installation diagram of a solar system with low-boiling liquid for hot water supply

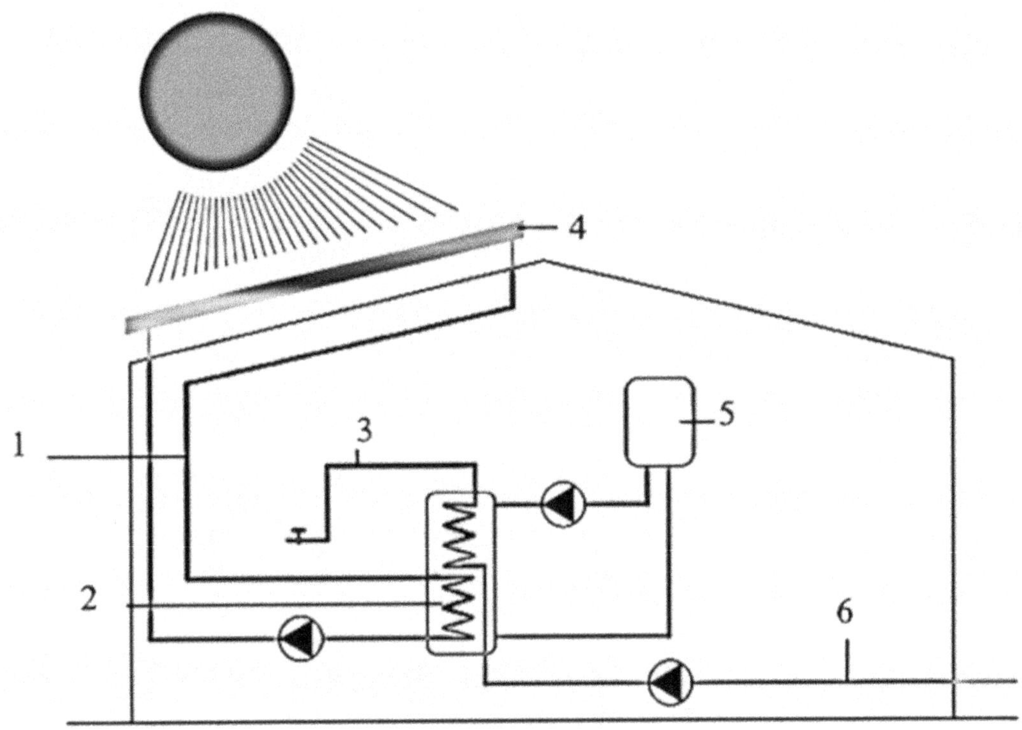

1 – collector circuit; 2 – buffer capacity of the heat accumulator of heating and hot water supply systems; 3 – hot water to consumers; 4 – solar collector; 5 – boiler with a heating circuit (without a DHW circuit); 6 – cold water from the tap

The third option involves using an indirectly heated tank that runs on fossil fuel. In late spring and summer, the boiler can be turned off, since the water will be heated by the collector. But in winter, the efficiency of solar systems in the northeastern regions of Russia is not very high, since the intensity of solar radiation is minimal. For this reason, the collector is used as a source of additional heating for heating systems. To ensure heating of the house in winter using only one solar collector, the dimensions of the entire structure must be at least 30–40% of the building area. But even in this case, we get the opportunity to use traditional energy resources more rationally, thereby reducing environmental risks.

1 – solar collector
2 – collector circuit
3 – indirect heating boiler for the DHW system;
4 – circulation pump
5 – hot water to consumers
6 – boiler with a heating circuit (without a DHW circuit)
7 – cold water from the tap

Figure 7. Connection diagram for solar heating system

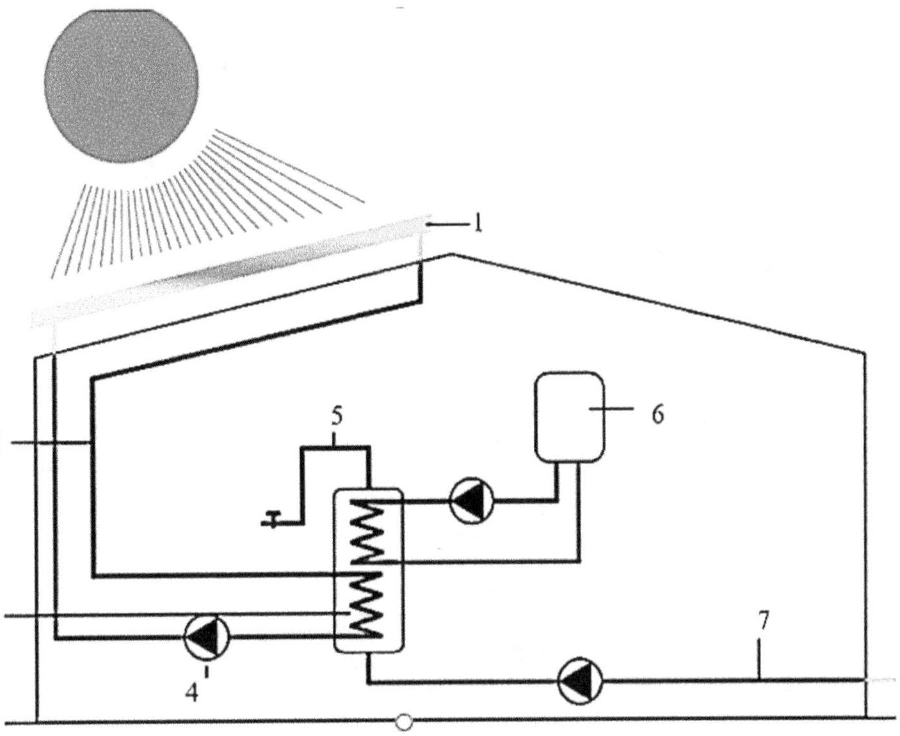

1 – solar collector; 2 – collector circuit;
3 – indirect heating boiler for the DHW system; 4 – circulation pump;
5 – hot water to consumers; 6 – boiler with a heating circuit (without a DHW circuit);
7 – cold water from the tap

Another option for connecting a solar system combines two options at once, that is, it is suitable for simultaneously organizing autonomous heating and hot water supply (Fig. 8). A double-circuit heat-storing tank is used here—in addition to the copper coil, an additional internal tank is also installed. This installation scheme makes it possible to separate technical liquid from drinking water. To automate the process of heating the coolant, a special solar collector controller is integrated into the system, which allows you to avoid excessive consumption of energy resources by controlling the temperature of the coolant in the solar system and the temperature of the water in the buffer.

1 – solar collector
2 – collector circuit
3 – storage boiler for indirect heating of the hot water supply system (right) and storage tank of the collector circuit (left)
4 – boiler with a heating circuit (without a hot water supply circuit)
5 – hot water to consumers

6 – cold water from the tap

Figure 8. Connection diagram for a solar system for heating and hot water supply

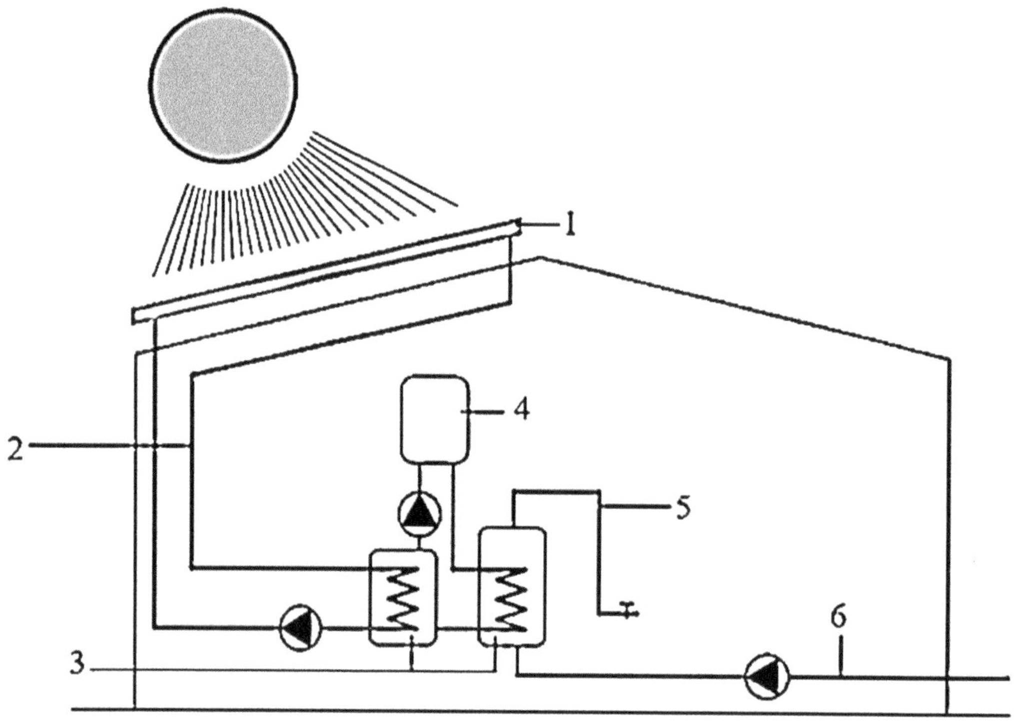

1 – solar collector; 2 – collector circuit;
3 – storage boiler for indirect heating of the hot water supply system (right) and storage tank of the collector circuit (left); 4 – boiler with a heating circuit (without a hot water supply circuit); 5 – hot water to consumers; 6 – cold water from the tap

And of course, swimming pools, which are found in almost every modern hotel. The proposed scheme (Fig. 9) is used when it is necessary to heat water in a portable outdoor pool. In this case, to ensure fluid circulation, it is possible to use a standard submersible pump. If there is a stationary swimming pool in the hotel complex, for greater convenience it is better to connect the equipment to a household automated pumping station.

Figure 9. Connection diagram for a solar system for heating a pool

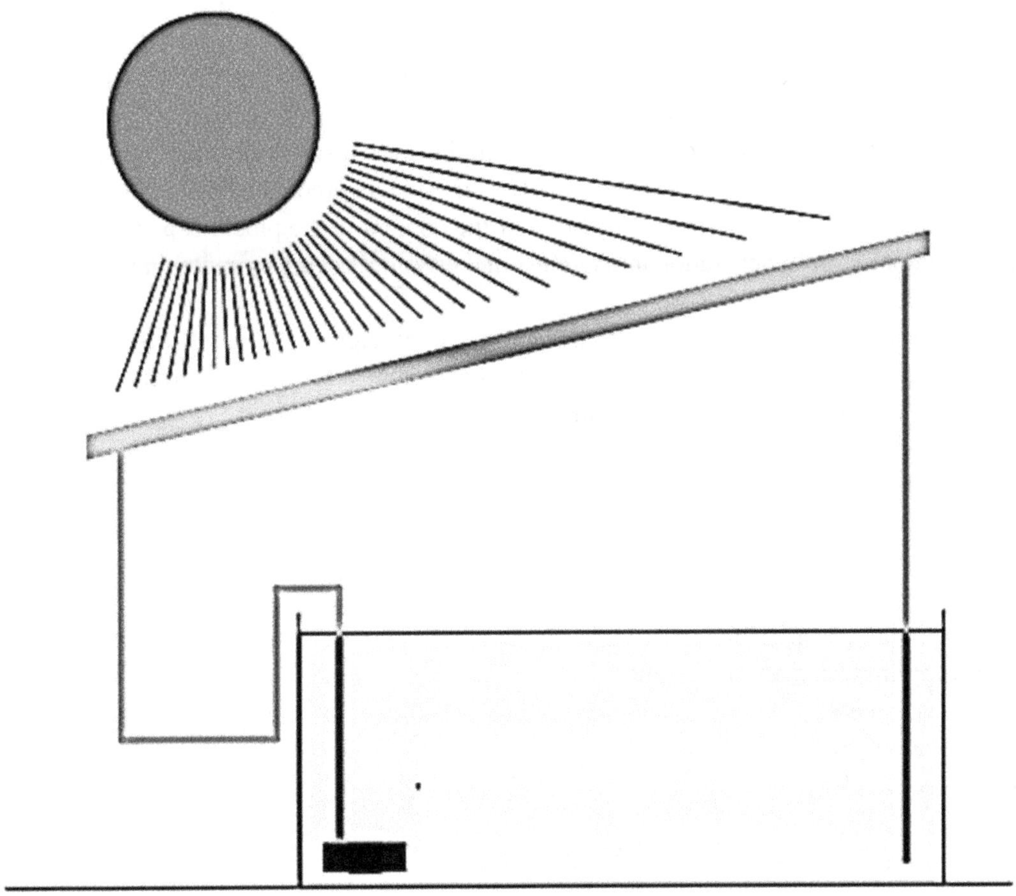

But in addition to heating and hot water supply, comfortable air temperature must be ensured in the premises of tourist sites. In the hot season, this is possible through the use of air conditioning systems. It is no coincidence that in the summer, electricity consumption increases rapidly, especially in hot resort areas. And if so, then fuel consumption at thermal power plants also increases, increasing environmental risks. But here, too, solar energy comes to the rescue.

Currently, there are quite a lot of inventions focused specifically on the use of the sun as a source of energy for refrigeration machines. In most cases, these devices operate on the principle of an absorption refrigerator using low-boiling components. The boiling point depends on the composition of the refrigerant components and the pressure in the system. Such solar refrigerators are simple in design and cheap to manufacture; as a rule, they have no moving parts, which ensures silent and reliable operation of such systems. The main feature of solar cooling is that refrigerators operate periodically and unevenly throughout the day and year. The good news is that solar cooling will produce its greatest performance during the hottest period of the day and year, when the need for space cooling is greatest. Therefore, it is advisable to use solar cooling to cool rooms during the hot season.

Some designs of solar refrigerators store energy during a hot day, and then release the stored energy in the form of cold. As a result, it is desirable to combine different solar cooling designs. Modern automation equipment is capable of maintaining a certain, specified cooling temperature by turning off or limiting the cooling performance. But, in the absence, especially for a long period, of sunlight, solar refrigerators cannot produce or accumulate cold for a long period (Slesarenko et al, 2010).

For example, a diagram of a combined solar heating and cooling system (Fig. 10), which includes a well-heat exchanger for collecting low-grade heat from the well, a solar collector, a heat pump, a heat accumulator consisting of an insulated housing, and a connected interconnected pipelines forming seven circulation circuits. For more intensive circulation of the coolant in the pipeline, a circulation pump is installed in front of the well. Separation into circuits in the system occurs by switching three-way valves (Varganova & Chinaeva, 2021).

The electronic control unit, using temperature sensors, continuously measures the temperature in the battery tank, heat exchanger well and solar collector and switches the taps depending on the operating mode to create the necessary circulation circuits.

1 – well – heat exchanger
2 – solar collector
3 – heat pump
4 – tank – battery
5 – step-up thermal transformer
6 – absorption refrigeration machine
7 – heat exchangers
8 – circulation pump
9 – three-way valves

Figure 10. Scheme of the solar-heat pump heating and cooling system

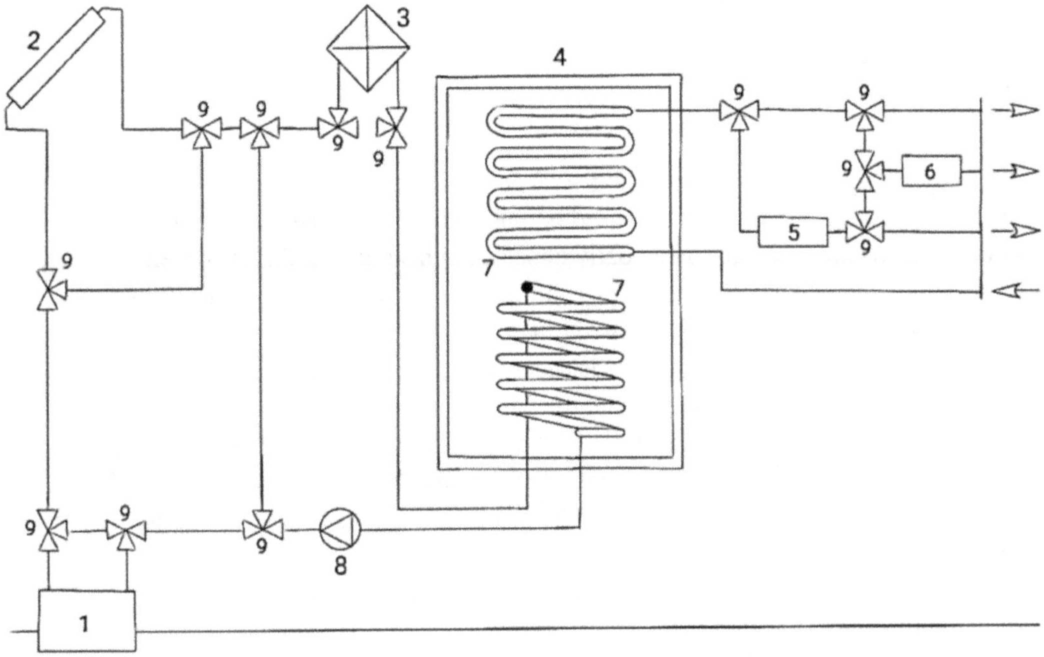

1 – well – heat exchanger; 2 – solar collector; 3 – heat pump; 4 – tank – battery;
5 – step-up thermal transformer; 6 – absorption refrigeration machine;
7 – heat exchangers; 8 – circulation pump; 9 – three-way valves

In winter, when solar radiation is low and the temperature of the coolant in the solar collector is below the set ground temperature, the electronic control unit switches the system to the second circulation circuit, which includes a well heat exchanger, a source heat exchanger and a circulation pump. The coolant, heated by low-grade heat in the well, enters the heat accumulator through a pipeline. The waste coolant is returned through the pipeline to the well.

In the summer, with low heat consumption, when temperature sensors indicate the temperature of the coolant in the solar collector and battery tank is higher than that required by the consumer, the electronic control unit switches the system to a third circulation circuit, including a well-heat exchanger, solar collector and circulation pump. The coolant, heated by the low-potential heat of the soil in the well, is supplied through a pipeline to the solar collector, in which it is heated by the energy of the sun, after which the coolant is returned through the pipeline back to the well, where it restores the temperature field around the well.

Under normal operating conditions, when the temperature in the thermal accumulator corresponds to the required parameters, the electronic control unit switches the pipeline to the fourth circulation circuit, including the consumer heat exchanger and pipelines. The coolant heated by the heat exchanger in the heat accumulator is supplied to the consumer, and then the waste coolant is returned back to the heat exchanger.

As we have shown, there are a wide variety of options for using solar systems for heating, hot water supply and air conditioning. But, of course, this is not all the possibilities of using solar energy. This study did not consider the issues of obtaining electricity through solar power plants.

RESULTS

Having analyzed various options for using solar installations, the authors believe that it is no coincidence that solar systems are increasingly used both in individual housing construction and to provide heat and hot water to entire villages using daily or seasonal heat accumulation systems.

But when deciding on the feasibility and possibility of using solar installations, it is necessary to take into account, consider and analyze the following main points.

First of all, an analysis of energy resources is necessary, that is, it is necessary to accurately determine the power of solar radiation, and then estimate the part that can be used.

Secondly, it is necessary to study the temporal characteristics of solar radiation. Energy requirements are usually not constant over time. For example, the demand for electricity is maximum in the morning and evening hours and minimum at night. The source must adjust to these fluctuations in demand. Thermal power plants can do this by regulating fuel consumption.

Thirdly, it is necessary to take into account the peculiarities of solar energy - sometimes the low density of incoming energy.

Fourthly, one should remember the determining role of a specific situation. Any renewable energy source cannot be universal, suitable for use in any situation, in any region. There are two main factors: the natural conditions of the area and the needs of society and the region.

Therefore, when planning the use of solar energy for heat supply, it is necessary to study the resources and possibilities of their use in a given area, as well as the structure and type of energy consumers.

When designing systems using solar power plants, it is necessary to have the results of long-term observations (monitoring) of the state of the environment in a given area (temperature, solar radiation intensity, wind speed, etc.). After analyzing the characteristics of consumers and potential sources of renewable energy, they need to be reconciled with each other.

As noted above, the demand for energy is constantly fluctuating, as well as its production from renewable energy sources. Therefore, the main coordination option is the use of energy storage devices, i.e. batteries. The simplest example of a heat storage device is a tank - an accumulator in a solar installation that provides hot water supply to the facility.

Another option is to connect the RES installation to a traditional larger power plant (heating water in front of the boilers in the boiler room). The use of storage devices in this case also increases the overall efficiency of the system.

One of the most important problems in the development of installations with renewable energy sources is the organization of control of the «power installation-consumer» system. The following options are possible.

1. The power plant is designed for maximum energy consumption. When it is reduced, only part of the energy coming from renewable energy sources is used. The rest is lost to the environment.
2. The installation includes energy storage devices (accumulators), which allows in some cases to design the installation not for maximum, but for some average energy consumption.

3. The system is designed with load regulation. The correspondence between demand and energy production is maintained by connecting or disconnecting the required number of consumers.

Analysis of the above points in the design of solar systems will not only reduce dependence on traditional fossil fuels, but will also reduce greenhouse gas emissions, that is, reduce environmental risks.

CONCLUSIONS

1. Energy consumption is a prerequisite for the existence of humanity. But extracting energy from natural resources has led to one of the greatest challenges of our time: climate change. To reduce environmental risks, the world must reduce its annual emissions by 22 billion tons of carbon dioxide equivalent in less than seven years.
2. At the moment, environmental risks are at a maximum permissible level, that is, their magnitude has not yet exceeded the level of irreversible processes of environmental degradation.
3. Environmental risks are caused by environmental hazards, that is, any change in the functioning parameters of natural, technical or natural-technical systems, leading to a deterioration in the quality of environmental components beyond the boundaries of established standards. And it is human activity in all forms of its manifestation that causes the highest degree of environmental danger.
4. The most destructive factor of environmental danger is the energy factor, since at the moment energy is the main driving factor in the development of all sectors of industry, public utilities and agriculture, and serves as the basis for increasing productivity and well-being of the population. Energy, for now, are precisely primary natural resources.
5. Today in the energy sector, one of the current areas of possible solutions for reducing environmental risks is the development of various methods of generating energy from renewable sources.
6. Renewable energy is the oldest and safest way to obtain energy. Today, as governments, businesses and consumers recognize global warming and climate change as one of the greatest threats of our time, the popularity and spread of renewable energy sources is steadily growing.
7. Solar energy is currently the most popular form of renewable energy. Based on the use of solar energy, the most widely available options are for obtaining thermal energy for heating, hot water supply and air conditioning, as well as obtaining electricity through solar power plants.
8. For efficient operation of solar installations, it is necessary to study the temporal characteristics of solar radiation, assess the quality of the energy source, as well as the structure and type of energy consumers. In addition, proper and clear organization of management of the "energy-installation-consumer" system is necessary. It is when these conditions are met that it will be possible to reduce the environmental risks of the energy type when using solar systems for heat supply.

REFERENCES

Butuzov, V. A., Bryantseva, E. V., & Butuzov, V. V. (2010). VACUUM TUBE SOLAR COLLECTORS. WORLD MANUFACTURERS AND PROSPECTS OF PRODUCTION IN RUSSIA. Альтернативная энергетика и экология, (5), 47-54.

. Butuzov V.A., Shetov V.Kh., Bryantseva E.V., Butuzov V.V., Gnatyuk I.S. (2009). Solar collectors. Trends in improving structures // Alternative energy and ecology. No. 10

Dmitriev, V. G. (2014). Environmental risk assessment. Analytical review of publications // AiS.. No. 14. URL: https://cyberleninka.ru/article/n/otsenka-ekologicheskogo-riska-analiticheskiy-obzor-publikatsiy (date of access: 10/23/2024).

Enikeeva, A. R., Sharipov, B. A., & Fedoseeva, E. A. (2019). Application of non-traditional energy sources to cover the load on heating tanks // Electrotechnical and information complexes and systems. No. 3. URL: https://cyberleninka.ru/article/n/primenenie-netraditsionnyh-istochnikov-energii-dlya-pokrytiya-nagruzki-na-podogrev-rezervuarov (date of access: 12/09/2024).

Fox, D. R., & Burgman, M. (2008). Ecological risk assessment. In Melnick, E., & Everitt, B. (Eds.), *Encyclopedia of Quantitative Risk Assessment and Analysis* (pp. 1600–1603). John Wiley & Sons Ltd. DOI: 10.1002/9780470061596.risk0268

Gaponov, V. L., & Samarskaya, N. S. Assessment of the environmental friendliness of burning organic fuels by enterprises of the fuel and energy complex // IVD. 2015. No. 3. URL: https://cyberleninka.ru/article/n/otsenka-ekologichnosti-szhiganiya-organicheskih-vidov-topliva-predpriyatiyami-toplivno-energeticheskogo-kompleksa (access date: 10/18/2024).

Grachev, V. A., & Plyamina, O. V. (2017). Global environmental problems, environmental safety and environmental efficiency of energy // Century of Globalization.. No. 1 (21). URL: https://cyberleninka.ru/article/n/globalnye-ekologicheskie-problemy-ekologicheskaya-bezopasnost-i-ekologicheskaya-effektiv-nost-energetiki (date of access: 08/08/2024).

Khanaev, V. V. (2015). Questions of the efficiency of using low-power solar systems // Systems for analysis and data processing. No. 2 (59). URL: https://cyberleninka.ru/article/n/voprosy-effektivnosti-primeneniya-geliosistem-maloy-moschnosti (date of access: 10/09/2024).

Makhmudov, B. M., & Kholboev, D. Zh. (2020). Solar collector devices and grouping // Economics and society. No. 12 (79). URL: https://cyberleninka.ru/article/n/gelio-kollektornye-ustroystva-i-gruppirovka (date of access: 10/08/2024).

Medvedeva, S. A. (2016). Environmental risk. General concepts, assessment methods // XXI century. Technosphere safety. No. 1. URL: https://cyberleninka.ru/article/n/ekologicheskiy-risk-obschie-ponyatiya-metody-otsenki (access date: 12/23/2024).

Mehdiyeva, V. Z. G. (2015). Sultanly Said Ismet Gizi Renewable energy sources in the context of global climate change // International scientific review.. No. 9 (10). URL: https://cyberleninka.ru/article/n/vozobnovlyaemye-istochniki-energii-v-usloviyah-globalnogo-izmeneniya-klimata (date of access: 12/08/2024).

Nikonova, A. A. (2017). Oil and gas resources of Russia: assessments and prospects for the development of the fuel and energy complex // Economic analysis: theory and practice.. No. 11 (470). URL: https://cyberleninka.ru/article/n/neftegazovye-resursy-rossii-otsenki-i-perspektivy-razvitiya-toplivno-energeticheskogo-kompleksa (date of access: 10/18/2024).

Podgornaya, T. I. (2015). Environmental Risk Management // Electronic Bulletin of the Rostov Social-Economic Institute. No. 3-4. URL: https://cyberleninka.ru/article/n/upravlenie-ekologicheskimi-riskami-1 (access date: 04/09/2024).

. Rabinovich M.D. (1986). Comparison of various methods for presenting climate-logical information when calculating the productivity of solar systems // Solar engineering, No. 3,

Revich, B. A. (2010). To assess the impact of fuel and energy complex activities on the quality of the environment and the health of the population // Problems of forecasting.. No. 4. URL: https://cyberleninka.ru/article/n/k-otsenke-vliyaniya-deyatelnosti-tek-na-kachestvo-okruzhayuschey-sredy-i-zdorovie-naseleniya (date of access: 02/09/2024).

Shcherban, A. V. (2021). Greenhouse effect and its impact on the environment // Economics and ecology of territorial entities. No. 2. URL: https://cyberleninka.ru/article/n/parnikovyy-effekt-i-ego-vozdeystvie-na-okruzhayuschuyu-sredu (date of access: 10/18/2024).

Shtebner, S. V. (2022). Assessment of environmental risks // Bulletin of science and practice.. No. 11. URL: https://cyberleninka.ru/article/n/otsenka-ekologicheskih-riskov-1 (date of access: 09/09/2024).

Sidorenko, A., & Goncharov, A. (2020). Renewable energy resources of the Black Sea coast of the Krasnodar Territory // Norwegian Journal of Development of the International Science.. No. 39-1. URL: https://cyberleninka.ru/article/n/resursy-vozobnovlyaemyh-istochnikov-energii-chernomorskogo-poberezhya-krasnodarskogo-kraya (date of access: 10/01/2024).

Sinyak, Y. V. Economic assessment of the potential of world oil and gas reserves // Problems of forecasting. 2015. No. 6. URL: https://cyberleninka.ru/article/n/ekonomicheskaya-otsenka-potentsiala-mirovyh-zapasov-nefti-i-gaza (access date: 10/18/2024).

Slesarenko, V. V., Kopylov, V. V., & Knyazhev, V. V. (2010). Assessing the efficiency of solar energy installations in heat supply systems // Bulletin of the Far Eastern Branch of the Russian Academy of Sciences. No. 3. URL: https://cyberleninka.ru/article/n/otsenka-effektivnosti-ustanovok-solnechnoy-energetiki-v-sistemah-teplosnabzheniya (date of access: 10/09/2024).

Varganova, T. N., & Chinaeva, T. I. Renewable energy sources: development trends // Russia: trends and development prospects. 2021. No. 16-1. URL: https://cyberleninka.ru/article/n/vozobnovlyaemye-istochniki-energii-tendentsii-razvitiya (access date: 10/08/2024).

Vishnyakov, Ya. D., & Radaev, N. N. (2008). *General theory of risks. M.* Academy.

Vorobyov, A. E., & Vorobyov, K. A. Distributed solar energy // Bulletin of Eurasian Science. 2019. No. 5. URL: https://cyberleninka.ru/article/n/raspredelennaya-solnechnaya-energetika (date of access: 10/08/2024).

Chapter 16
Mud Volcanoes of the Taman Peninsula as Objects of Excursion Display

Yulia Ivanovna Zakharchenko
https://orcid.org/0009-0007-1736-5275
Kuban State University, Russia

Evgenia Ivanovna Zakharchenko
https://orcid.org/0009-0009-4172-4244
Kuban State University, Russia

Natalia Gennadievna Andreiko
https://orcid.org/0009-0008-3601-6115
Kuban State Technological University, Russia

Musabek Abdusharipovich Duvanakulov
Osh Technological University, Kyrgyzstan

Mikhail Andreevich Zakharchenko
https://orcid.org/0009-0004-7104-7014
Kuban State University, Russia

ABSTRACT

Mud volcanism is a rather rare, complex, and in many ways still mysterious geological phenomenon. Currently, the Taman Peninsula is the region with the greatest development of mud volcanoes in the Russian Federation. The Taman Peninsula belongs to one of the youngest land areas. Back in the first century AD, this area was an archipelago of several islands. However, due to the high dynamism of the Kuban River delta, the high intensity of sediment coming from the river (up to several million tons per year) and the rise of tectonic plates, the islands joined together. Currently, the Taman Peninsula is the region with the greatest development of mud volcanoes in the Russian Federation. All known forms of mud volcanic manifestations (continental and marine, buried and open, extinct and active) are found here. Many of them are highly active and are located in close proximity to populated areas, so the mud volcanoes of the Taman Peninsula are currently considered as objects of sightseeing display.

DOI: 10.4018/979-8-3693-8232-5.ch016

RELEVANCE OF THE STUDY

Mud volcanism is a rather rare, complex, and largely still mysterious geological phenomenon. Currently, there are about 2000 mud-volcanic structures on earth, both land-based and underwater, grouped into mud-volcanic provinces. Mud volcanoes are often concentrated in oil and gas basins of the Earth's mobile belts.

The largest provinces of mud volcanoes, numbering more than 100 mud volcanoes, are located in the south-eastern and north-western parts of the Caucasus, within Azerbaijan, on the Kerch and Taman Peninsulas.

Smaller provinces, including several dozen mud volcanoes, are located on Sakhalin Island, in Northern Italy (Po River Valley, Sicily), in southwestern Albania, in Transylvania, on the Lower Danube Plain in Romania, in Western Turkmenistan, in the Kerch Strait, in the northeast of the Black Sea, in the south-in the east of the Sea of Azov, in the Stavropol Territory, in Eastern Georgia, in the west and east of Azerbaijan, in the southern part of the Caspian Sea, in the northern part of the Arabian Sea, within the Gorgan Plain of Iran, in Eastern and Northern Baluchistan (Pakistan), in Punjab province (India), in China (in Dzungaria), in the western regions of Burma, on the Andaman and Nicobar Islands, on the islands of Malaysia and Indonesia, in the west of the island of Taiwan, in the northwest of the island of New Guinea, Japan (Honshu and Hokkaido Islands), New Zealand, California, Greenland and Iceland.

In the Western Hemisphere, mud volcanoes are found on the island of Trinidad, in northeastern Mexico, in the Caribbean Sea, in northeastern Venezuela, in northern Colombia, on the Pacific coast of Ecuador.

Underwater mud-volcanic provinces are found within the South Caspian Basin, the largest province (136 structures), the Black Sea (25 structures) and the Mediterranean Sea (16 mud-volcanic structures). Underwater mud volcanoes are found on the shelves of the oceans and outer seas.

The development of mud volcanoes all over the world occurs in zones associated with foothill and inter-mountain depressions; accumulation of clay deposits with ultrahigh (abnormally high) reservoir pressures; confined to oil and gas basins; and elision systems.

All recorded areas of mud volcanism development are localized within the Alpine, Himalayan, and Pacific mobile belts. There are also a number of possible mud-volcanic occurrences associated with permafrost, magma chambers, and the like (Shnyukov et al, 1986).

Currently, the Taman Peninsula is the region with the greatest development of mud volcanoes in the Russian Federation. The Taman Peninsula belongs to one of the youngest land areas. Even in the first century AD, this area was an archipelago of several islands. However, due to the high dynamism of the Kuban River Delta, the high intensity of sediment coming from the river (up to several million tons per year) and the rise of tectonic plates, the islands were connected to each other.

Currently, the Taman Peninsula stretches from north to south for almost 40 km, from east to west-for 60 km. Its area is almost 2000 km^2, of which 800 km^2 are estuaries and lakes. On the territory of the peninsula are located: Temryuksky district, most of the Anapa resort region, partially Krymsky district. The peninsula is bordered by the Sea of Azov to the north and the Black Sea to the south. Almost all of its land is occupied by farmland.

The relief of the peninsula is peculiar: flat areas alternate with gentle ridges and cone-shaped hills. The highest mountain (Komendantskaya) has a height of 164 m above sea level. Land covers about 55% of the territory. The peninsula has a large number of mud volcanoes.

Geotectonically, the Taman Peninsula is a heterogeneous formation located at the junction of the mountain ranges of the Crimea and the Caucasus. This is precisely what determined the identification of several large first-order tectonic structures in the region: the closure of the meganticlinories of the Mountainous Crimea and the Greater Caucasus, the Indo-Kuban trough, the northwestern part of the platform side of the West Kuban forward trough, and the inter periclinal Kerch-Taman transverse trough. The general scheme of structural and tectonic disturbances of the Taman Peninsula is described by E. F. Shnyukov (1986).

According to estimates by Naumenko P. I. (1963), the total mass of sopochny sediments ejected by mud volcanoes reaches 40 billion m^3. In order to release such a mass of sopochny silt and breccia from the subsurface to the surface, it is necessary to eruption many trillions of cubic meters of deep gases. This indicates a huge potential energy stored in the bowels of the earth.

In the deep structure of the Taman Peninsula, five structural floors are distinguished (Shnyukov et al, 2006):

- neogene.
- maikop district.
- eocene-Upper Cretaceous.
- lower Cretaceous.
- jurassic.

The studied area of the Taman Peninsula, which includes a large mud-volcanic province, is located in the northwestern part of the platform side of the West Kuban forward trough. It began to form as a compensatory trough in the Oligocene before the growing massif of mountain-folded structures in the Northwestern Caucasus (Alpine folding zone). The thickness of the sedimentary cover in the trough in some places reaches 14 km. The area is confined to the Azov-Kuban oil and gas basin. The section contains clay strata of the maikop series with a thickness of up to 3–4 km. It is with them that the foci of volcanic activity in the Kerch-Taman region are associated (Shnyukov et al, 1986).

Currently, the Taman Peninsula is the region with the greatest development of mud volcanoes in the Russian Federation. All known forms of mud volcanics occur here (continental and marine, buried and open, extinct and active). Many of them are highly active, located in close proximity to populated areas, so the mud volcanoes of the Taman Peninsula are currently considered as objects of excursion display.

RESEARCH METHODS

The following research methods were used in the preparation of this article:

1) general scientific methods (theoretical):
    - method of analysis of literary and stock sources;
    - the method of generalization;
    - synthesis method;
2) empirical methods (collection of specific facts):
    - measurement method;
    - the method of observation;

– the method of comparison.

INTRODUCTION

Mud volcanism is a set of processes that cause the rise and release of clay masses, hard rock fragments, and hydrocarbon gases softened by reservoir waters to the surface, which leads to the formation of unique geological bodies from the products of eruptions – mud volcanoes.

The mechanism of action of mud volcanoes is similar to that of a magma volcano. The causes of mud volcanoes are tectonic movements of the earth's crust, as a result of which clay mass, water and gases are squeezed to the surface where "breaks" were formed during tectonic movements.

Mud volcanoes are common in both tectonic compression and extension regions. When moving up the channel, the liquefied pulp is a mixture of water and gas, fine clay-siltstone material, and fragments of various rocks that make up the sedimentary complex. The main cause of mud volcanism is the pressure of oil gases (figure 1). Combustible gases released from oil deposits rise along tectonic fractures to the surface and, meeting clay breccias liquefied by pressure waters, bring them to the surface. Gases that play a major role in the mechanism of eruptions of mud volcanoes consist of various hydrocarbons, mainly methane, nitrogen, carbon, and in small amounts – hydrogen sulfide. The release of hydrocarbon gases is mainly characteristic of those mud volcanoes whose mud has a low temperature during an eruption.

Figure 1. Mud volcano structure diagram (Shnyukov et al, 2006)

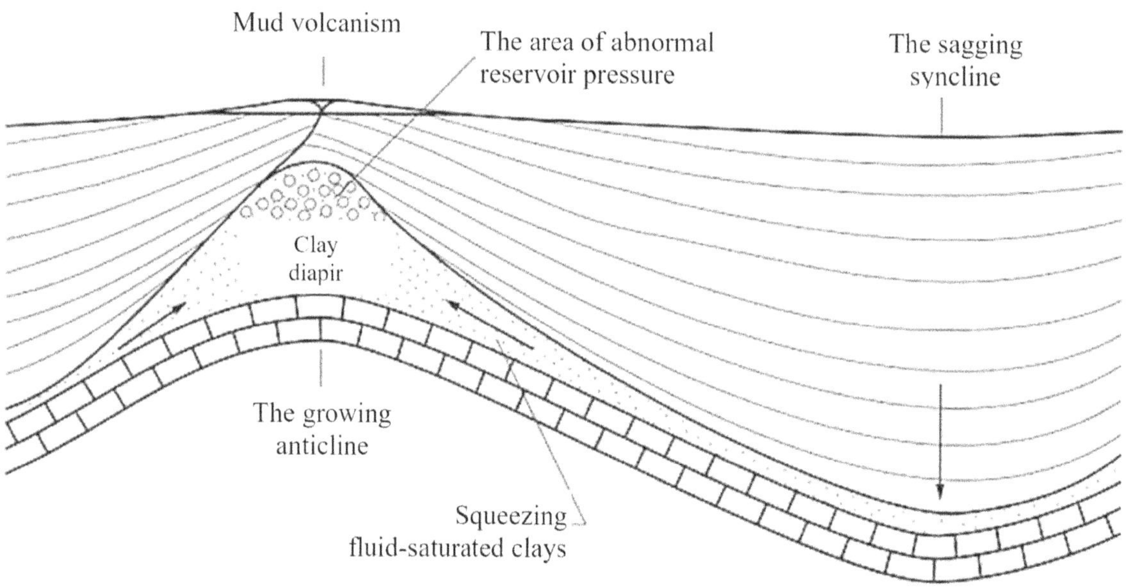

The large scale distribution of mud volcanism on the Taman Peninsula and the unusual nature of the phenomena attracted public attention, and geologists studying mud volcanoes also set a popularizing task: to present the mud volcanoes of the Taman Peninsula as objects of excursion display, to show the nature of mud volcanism, its imaginary and real dangers, the peculiar beauty of the hills, their geological and search significance, importance of accompanying geological manifestations, local lore opportunities for studying mud volcanism, connection with the traditions and history of the Krasnodar Territory.

It is not by chance that Homer wrote about the mud volcanoes of Taman in his Odyssey. It was on Taman that Homer placed the entrances to Pluto's subterranean realm, «in a sad, bare area» – in mud volcanoes. According to the authors ((Shnyukov et al, 1986; Lagunova & Gemp, 1978), one of the verses of the Odyssey tells about the Argonauts visit to the Kerch-Taman region, or rather the Kerch Strait, and about the Argo's sailing along the «fire-breathing» mud volcanoes: «Not a single human vessel has yet passed through here, but the sea waves and the raging sea together, the planks of ships and the corpses of men carry away the destructive flames. Only one seaworthy ship, the well-known Argo, sailed here on its return journey».

The authors believe that if indeed the Argonauts passed through the Strait, then, obviously, the path of the «Argo» passed not along the modern fairway of the Kerch Strait, but along the old channel of the Kuban and the Taman Bay, along the Akhtanizovsky estuary and the Kuban River, which flows into the Sea of Azov near Temryuk. Only in this case, «Argo» sailed through a chain of such «exploding» mud volcanoes as Gorelaya Mountain, Karabetova Mountain, Tsymbaly, mud volcanoes as Boris and Gleb, Akhtanizovskaya Sopka and others.

Basic concepts of mud volcanism and the mechanism of action of mud volcanoes

Currently, there is no unified terminology for determining the morphological forms of mud volcanism. Different researchers formulate these concepts differently.

In the Geological Dictionary (1973), the following definitions are given: «Mud volcano is a large hill composed entirely or only from the surface of hill deposits and has a funnel-shaped crater at the top and a channel extending to the depth, from which gas, water, sometimes with oil films, rock fragments, and hill mud are periodically or continuously released; the latter spreads out along the slope of the hill, increasing the hill cone. At certain intervals, eruptions occur, accompanied by very violent releases of gases (explosions) and ejections of rock fragments to a considerable height».

Currently, experts (Shnyukov et al, 1986; Shnyukov et al, 2006) have proposed a more concise formulation. A mud volcano is an eruptive unit of gas volcanism, whose emissions (hill breccia, water, gases, solid debris) most often form positive landforms, sometimes, depending on local conditions and the consistency of hill emissions, depressions filled with liquid mud. As a rule, mud volcanoes are quite significant in size — at least hundreds of meters. Small mud volcanoes are called mud hills (salzas). Small secondary eruptive apparatuses (griffins) develop on hills or volcanoes. Very often, the tops of mud volcanoes are crowned with subsidence calderas. Sometimes volcanoes or hills are located separately, but most often in groups that have a common deep eruptive center. It is advisable to call a group of volcanoes or salzas a mud-volcanic focus. Figure 2 shows the morphological elements of mud volcanoes, distinguished by their shape, size, and subordination.

I – mud volcano
a – positive shape
b – negative shape.

II – sopka (salza)
a – positive shape
b – negative shape
III – secondary geomorphological elements
a – positive form (gryphon)
b – negative form

Figure 2. Main geomorphological elements of mud volcanoe

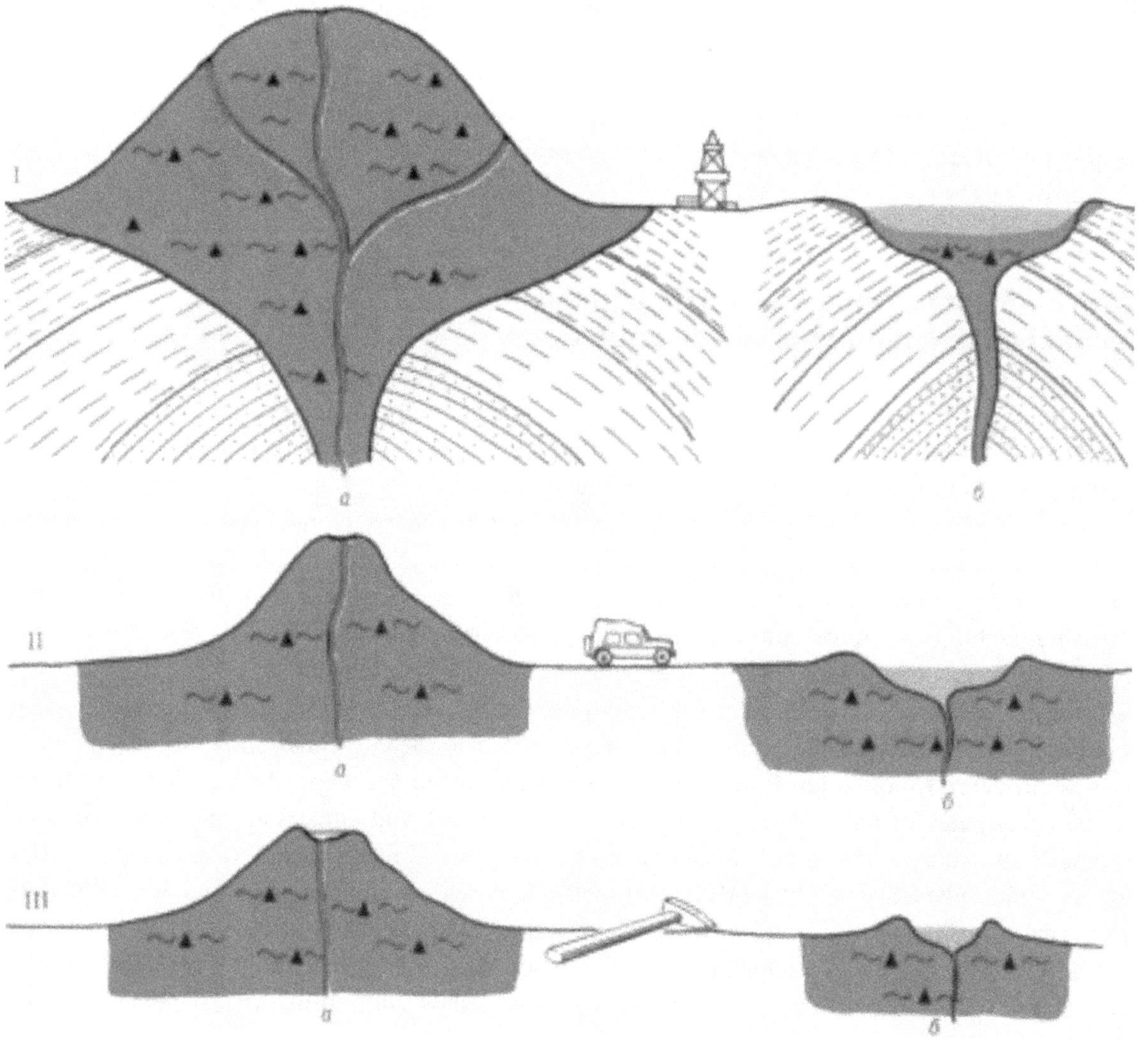

(Shnyukov et al, 1986)

Mud-volcanic occurrences should be perceived in their geological development, understanding the modern section as a characteristic of the observed moment in the geological history of the mud-volcanic focus. As a rule, modern mud volcanoes are only the upper floor of a complex multi-tiered building, the lower floors of which were formed in the geological past.

Often, the mud-volcanic focus is no longer functioning and is recorded only by fossilized hill deposits. The entire mass of sopochny deposits (modern or fossil) forms a «sopochny field». The fossil caldera of subsidence, filled with a layer of sedimentary rocks, received a local name-the depressed syncline.

Mud volcanoes are characterized by many other morphological features. Thus, on satellite and aerial photographs, significant ring structures are clearly visible. Smaller-scale ring disturbances can be observed within the upland foci. They are formed during catastrophic eruptions of mud volcanoes accompanied by an earthquake. This also creates a network of radially divergent cracks, sinking funnels, and vertical rock displacements.

The slopes of mud hills are often cut by deep gullies-barrancos, at the bottom of which clastic material accumulates. Large mud-volcanic structures with an annular rampart around the caldera are characterized by the presence of a washout-a washout in the body of the rampart, through which sopochny and atmospheric water flows out of the inner sopochny field. Hill fields are also characterized by numerous white spots of discoloration (boron minerals, halides, etc.), oil black spots and streams.

There are underwater mud volcanoes. They are usually found in shallow areas of the bottom, they are confined to diapir structures and can serve as a sign of the oil and gas potential of this section of the shelf. In a calm state, they emit gas, bubbles of which rise to the surface along with turbidity, revealing the location of volcanoes. Eruptions of these volcanoes often lead to the formation of islands, which are quickly eroded by waves. Instead of eroded mud-volcanic islands, detrital material remains on the bottom and accumulates over time. The wreckage is overgrown with seaweed and seashell – and marine underwater banks of mud-volcanic origin are formed. Areas of the sea where mud volcanoes are located are dangerous for navigation. They are plotted on maps and indicated in lots. In the Sea of Azov, for example, these are the Temryuk and Golubitskaya banks.

Mud-volcanic structures were discovered by geological and geophysical methods in the Kerch Strait and the north-eastern part of the Black Sea. They are located at great depths and are not clearly defined in the modern bottom relief. Currently, mud volcanoes associated with these structures are in a state of partial dormancy.

Mud volcanoes are formed under strictly defined conditions, including favorable tectonic conditions, the presence of diapir structures, a thick clay layer-a cushion that contributes to the appearance of abnormally high reservoir pressures of gases in the subsurface, a system of deep-seated discontinuous disturbances that facilitate the breakthrough of gases through a thick sedimentary layer, aquifers (Shnyukov et al, 1971). Various opinions were expressed about the depth of mud volcanoes. Thus, some authors defined them from 5–7 km to 8–10 km. Almost all these data are conjectural and do not exclude even greater depths of volcanoes, since they are usually confined to geosynclinal zones with sedimentary thicknesses up to 13–15 km. The basic scheme of action of mud volcanoes is as follows. If there are gas deposits with abnormally high reservoir pressures and aquifers in the subsurface, discontinuous disturbances are necessary for the emergence of a mud volcano, which are channels for upward migration of mud volcanism products. Gases and water carry along clayey rocks that have been processed due to tectonic processes in the breccia. The latter, in the process of moving up, disintegrate, roll over, turning into typical breccia-shaped hill clays, pelites containing various amounts of large fragments of hard rocks.

The duration and speed of gas movement to the surface, their volume, the degree of waterlogging of rocks along the path of movement, abnormally high reservoir gas pressures in the source reservoir-these are some factors that determine the volume of clay material and solid debris ejected during the eruption of a mud volcano. Three types of volcanoes are distinguished by the nature of their activity: Lokbatansky, Bulganaksky, and Shuginsky.

The activity of volcanoes of the Lokbatan type is reduced to the following (Kalinko, 1964). If the intensity of the initial eruption was high, only gas accumulated in the eruptive channel, the pressure in the gas deposit sharply decreased, the dynamic pressure became lower than hydrostatic, and the channel was blocked by plastic clays, which provided almost complete tightness. The ongoing accumulation of gases, which restores the anomalously high pressure in the deposits, created an unstable equilibrium, easily disturbed by any external influences (tectonic movements, earthquakes), which ultimately leads to repeated, often explosive eruptions of the mud volcano. Explosive, Lokbatan, and mud volcanoes on the Taman Peninsula include Gorelaya Mountain, Boris and Gleb volcanoes, Severo-Akhtanizovsky volcanoes, and Jau–Tepe on the Kerch Peninsula.

When the pressure difference is relatively low, the gases move up slowly, through a kind of overflow. In this case, volcanoes of the Bulganak type appear, characterized by a constant action, systematically releasing gases and hill mud with a small content of coarse-grained material for a long time (figure 3). This type includes most of the mud volcanoes of the Kerch-Taman region, in particular Akhtanizovskaya Sopka, Gnylaya Gora and others.

Figure 3. Bulganak type eruption – mud volcano Akhtanizovskaya Sopka

(photo by E.I. Zakharchenko, 2024)

Mud volcanoes of the Shuginsky type are characterized by activity typical of volcanoes of the Bulganak type, which is periodically disturbed by significant emissions, resembling the primary eruptions of volcanoes of the Lokbatan type. This is due to the presence of large pockets of gas accumulation in porous reservoirs on the gas migration routes, which create additional excess pressure (Kalinko, 1964). Mud volcanoes of the Shuginsky type include the Karabetova Gora (figure 4), Miska, Shugo, Voskhodovsky (Dzharjava) volcanoes on the Kerch Peninsula, and the Sea of Azov volcanoes.

Figure 4. Shuginsky type eruption – mud volcano Karabetova Gora (May 2000)

This division of volcanoes into three types according to the nature of their activity is quite conditional. Mud volcanoes probably go through various stages in their long geological history, both violent and calm.

Of considerable interest is the mechanism of formation of superimposed negative structures – depressed synclines formed as a result of mud-volcanic activity.

The completion of the anticlinal structure formation and the manifestation of mud volcanism create the initial prerequisites for the occurrence of a compensatory trough (Shnyukov et al, 1971; Prokopov, 1931). The activity of a mud volcano inevitably forms a depression in any paleogeographic conditions, in a marine or subaerial environment, and leads to bends in the rock layers that make up the arched part of the anticline.

In the marine basin, during underwater eruptions, the sopochnaya breccia ejected by the mud volcano, together with the accumulation of normal marine sediments, creates a heavy pressure, which additionally accelerates the subsidence and sinking of sections of anticlinal structures lying above mud volcanic foci. In a subaerial environment, the process of compensatory deflection formation is inhibited, its scale and amplitude decrease (Shnyukov et al, 1986).

When the marine regime is maintained, as precipitation and breccia accumulate, the layers in the anticline arch sink, and their continuity is disrupted. The resulting compensation structure is most often a rounded crater delineated by an annular discharge formed by a system of smaller disturbances, that is, a kind of rounded graben filled with sediments and breccia. In some places, the volcano erupts directly into the compensation syncline, sometimes along an annular discharge, and often the center of the volcano is shifted away from the compensation syncline. After the sea retreats, the depressed synclines are prepared by weathering processes and in most cases form positive landforms – low hills, rocks, and the like (Shnyukov et al, 2006).

In fact, the development of depressed synclines was immeasurably more complex: the sections of many of them represent an extremely complex «layer cake» of various sedimentary rocks and hill breccias. Reef limestones developed along the periphery of the structures, which turned to the center as they sank further. In land conditions, further activity of the mud volcano leads to continued lowering and leveling of the terrain. With the deepening of the sea and the cessation of mud-volcanic activity, the depressed synclines are buried.

GENERAL INFORMATION ABOUT THE KERCH-TAMAN MUD VOLCANO REGION

The Kerch-Taman region of mud volcanic activity is a rather extensive region at the junction of the Crimean and Caucasian folding, potentially oil-and-gas bearing, with a complex multi-tiered geological structure, frequent mismatch of structural plans of different structural floors, the presence of thick thicknesses of clay deposits – maikop and cretaceous, the widest development of discontinuous tectonics of different levels-from deep and regional faults to small discontinuous dislocations, with certain folding systems in the upper structural floor, and a wide development of diapirism.

The described mud-volcanic area extends far beyond the Kerch and Taman Peninsulas in its area (Shnyukov et al, 1986). In this regard, its name should be considered somewhat conditional. The region should include areas of the Sea of Azov corresponding to the Indo-Kuban trough or at least its southern part, adjacent areas of the Kerch-Taman shelf and the continental slope in the Black Sea. The boundaries of the mud-volcanic region are quite mobile. It is possible that abnormally high pressures of gas deposits in the subsurface, as one of the reasons for the action of mud volcanism, decrease away from the geological core of the region, and in this regard, volcanism subsides Obviously, it is no coincidence that the largest modern mud volcanic centers, such as Jau-Tepe, Bulganak, Gnilaya Gora, Karabetova Gora, and Shugo, are located not on the periphery of the region, but in its axial part. Perhaps in the geological past there was a different situation, the boundaries of the region were different, but now this trend can be traced quite clearly.

The Kerch-Taman mud volcanic region generally has its own geological and historical peculiarity in comparison with other areas of mud volcanism: the maximum activity of mud volcanism occurs here in the geological past, most likely in the chokrak and sarmatian times. Now the level of mud volcanism activity in the Kerch-Taman region is lower than in the geological past (Shnyukov et al, 2006). This is the difference between the studied region and the Azerbaijani or Turkmen mud-volcanic regions. This may also explain the smaller scale of individual expression of modern mud volcanoes, the smaller size of modern mud volcanic structures on the Kerch and Taman peninsulas than in Azerbaijan, along with the grand scale of volcanic activity in the past. That is why there are now fewer active volcanoes in the Kerch-Taman region than was known in the geological past.

Probably, this is also due to the lower development of catastrophic, explosive eruptions of mud volcanoes recorded in human memory. The main core of the Kerch-Taman region is the Kerch and Taman Peninsulas. At the same time, if the western contour of the region ends rather sharply at the borders of the Kerch Peninsula, then the eastern border of the mud-volcanic region is less defined.

The main mass of mud volcanoes is located within the Taman Peninsula. At the same time, a number of mud volcanoes are developed at the northwestern end of the Caucasian anticlinorium. This series begins with the Raznokolsky volcano and continues with the Shugo, Gladkovsky, Kievsky, Semyonovsky and finally Shapsugsky volcanoes. It is very likely that the Shapsugsky volcano is not yet an extreme mud volcanic focus, and other, more eastern hill manifestations are also possible (Slavyanov, 1928). All these volcanoes are localized along the northern slope of the Caucasian anticlinorium.

On the southern slope of the extreme eastern volcano – Semigorsky near the village of Natukhaevskaya (district of Anapa). The westernmost mud volcano in the region, reliably identified and studied, is Vladislavovsky, located near the Tush-Oba hill and Lake Achi (Kerch Peninsula).

CHARACTERISTIC FEATURES OF THE MODERN ACTIVITY OF MUD VOLCANOES

The Kerch-Taman region is a zone of decaying mud volcanism. Mud-volcanic activity flourished in the past geological epochs, mainly in the neogene. At the end of the paleogene, in the neogene, anthropogenic period, huge volumes of sediments, combustible and other gases, and water were erupted. The total mass of the ejected sopochny sediments reaches, according to the calculationsof some specialists (Shnyukov et al, 1986; Naumenko, 1963; Shnyukov et al, 2006), 40 billion m^3. In order to throw such a mass of sopochny breccia from the bowels to the surface, it was necessary to eruption of many trillions of cubic meters of combustible and other gases from the depths of the Kerch-Taman region.

As a result of intensive mud-volcanic activity in the geological past and in the present, numerous existing and extinct mud-volcanic foci and their derivatives – depressed synclines-have emerged in the Kerch-Taman region. On the Kerch Peninsula, there are more than 50 mud volcanic foci, the main activity of which is confined to the geological past, many of them continue to function. At least 50 mud volcanoes have been found in Taman and the Northwestern Caucasus, and most of them are still active today.

Over the past forty to fifty years, our knowledge of mud volcanism in the Kerch-Taman region has significantly expanded. For a long time, there was an opinion about the mud volcanoes of the Kerch-Taman region as quietly developing, in their overwhelming majority. The main mass of mud volcanoes in the Kerch-Taman region does function relatively calmly. There is a constant release of combustible gases, hill silts, breccias, water, and in some volcanoes – petroleum products.

However, the analysis of literary, archival and field expedition materials leads the authors to the conclusion that explosive phenomena often occur in the volcanoes of the Kerch-Taman region.

The largest explosion was recorded on the Taman Peninsula, on Goreloi Mountain, where about 1 million m^3 of dairy products were ejected during one eruption m^3; other major explosive centers are the Jau-Tepe and Dzharjava (Voskhodovsky) volcanoes on the Kerch Peninsula. Most often, large breccia emissions are accompanied by a hum, earth tremors, fire emissions, smoke, and sometimes steam. Dirt and debris are sometimes thrown up to a height of 50–100 m. Most often erupted with explosive phenomena documented in the literature, the Golubitsky mud underwater volcano in the Sea of Azov near the village of Golubitskaya, Temryuk district. After explosive eruptions of the sopochnaya breccia accompanied by fiery phenomena, burnt annular or elongated zones composed of black and red cinder-like formations remain around the explosion centers. The total number of explosive eruptions that have been preserved in the literature, human memory, and on the ground is relatively small: in the Kerch-Taman region, it reaches 58. This is undoubtedly an understated figure, since it is not possible to take into account all eruptions without a systematically established monitoring service.

Of the 58 eruptions, 12 occurred on the Kerch Peninsula, 14 in the Strait and the Sea of Azov, and 32 in Taman and the northwestern Caucasus. Explosive, lokbatan-type mud volcanoes in Taman and the North Caucasus include Gorelaya Mountain, Boris and Gleb volcano, Severo-Akhtanizovsky volcano, and Jau-Tepe on the Kerch Peninsula. Mud volcanoes of the Shuginsky type include the volcanoes Karabetova Gora, Miska, Shugo, Voskhodovsky (Dzharjava) on the Kerch Peninsula, and the volcanoes of the Sea of Azov (Shnyukov et al, 2006).

During catastrophic eruptions accompanied by explosive phenomena and gas combustion, the sopochnaya breccia undergoes various stages of roasting directly at the site of gas jet combustion.

PRODUCTS OF MUD VOLCANISM: HILL BRECCIAS

Along with hill gases and waters, the main product of mud-volcanic activity is hill breccia – a specific rock that has no analogues among sediments resulting from various geological processes, and consists of fragments, mainly clay rocks (figure 5). According to the size of the fragments, there are hill pelites, siltstones, sandstones, gravelites, and conglobreccias.

Figure 5. Surface of the sopochny field of the Rotten Mountain mud volcano

(photo by E.I. Zakharchenko, 2023)

Often, there is an alternation of different types of hill breccias, starting from conglobreccias and ending with sands, which indicates a complex section and heterogeneity in area.

This distribution of different lithological types is due to the introduction of liquefied hill breccias localized within the depressed syncline of the reservoir, which experience repeated washing and redeposition.

Classification of hill deposits is debatable. The works of E.F. Shnyukov and others (1971), N.S. Shatsky and others (1929), V.E. Hain and others (1953), A.D. Arkhangelsky (1930), P.P. Avdusin (1948) are devoted to it.

Lithologically, the sopochnaya breccia is a clastic rock consisting of fragments of various, mainly clay rocks. Based on the formal generally accepted classifications, the term hill breccia is to a certain extent conditional, since this concept includes, by the nature of rolling and size, not only uncoated fragments of pebble-gravel dimensions, but also gravel, sand and siltstone fragments of various degrees of rolling, from angular to well rounded, that is, among the hill deposits, the dimensions of the fragments are distinguished hill pelites, siltstones, sandstones(sands), gravelites, breccias, conglobreccias (Grechishkin, 1936; Avdusin, 1939; Avdusin, 1948; Beloussov & Yarotsky, 1936; Popov, 1938; Prokopov, 1931).

The size of the debris in the sopochnaya breccia is quite different. The composition of large fragments within the sopochny fields is very diverse. They are mainly represented by conglomerates, conglobreccias, sandstones, siltstones, limestones, dolomites, and siderites. Fragments of phosphorites, iron ores, marls, and igneous rocks occur sporadically (Shnyukov et al, 1986; Shnyukov et al, 2006; Shnyukov et al, 1971).

Carbonate rocks in the form of fragments in hill deposits are very widespread and occur almost within all mud volcanic foci.

Fragments of igneous rocks in the sopochnaya breccia are extremely rare. The interpretation of their genetic affiliation is ambiguous and very contradictory.

A peculiar rock was found in the solid ejections of mud volcanoes of Karabetova Gora, Gorelaya Gora, and the Azov Inferno. It is a dense dark gray lithiated hill breccia, most likely ancient, ejected by a mud volcano from strata submerged to a considerable depth. Microscopic examination of the rock showed that it consists of fine-grained siderite, recrystallized in parts, containing fragments of clays and sandstones. At the same time, siderite completely replaced the clay cement of the hill breccia and, in part, often clay fragments.

PRODUCTS OF MUD VOLCANISM: HILL GASES

Gases from mud volcanoes, along with sopochny waters, are the main agent determining mud volcanic activity in the region. The chemistry of the gas component and the isotopic composition of gases have been studied by many experts (Lagunova & Gemp, 1978; Shnyukov et al, 1971; 22. Beloussov & Yarotsky, 1936; Burkser & Bronstein, 1933; Gemp & Lagunova, 1978; Gemp et al, 1970; Zaitsev, 1965; Kudryavtsev, 1963; Kurishko et al, 1968; Lukashuk, 1933; Subbota, 1964; Shternov, 1968).

The daily gas flow rate of active mud volcanoes in the Kerch-Taman zone is estimated by the authors (Beloussov & Yarotsky, 1936) to be about 350–400 m^3 /day. Some researchers (Kudryavtsev, 1963) believe that a mud volcano emits 0,5–40 billion m^{3}of water per eruption others (Shternov, 1968) – estimated that, assuming the daily gas release of Kerch volcanoes equal to 520 m^3 in a calm state and 1500–2000 m^3 during recovery, the volume of gases released over 20 million years will be 11–15 trillion m^3 for the Kerch Peninsula.

The gases of mud volcanoes in the Kerch-Taman zone contain methane, carbon dioxide, heavy hydrocarbons, nitrogen, argon, helium, and sometimes hydrogen, hydrogen sulfide, and carbon monoxide. The main ones are methane and carbon dioxide. The methane content in most cases is 80–99%, sometimes decreasing to one and a half dozen percent and below – until it is completely absent. Carbon dioxide is contained in the amount of fractions of a percent to 91,7%, in the vast majority-up to 10%. Nitrogen is constantly contained in small amounts from 0,2% to 5,0% for most volcanoes. According to the ratio of the gas component, six types of exhaust gases are distinguished (Shnyukov et al, 1986; Shnyukov et al, 2006):

- methane gas,
- methane-carbon dioxide,
- carbon dioxide,
- nitrogen,
- heavy-hydrocarbon,
- carbon dioxide-nitrogen-methane.

Most of the volcanoes of the Taman Peninsula, with the exception of mud volcanoes: Karabetova Gora, Gnilaya Gora, and western gryphons of the Shugo volcano, belong тоервому the erw omu type and u-methaneomu. This type is characterized by a methane content of about 80–99%. Small amounts contain carbon dioxide, nitrogen, heavy hydrocarbons, and other gases.

All these types of exhaust gases are characterized by constant chemistry, which was studied at different times. Some mud volcanoes (Tsentralnaya Sopka of the Bulganak group, Podmayachnaya sopka of the Yenikalsky volcano, Karabetova Gora, Gnilaya Gora, and Zapadnye gryphony of the Shugo volcano) are characterized by changes in the ratio of gas components over time (Kurishko et al, 1968).

The other volcanoes listed above are characterized by a change in the type of cap gases over time from methane to methane-carbon dioxide. A similar phenomenon was observed in the same years for mud volcanoes: Rotten Mountain, Karabetova Mountain and Shugo (Lagunova, 1974). When considering the distribution of volcanoes with different gas compositions within the structures of the neogene-quaternary structural floor, we note the absence of any regularities.

In all likelihood, the specific distribution is determined by the structural and tectonic features of the lower structural floors lying at significant depths, the duration, activity and intensity of mud volcanoes with different depths of occurrence. The reason for the development of hills with different gas compositions within a single hill field is that the supply channels are confined to fault zones that are fixed at different depths in the lower structural floors, that is, the roots of mud hills in the vertical section are located at different depths.

Isotopic composition of mud volcanoes in the studied region, undoubtedly, indicates that their composition includes products of degassing of the Earth's deep interior. In general, the gas composition of mud volcanoes in the Kerch-Taman zone is determined by a number of factors, such as tectonic features of various structural floors, the depth of occurrence and activity of fault zones (including seismicity (Shnyukov et al, 1971), the thickness of the sedimentary cover and the presence of oil and gas deposits at various stratigraphic levels.

MUD VOLCANISM PRODUCTS: SOPOCHNYE VODY

Hydrochemical features of the waters of mud volcanoes in the Kerch-Taman zone have been studied by many scientists (Lagunova &, 1978; Shnyukov et al, 1971; Gemp & Lagunova, 1978; Gemp et al, 1970; Kurishko et al, 1968; Beloussov & Yarotsky, 1936; Albov, 1956; Albov, 1971; Albov, 1973; Sulin, 1939; Nesterov, 1979).

The water flow rate of mud volcanoes is generally insignificant. The most complete analytical data on the hydrochemical characteristics of sopochny waters, as well as their geological interpretation, are given in the works (Shnyukov et al, 1971; Gemp & Lagunova, 1978; Gemp et al, 1970; Kurishko et al, 1968). Analyses performed in different years showed a good match of the results.

The waters of mud volcanoes are characterized by a relatively low salinity-from 3,84 g/l to 23,36 g/l. In terms of composition, they are quite monotonous, belonging to the chloride-bicarbonate-sodium, bicarbonate-chloride-sodium, occasionally to the sulfate-chloride-sodium, chloride-bicarbonate-sodium, and sodium chloride types.

S.D. Gemp and I.A. Lagunova (1978) distinguish three hydrodynamic zones within the Kerch-Taman region: the western (southwestern part of the Kerch Peninsula), the central (northeastern part of the Kerch and western part of the Taman Peninsula), and the eastern (rest of the Taman Peninsula). The allocation of these zones is conditional, since often within the same large geotectonic unit, and often within the same sopochny field, the waters of volcanoes and griffins have different hydrochemical characteristics.

According to the hydrochemical characteristics of water within the Kerch-Taman region, there are six types of volcanoes:

- Type I-sodium chloride-bicarbonate;
- Type II-bicarbonate-sodium chloride;
- Type III-sodium chloride-carbonate;
- Type IV-sulfate-chloride-sodium;
- Type V-sodium chloride-bicarbonate-sulfate;
- Type VI-sodium chloride.

The study of the oxygen isotope composition of the hill waters of the Kerch and Taman peninsulas showed that the values of γO^{18} range from +3,8 to +6,0, which brings the hill waters closer to both oil and fumarolic waters in this indicator.

So, the balneological wealth, hydrogen sulfide sources of good flow, the proximity of the sea and sandy beaches, a favorable climate – all this remains practically unused so far if there are needs for the treatment and rehabilitation of vacationers.

MUD VOLCANOES ARE ONE OF THE MAIN ATTRACTIONS OF THE TAMAN PENINSULA

Within the Kerch-Taman mud volcanic region, the total number of mud volcanoes reaches almost one hundred, and half of them are localized on the Kerch Peninsula. The general pattern of volcanism development in the Kerch and Taman Peninsulas has already been noted above. Kerch mud volcanoes were more active in the geological past than at the present moment of geological history. Taman volcanoes, on the contrary, are more active now and there are almost no traces of their geological activity in the past.

Mud volcanoes are one of the main attractions of the Taman Peninsula. Most of them are located in the vicinity of the city of Temryuk and in the Temryuk district. Despite their small size, many of the volcanoes of the Taman region are active and still regularly erupt (Zakharchenko & Korneva, 2023).

The main activity of mud volcanoes in the geological history of the Kerch-Taman region is the late maikop, chokrak, karagan, konka, meotis, pontus, cimmerian, kuyalnik, and anthropogenic; the most active mud volcanoes are in chokrak, sarmatian, and cimmeria, which has a definite tectonic explanation.

On the territory of the peninsula there are many mud volcanoes that form negative (mud lakes with silt deposits) and positive forms (mud volcanic hills with heights from 15–20 m to 164 m), as well as underwater marine volcanoes. Each volcano is a unique geological feature of high scientific significance (Zakharchenko & Korneva, 2023).

Mud volcanoes of the Taman Peninsula look like real large volcanoes, only they are smaller and they do not spew lava, but a liquefied hill breccia that rises to the surface with water and gases. Its temperature is usually constant and rarely exceeds 15°C, even in the hottest summer.

Mineral hill mud of the Taman Peninsula has a curative effect, and silt hydrogen sulfide mud is even more curative. Volcanic mud, in addition to hydrogen sulfide, contains a lot of iodine, boron, bromine and other trace elements and has valuable balneological properties.

Below you will find basic information about the most interesting mud volcanoes of the Taman Peninsula for tourism development.

MUD VOLCANO AKHTANIZOVSKAYA SOPKA

Location griazevogo volcano and Akhtanizovskaya Sopka-Krasnodar Territory, Temryuksky district. The Akhtanizovskaya Sopka mud volcano is located near the southwestern edge of the village of Akhtanizovskaya (figure 6).

Figure 6. View of the top of the mud volcano Akhtanizovskaya Sopka near the village of Akhtanizovskaya

(photo by E.I. Zakharchenko, 2023)

A dirt road leads to the top of the mud volcano (figure 7).

Figure 7. View of the dirt road to the top of the mud volcano Akhtanizovskaya Sopka

(photo by E.I. Zakharchenko, 2022)

The Akhtanizovskaya Sopka mud volcano has been active for many centuries. In the 19th century, it was an active volcano. In the past, the Akhtanizovskaya Sopka volcano was distinguished by violent eruptions during certain periods. Strong eruptions were recorded in 1818, 1842 and 1853. From the gryphon located on the top of the mountain, hill mud constantly poured out in the form of bizarre tongues descending from the pointed peak. Since then, the volcano has been functioning relatively calmly (Zakharchenko & Korneva, 2023).

Back in 1900, researchers described this volcano as two almost level peaks separated by a depression. In 1936, the volcano was a regular cone, truncated at the very top, about 30 m high and with a base diameter of 200 m. In the crater there was a lake filled with liquid mud with a diameter of 2 m. Sometimes it flowed in a southwesterly direction along a narrow and shallow ditch. In the crater lake, gas was released in bubbles.

In February 2002, an outflow of a significant mass of sopochnaya breccia from the interior of the volcano was recorded. The last powerful eruption was recorded in 2008. Prior to 2008, eruptions occurred at intervals of 8–10 years. Since then and until now (2024), the mud volcano Akhtanizovskaya Sopka is in a slightly active state. Currently, the activity of the volcano is weakened, eruptions occur continuously through the side gryphons (depending on the activity of the volcano, eruptions occur through 1 or 2 side gryphons) (figure 8) (Zakharchenko & Korneva, 2023).

Figure 8. Eruption of the mud volcano Akhtanizovskaya Sopka through the side gryphon

(photo by E.I. Zakharchenko, 2024)

The Akhtanizovskaya Sopka mud volcano is a continuation of the eastern part of the Tsymbala ridge (from the top of the volcano you can see the neighboring cones of the Tsymbala, Boris and Gleb mud volcanoes, and others) and includes two conical parts. The main cone is

67 m above sea level, and the steepness of the slopes reaches 15–30 degrees. The base of the hill is made of clay and dark gray limestone. In front of the main cone, there is a small plateau formed by eruptions of the lower cone on the eastern side (figure 9).

Figure 9. East side view of the plateau in front of the main cone and the main cone Akhtanizovskaya Sopka volcano

(photo by E.I. Zakharchenko, 2022)

The main crater of the volcano is active and has the shape of an ellipse. The crater at the top has a size of 3,5×13,0 m. In the southwestern part, there is an active gryphon vent, from which a dark gray liquid sopochnaya breccia erupts (figure 10). Its temperature does not exceed 20^{0}C (Zakharchenko & Korneva, 2023).

Figure 10. Eruption of the mud volcano Akhtanizovskaya Sopka through the side gryphon

(photo by E.I. Zakharchenko, 2022)

The top surface and slopes of the mud volcano Akhtanizovskaya Sopka are covered with dried-up hill breccia of light gray color. Tongues of dry hill silt are visible on the slopes, the breccia surface is broken by cracks, covered with whitish discolorations of salts (figure 11).

Figure 11. Eruption of the mud volcano Akhtanizovskaya Sopka through the side gryphon

(photo by E.I. Zakharchenko, 2021)

Overgrown with grass tongues of dried hill silt in the lower part of the cone create a bizarre landscape. The slopes are strewn with numerous fragments, among which there are often sandstones and fragments of mud-volcanic breccia of different stages of firing-from gray to red and dark brown.

On the Akhtanizovskaya Hill at an altitude of 20–60 meters, layers with fossilized animal remains were found, indicating a relatively recent uplift of the territory.

The Akhtanizovskaya Sopka mud volcano is surrounded by unique flora and fauna that have managed to survive and adapt to the difficult conditions of mud volcanic activity. Here you can see representatives of rare bird species that feed on the inhabitants of volcanic lakes: flamingos, white-tailed eagles and others.

The local soil is unique, as are the grasses and shrubs that grow on it: they had to adjust to the temperature and degree of mineralization of the surface. This unique ecosystem, which combines wildlife and rare natural phenomena, has become an interesting object for research and conservation of the diversity of biological species.

MUD VOLCANO MISKA (MYSKA)

Location griazevogo volcanoa Miska – Myska – Krasnodar Territory, Temryuksky district, city of Temryuk. Mud volcano Miska is located on the right bank of the Kuban River, in the eastern part of the city of Temryuk. The name of the volcano is associated with the shape of its top, which resembles a giant bowl (figure 12).

Figure 12. Mud volcano Miska crater filled with lake waters

Catastrophic eruptions of the mud volcano Miska occurred in 1812, 1842 and 1860, accompanied by emissions of sopochnaya breccia, vapors, smoke and flames.

It is believed that Miska is a dormant volcano, since the last eruption was observed in 1924. Currently, the volcano is in a state of relative rest, but within the museum site with military equipment, a hill breccia periodically comes to the surface through small griffins (figure 13) (Zakharchenko & Korneva, 2023).

Figure 13. Side gryphon of the mud volcano Miska

At its base, the mud volcano hill is composed of clays and sands formed in the Paleogene and Neogene. During the Alpine folding, the layers of these rocks were domed.

The solidified hill breccia lies in the form of a giant loaf, the base of which drops 150 m below sea level. The maximum height of the hill of the mud volcano Miska is 75.3 m (Figure 14).

Figure 14. Top of the hill of mud volcano Miska

Mud volcano rises in the central part of the Temryuk ridge. Its cone is composed mainly of Maikop clays. Drilling revealed sands. The central part of the volcano's cone along the vertical axis (from the crater and below) is composed of hill breccia, which would have been drilled to 225 m (Zakharchenko & Korneva, 2023).

The crater of the volcano is very large: its area reaches 22 hectares. In plan, it has the shape of a weakly pronounced ellipse with a long axis stretching from west-northwest to east-southeast by 625 m. The small axis is shorter by 100 m. It stretches in a direction from south-southwest to north-northeast for 525 m.

The caldera wall surrounding the crater is composed of hill breccia and rises 10–15 m and above the bottom of the latter (figure 15), which is occupied by a shallow (1 m deep) lake, which resembles the head of a wrench in its contours. The area of the lake is 6,5 ha, length 500 m, width 350 m (figure 16) (Zakharchenko & Korneva, 2023).

Figure 15. External view of the mud volcano Miska caldera shaft

Figure 16. Mud volcano Miska crater filled with lake waters

Subsidence of hill breccia and outpouring of hill silt under asphalt lead to deformations of the roadbed passing in the northern part of the crater.

MUD VOLCANO KARABETOVA GORA

Location griazevogo volcanoand Karabetova Gora-Krasnodar Territory, Temryuksky district, district of the village of Taman.

The Karabetova Gora mud volcano is the largest active mud volcano on the Taman Peninsula. It is located 5 km east-southeast of the village of Taman (figure 17).

Figure 17. General view of the mud volcano Karabetova Gora

The mud volcano Karabetova Gora is a natural monument. It is an кoval-shaped crater plateau, stretching from west to east, with a length of 2726 meters, and from north to south by 1718 meters (the most remote points), composed of hill breccia. The absolute elevation of the plateau is 152 m above sea level. The relief of the mud volcanic plateau is formed as a result of repeated catastrophic eruptions of a mud volcano and, as a result, has a very heterogeneous character. The mud volcano is a grandiose system of hills, salzas, and crater lakes (figure 18, 19) (Zakharchenko & Korneva, 2023).

Figure 18. Active crater of the Karabetova Gora mud volcano

Figure 19. Gryphon on the eastern slope of the mud volcano Karabetova Gora

The Karabetova Gora mud volcanois located within the Karabetovskaya anticline zone, which stretches in a northeasterly direction across the entire Taman Peninsula from the Kerch Strait to the Kuban River. This zone also includes the mud volcanoes of Mount Chirkov, Severo-Neftyanoy and Dubovy Rynok.

The most powerful mud volcano Karabetova Gora erupted in 1876. The eruption was accompanied by explosions and ejections of gas and rocks with clouds of black smoke and flames. From 1818 to 1974, at least 19 explosive eruptions of this volcano are known. Many eruptions were accompanied by underground impacts, explosions, flames and clouds of black smoke, the formation of cracks on the surface of the earth. At the same time, powerful volcanic breccia emissions were observed, covering the mountain top with a layer of up to 2 m. The revival of mud volcanic activity was also noted in the early 1980s. The eruption events of 1968, 1980, 1981, 1982, 2001, and 2004 were large bursts of hill breccia accompanied by rumbling, tremors, and fire events.

According to observations, the volcano also slightly erupts with a frequency of 2 times a year-in spring and autumn. The solid ejecta of this volcano are dominated by rocks of the Maikop and, to a lesser extent, Sarmatian deposits (Zakharchenko & Korneva, 2023).

On the north side of the mud volcano there are three main large cones with different diameters. To the northwest is a large dormant lateral cone of a mud volcano with a flat and dense crater field (figure 20).

Figure 20. Central hill fieldof the Karabetova Gora mudvolcano Гора

Due to its very long-standing activity and under the influence of the forces of the elements (mainly water), it is difficult to find this side cone of the volcano, since it rises only 15–20 centimeters above the general level of the plateau, and its edges are very flat. To the north-east and above the dormant side cones of the volcano is the side cone, which was the last to stop erupting (figure 21). The last eruption of the sopochnaya breccia was observed in 2023.

Figure 21. Extinct lateral crater of the mud volcano Karabetova Gora

At the foot of the volcano there is a large, 100×20 meter freshwater lake (figure 22).

Figure 22. Lake located near the foot of the mud volcano Karabetova Gora

The height of the hill of the mud volcano Karabetova Gora is 152 meters. Externally – it is a large hill with a flat top and calm outlines, planted on one of the anticlinal ridges of the peninsula. The crater of the volcano has the shape of an oval (Zakharchenko & Korneva, 2023). The entire surface is covered with cones of mud hills (figures 23, 24). The most active part of the volcano is the south-eastern slope. A mud lake is also located here (figure 25).

Figure 23. One of the active hills of the mud volcano Karabetova Gora

Figure 24. One of the active hills of the mud volcano Karabetova Gora

Figure 25. Mud lake located on the south-eastern slope of the mud volcano Karabetova Gora

The Karabetovskaya anticline is composed of rocks of the maikop series, chokrak, karagan, sarmatian, meotic, pontic, and cimmerian stages. The mud volcano is confined to the Maikop core of the fold and is composed of a powerful mass of sopochnaya breccia, which in turn is erupted by new arrivals of mud volcanic products.

Numerous oil occurrences are known on the mud volcano oil events – the water surface of almost all active sources is covered with iridescent oil films (figure 26). In some griffins, the release of liquid or plastic sludge is noted.

Figure 26. Oil occurrences on the active hill of the mud volcano Karabetova Gora

The surface of the sopochny field is almost entirely covered with fragments of sopochny breccia of various stages of firing: from the initial fresh light gray color to brick-red (firing temperature 900–950°C) and black slag-like (above 1050°C) (Zakharchenko & Korneva, 2023).

On the surface of the volcano, tourists can easily find rock fragments and fossilized pieces of ancient plants. The soil composition is dominated by sand and clay, fossilized limestone is found. Due to the poor soil composition, the plant world is represented exclusively by soleros grasses (figure 27).

Figure 27. Soleros growing on the slopes of the mud volcano Karabetova Gora

The formation of the fertile layer occurs very slowly, this is facilitated by wind and water erosion. In spring, the surface of the volcano turns bright green, and in summer and autumn it turns yellow. The gullies are dominated by communities of wheatgrass, on the peaks of fescue, cornflowers, sage, and feather grass, and various types of wormwood are common.

MUD VOLCANO ROTTEN MOUNTAIN (HEPHAESTUS)

Location ofthero Rotten Mountain volcanoand the Rotten Mountain (Hephaestus) – Krasnodar Territory, Temryuksky district, the city of Temryuk.

The Rotten Mountain mud volcano is active, located on the eastern outskirts of Temryuk, and has been in the griffin stage for a long time. Currently, a marketing ploy is being used – the name of the mud volcano Rotten Mountain in the media is being changed to Hephaestus.

Rotten Mountain mud volcano is known for periods of high activity. Several intense eruptions took place in the first half of the 19th century. Asimilar eruption was observed in October 1978. Recent eruptions of the volcano occurred in 2001-2002. An active eruption of the sopochnaya breccia with a gas discharge on the Rotten Mountain mud volcano was recorded in 2001. In February 2002, a powerful explosion at the Rotten Mountain volcano was repeated, coinciding with the eruption of the Western Tsymbaly mud volcano. The last major eruption occurred in April 2010.

The Rotten Mountain mud volcano is a low, flat hill (plateau) of ellipsoid shape, with a chain of mud ramparts (Figure 28). The plateau has a slightly hilly surface in the form of a truncated cone, rises 13–15 m above the adjacent plain, and 32 m above the Sea of Azov level. The mud volcano measures about 500 m across (the large axis of the oval is about 550 m, the small axis is about 500 m) (Zakharchenko & Korneva, 2023).

Figure 28. Top view of the Rotten Mountain mud volcano (Hephaestus)

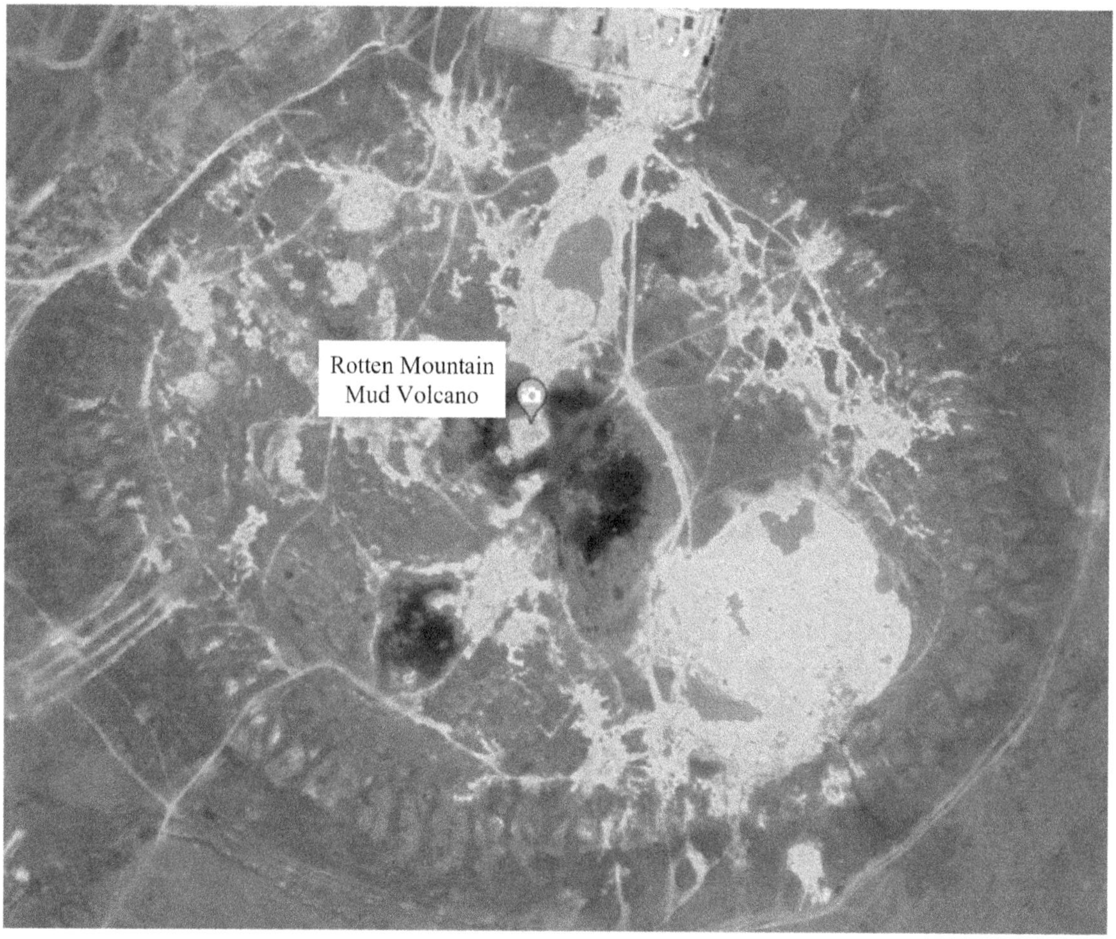

A large number of small and very small griffins of various shapes are located on the surface of the sopochny field. Among the numerous griffins, there are both temporarily stopped activities and active ones. In different years, there are from 75 to 200 of them. The height of the largest griffins is 2–3 m (Zakharchenko & Korneva, 2023).

The entire surface of the hill field of the Rotten Mountain mud volcano is covered with a hill breccia. Stand – out breccia is gray, plastic, odorless. Zones of recently overflowed dark gray hill breccia are interspersed with areas of light sand of Aeolian origin (figure 29).

Figure 29. Active volcano Rotten Mountain

(photo by E.I. Zakharchenko, 2023)

In the central part of the sopochny field, there is a 100x50 m lake with brackish water (figure 30).

Figure 30. Brackishlake on the sopochny field of the Rotten Mountain mud volcano

Also within the sopochny field there are several other small craters that represent mud lakes (figure 31), theirnumber may vary (Zakharchenko & Korneva, 2023).

Figure 31. One of the craters of the Rotten Mountain mud volcano, representing a mud lake

(photo by E.I. Zakharchenko, 2024)

Larger griffins are located on the periphery of the sopochny field (figure 32).

Figure 32. Fragment of the sopochny field of the Gnylaya Gora mud volcano

(photo by E.I. Zakharchenko, 2023)

They are grouped together, often arranged in chains. Gryphons constantly secrete hill silt that flows down the slopes of gryphon cones (figure 33).

Figure 33. Streams of fresh hill mud flow down the slopes of gryphons located on the periphery of the hill field (photo may 2011)

The largest gryphon of the Rotten Mountain mud volcano is shown in figure 34.

Figure 34. Eruption of the sopochnaya breccia from a large gryphon

The intensity of activity and consistency of the breccia substance released by griffins varies from water and liquid silt to plastic clay (figure 35).

Figure 35. Fragment of the sopochny field of the Gnylaya Gora mud volcano with streams of hill breccia

Numerous microgryphons are widely developed within the sopochny field among fresh, sometimes already dried-up sopochny breccia (figure 36) (Zakharchenko & Korneva, 2023).

Figure 36. Fragment of the hill field of the Rotten Mountain mud volcano with a micro-gryphon

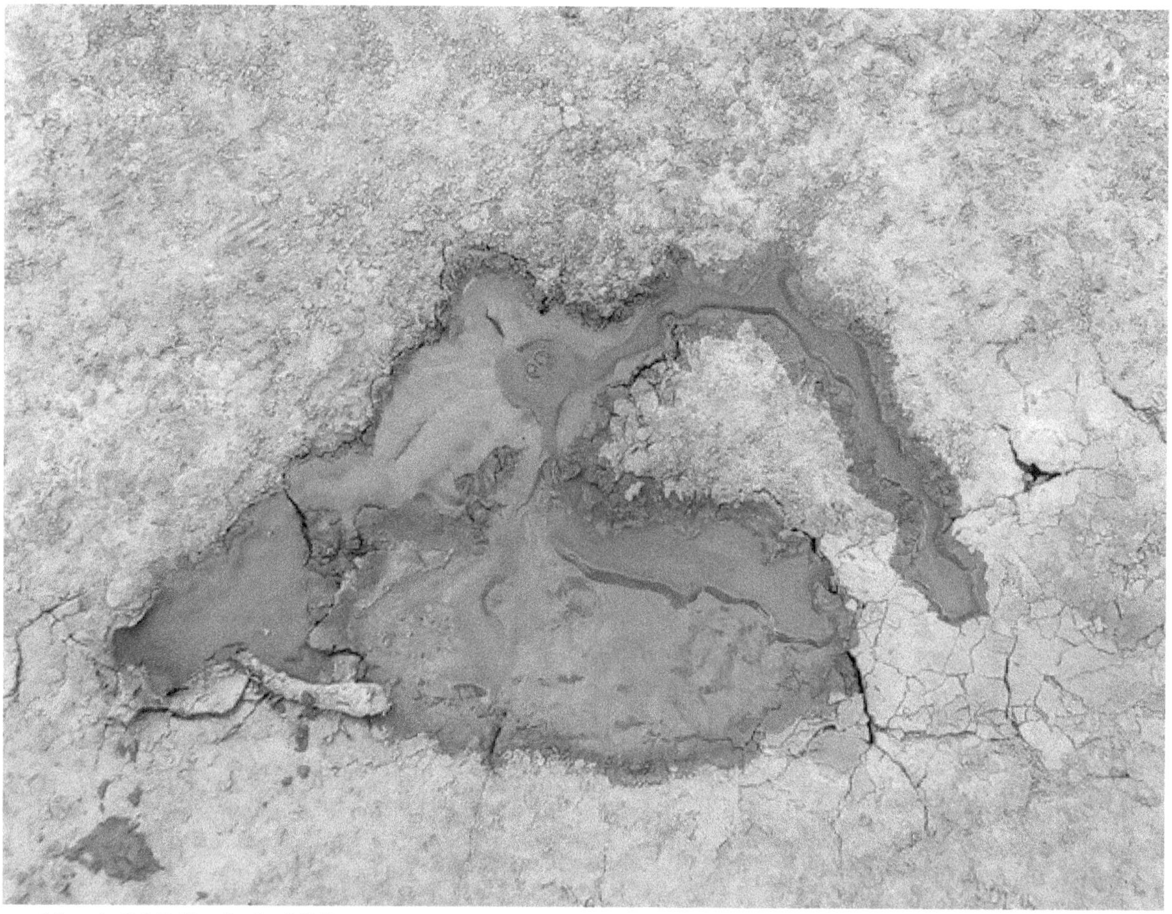

(photo by E.I. Zakharchenko, 2024)

On the slopes of gryphons, tongues of dry hill silt are visible, the surface of the breccia is broken by cracks, covered with whitish discolorations of salts (figure 37). Drying cracks form very quickly in the hill breccia (figure 38).

Figure 37. Gryphon of the Rotten Mountain mud volcano

(photo by E.I. Zakharchenko, 2023)

Figure 38. Fragment of the sopochny field of the Gnylaya Gora mud volcano

(photo by E.I. Zakharchenko, 2024)

In some areas, griffins break through the surface of the hill field, which is already overgrown with grass (figure 39). Whitish discolorations of mineral salts are visible everywhere on the surface of the hill breccia.

Figure 39. Small griffins among red-flowered soleros plants

(photo by E.I. Zakharchenko, 2023)

Due to the poor soil composition, the plant world is represented exclusively by soleros grasses (figure 40).

Figure 40. Soleros growing on the sopochny field Rotten Mountain mud volcano

(photo by E.I. Zakharchenko, 2023)

BLUE BALKA MUD VOLCANO (TIZDAR)

Location griazevogo volcanoa Sinyaya balka (Tizdar) – Krasnodar Krai, Temryuksky district, village. For the Motherland.
The Blue Balka mud volcano (Tizdar) is an active mud volcano of the Bulganak type. It is a negative-shaped mud volcano. It is located on the coast of the Sea of Azov, on the territory of the village of Za Rodina, 4 km west of the village of Peresyp.
Currently, a marketing ploy is being used – the name of the Blue Balka mud volcano is being changed to Tizdar in the media.
The first mention of the Sinyaya Balka mud volcano dates back to 1799, when emissions from the ground were observed by Cossack Cossacks who had recently moved to Taman.
The crater of the Blue Beam mud volcano has a regular funnel shape measuring about 30 m across. The height of the volcano above sea level is 74 meters. The crater is filled with a mud lake, which is a bluish-gray liquefied hill breccia (figure 41). A well-defined caldera wall covered with reeds is observed around the mud lake (figure 42). Its height is 7–10 m.

Figure 41. Blue Beam mud volcano crater filled with mud lake

Figure 42. Pronounced caldera wall around the crater of the Blue Beam mud volcano

Currently, volcanic mud is extracted and used for recreational purposes (figure 43). «Tizdar Family Resort & Spa 5*» is located near the mud volcano crater. There is a recreation area, equipped with tanks for taking mud baths, near the Azov coast with a natural sandy beach.

Figure 43. Top view of the Blue Beam mud volcano crater

Due to the recreational use of the hill breccia of the Sinyaya Balka mud volcano, it is one of the most studied. The mud of the volcano is renewable, the daily flow rate of new portions of breccia is 2,5 m^3 per day, and in the very center of the crater, insignificant outflows of hill breccia with gas are continuously observed.

Sopochnye mud belongs to mineral mud type clays, they have a light gray color, rich in biologically active elements: iodine, bromine, boron, do not contain hydrogen sulfide and organic substances, has antimicrobial properties. Heat convection in the hill mud is insignificant, so it has a high heat retention capacity.

The crater is oftensurrounded by fragments of hill breccia of various stages of roasting: from the initial fresh light gray color to brick-red (roasting temperature 900–950°C).

MUD VOLCANO AZOVSKOE PEKLO

Location griazevogo volcanoand Azovskoe Peklo-Krasnodar Territory, Temryuksky district, village of Kuchuguri.

PIt is located on the coast of the Sea of Azov, 2 km west of the village of Kuchugury. The Azovskoe Peklo mud volcano is an active mud volcano of the Bulganak type. It is a negative-shaped mud volcano (figure 44).

Figure 44. Azovskoe Peklo mud volcano crater

(photo by E.I. Zakharchenko, May 2024)

The size of the mud lake that fills the crater of the volcano varies from 5 meters to 12 meters in diameter at different times (figures 45, 46).

Figure 45. Azovskoe Peklo mud volcano crater (photo from 2013)

Figure 46. Azovskoe Peklo mud volcano crater

(photo by E.I. Zakharchenko, May 2024)

The Azovskoe Peklo mud volcano is located in a cirrus-shaped depression within the anticline of Cape Peklo, which is fixed at its top by a sharp vertical cliff that has arisen as a result of landslide phenomena. The hill field of this volcano is located on the body of a landslide and, in accordance with its slope (figure 47), also has a slight slope towards the Sea of Azov (figure 48).

Figure 47. Crater and sopochnoe field of the mud volcano Azovskoe Peklo (side view)

(photo by E.I. Zakharchenko, May 2024)

Figure 48. The lower part of the landslide body has a slope towards the Sea of Azov

(photo by E. I. Zakharchenko, May 2024)

The size of the sopochny field varies due to different volcanic activity and varies in recent years from 30–60 m in width to 80–120 m in length. The main flow of mud-volcanic deposits is directed towards the sea to the northeast (figures 49, 50).

A high concentration of minerals is observed in the sopochnaya breccia, so when the liquid evaporates, a loose white crust forms on the surface of the sopochnaya field (figure 51).

Figure 49. Hill field of the mud volcano Azovskoe Peklo

(photo by E.I. Zakharchenko, May 2024)

Figure 50. Hill field of the mud volcano Azovskoe Peklo

(photo by E.I. Zakharchenko, May 2024)

Figure 51. Mineral substances observed in the sopochny field mud volcano Azovskoe Peklo

(photo by E.I. Zakharchenko, May 2024)

Directly at the cliff, among the constantly pouring hill mud, there are outcrops of loess-like rocks of Quaternary age, lying higher up the slope. Numerous small lateral griffins (figures 52), including hydrogen sulfide ones, which operate continuously and are located mainly in the marginal parts of the volcano, further dilute the hill breccia on the surface, which, combined with the slope of the terrain towards the sea, creates an amazing picture of the movement of mud volcanic tongues (figures 53).

Figure 52. Side Gryphon crater located in the edge part mud volcano Azovskoe Peklo

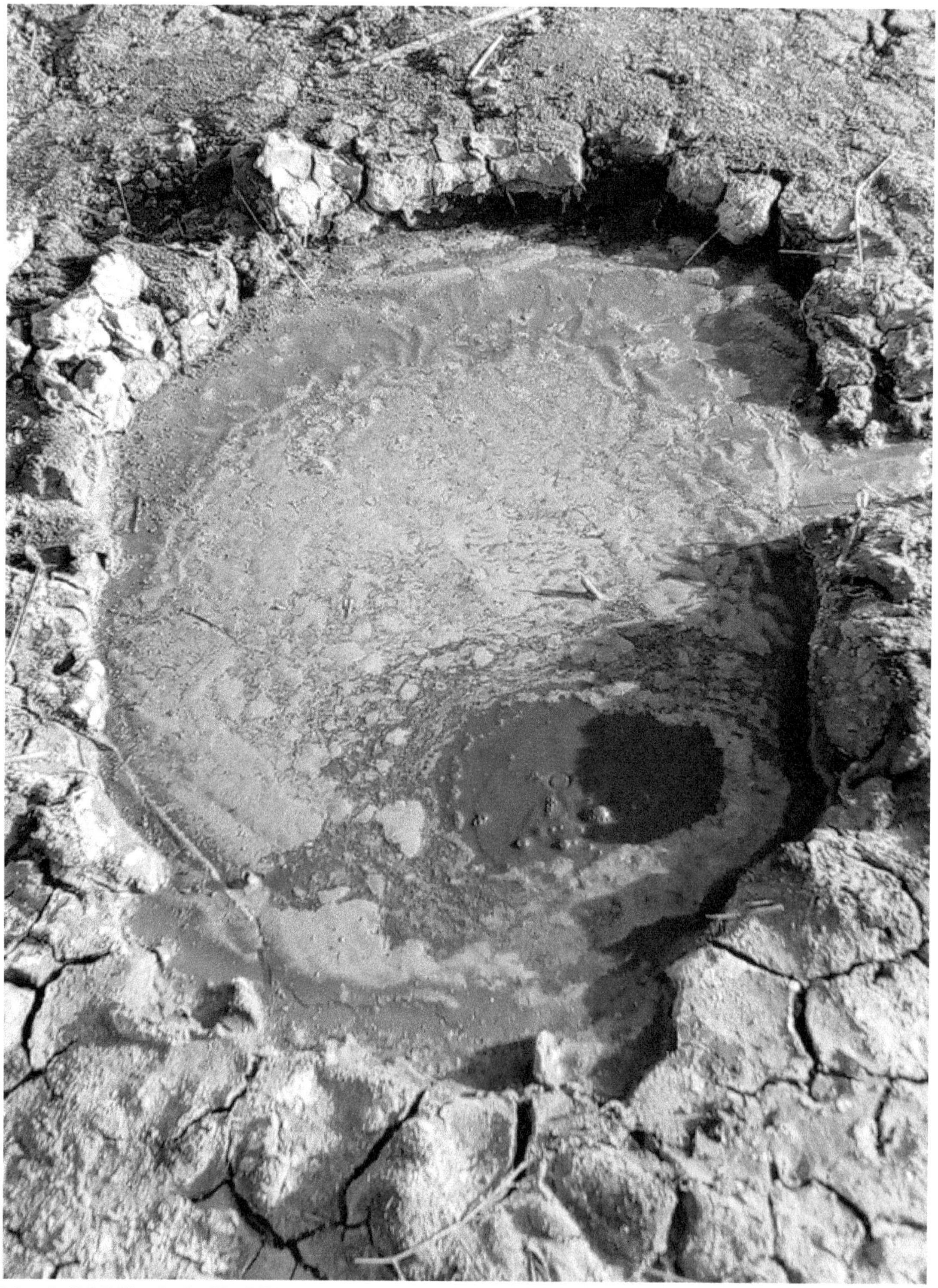

(photo by E.I. Zakharchenko, May 2024)

Figure 53. One of the side griffins in the edge part mud volcano Azovskoe Peklo

(photo by E.I. Zakharchenko, May 2024)

Just below the sopochny field, to the north of it, there is a small lake reservoir, which was formed in the hilly-zapadinnaya pressure part of the landslide and is fed by water from a mud volcano (figure 54).

Figure 54. Lake in the lower part of the sopochny field of the Azovskoe Peklo volcano

The lake has dimensions of approximately 40 m by 25 m, which can be quite variable in different periods. The lake's waters have a complex chemical composition and high mineralization.

MUD VOLCANO SHUGO

Location griazevogo volcanoa Shugo-Krasnodar Territory, Krymsky district, district of the village of Varenikovskaya.

Shugo Mud Volcano is a geological nature monument of regional significance, established in 1988.

Mud volcano Shugo-active, one of the most active and picturesque volcanoes of the Kerch-Taman mud volcanic province, as it is located in a mountainous and wooded area and is located on the right bank of the Shugo River, 5 km northeast of the village. Fadeyevo and 6 km south of Varenikovskaya station (figure 55).

Figure 55. Top view of the hill field of the Shugo mud volcano (from a height of 100 m)

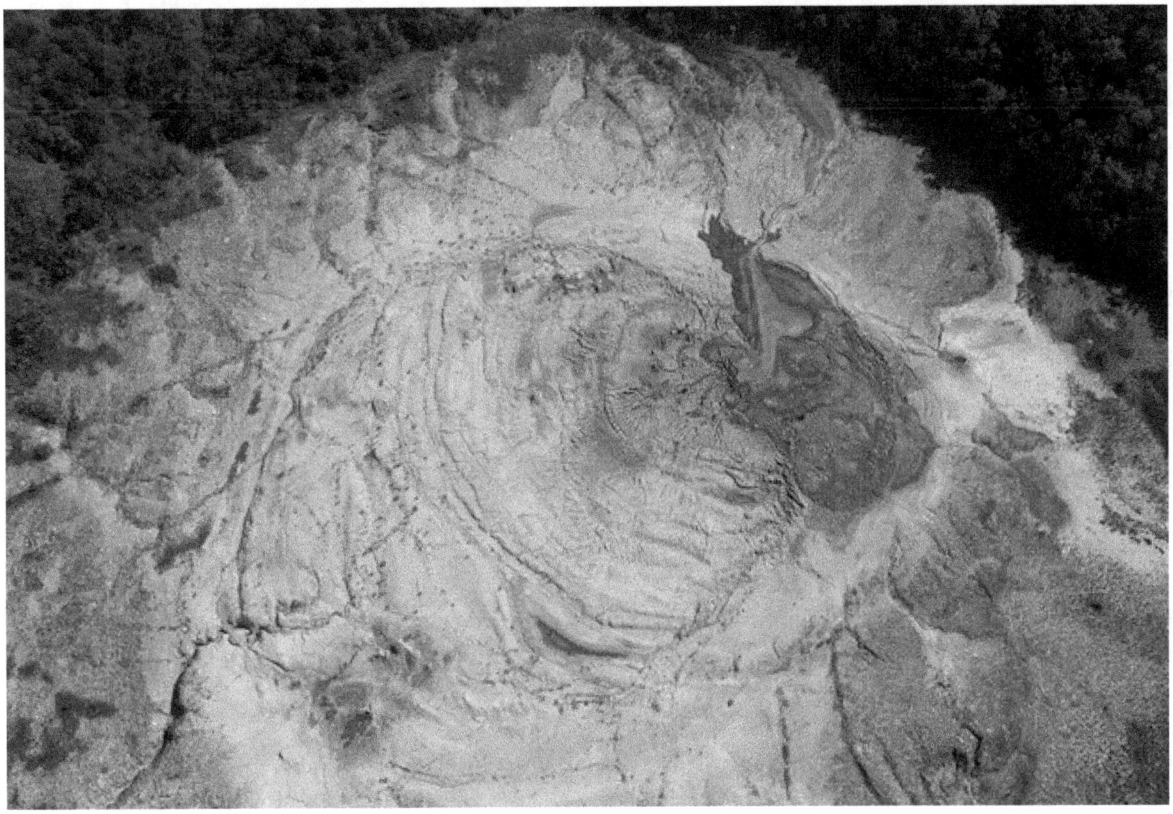

The volcano historically had several names: Divnaya Gora, Iezheuasa (Circassian) – Ash Mountain, Rotten Mountain on Shuga (the name given by the Kuban Cossacks). Currently, the generally accepted name is Shugo.

In the past, strong eruptions of the Shugo mud volcano were known, accompanied by explosions, columns of flame, strong noise and powerful ejections of hill breccia to a height of several meters (1866, 1903, 1907, 1908, 1912, 1959, 2006). Thedischarge of the sopochnaya breccia on the Shugo mud volcano occurs constantly, as it is currently in the griffin stage, daily spewing dozens of cubic meters of gas, a large amount of sopochnaya breccia. It belongs to the volcanoes of the Bulganak type. The last major eruption of the Shugo mud volcano occurred in May 2023.

As you approach the top of the volcanic structure, which is densely wooded, you can see the almost devoid of vegetation sopochnoe field and the main eruptive center (figure 56).

Figure 56. Top view of the sopochny field of the Shugo mud volcano

The activity of the Shugo volcano, located within the Akhtyr flexure-rupture zone, is characterized by a constant change of active (figure 57) and calm (figure 58) periods over a long period of time, systematic emission of gases and hill breccia with a low content of coarse-grained material.

Figure 57. Shugo Mud Volcano Eruption (2023)

Figure 58. Griffin stage of the Shugo mud volcano eruption (2024)

Tectonically, the eastern part of the mud volcano province is located in the area of submergence of orogenic folded structures of the meganaticlinorium of the Greater Caucasus under the weakly dislocated Cenozoic cover of the Taman and West Kuban troughs (Sobisevich et al, 2014; Gorbatikov et al, 2008; Kholodov, 2002a; Kholodov, 2002b).

The anticlinal ridges of the Taman Peninsula in the direction from west to east change their strike from north – north-east to sublatitudinal, then to north-west, bordering the periclinal closure of the Caucasus. The northern wing of the meganticlinorium of the Northwest Caucasus, represented by the Akhtyr flexure-discontinuity zone, forms a structural stage that serves as a transition to the depression of the West Kuban trough. The upper cretaceous, paleogene, and maikop series of sediments sink steeply toward the foothill trough (Sobisevich et al, 2014; Gorbatikov et al, 2008; Kholodov, 2002a; Kholodov, 2002b). Disjunctive complications of the flexure of the northern slope are faults identified in the course of geological work in the Caucasus by various researchers and studied in detail by geological and geophysical methods (Lagunova & Gemp, 1978; Geological dictionary, 1973).

The volcano is confined to the top of the anticlinal fold of the same name, composed of Cretaceous and Cenozoic deposits. The mud-volcanic structure of Shugo is associated with a north-eastern strike-slip fault. This fault is expressed in relief and affects exogenous processes, its south-eastern wing is raised, and the right-hand thrust component is manifested in the displacement of the ring shaft that limits the caldera of the volcano. The rate of horizontal displacements measured by geomorphological methods in the Holocene is 3.7–5 mm/year, and the rate of vertical displacements is 0.7–1 mm / year (Lagunova & Gemp, 1978; Geological dictionary, 1973; Shnyukov et al, 1971).

On the ground, the Shugo mud volcano is a wooded hill with a relative height of 30 m and dimensions of 1.5×2.5 km (figure 59). At the top of the hill, which looks like a truncated cone, there is a crater. It is round in shape, reaches 230–280 m, and is surrounded by an annular shaft 20–30 m high. The rampart was breached in several places by flows of sopochnaya breccia in the east and south-east, where a natural gully was formed, which ensures the demolition of accumulated igneous material.

Figure 59. Top view of the sopochny field of the Shugo mud volcano

The interior of the crater is mostly filled with fresh light gray hill breccia with numerous clay streams and a number of active cones spewing liquefied breccia and gas (figures 60).

Figure 60. Griffins of the Shugo Mud volcano

In the crater there are several active and extinct griffins, volcanic cones or hills (salzas), composed of hill breccia. The number of active griffins varies from 15 to 40 in different periods. The height of the cones reaches up to 2.5 m (figure 61).

Figure 61. Griffins of the Shugo mud volcano

Volcanic cones (or hills) are broken by a system of concentric fold-break structures and radial breaks. Concentric ramparts surrounding the perimeter of volcanic structures are complicated by discharges, expressed in relief by ledges 1–1.5 m high (figure 62).

Figure 62. Crater field of the mud volcano Shugo, complicated by discharges

On the western periphery of the main crater field, there are 3 largest cone-shaped gryphons. The lower diameters of griffins are 4–6 m, the upper ones 1.3 m. The vents are filled with semi-liquid hill breccia, which flows down the slopes of griffin cones and forms crusts. The breccia is cold and often covered with an oil film (figure 63). Gas escapes from the griffins. On the same line there is a small rounded lake, 1.5 m in diameter, strongly carbonating (figure 64).

Figure 63. Eruption of a sopochnaya breccia with oil from Gryphon

Figure 64. Griffins of the Shugo Mud volcano

The second group of griffins is located on the southern periphery of the crater field, 65 m east of the previous group. The nature of activity and appearance of griffins differs from the first. The hill breccia becomes more viscous, plastic, and light gray. The flat, saucer-shaped surface with a diameter of 6 m is covered with cracked breccia, which sharply contrasts with the whitish background of the crater field. Columnar wheels spring underfoot on a fluid base. In the center of the gryphon with a diameter of 1.1 m, pieces of hill silt periodically erupt, accompanied by noise and gas breakthrough (figure 65). Griffins act after a short period of time, one after the other. Gas also escapes noisily through drying cracks. Small griffins squeeze out gray plastic silt. Sometimes it's just spots of fresh portions of hill breccia.

Figure 65. Hill field of Shugo mud volcano

The third group of griffins is located at the northern periphery of the sopochny field. Three large cones are represented with a low-viscosity hill breccia, which forms a boning at the top and cracks at the bottom (figure 66). Gas and water are also intensively released.

Figure 66. Gryphon Crater of Shugo Mud Volcano

Along with active gryphons, there are a number of temporarily inactive gryphons. A high concentration of minerals is observed in the sopochnaya breccia, so when the liquid evaporates, a loose white crust forms on the surface of the sopochnaya field (figure 67). Along the perimeter of the basin there are thickets of reeds.

Figure 67. Mineral substances observed in the sopochny field mud volcano Shugo

The eruptions of the Shugo mud volcano contain various detrital material mixed with fine clay, which forms a stable emulsion with water. Detrital material of various ages and compositions – marls, glauconite sandstones, strong crinoid limestones, concretion and slab siderites, and shale clays form placers in the depressions and hollows of the runoff. The linear localization of individual griffins and clusters is due to the existence of radial and annular cracks in the diapir core.

The waters of Shugo volcano belong to the chloride-bicarbonate-sodium type. Vulcan mud is recommended for application treatment. Its reserves amount to 36 thousand m^3. The sopochnaya breccia has a high content of iodine and bromine, as well as a small amount of oil.

The mud volcano Shugo has its roots presumably associated with a regional fault that runs along the axis of the Verkhne-Chekup structure. The thickness of the hill deposits on the mud volcano is 75–100 m and they overlap the deposits of the supra-ore layers.

Shugo mud volcano belongs to the category of specially protected natural objects.

CONCLUSIONS

Mud volcanoes of the Taman Peninsula are of the greatest interest as objects of excursion display. The Taman Peninsula is one of the youngest land areas and is the region with the greatest development of mud volcanoes in the Russian Federation. Even in the first century AD, this area was an archipelago of several islands. However, due to the high dynamism of the Kuban River Delta, the high intensity of sediment coming from the river (up to several million tons per year) and the rise of tectonic plates, the islands were connected to each other.

Currently, the Taman Peninsula stretches from north to south for almost 40 km, from east to west-for 60 km. Its area is almost 2000 km^2, of which 800 km^2 are estuaries and lakes. On the territory of the peninsula are located: Temryuksky district, most of the Anapa resort region, partially Krymsky district. The peninsula is bordered by the Sea of Azov to the north and the Black Sea to the south. Almost all of its land is occupied by farmland.

The relief of the peninsula is peculiar: flat areas alternate with gentle ridges and cone-shaped hills. The highest mountain (Komendantskaya) has a height of 164 m above sea level. Land covers about 55% of the territory. The peninsula has a large number of mud volcanoes.

Geotectonically, the Taman Peninsula is a heterogeneous formation located at the junction of the mountain ranges of the Crimea and the Caucasus. This is precisely what determined the identification of several large first-order tectonic structures in the region: the closure of the meganticlinories of the Mountainous Crimea and the Greater Caucasus, the Indo-Kuban trough, the northwestern part of the platform side of the West Kuban forward trough, and the interpericlinal Kerch-Taman transverse trough. According to researchers, the total mass of sopochny sediments ejected by mud volcanoes reaches 40 billion m^3. In order to release such a mass of sopochny silt and breccia from the subsurface to the surface, it is necessary to eruption many trillions of cubic meters of deep gases. This indicates a huge potential energy stored in the bowels of the earth.

In the deep structure of the Taman Peninsula, five structural floors are distinguished: neogene; maikop; eocene-upper cretaceous; lower cretaceous; jurassic. The studied area of the Taman Peninsula, which includes a large mud-volcanic province, is located in the northwestern part of the platform side of the West Kuban forward trough. It began to form as a compensatory trough in the Oligocene before the growing massif of mountain-folded structures in the Northwestern Caucasus (Alpine folding zone). The thickness of the sedimentary cover in the trough in some places reaches 14 km. The area is confined to the Azov-Kuban oil and gas basin. The section contains clay strata of the maikop series with a thickness of up to 3–4 km. It is with them that the foci of volcanic activity in the Kerch-Taman region are associated.

Currently, the Taman Peninsula is the region with the greatest development of mud volcanoes in the Russian Federation. All known forms of mud volcanics occur here (continental and marine, buried and open, extinct and active). Many of them are highly active, located in close proximity to populated areas, so the mud volcanoes of the Taman Peninsula are currently considered as objects of excursion display.

There are about 50 mud volcanoes on the peninsula. They form hills up to 164 m high.

In fact, there are many more areas of volcanic activity:

1) areas of mud volcanism are quickly leveled by slope processes and are subsequently detected only during detailed drilling of wells;
2) some of the mud volcanoes are located in the marine zone.

Each volcano on the Taman Peninsula is a unique geological feature. There are mud lakes with silt deposits, underwater sea volcanoes, and land-based hills. The mud volcanoes of the Taman Peninsula look like real large volcanoes, only they are smaller.

Mud volcano – a geological formation that is a hole (or depression) on the earth's surface or a cone-shaped elevation with a crater (mud hill), from which mud masses, water and gases, often accompanied by oil, are constantly or periodically spewed to the Earth's surface.

As a rule, mud volcanoes are quite significant in size – at least a hundred meters.

The Taman Peninsula is home to the most diverse complex of mud volcanics.

There are several morphological classifications of mud volcanism of the Taman Peninsula, the most generalized of which includes:

1) the so-called «real» volcanoes, that is, having at the base of any hill composed of ordinary rocks;
2) bulk volcanoes (consisting entirely of hill breccia).

Classification with the allocation of a number of varieties (the allocation took into account both the features of the internal structure of mud massifs and the nature of their expression in the modern relief):

1. Powerful bulk mud volcanoes of the «iceberg» type, confined to the near-water zones of anticlinal structures, occupy significant areas and are characterized by the maximum capacity of sopochnaya breccia for the peninsula (the most characteristic representatives of this variety are the mud volcanoes of Goreloi Mountain, Karabetovaya Sopka Mountain, and others).

A distinctive feature of their internal structure is that the hill breccia not only has a completely visible mud-volcanic cone, but also opens up at a very significant depth (up to 150–200 m or more). under it. At the same time, sedimentary deposits involved in the structure of the arches of anticlinal structures to which the pseudovolcanic massif is confined form under the latter superimposed synclinal bends of compensatory genesis, due to the deflection of sediments over the zones of mud mass intake. This deflection increases as a result of the additional load created by a powerful mud mass from above, which increases as it accumulates. Such complications are commonly known as «depressed synclines».

2. Impaled mud volcanoes are a common variety and are most characteristic of the Taman Peninsula. They are generally confined to the near-axial zones of anticlinal structures and differ from the former by a much smaller thickness of hill breccias and the absence of compensatory depressions at their base (the most characteristic representatives of this variety are the mud volcanoes of Mount Shopurskaya and Mount Yanovskaya within the Fanagoria anticlinal ridge and others).
3. Reservoir mud-volcanic deposits.

4. Mud-volcanic rods are small in size, usually steeply falling, rounded or irregular hill bodies. They are usually confined to watershed zones and, sometimes, to the slopes of hilly ridge elevations (Central Taman Ridge, Tsymbaly Ridge, eastern part of the Fontalovskaya ridge). Their diameter is relatively small and does not exceed several tens of meters.

On the surface of active or recently extinct mud volcanoes, a whole complex of numerous small forms associated with their activity is observed.

Small craters, whose diameter can reach several meters, usually complicate the surfaces of crater plateaus of the first order and are rounded formations rounded by a shaft of hardened breccia. The interior of the active craters is filled with a silt mass of a liquid-flowing consistency, with the usual admixture of small fragments of solid sedimentary rocks; episodic water and gas occurrences are noted here.

Hill lentils (tortillas) and streams are usually flat, flat-shaped (on the subhorizontal surfaces of large craters) or elongated tongue-shaped (on more or less steep slopes) formations. The size of tortillas is up to 100–150 m, with a height of up to 1–1,5 m, they are composed of fluid-plastic «breccias».

The height and generally all flow parameters depend on the consistency of mud and vary widely. Sufficiently thin streams form on the slopes of volcanoes radially diverging from the crater downwards furrows of the barrancos type, stretching from the top to the foot of the volcanoes. On the slopes of large volcanoes, such forms are deepened by the erosion of temporary watercourses and sometimes turn into real ravines.

Small mud hills, which are parasitic centers of outpouring of hill breccia, have the appearance of small, regular cones. They are composed of solidified mud with numerous fragments of harder rocks (marls, limestones, siderite nodules, etc.). Their height reaches 6–10 m, less often more, the steepness of the slopes – up to 40–50°.

Salzas are underdeveloped, poorly defined cones filled with creamy mud, distributed mainly on the flat surfaces of crater plateaus. Their diameters do not exceed 10–12 m, the height is only 1–2 m, less often up to 4–6 m.

Griffins are small (no more than 2 m in diameter) foci of mud, water and gas release, in many cases also oil, and are widely distributed.

Lakes of mud-volcanic origin, filled with water or liquid mud, are found on the crater plateaus of Mount Karabetova Gora, Mount Miska, on Rotten Mountain. Lake baths are usually rounded or oval in shape and are surrounded by a rampart of hill breccia. The size of lake baths is very diverse-from 1 m to 40–50 m in diameter.

Currently, the Taman Peninsula is the region with the greatest development of mud volcanoes in the Russian Federation.

And since many of the volcanoes are located in close proximity to populated areas (for example, the volcano Akhtanizovskaya Sopka is located about 500 meters from the outskirts of the village of Akhtanizovskaya, the crater of the mud volcano Azovskoe Peklo is located 1.5 km from the outskirts of the village of Kuchugury, the crater field of the mud volcano Shugo is 5 km from the village of Varenikovskaya, the volcanoes Rotten Mountain and Miska are located in the center of Temryuk) and show high activity (some mud volcanoes have year-round eruptions of sopochnaya breccia, sopochnaya waters and gases) – they are unique geological objects of the excursion show.

That is why in July 2022 in the Temryuksky district of the Krasnodar Territory on an area of 1500 hectares was created «Natural Park «Volcanoes of Taman».

REFERENCES

Albov, S. V. (1956). *Mineral springs of Crimea*. Krymizdat Publishing House.

Albov, S. V. (1971). On the Kerch-Taman hydrogeochemical and mud volcanic region. [USSR Academy of Sciences Publishing House.]. *Reports of the USSR Academy of Sciences*, *197*(1), 175–177.

Albov, S. V. (1973). On the geochemical environment in the Kerch-Taman region and nearby areas. [USSR Academy of Sciences Publishing House.]. *Reports of the USSR Academy of Sciences*, *208*(1), 184–187.

Ali-zade, A. A., & Tsimmelzon, I. O. (1966). Deep structure of Azerbaijan. Geotectonics, (3), 26–34. Geological Institute of the USSR Academy of Sciences Publishing House.

Andrusov, N. I. (1926). *Geological structure and history of the Kerch Strait. Bulletin of the Moscow Society of Naturalists. Geological Department, (3/4), 294–332.* GosIzdat Publishing House.

Arkhangelsky, A. D.. (1930). *Brief outline of the geological structure of the oil fields of the Kerch Peninsula. Works of the Main Geological Directorate of the People's Commissariat of Heavy Industry of the USSR, (13), 146.* Main Geological Directorate Publishing House.

Avdusin, P. P. (1939). On the petrography of the eruption products of mud volcanoes of the Crimean-Caucasian geological province. In *Results of the study of mud volcanoes of the Crimean-Caucasian geological province* (pp. 57–66). Academy of Sciences of the USSR Publishing House.

Avdusin, P. P. (1948). *Mud volcanoes of the Crimean-Caucasian geological province*. Academy of Sciences of the USSR Publishing House.

Avdusin, P. P. (1948). *Mud volcanoes of the Crimean-Caucasian geological province*. Academy of Sciences of the USSR Publishing House.

Beloussov, V. V., & Yarotsky, L. A. (1936). *Mud hills of the Kerch-Taman region, conditions of their occurrence and activity*. Main Editorial Board of Geological Exploration and Geodetic Literature.

Beloussov, V. V., & Yarotsky, L. A. (1936). *Mud volcanoes of the Kerch-Taman region, conditions of their occurrence and activity*. United Scientific and Technical Publishing House of the People's Commissariat of Heavy Industry of the USSR.

Blagovolin, N. S. (1960). *Origin and history of the development of the Kerch Strait. Bulletin of the USSR Academy of Sciences. Geographical Series, (2), 105–109.* USSR Academy of Sciences Publishing House.

Burkser, E. S., & Bronstein, K. G. (1933). Gases of Crimea and the Kerch Peninsula. In Natural gases of the USSR (pp. 2–12). Goskhimtekhizdat Publishing House, (13).

Fedorov, A. Y. (1931). *Geological description of the Dzhiginsko-Varenikovsky oil-bearing region: Report on the work of 1928–1929. Works of the Oil Geological Prospecting Institute, (4), 34.* Main Editorial Board of Mining and Fuel Literature.

Gemp, S. D., Dubrova, N. V., & Nesmelova, Z. N. (1970). *Isotopic composition of carbon in carbon-containing gases (CH4 and CO2) of mud volcanoes of the Kerch-Taman region. Geochemistry, (2), 243–247.* USSR Academy of Sciences Publishing House.

Gemp, S. D., & Lagunova, I. A. (1978). Relationship of mud volcanism with endogenous processes. In The influence of endogenous factors on the formation of oil and gas deposits (pp. 75–97). Transactions of the All-Union Oil Research Geological Prospecting Institute (VNIGRI). State Scientific and Technical Publishing House of Oil and Mining and Fuel Literature.

Glinsky, B. M., Sobisevich, A. L., & Khairetdinov, M. S. (2007). Experience of active monitoring of complex geological structures (using the Shugo mud volcano as an example). [Nauka Publishing House.]. *Reports of the Academy of Sciences*, *413*(3), 398–402.

Gorbatikov, A. V., Sobisevich, A. L., & Ovsyuchenko, A. N. (2008). Development of the model of the deep structure of the Akhtyr flexural-fault zone and the Shugo mud volcano. [Nauka Publishing House.]. *Reports of the Academy of Sciences*, *421*(5), 670–674.

Grechishkin, L. A. (1936). Study of oil fields in the northern and central parts of the Kerch Peninsula. In Geological exploration studies in the oil-bearing regions of the Kerch Peninsula in 1926 (pp. 49–71). Works of the Main Geological Directorate of the People's Commissariat of Heavy Industry of the USSR, (39). Geologizdat Publishing House.

Kalinko, M. K. (1964). *Basic patterns of oil and gas distribution in the earth's crust*. Nedra Publishing House.

Khain, V. E., Zhabree, Y. Y., & Shapiro, A. R.. (1953). *Types of tectonic and mud volcanic breccias of Kobystan. Azerbaijan Oil Industry, (6), 5–7.* Azneft Publishing House.

Kholodov, V. N. (2002a). *Mud volcanoes: Patterns of placement and genesis. Mud volcanic provinces and morphology of mud volcanoes. Lithology and Useful Minerals, (1), 227–241.* Nauka Publishing House.

Kholodov, V. N. (2002b). On the nature of mud volcanoes. Priroda, (11), 47–58. Nauka Publishing House.

Kudryavtsev, Y. A. (1963). *On the quantitative relationships between hydrocarbons of sedimentary rocks and oil. Geology of Oil and Gas of the USSR, (9), 45–49.* State Scientific and Technical Publishing House of Oil and Mining and Fuel Literature.

Kurishko, V. A., Mesyats, I. A., & Terdovidov, A. S. (1968). Hydrogeology of mud volcanism of the Kerch Peninsula. [Naukova Dumka Publishing House.]. *Geological Journal*, *28*(1), 49–59.

Lagunova, I. A. (1974). *On the genesis of CO2 in gases of mud volcanoes of the Kerch-Taman region. Geochemistry, (11), 1711–1716.* Institute of Geochemistry and Analytical Chemistry, Russian Academy of Sciences Publishing House.

Lagunova, I. A., & Gemp, S. D. (1978). *Hydrochemical features of mud volcanoes. Soviet Geology, (8), 108–124.* Nedra Publishing House.

Lukashuk, A. I. (1933). *Gases of the Kerch volcanoes. Mineral spring Adzhi-Su. Results of analyses. Natural Gases of the USSR, (13), 11–18.* Goskhimtekhizdat Publishing House.

Naumenko, P. I. (1963). Modern activity of mud volcanism of the Kerch Peninsula. In Erofeev, B. N., & Shcherbakov, D. I. (Eds.), *Materials on mineralogy, petrography and geochemistry of sedimentary rocks and ores* (p. 459). USSR Academy of Sciences Publishing House.

Nesterov, K. Y. (1979). Geochemistry of mineral waters of the western part of the Indolo-Kuban trough. In *Geochemistry of Landscapes and Groundwater* (pp. 80–98). North Caucasian Scientific Center Publishing House.

Ovsyuchenko, A. N. (2005). *Manifestations of young tectonic activity in the structure of mud volcanoes of Taman and the North-West Caucasus. Modern Methods of Geological and Geophysical Monitoring of Natural Processes in the Territory of the North Caucasus*. Institute of Physics of the Earth RAS Publishing House.

Paffenholz, K. N. (Ed.). (1973). *Geological dictionary: In 2 volumes*. Nedra Publishing House.

Popov, S. Y. (1938). *Mineralogy of Crimea*. Academy of Sciences of the USSR Publishing House.

Prokopov, K. A. (1931). *Complications in the anticlines of the Kerch Peninsula. Works of the Main Geological Directorate of the People's Commissariat of Heavy Industry of the USSR, (38), 37–83*. Main Geological Directorate Publishing House.

Prokopov, K. A. (1931). *Complications in the anticlines of the Kerch Peninsula. Works of the Main Geological Administration of the People's Commissariat of Heavy Industry of the USSR, (38), 37–83*. Geologizdat Publishing House.

Shardanov, A. N., Malyshek, V. G., & Peklo, V. P. (1962). On the roots of mud volcanoes of the Taman Peninsula. Transactions of the All-Union Research Institute of Oil, (10), 53–66. VNIIneft Publishing House.

Shatskiy, N. S., Zhukov, M. M., & Milanovsky, E. V.. (1929). *Dislocation breccias and mud volcanoes in Azerbaijan. Bulletin of the Moscow Society of Naturalists. Geological Department, (1/2), 97–161*. GosIzdat Publishing House.

Shnyukov, E. F., Naumenko, P. I., & Lebedev, Y. S.. (1971). *Mud volcanism and ore formation*. Naukova Dumka Publishing House.

Shnyukov, E. F., Sheremetyev, V. M., & Maslakov, N. A.. (2006). *Mud volcanoes of the Kerch-Taman region*. GlavMedia Publishing House.

Shnyukov, E. F., Sobolevsky, Y. V., & Gnatenko, G. I.. (1986). *Mud volcanoes of the Kerch-Taman region*. Naukova Dumka Publishing House.

Shternov, A. G. (1968). Geological conditions of formation of natural gas deposits in the Kerch mud volcanic region. *Abstracts of Reports of the IV Scientific Conference on the Study of Useful Mineral Deposits of the Sedimentary Cover of the South of* Ukraine. Kyiv University Publishing House.

Slavyanov, Y. Y. (1928). *On some little-known mineral springs of the Kuban region. Materials on General and Applied Geology, (82), 15–25*. Geological Committee Publishing House.

Sobisevich, A. L., Sobisevich, L. E., & Tveritinova, T. Y. (2014). On mud volcanism in the late Alapian folded structure of the Northwestern Caucasus (based on the study of the deep structure of the Shugo mud volcano). [State Scientific Institution Department of Marine Geology and Sedimentary Ore Formation of the National Academy of Sciences of Ukraine.]. *Geology and Useful Minerals of the World Ocean*, *2*(36), 80–93.

Subbota, M. I. (1964). *New data on gases and types of mud volcanoes in the Western Kuban region. Works of the All-Union Oil Research Geological Prospecting Institute, (41), 156–178*. State Scientific and Technical Publishing House of Oil and Mining and Fuel Literature.

Sulin, V. A. (1939). Objectives and some results of hydrogeological and geochemical study of mud volcanoes of the Crimean-Caucasian province. In *Results of Studies of Mud Volcanoes of the Crimean-Caucasian Geological Province* (pp. 93–102). Academy of Sciences of the USSR Publishing House.

Sungurov, A. M. (1958). *Features of the structure of mud volcanoes in Southwestern Turkmenistan. News of Oil Technology. Geology Series, (9), 12–14*. Central Research Institute for Technical Information and Economics of the Oil Industry Publishing House.

Zaitsev, A. V. (1965). Mud volcanoes of the Azov region: Abstract of the dissertation of a candidate of geological and mineralogical sciences. Rostov State University Publishing House.

Zakharchenko, E. I., & Korneva, O. R. (2023). Mud volcanism as an object of study in educational geophysical practice. Proceedings of the II All-Russian Scientific and Practical Conference "Student Practice: From Educational to Pre-Graduation", 87–90. Kuban State University Publishing House.

Chapter 17
Double Jeopardy:
Unveiling the Synergistic Impacts of Climate Change and Tourism on Biodiversity Loss

Suraj Jaywant Yadav
https://orcid.org/0000-0002-2687-4852
D.Y. Patil Education Society, Kolhapur, India

ABSTRACT

This study investigates the synergistic effects of climate change and tourism on biodiversity loss, focusing on the complex interactions between these two factors and their collective impact on various ecosystems. The research examines the role of tourism in exacerbating environmental degradation and the additional pressures placed on biodiversity by changing climate conditions. By analyzing data from different regions and ecosystems, the study identifies critical areas where the intersection of tourism and climate change is most pronounced. The findings highlight the urgent need for integrated conservation strategies that address both climate change mitigation and sustainable tourism practices. Through a detailed examination of various case studies and empirical data, this research contributes to the understanding of how combined human activities threaten biodiversity and offers recommendations for policy and management interventions to mitigate these impacts.

1.INTRODUCTION

Numerous ecosystems' biodiversity and structural integrity are being impacted by climate change. This chapter focuses on how it affects forests. We will concentrate on the potential effects of climate change on forest structure and biodiversity and how these effects could affect the ecosystem services that forests offer(Imbert et al., 2021). As the most significant global environmental and development concern of our time, climate change has become a hot topic in the field of tourism studies. Tourism in the twenty-first century is expected to be significantly impacted by climate change, which is already having an impact on the sector. These effects include consumer holiday choices, demand for tourism in specific geographic areas, destination sustainability and competitiveness, and tourism's role in global development(D. Scott, Hall, et al., 2012). Natural capital such as biodiversity and ecosystems support

DOI: 10.4018/979-8-3693-8232-5.ch017

human society through cultural, regulating, provisioning, and other functions, and they are essential to maintaining human well-being. However, ecosystems and biodiversity are facing widespread and grave global threats due to climate change. According to the sixth assessment report of the United Nations Panel on Climate Change (IPCC), the average temperature change predicted for the next 20 years indicates that the global temperature rise will reach or surpass 1.5 °C; this means that the global surface temperature will continue to rise until the middle of this century at the latest(Yang et al., 2021). As environmental change occurs, people must negotiate the trade-offs between various ecosystem contributions to adaptation, or adaptation services (AS), and can maximize their synergies and co-benefits along the way. Therefore, analyzing how people co-produce AS is necessary to understand trade-offs and cobenefits of AS and to support social adaptation(Lavorel et al., 2020). Aesthetic values, cultural heritage, spiritual and religious values, educational values, social relations, tourism and recreation, and ecotourism are examples of cultural services that will disappear. It has become essential to comprehend how climate change affects ecosystem services in order to adapt and mitigate its effects. Therefore, ecosystem management that is appropriate, conservation, protection, and restoration are necessary for both adaptation and mitigation of the effects of climate change(Bakure et al., 2022). It is vital and difficult to manage many ecosystem services (ESs) in a win-win way. Unfortunately, especially in mountainous regions, our knowledge of the spatial interactions among ESs is still rather limited(Lin et al., 2018). It is imperative to comprehend the ecological dynamics underlying these climate impacts, pinpoint hotspots of susceptibility and adaptability, and identify management strategies that could bolster the biosphere's ability to withstand climate change. Ecosystems can also help with climate change adaptation and mitigation at the same time(Malhi et al., 2020). When multiple factors act simultaneously and have a larger overall effect than the sum of their individual effects, this is referred to as having a synergistic effect. The specific roles of various drivers are frequently overlooked, and our understanding of how synergistic effects arise is still rather limited. Therefore, in order to reduce the loss of biodiversity, it becomes extremely difficult to manage several dangerous processes at once(Chen et al., 2015). Numerous people could be impacted by one climate-related phenomenon. Because of the indirect economic effects of climate change, the direct costs related to climate-related factors also affect many other climate-related costs(Halkos & Tsilika, 2014). Over time, intermediate and final ecosystem goods and services are generated by ecosystems, which can be thought of as a fund of natural capital stocks that are disrupted by the aforementioned scenarios. Resources including biotic, geologic, water, atmosphere, and land resources are included in these natural capital stocks, both renewable and non-renewable(Rinawati et al., 2013). Building practical solutions and integrated institutional capacity in developing nations to address climate change may start with acknowledging how other development priorities are likely to be impacted by climate change. Additionally, developing nations may be able to take advantage of synergies between policies addressing climate change and others, like those aimed at preventing desertification and protecting biodiversity, by utilizing regional economic organizations to help with the design of integrated responses(Beg et al., 2002). The basis for intervention cannot and should not be limited to the quantifiable effects of tourism on biodiversity(van der Duim & Caalders, 2002). Because forests are essential for maintaining the natural habitat of plants and animals, they store billions of tons of carbon, act as a buffer against floods and droughts, stabilize soils, impact climate change, and provide food and shelter for indigenous people, the world community has decided to safeguard them from future destruction. Additionally, local communities rely heavily on sustainable tourism as a source of employment and income, which strongly encourages biodiversity protection(Din et al., 2014). Aviation, rural and ecotourism, the destination's isolation from cities and other well-known destinations, the degree of animal-human interaction growing, environmental change (e.g. G. roadway construction), as well as interactions between people and disease-carrying organisms (e.g. G. both mosquitoes and flies). Baggage, food, clothing, shoes, and other items, as well as vehicles, are common biohazard vectors and pathways. Managing biological invasions and other biosecurity issues are covered in this chapter as well(Hall, 2019).

2. RESEARCH OBJECTIVES

2.1. To Analyze the Combined Effects of Climate Change and Tourism on Habitat Destruction and Biodiversity Loss.
2.2. To Evaluate the Impact of Tourism Activities on Marine Ecosystems in the Context of Climate Change.
2.3. To Investigate the Challenges of Wildlife Conservation in Tourist-Dominated Regions Affected by Climate Change.
2.4. To Identify and Assess Mitigation and Adaptation Strategies for Minimizing the Synergistic Impacts of Tourism and Climate Change on Biodiversity.
2.5. To Develop a Framework for Integrating Biodiversity Conservation into Tourism Planning and Climate Policy.

3. THEORETIC REVIEW

Both direct and indirect goods and services—such as crop pollination, waste management, waste water and air purification, fish and game—are provided by ecosystems to humans. Without these services, life as we know it would not be possible(Lopes et al., 2015). When ecological structure and functions change, ecosystems can support social adaptation to environmental change by shielding people from the effects of climate change and offering options for maintaining both material and intangible benefits(Lavorel et al., 2020). Many theories concerning the definitions, effects of introduction, damage causes, population dynamics, mode of propagation, prevention, adaptability, and effects of invasive species on native species—both directly and indirectly—by changing the dynamics of ecosystems have been proposed(Chakraborty, 2019). Ecosystems are being impacted by climate change due to variations in average weather and climate variability, as well as other related changes like elevated atmospheric carbon dioxide levels and ocean acidification. Additionally, it interacts with other stresses on ecosystems, such as fragmentation, defaunation, and degradation(Halkos & Tsilika, 2014). One of three possible outcomes for a species impacted by climate change is to adapt, relocate, or perish. A rapidly deteriorating ecosystem or local species extinctions may approach their respective "tipping points," depriving human society of their services and potentially resulting in a global crisis. A variety of scenarios involving the impacts of climate change on biodiversity, particularly in tropical regions, require the consideration of appropriate and urgent action(Rinawati et al., 2013). Stress is already being placed on ecosystems' ability to provide vital services to humanity. Significant adaptation will be needed to handle the additional stresses brought on by climate change in the upcoming years. In order to maintain and restore robust biological and social systems, we must monitor the state of ecosystems as they change, expand our knowledge of the biological principles underlying ecosystem service delivery, and create new instruments and methods(Yang et al., 2021). Developing nations' economic and environmental policy agendas do not currently place a high priority on climate change. However, data indicates that developing nations will experience some of the worst effects of climate change, as their populations are the most susceptible and unlikely to adapt to it readily. Additionally, climate change will have an impact on these nations' ability to develop(Beg et al., 2002). The conservation and decline of biodiversity are influenced by tourism. In addition to outlining some of the key themes in the relationship, the paper emphasizes the need for additional research on biological invasion, extinction and habitat loss, the impact of tourism on biodiversity, climate change, and conservation tactics(Hall, 2010). A significant feature of climate change is the erosion of ecological integrity caused by the development of more and more landscapes, which leaves a patchwork of intact refuges and degraded areas that might not be enough to sustain biodiversity. While changes in the distribution and growth of the human population have an impact on land use, transient human migration can have significant effects on biodiversity and conservation(Burger, 2000). Travelers' recreational activities both inside and outside of lodging structures, as well as tourism-related transportation and activities carried out at the level of those structures, can all negatively impact the environment, depending on how intensely they occur over time and space. We outline the effects of tourism-related factors on the environment below(tefănică et al., 2021).

Figure 1. Cascading effect of Climate Change(Danylchuk et al., 2023)

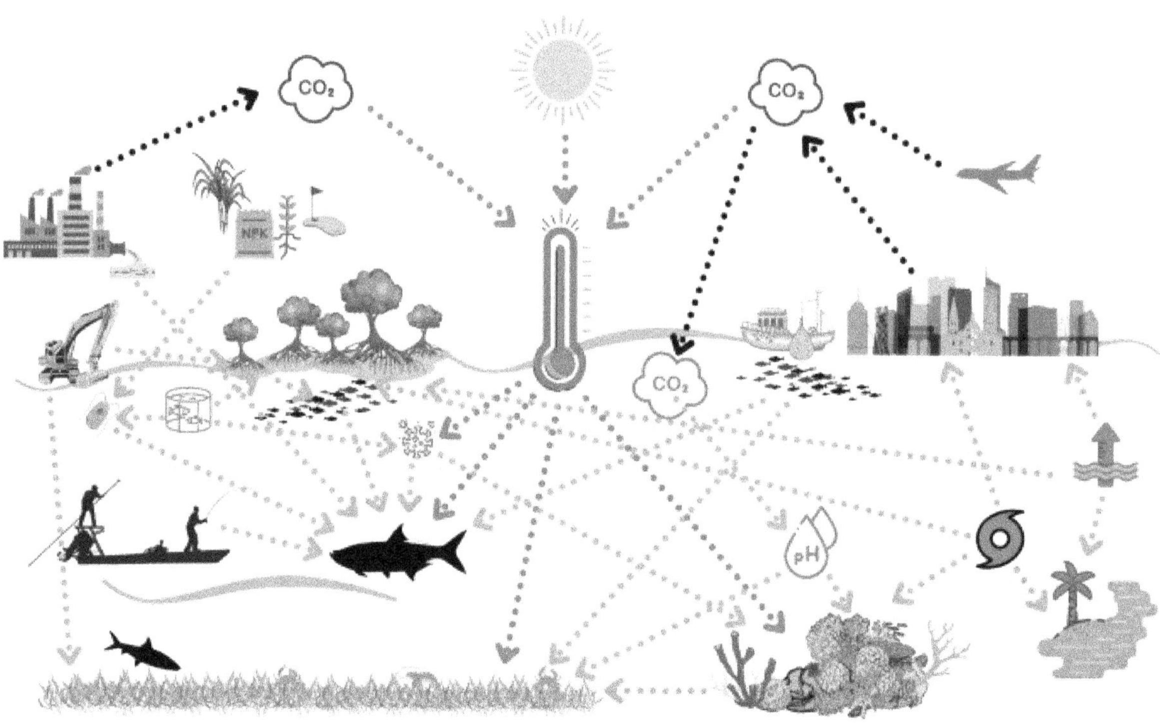

3.1 Climate Change and Habitat Destruction: An Overview

Two of the biggest threats to biodiversity worldwide are habitat destruction and climate change. The response of hypothetical species with various traits to habitat loss has been studied using lattice models(Travis, 2003). Patch occupancy stays high up to a critical rate of climate change, beyond which species extinction becomes likely. This threshold for the rate of climate change is similar to that found for habitat loss. The least able to adapt to climate change are specialists in particular habitats, particularly those with comparatively low colonization rates. The potential consequences of combining habitat loss and climate change could be catastrophic. The habitat threshold happens earlier due to climate change. Similarly, in a fragmented habitat, species are more negatively impacted by climate change(Travis, 2003). Empirical changes in habitat connectivity for certain species have also been brought about by climate-related changes in movement behavior and the quantity, quality, and distribution of habitat. The effects of climatic processes on habitat dynamics on species that inhabit dynamic habitats is a topic that has not received enough attention(Holyoak & Heath, 2016). Two major threatening factors contributing to the global decline in biodiversity are habitat loss and climate change. The intricacy of both processes means that little is known about their synergistic effects on biological populations. The effectiveness of current conservation management strategies may be compromised if the combined effects of habitat loss and climate change outweigh the effects of each threat separately(Mantyka-pringle et al., 2012). Because of a variety of activities, including deforestation, urbanization, roads, the energy sector (coal and renewable), mining, and climate change, humans have a negative influence on natural habitat(Scanes, 2018). These ecological effects are connected to both local and global species extinctions as well as recent population declines. Even though it is impossible to link these ecological effects to climate change, these findings have significant conservation biology ramifications. It is no longer reasonable to believe that a species' historical range still applies to its whole range. These studies highlight the significance of climate change as a contemporary threat to species and the necessity of taking climate change into account in both in situ conservation and reintroduction efforts(McCarty, 2001). The fragmentation of habitats and climate change are two major factors affecting biodiversity. We investigate the possible synergistic effects between these factors in this paper. We contend that interactions occur between processes operating at the species range level and the metapopulation level of spatial scale(Opdam & Wascher, 2004). Loss of biodiversity is mostly caused by human expansion into natural habitats. Original flora and fauna communities in wooded areas are being disrupted by this process of transformation, which signifies a significant change. Many animals, particularly parrots (Psittaciformes), whose extinction is a threat to nearly one-third of their species, rely heavily on forests(Vergara-Tabares et al., 2020).

3.2 Tourism-Driven Environmental Degradation

Although it is a double-edged sword, the contribution of tourism to economic development is widely acknowledged. Degradation of the environment in the absence of necessary environmental protection measures is the opportunity cost for this substantial share(Ahmad et al., 2019). As environmental degradation worsens over time, there will be a long-term negative impact on tourism demand, making this relationship significant(Amir et al., 2022). Long-term economic growth is made possible by ecotourism, but there is a social cost to the welfare of the current generation. Lastly, the presence of unbalanced effects between environmental quality and tourism infrastructures offers fresh perspectives on the dynamics of transitional tourism destinations(Ouattara et al., 2019). The environment is seen by the tourism industry as a resource and an opportunity rather than a limitation. Not only are tourism and the environment connected, but they also depend on each other. Instead of being at odds with environmental preservation, the viability of tourism actually depends on it; otherwise, visitor satisfaction will suffer as the natural allure of the tourism destination is undermined(Pigram, 1980). One crucial component of business plans to preserve and improve destination competitiveness is a focus on "tourism yield.". The concept of "tourism yield" ought to encompass not only the economic benefits of tourism to a destination, but also its environmental and social implications(Lundie et al., 2007). Environmental management is divided into four categories: self-developed environmental practices, ethical codes of conduct, best practices that have been recognized or awarded, and accreditation programs. Evaluation is done on their value for competitiveness and environmental destination management(Mihalič, 2000). The primary financial gains from tourism, its effects on the environment, and the evolution of the global agenda for sustainable tourism. This article will concentrate on the (economic) growth of the tourism industry in developing nations, even though the developed world accounts for a large portion of global tourism activity(Neto, 2003).

3.3 The Intersection of Climate Change and Tourism on Marine Ecosystems

The main contention is that, with the onset of global climate change, long-term adaptation strategies must take into account both the material and social aspects. The results provide insight into the challenges associated with adapting tourism destinations to changes in geomorphology and climate, as well as the contested meanings that result from interactions between humans and their environment(Buzinde et al., 2010). Caribbean small island developing states' coastal and marine environments are extremely susceptible to both present-day and long-term climate change. Societies negotiate these shifts in part because of their institutional frameworks and governance procedures. The notion of institutional adaptive capacity is employed to investigate the potential for modifying governance procedures and institutional configurations to align with the magnitude and scope of climate change(Pittman et al., 2015). Millions of tourists visit coastal and marine environments each year, and tourism plays a major role in the economies of many regions, such as the Mediterranean and the Caribbean. The tourism industry depends on the integrity of coastal resources, such as clean beaches and waters, because a significant portion of travel is focused on the "sun, sand, and sea."(Gössling et al., 2018). There is a significant degree of variability in the results, with estimates of the socioeconomic and physical effects of climate change varying greatly amongst destinations and methodologies. Furthermore, most recent research has focused on a small number of the most significant effects; therefore, future studies should focus on overlooked indicators and relationships, which are crucial for creating efficient climate policies at tourist destinations(Arabadzhyan et al., 2021). Numerous risks related to climate change affect coastal and marine tourism. Destinations may suffer greatly from the deterioration of marine ecosystems, particularly if marine activities are the primary draw. Assessing the risk of marine habitat degradation to coastal and marine tourism due to climate change is the goal of this study(González Hernández et al., 2023). Oceanic effects of human-caused climate change are already apparent from the equator to the poles and from the surface to the deepest recesses. Although more effects of climate change are unavoidable, they can be mitigated to a great extent if greenhouse gas emissions are quickly cut. This will protect marine ecosystems and organisms as well as the services they provide(Williamson & Guinder, 2021).

Figure 2. Synergies Between Climate Change, Biodiversity, Ecosystem Function and Services, Indirect Drivers of Change and Human Well-Being in Forests(Imbert et al., 2021)

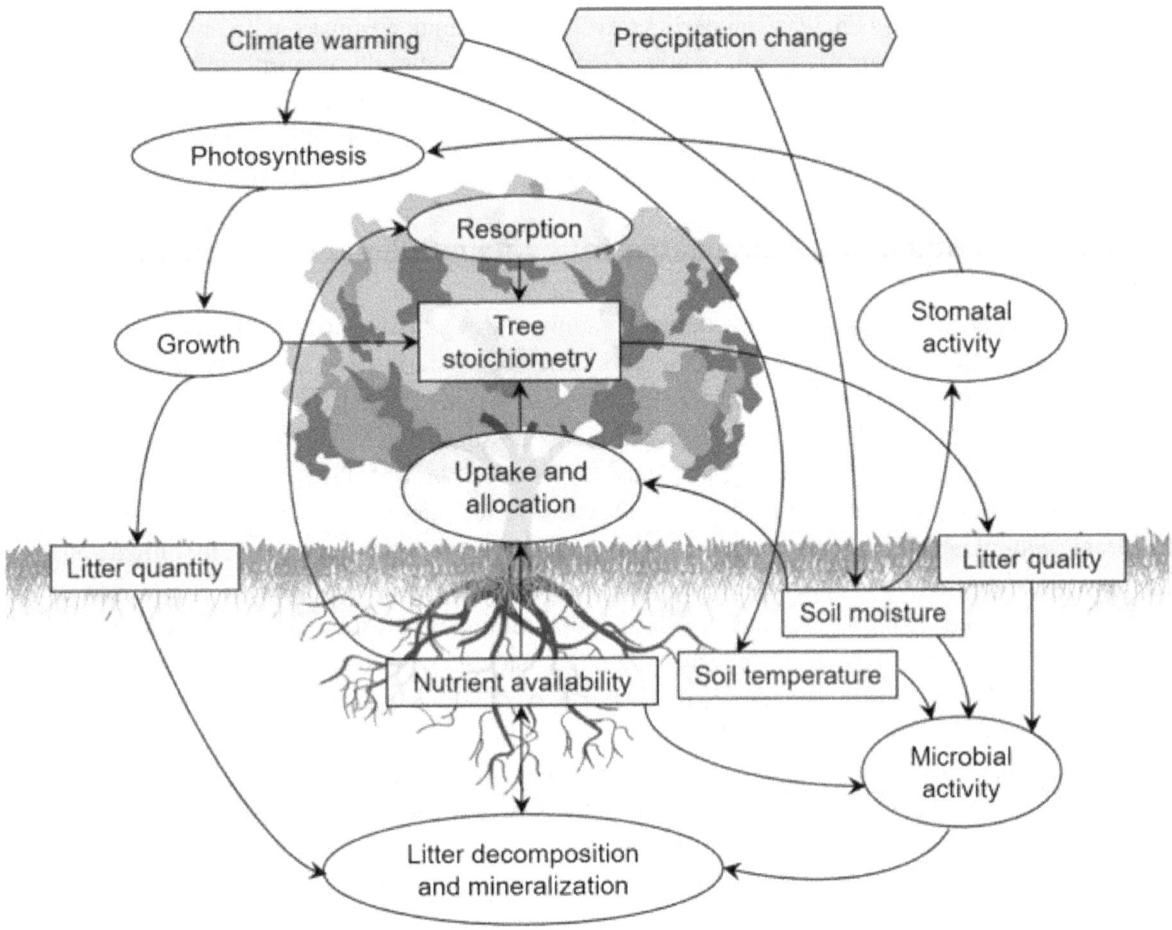

3.4 Wildlife Conservation Challenges in Tourism-Heavy Regions

One of the main causes of the decline in biodiversity worldwide is biological invasions, which also compromise the structure and functionality of ecosystems. A major scientific project at the intersection of conservation biology and ecology is to underpin the mechanisms and patterns of biological invasions, specifically how exotic invasions and native-weed expansion differ across different habitats and variable degrees of human-induced disturbances(Aththanayaka et al., 2023). The idea of sustainable tourism is closely related to the aspects of the economy, environment and socio-cultural. In short, sustainable tourism aims to economically empower the people in potential tourism areas without disturbing the balance of natural biodiversity and the established social relations among them. Along with sustainable tourism development, governments are also eager to speed up the use of renewable energy in tourism sites(Rudiany et al., 2023). In recent years, zoos and aquariums have changed their emphasis, becoming much more involved in the conservation of wildlife and encouraging visitors to learn about conservation. Studies conducted in these environments offer a significant basis for the developing domain of non-captive wildlife tourism(Ballantyne et al., 2007). When illegal activities are becoming the backdrop for wildlife tourism, it is imperative that innovative technologies be applied to safeguard and protect wildlife(Hassan & Sharma, 2017). Comparing travel to last-minute destinations to other domestic and international travel, there are two main findings: first, new philanthropic support is influenced by other important factors such as first-time visitation and/or a stronger sense of identity with nature. In addition to its scholarly contributions, this study offers practical recommendations for promoting charitable behavior when collaborating with non-profits and tour operators(Hehir et al., 2023).

3.5 Mitigation Strategies: Sustainable Tourism and Climate Adaptation

The use of strategies that can quickly reduce vulnerability and eventually achieve sustainable development is required due to the climate change vulnerability of beaches, coastal tourism facilities, and communities that depend on tourism(Mycoo, 2014). This opinion piece contends that tourism's expanding engagement with climate change, as it is currently unfolding, is not necessarily conducive to the interests of tourism sustainability. Inherent unpredictability, long-term timeframes, lack of directly tangible consequences or clearly identifiable villains, issues with credibility and vested interests and cost implications in an era of chronic economic uncertainty all combine to increase the likelihood of unsuccessful climate change policies and strategies(Weaver, 2011). To adjust in a way that is sustainable in terms of the environment, society, and economy to climate change. In addition to outlining the various uses of meteorological and climate data in the travel and tourism industry, this paper makes recommendations for future research on climate services that will help the industry manage weather risk and adapt to climate change(D. J. Scott et al., 2011). Not only does tourism contribute to climate change, but it also puts travelers at risk from the shifting weather patterns. Among the many direct and indirect effects that could endanger the tourism sector are sea level rise, extreme weather, biodiversity loss, and water stress(Njoroge, 2014). Weaver's argument that the growing involvement of sustainable tourism in addressing climate change may not always be in the best interests of sustainable tourism is examined in this opinion piece. The opinion put forth by Weaver is supported by a critical examination and response to seven interconnected issues(D. Scott, 2011). As one of the most climate-sensitive industries, tourism is not an exception when it comes to policymakers' biggest concerns in the twenty-first century, both at the national and international levels. To create a variety of useful policy recommendations, tourism researchers have persisted in examining the connection between climate change and sustainable tourism(Farid et al., 2016). In order to differentiate their business as "climate friendly" and maintain business competitiveness, tourism operators are primarily motivated to implement these carbon actions due to their ecological responsibility and desire to save costs. These goals coincide with the overarching business concepts that underpin businesses' adoption of environmentally and socially conscious practices(Zeppel & Beaumont, 2014). Because of the tourism system, all locations must adjust to the opportunities and risks presented by climate change and climate policy(D. Scott, Gössling, et al., 2012).

4. DATA COLLECTION

Data collection involved a mixed-methods approach, utilizing both quantitative and qualitative data from multiple sources(S. Yadav, 2021a). Respondents included local community members, tourists, environmental experts, and wildlife conservationists from regions heavily impacted by tourism and climate change. Surveys and structured interviews were conducted to gather firsthand information on observed changes in biodiversity and the perceived impacts of tourism and climate-related factors. Archival data from environmental reports and tourism statistics were also reviewed. Several tests were conducted to analyse the data: regression analysis was used to examine the relationship between tourism intensity and biodiversity loss; ANOVA tests (S. J. Yadav, 2024b) compared biodiversity indicators across different regions with varying levels of tourism; and thematic analysis was employed to identify recurring themes and insights from the qualitative data(S. Yadav, 2021b). This comprehensive approach ensured a robust analysis of the synergistic impacts of climate change and tourism on biodiversity.

5. DATA ANALYSIS

The data presented in this study provides a comprehensive analysis of the synergistic impacts of climate change and tourism on biodiversity loss across various regions. Through a combination of demographic surveys, quantitative measurements, and qualitative assessments, the study examines the relationship between increasing tourism activities(S. Yadav, 2021a), climate changes, and the subsequent effects on local ecosystems. It explores key factors such as changes in species diversity, the effectiveness of existing mitigation strategies, and public perceptions of environmental issues(S. J. Yadav, 2024b). The findings highlight significant regional differences in biodiversity indicators, the severity of climate change impacts, and the effectiveness of policy measures, offering critical insights into the challenges and opportunities for sustainable environmental management.

Table 1. Descriptive Statistics of Respondent Demographics

Variables	N	Mean	SD	Min	Max
Age	150	34.5	10.2	18	65
Gender (Male %)	150	55%			
Occupation (Tourism-Related %)	150	35%			
Residence Duration (Years)	150	12.3	8.4	1	40

This table provides an overview of the demographic characteristics of the respondents who participated in the study(D. Scott, Gössling, et al., 2012). The average age of the respondents was 34.5 years, with a standard deviation of 10.2 years. The youngest respondent was 18 years old, while the oldest was 65 years old. The gender distribution showed that 55% of the respondents were male. Regarding occupation, 35% of the respondents were engaged in tourism-related professions. The average duration of residence in the study regions was 12.3 years, with a range from 1 to 40 years. This demographic data helps contextualize the perspectives and experiences reported by the respondents(S. J. Yadav, 2024a).

Table 2. Tourism Intensity by Region

Region	Number of Tourists (Annual)	Tourism Revenue (Million USD)	Tourism Growth Rate (%)
Region A	500,000	150	5.2
Region B	300,000	90	3.8
Region C	750,000	225	6.5

This table illustrates the level of tourism activity across different regions, highlighting the number of tourists, tourism revenue, and the annual growth rate in tourism. Region A received approximately 500,000 tourists annually, generating 150 million USD in revenue and experiencing a growth rate of 5.2%. Region B saw around 300,000 tourists, bringing in 90 million USD with a 3.8% growth rate. Region C was the most visited, attracting 750,000 tourists, generating 225 million USD, and growing at a rate of 6.5%. These figures indicate the varying degrees of tourism development and economic impact across the regions.

Table 3. Biodiversity Indicators by Region

Region	Species Richness (Number of Species)	Endemic Species (%)	Threatened Species (%)
Region A	250	15	10
Region B	180	12	14
Region C	300	20	8

This table presents key indicators of biodiversity, such as species richness, the percentage of endemic species, and the percentage of threatened species in each region. Region A was home to 250 species, with 15% being endemic and 10% classified as threatened. Region B contained 180 species, with 12% endemic and 14% threatened species. Region C exhibited the highest biodiversity, with 300 species, 20% of which were endemic, and 8% threatened. These indicators help assess the biodiversity health and conservation needs of each region.

Table 4. Changes in Average Temperature Over the Past Decade (°C)

Region	2005-2010 Average	2011-2015 Average	2016-2020 Average	Change
Region A	15.2	15.5	16	0.8
Region B	22.1	22.4	22.9	0.8
Region C	18.4	18.6	19.1	0.7

This table tracks the changes in average temperatures in each region over three periods: 2005-2010, 2011-2015, and 2016-2020. Region A experienced an increase in average temperature from 15.2°C to 16.0°C, a total rise of 0.8°C over the decade. Similarly, Region B saw an increase from 22.1°C to 22.9°C, also a rise of 0.8°C. Region C had a temperature increase from 18.4°C to 19.1°C, marking a 0.7°C rise. These temperature changes are indicative of the broader climate change trends affecting these regions.

Table 5. Perceived Impact of Climate Change on Local Biodiversity (Survey Results)

Impact Level	Frequency (N)	Percentage (%)
No Impact	10	6.7
Mild Impact	40	26.7
Moderate Impact	60	40
Severe Impact	40	26.7

This table summarizes the respondents' perceptions of the impact of climate change on local biodiversity. A small percentage of respondents, 6.7%, felt that there was no impact, while 26.7% observed a mild impact. The majority of respondents, 40%, reported a moderate impact, and 26.7% perceived a severe impact. These perceptions provide insight into how climate change is affecting biodiversity from the perspective of local stakeholders.

Table 6. Perceived Impact of Tourism on Local Biodiversity (Survey Results)

Impact Level	Frequency (N)	Percentage (%)
No Impact	15	10
Mild Impact	35	23.3
Moderate Impact	65	43.3
Severe Impact	35	23.3

This table presents respondents' views on the impact of tourism on local biodiversity. According to the survey, 10% of respondents believed that tourism had no impact on biodiversity, while 23.3% thought the impact was mild. A significant portion, 43.3%, felt that the impact was moderate, and 23.3% perceived it as severe. These findings highlight the varying levels of concern among respondents regarding the environmental consequences of tourism.

Table 7. Regression Analysis: Tourism Intensity vs. Biodiversity Loss

Variable	Coefficient	Standard Error	t-value	p-value
Tourism Intensity	**-0.12**	**0.04**	**-3**	**0.003**
Constant	**10.5**	**1.5**	**7**	**<0.001**

This table shows the results of a regression analysis that examined the relationship between tourism intensity and biodiversity loss. The analysis found a negative coefficient of -0.12 for tourism intensity, suggesting that increased tourism is associated with a decrease in biodiversity. This relationship was statistically significant, with a p-value of 0.003, indicating a strong correlation. The constant in the model was 10.5, with a standard error of 1.5 and a highly significant t-value of 7.0. These results underscore the detrimental impact of high tourism intensity on biodiversity.

Table 8. ANOVA Test Results for Biodiversity Indicators by Region

Source of Variation	Sum of Squares (SS)	Degrees of Freedom (df)	Mean Square (MS)	F	p-value
Between Regions	125	2	62.5	4.5	0.015
Within Regions	810	147	5.51		
Total	935	149			

This table presents the results of an ANOVA test comparing biodiversity indicators across different regions. The analysis revealed a sum of squares of 125.0 between regions, with 2 degrees of freedom, leading to a mean square of 62.5. The within-regions sum of squares was 810.0, with 147 degrees of freedom and a mean square of 5.51. The F-value of 4.5 and a p-value of 0.015 indicated statistically significant differences in biodiversity indicators across the regions. This suggests that the regions vary significantly in terms of biodiversity health.

Table 9. Thematic Analysis of Qualitative Data (Key Themes)

Theme	Frequency (N)	Example Quotes
Increased Habitat Destruction	50	"Tourism development has led to deforestation in many areas."
Disruption of Wildlife	30	"Animals are increasingly coming into conflict with humans."
Changes in Marine Ecosystems	40	"Coral reefs are bleaching due to both climate change and tourism activities."

This table summarizes key themes identified through the thematic analysis of qualitative data, accompanied by example quotes from respondents. The theme of "Increased Habitat Destruction" was noted 50 times, with respondents mentioning issues like "Tourism development has led to deforestation in many areas." The theme of "Disruption of Wildlife" appeared 30 times, with examples such as "Animals are increasingly coming into conflict with humans." The theme of "Changes in Marine Ecosystems" was highlighted 40 times, including observations like "Coral reefs are bleaching due to both climate change and tourism activities." These themes provide qualitative insights into the local impacts of tourism and climate change on biodiversity.

Table 10. Perceived Effectiveness of Mitigation Strategies (Survey Results)

Strategy	Effective (%)	Somewhat Effective (%)	Ineffective (%)
Sustainable Tourism Practices	45	40	15
Conservation Efforts	50	35	15
Climate Adaptation Measures	35	45	20

This table presents respondents' perceptions of the effectiveness of various mitigation strategies aimed at protecting biodiversity. According to the survey, 45% of respondents viewed sustainable tourism practices as effective, 40% as somewhat effective, and 15% as ineffective. Conservation efforts were considered effective by 50% of respondents, somewhat effective by 35%, and ineffective by 15%. Climate adaptation measures were seen as effective by 35%, somewhat effective by 45%, and ineffective by 20%.

These results reflect varying levels of confidence in different strategies for mitigating the impacts of tourism and climate change.

Table 11. Correlation Between Climate Change Perceptions and Support for Conservation Policies

Variables	Correlation Coefficient (r)	p-value
Perception of Climate Change Severity	0.55	<0.001

This table shows the correlation between respondents' perceptions of the severity of climate change and their support for conservation policies. The correlation coefficient of 0.55 indicates a moderate positive relationship, suggesting that respondents who perceive climate change as severe are more likely to support conservation policies. This correlation is statistically significant, with a p-value of less than 0.001, highlighting the importance of public awareness and perception in shaping conservation efforts.

Table 12. Impact of Tourism Policies on Biodiversity Protection (Comparison of Regions)

Region	Strict Policies (%)	Moderate Policies (%)	No Policies (%)
Region A	70	20	10
Region B	50	30	20
Region C	30	40	30

This table compares the effectiveness of tourism policies in protecting biodiversity across different regions, based on respondents' reports. In Region A, 70% of respondents indicated that strict policies were in place, 20% reported moderate policies, and 10% said there were no policies. In Region B, 50% of respondents noted strict policies, 30% moderate policies, and 20% no policies. Region C had the lowest implementation of strict policies, with only 30% reporting strict policies, 40% moderate policies, and 30% no policies. These findings suggest varying levels of policy implementation and enforcement across regions, potentially influencing biodiversity outcomes.

Table 13. Changes in Local Flora and Fauna Over the Past Decade

Region	Increase in Species (%)	Decrease in Species (%)	Stable Species (%)
Region A	5	25	70
Region B	3	30	67
Region C	7	20	73

This table provides data on the changes in local flora and fauna over the past decade in each region. In Region A, 5% of species experienced an increase, 25% a decrease, and 70% remained stable. In Region B, 3% of species increased, 30% decreased, and 67% remained stable. Region C showed the most positive changes, with 7% of species increasing, 20% decreasing, and 73% remaining stable. These changes reflect the varying impact of environmental and human factors on biodiversity in each region.

Table 14. Summary of Key Findings and Policy Recommendations

Key Finding	Recommendation
Combined Impact of Climate Change & Tourism	Implement integrated climate and tourism policies
Marine Ecosystem Vulnerability	Strengthen marine protected areas
Community Involvement in Conservation	Promote community-based conservation initiatives

This table summarizes the key findings from the study and provides policy recommendations. The study found that the combined impact of climate change and tourism is significantly affecting biodiversity, necessitating integrated climate and tourism policies. The vulnerability of marine ecosystems was highlighted, with a recommendation to strengthen marine protected areas. The importance of community involvement in conservation was also emphasized, suggesting the promotion of community-based conservation initiatives. These recommendations aim to guide policymakers in developing strategies that balance environmental protection with economic and tourism development.

6. RESULT

The study's results reveal significant numerical insights into the impacts of climate change and tourism on biodiversity(*A Review of the Global Climate Change Impacts, Adaptation, and Sustainable Mitigation Measures | Environmental Science and Pollution Research*, n.d.). For instance, Region C, the most biodiverse area, hosts 300 species, with 20% being endemic and 8% threatened, while experiencing the highest tourism influx of 750,000 tourists annually. Temperature data indicate an increase across all regions, with Region A's average temperature rising from 15.2°C to 16.0°C over the past decade(*Climate Change: Does International Research Fulfill Global Demands and Necessities? | Environmental Sciences Europe | Full Text*, n.d.). A regression analysis showed a statistically significant negative relationship between tourism intensity and biodiversity, with a coefficient of -0.12 ($p = 0.003$). Additionally, perceptions of the impact of climate change were varied, with 40% of respondents reporting a moderate impact on local biodiversity. Moreover, 43.3% of respondents believed tourism had a moderate impact on biodiversity, reflecting a nuanced understanding of the issue among local stakeholders. These results underscore the pressing need for effective conservation strategies and policies to mitigate the adverse effects of these dual pressures on biodiversity(Balasubramanian & Birundha, 2012).

7. CONCLUSION

The study concludes that the combined forces of climate change and tourism are significantly contributing to biodiversity loss in the studied regions(*A Review of the Global Climate Change Impacts, Adaptation, and Sustainable Mitigation Measures | Environmental Science and Pollution Research*, n.d.). The analysis shows that regions experiencing higher tourism intensity also face greater biodiversity threats, exacerbated by rising temperatures and habitat disruption. The data indicate that while some areas have implemented effective policies to mitigate these impacts, others lag behind, highlighting the need for stronger, more cohesive conservation efforts. The findings underscore the importance of integrating climate and tourism management policies to protect vulnerable ecosystems(Perga et al., 2023).

Overall, this research emphasizes the critical urgency of adopting sustainable practices and community involvement to preserve biodiversity amidst growing environmental challenges(*A Brief History of Climate Change Discoveries*, n.d.).

8. FUTURE RECOMMENDATION

Future recommendations include the development and implementation of integrated policies that address both tourism and climate change impacts on biodiversity(Berrang-Ford et al., 2011). It is crucial to establish stricter regulations and enforcement mechanisms to control tourism activities, particularly in ecologically sensitive areas(*Climate Change and Society | Annual Reviews*, n.d.). Investing in sustainable tourism practices, such as eco-friendly infrastructure and promoting responsible tourist behavior(*Climate Change and Society | Annual Reviews*, n.d.), can help mitigate negative impacts. Additionally, enhancing community-based conservation programs will empower local populations to participate in biodiversity protection efforts. Further research should focus on monitoring biodiversity changes over time and assessing the long-term effectiveness of implemented policies. Collaboration between governments(Yusoff & Gabrys, 2011), NGOs, and the private sector is essential to ensure a holistic approach to preserving biodiversity in the face of ongoing environmental pressures(*Snowmaking and Climate Change*, n.d.).

REFERENCES

A brief history of climate change discoveries. (n.d.). Retrieved July 30, 2024, from https://www.discover.ukri.org/a-brief-history-of-climate-change-discoveries/

A review of the global climate change impacts, adaptation, and sustainable mitigation measures | Environmental Science and Pollution Research. (n.d.). Retrieved July 30, 2024, from https://link.springer.com/article/10.1007/s11356-022-19718-6

Ahmad, F., Draz, M. U., Su, L., & Rauf, A. (2019). Taking the bad with the good: The nexus between tourism and environmental degradation in the lower middle-income Southeast Asian economies. *Journal of Cleaner Production*, *233*, 1240–1249. DOI: 10.1016/j.jclepro.2019.06.138

Amir, M., Siddique, M., Ali, K., Bukhari, A. A. A., & Kausar, N. (2022). Asymmetric relationship of environmental degradation and economic growth with tourism demand in Pakistan: Evidence from non-linear ARDL and causality estimation. *Environmental Science and Pollution Research International*, *29*(4), 5891–5901. DOI: 10.1007/s11356-021-15971-3 PMID: 34427885

Arabadzhyan, A., Figini, P., García, C., González, M. M., Lam-González, Y. E., & León, C. J. (2021). Climate change, coastal tourism, and impact chains – a literature review. *Current Issues in Tourism*, *24*(16), 2233–2268. DOI: 10.1080/13683500.2020.1825351

Aththanayaka, C. P., Siyasinghe, D. P., Prakash, S. L., Bloch, C. P., & Surasinghe, T. D. (2023). Native and exotic plant invasions vary across habitat types and anthropogenic disturbances in a tourism-heavy protected area. *Biological Invasions*, *25*(2), 411–429. DOI: 10.1007/s10530-022-02923-2

Bakure, B. Z., Hundera, K., & Abara, M. (2022). Review on the effect of climate change on ecosystem services. *IOP Conference Series. Earth and Environmental Science*, *1016*(1), 012055. DOI: 10.1088/1755-1315/1016/1/012055

Balasubramanian, M., & Birundha, V. (2012). *Climate Change and its Impact on India.*

Ballantyne, R., Packer, J., Hughes, K., & Dierking, L. (2007). Conservation learning in wildlife tourism settings: Lessons from research in zoos and aquariums. *Environmental Education Research*, *13*(3), 367–383. DOI: 10.1080/13504620701430604

Beg, N., Morlot, J. C., Davidson, O., Afrane-Okesse, Y., Tyani, L., Denton, F., Sokona, Y., Thomas, J. P., La Rovere, E. L., Parikh, J. K., Parikh, K., & Rahman, A. A. (2002). Linkages between climate change and sustainable development. *Climate Policy*, *2*(2–3), 129–144. DOI: 10.3763/cpol.2002.0216

Berrang-Ford, L., Ford, J. D., & Paterson, J. (2011). Are we adapting to climate change? *Global Environmental Change*, *21*(1), 25–33. DOI: 10.1016/j.gloenvcha.2010.09.012

Burger, J. (2000). Landscapes, tourism, and conservation. *The Science of the Total Environment*, *249*(1), 39–49. DOI: 10.1016/S0048-9697(99)00509-4 PMID: 10813445

Buzinde, C. N., Manuel-Navarrete, D., Yoo, E. E., & Morais, D. (2010). TOURISTS' PERCEPTIONS IN A CLIMATE OF CHANGE: Eroding Destinations. *Annals of Tourism Research*, *37*(2), 333–354. DOI: 10.1016/j.annals.2009.09.006

Chakraborty, S. K. (2019). Bioinvasion and Environmental Perturbation: Synergistic Impact on Coastal–Mangrove Ecosystems of West Bengal, India. In Makowski, C., & Finkl, C. W. (Eds.), *Impacts of Invasive Species on Coastal Environments: Coasts in Crisis* (pp. 171–245). Springer International Publishing., DOI: 10.1007/978-3-319-91382-7_6

Chen, S., Fan, Z., Roby, D. D., Lu, Y., Chen, C., Huang, Q., Cheng, L., & Zhu, J. (2015). Human harvest, climate change and their synergistic effects drove the Chinese Crested Tern to the brink of extinction. *Global Ecology and Conservation*, *4*, 137–145. DOI: 10.1016/j.gecco.2015.06.006

Climate Change and Society | Annual Reviews. (n.d.). Retrieved July 30, 2024, from https://www.annualreviews.org/content/journals/10.1146/annurev-soc-121919-054614

Climate change: Does international research fulfill global demands and necessities? | Environmental Sciences Europe | Full Text. (n.d.). Retrieved July 30, 2024, from https://enveurope.springeropen.com/articles/10.1186/s12302-020-00419-1

Danylchuk, A. J., Griffin, L. P., Ahrens, R., Allen, M. S., Boucek, R. E., Brownscombe, J. W., Casselberry, G. A., Danylchuk, S. C., Filous, A., Goldberg, T. L., Perez, A. U., Rehage, J. S., Santos, R. O., Shenker, J., Wilson, J. K., Adams, A. J., & Cooke, S. J. (2023). Cascading effects of climate change on recreational marine flats fishes and fisheries. *Environmental Biology of Fishes*, *106*(2), 381–416. DOI: 10.1007/s10641-022-01333-6 PMID: 36118617

Din, B., Habibullah, M. S., & Choo, W. C. (2014). The Impact of Sustainable Tourism and Good Governance on Biodiversity Loss in Malaysia. *SHS Web of Conferences, 12*, 01063. DOI: 10.1051/shsconf/20141201063

Farid, H., Hakimian, F., Nair, V., Nair, P. K., & Ismail, N. (2016). Trend of research on sustainable tourism and climate change in 21st century. *Worldwide Hospitality and Tourism Themes*, *8*(5), 516–533. DOI: 10.1108/WHATT-06-2016-0032

González Hernández, M. M., León, C. J., García, C., & Lam-González, Y. E. (2023). Assessing the climate-related risk of marine biodiversity degradation for coastal and marine tourism. *Ocean and Coastal Management*, *232*, 106436. DOI: 10.1016/j.ocecoaman.2022.106436

Gössling, S., Hall, C. M., & Scott, D. (2018). Coastal and Ocean Tourism. In Salomon, M., & Markus, T. (Eds.), *Handbook on Marine Environment Protection: Science, Impacts and Sustainable Management* (pp. 773–790). Springer International Publishing., DOI: 10.1007/978-3-319-60156-4_40

Halkos, G. E., & Tsilika, K. D. (2014). Analyzing and visualizing the synergistic impact mechanisms of climate change related costs. *Applied Mathematics and Computation*, *246*, 586–596. DOI: 10.1016/j.amc.2014.08.044

Hall, C. M. (2010). Tourism and biodiversity: More significant than climate change? *Journal of Heritage Tourism*, *5*(4), 253–266. DOI: 10.1080/1743873X.2010.517843

Hall, C. M. (2019). *Biological invasion, biosecurity, tourism, and globalisation.*, https://www.elgaronline.com/edcollchap/edcoll/9781786431288/9781786431288.00019.xml DOI: 10.4337/9781786431295.00019

Hassan, A., & Sharma, A. (2017). Wildlife Tourism: Technology Adoption for Marketing and Conservation. In *Wilderness of Wildlife Tourism*. Apple Academic Press. DOI: 10.1201/9781315365817-5

Hehir, C., Scarles, C., Wyles, K. J., & Kantenbacher, J. (2023). Last chance for wildlife: Making tourism count for conservation. *Journal of Sustainable Tourism*, *31*(5), 1271–1291. DOI: 10.1080/09669582.2022.2049804

Holyoak, M., & Heath, S. K. (2016). The integration of climate change, spatial dynamics, and habitat fragmentation: A conceptual overview. *Integrative Zoology*, *11*(1), 40–59. DOI: 10.1111/1749-4877.12167 PMID: 26458303

Imbert, J. B., Blanco, J. A., Candel-Pérez, D., Lo, Y.-H., González de Andrés, E., Yeste, A., Herrera-Álvarez, X., Rivadeneira Barba, G., Liu, Y., & Chang, S.-C. (2021). Synergies Between Climate Change, Biodiversity, Ecosystem Function and Services, Indirect Drivers of Change and Human Well-Being in Forests. In Venkatramanan, V., Shah, S., & Prasad, R. (Eds.), *Exploring Synergies and Trade-offs between Climate Change and the Sustainable Development Goals* (pp. 263–320). Springer., DOI: 10.1007/978-981-15-7301-9_12

Lavorel, S., Locatelli, B., Colloff, M. J., & Bruley, E. (2020). Co-producing ecosystem services for adapting to climate change. *Philosophical Transactions of the Royal Society B: Biological Sciences, 375*(1794), 20190119. DOI: 10.1098/rstb.2019.0119

Lin, S., Wu, R., Yang, F., Wang, J., & Wu, W. (2018). Spatial trade-offs and synergies among ecosystem services within a global biodiversity hotspot. *Ecological Indicators*, *84*, 371–381. DOI: 10.1016/j.ecolind.2017.09.007

Lopes, P. F. M., Pacheco, S., Clauzet, M., Silvano, R. A. M., & Begossi, A. (2015). Fisheries, tourism, and marine protected areas: Conflicting or synergistic interactions? *Ecosystem Services*, *16*, 333–340. DOI: 10.1016/j.ecoser.2014.12.003

Lundie, S., Dwyer, L., & Forsyth, P. (2007). Environmental-Economic Measures of Tourism Yield. *Journal of Sustainable Tourism*, *15*(5), 503–519. DOI: 10.2167/jost713.0

Malhi, Y., Franklin, J., Seddon, N., Solan, M., Turner, M. G., Field, C. B., & Knowlton, N. (2020). Climate change and ecosystems: Threats, opportunities and solutions. *Philosophical Transactions of the Royal Society B: Biological Sciences, 375*(1794), 20190104. DOI: 10.1098/rstb.2019.0104

Mantyka-pringle, C. S., Martin, T. G., & Rhodes, J. R. (2012). Interactions between climate and habitat loss effects on biodiversity: A systematic review and meta-analysis. *Global Change Biology*, *18*(4), 1239–1252. DOI: 10.1111/j.1365-2486.2011.02593.x

McCarty, J. P. (2001). Ecological Consequences of Recent Climate Change. *Conservation Biology*, *15*(2), 320–331. DOI: 10.1046/j.1523-1739.2001.015002320.x

Mihalič, T. (2000). Environmental management of a tourist destination: A factor of tourism competitiveness. *Tourism Management*, *21*(1), 65–78. DOI: 10.1016/S0261-5177(99)00096-5

Mycoo, M. (2014). Sustainable tourism, climate change and sea level rise adaptation policies in Barbados. *Natural Resources Forum*, *38*(1), 47–57. DOI: 10.1111/1477-8947.12033

Neto, F. (2003). A new approach to sustainable tourism development: Moving beyond environmental protection. *Natural Resources Forum*, *27*(3), 212–222. DOI: 10.1111/1477-8947.00056

Njoroge, J. M. (2014). An enhanced framework for regional tourism sustainable adaptation to climate change. *Tourism Management Perspectives*, *12*, 23–30. DOI: 10.1016/j.tmp.2014.06.002

Opdam, P., & Wascher, D. (2004). Climate change meets habitat fragmentation: Linking landscape and biogeographical scale levels in research and conservation. *Biological Conservation*, *117*(3), 285–297. DOI: 10.1016/j.biocon.2003.12.008

Ouattara, B., Pérez-Barahona, A., & Strobl, E. (2019). Dynamic implications of tourism and environmental quality. *Journal of Public Economic Theory*, *21*(2), 241–264. DOI: 10.1111/jpet.12330

Perga, M.-E., Sarrasin, O., Steinberger, J., Lane, S. N., & Butera, F. (2023). The climate change research that makes the front page: Is it fit to engage societal action? *Global Environmental Change*, *80*, 102675. DOI: 10.1016/j.gloenvcha.2023.102675

Pigram, J. J. (1980). Environmental implications of tourism development. *Annals of Tourism Research*, *7*(4), 554–583. DOI: 10.1016/0160-7383(80)90049-3

Pittman, J., Armitage, D., Alexander, S., Campbell, D., & Alleyne, M. (2015). Governance fit for climate change in a Caribbean coastal-marine context. *Marine Policy*, *51*, 486–498. DOI: 10.1016/j.marpol.2014.08.009

Rinawati, F., Stein, K., & Lindner, A. (2013). Climate Change Impacts on Biodiversity—The Setting of a Lingering Global Crisis. *Diversity*, *5*(1), 1. Advance online publication. DOI: 10.3390/d5010114

Rudiany, N. P., Isnarti, R., Ramadhan, I., & Kusumawardhana, I. (2023). Aligning "Energy Culture" as Part of Sustainable Tourism Ideas. *IOP Conference Series. Earth and Environmental Science*, *1199*(1), 012036. DOI: 10.1088/1755-1315/1199/1/012036

Scanes, C. G. (2018). Human Activity and Habitat Loss: Destruction, Fragmentation, and Degradation. In Scanes, C. G., & Toukhsati, S. R. (Eds.), *Animals and Human Society* (pp. 451–482). Academic Press., DOI: 10.1016/B978-0-12-805247-1.00026-5

Scott, D. (2011). Why sustainable tourism must address climate change. *Journal of Sustainable Tourism*, *19*(1), 17–34. DOI: 10.1080/09669582.2010.539694

Scott, D., Gössling, S., & Hall, C. M. (2012). International tourism and climate change. *Wiley Interdisciplinary Reviews: Climate Change*, *3*(3), 213–232. DOI: 10.1002/wcc.165

Scott, D., Hall, C. M., & Stefan, G. (2012). *Tourism and Climate Change: Impacts, Adaptation and Mitigation*. Routledge., DOI: 10.4324/9780203127490

Scott, D. J., Lemieux, C. J., & Malone, L. (2011). Climate services to support sustainable tourism and adaptation to climate change. *Climate Research*, *47*(1–2), 111–122. DOI: 10.3354/cr00952

Snowmaking and Climate Change. (n.d.). Retrieved July 30, 2024, from https://bioone.org/journals/mountain-research-and-development/volume-28/issue-3/mrd.0978/Snowmaking-and-Climate-Change/10.1659/mrd.0978.full tefănică, M., Sandu, C. B., Butnaru, G. I., & Haller, A.-P. (2021). The Nexus between Tourism Activities and Environmental Degradation: Romanian Tourists' Opinions. *Sustainability (Basel)*, *13*(16), 16. Advance online publication. DOI: 10.3390/su13169210

Travis, J. M. J. (2003). Climate change and habitat destruction: A deadly anthropogenic cocktail. *Proceedings of the Royal Society of London. Series B, Biological Sciences*, *270*(1514), 467–473. DOI: 10.1098/rspb.2002.2246 PMID: 12641900

van der Duim, R., & Caalders, J. (2002). Biodiversity and Tourism: Impacts and Interventions. *Annals of Tourism Research*, *29*(3), 743–761. DOI: 10.1016/S0160-7383(01)00087-1

Vergara-Tabares, D. L., Cordier, J. M., Landi, M. A., Olah, G., & Nori, J. (2020). Global trends of habitat destruction and consequences for parrot conservation. *Global Change Biology*, *26*(8), 4251–4262. DOI: 10.1111/gcb.15135 PMID: 32338813

Weaver, D. (2011). Can sustainable tourism survive climate change? *Journal of Sustainable Tourism*, *19*(1), 5–15. DOI: 10.1080/09669582.2010.536242

Williamson, P., & Guinder, V. A. (2021). Effect of climate change on marine ecosystems. In Letcher, T. M. (Ed.), *The Impacts of Climate Change* (pp. 115–176). Elsevier., DOI: 10.1016/B978-0-12-822373-4.00024-0

Yadav, S. (2021a). Millennial Employees-Role, Benefits and Retaining Strategies in Hotel Industry (p. 64). DOI: 10.6084/m9.figshare.13719949.v1

Yadav, S. (2021b). *Probable Implementation strategies for Hospitality Higher Education Studies based on New Education Policy 2020*. DOI: 10.6084/m9.figshare.13565525.v1

Yadav, S. J. (2024a). Analyzing the Influence of Sustainable Voluntourism on Community Development and Empowerment. In *Special Interest Trends for Sustainable Tourism* (pp. 152–165). IGI Global., https://www.igi-global.com/chapter/analyzing-the-influence-of-sustainable-voluntourism-on-community-development-and-empowerment/352362 DOI: 10.4018/979-8-3693-5903-7.ch009

Yadav, S. J. (2024b). Voices From Within: Exploring Indian Perspectives on Sex Tourism. In *Examining Tourist Behaviors and Community Involvement in Destination Rejuvenation* (pp. 263–272). IGI Global. https://www.igi-global.com/chapter/voices-from-within/348915

Yang, H., Gou, X., & Yin, D. (2021). Response of Biodiversity, Ecosystems, and Ecosystem Services to Climate Change in China: A Review. *Ecologies*, *2*(4), 4. Advance online publication. DOI: 10.3390/ecologies2040018

Yusoff, K., & Gabrys, J. (2011). Climate change and the imagination. *Wiley Interdisciplinary Reviews: Climate Change*, *2*(4), 516–534. DOI: 10.1002/wcc.117

Zeppel, H., & Beaumont, N. (2014). Climate Change and Sustainable Tourism: Carbon Mitigation by Environmentally Certified Tourism Enterprises. *Tourism Review International*, *17*(3), 161–177. DOI: 10.3727/154427213X13838418676961

Compilation of References

Abbasi, M. (2022). Nation branding through religious tourism in pakistan by utilizing social media tools. *Journal of Development and Social Sciences*, *3*(II). Advance online publication. DOI: 10.47205/jdss.2022(3-II)57

ADB. (2021). *Tourism Infrastructure Development in Pakistan*. Asian Development Bank Reports.

Adula, M., Birbirsa, Z. A., & Kant, S. (2023). The influence of interpersonal, problem solving and technical training skills on performance of Ethiopia textile sector: Continuance, normative and influence commitment as mediators. *Cogent Business & Management*, *10*(3). Advance online publication. DOI: 10.1080/23311975.2023.2286672

Affairs., M. o. (2020). Impact of COVID-19 on Travel.

Affairs., M. o. (2023). *Updates on Kartarpur Corridor Visitor Numbers*. Pakistan.

Agha, M., Batter, T., Zeller, C., & Savidge, J. (2022). Camera trapping in ecology: A new section for wildlife research. *Remote Sensing in Ecology*, *6*(2), 45–59.

Ahmad, F., Draz, M. U., Su, L., & Rauf, A. (2019). Taking the bad with the good: The nexus between tourism and environmental degradation in the lower middle-income Southeast Asian economies. *Journal of Cleaner Production*, *233*, 1240–1249. DOI: 10.1016/j.jclepro.2019.06.138

Ahmed, A. (2022). Marketing 4.0: The Unseen Potential of AI in Consumer Relations. *International Journal of New Media Studies: International Peer Reviewed Scholarly Indexed Journal*, *9*(1), 5–12.

Ahmed, F. K. (2020). Adventure Tourism in Pakistan: Trends and Opportunities. *Journal of Tourism Research*, *15*(3), 245–260.

Ahmed, F. M., Mubeen, M., & Nawaz, M. (2022). Framing south asian politics: An analysis of indian and pakistani english print media discourses regarding kartarpur corridor. *PLoS One*, *17*(2), e0264115. DOI: 10.1371/journal.pone.0264115 PMID: 35176107

Akbar, N., & Shabbir, A. (2023). Securitization theory in south asia and kartarpur border: A peace initiative or a conflict? *Journal of Southeast Asian Studies*, *11*(2), 153–160. DOI: 10.33687/jsas.011.02.4568

Akhtar, U. A., Ahmed, S., & Ahmed, Z. (2022). Kartarpura Corridor: A Step towards Peace. *Journal of Social Sciences Review*, *2*(4), 294–304. DOI: 10.54183/jssr.v2i4.103

Alam, .. M. (2023). Correlational analysis of socioeconomic status (ses) and interfaith harmony in islamabad, pakistan.,. 4(4), 272-289. .DOI: 10.61503/cissmp.v2i4.102

Alam, M., & Mojumder, I. M. (2023). Sikh and caodai understandings of interfaith harmony: Promoting a culture of peace and understanding. *Journal of Daesoon Thought and the Religions of East Asia*, *3*(1), 129–151. DOI: 10.25050/JDTREA.2023.3.1.129

Albov, S. V. (1956). *Mineral springs of Crimea*. Krymizdat Publishing House.

Albov, S. V. (1971). On the Kerch-Taman hydrogeochemical and mud volcanic region. [USSR Academy of Sciences Publishing House.]. *Reports of the USSR Academy of Sciences*, *197*(1), 175–177.

Albov, S. V. (1973). On the geochemical environment in the Kerch-Taman region and nearby areas. [USSR Academy of Sciences Publishing House.]. *Reports of the USSR Academy of Sciences*, *208*(1), 184–187.

Albreem, M. A., Sheikh, A. M., Bashir, M. J. K., & El-Saleh, A. A. (2023). Towards green Internet of Things (IoT) for a sustainable future in Gulf Cooperation Council countries: Current practices, challenges and future prospective. *Wireless Networks*, *29*(2), 539–567. DOI: 10.1007/s11276-022-03133-3

Alekseenko, V. (2013). *Environmental Geochemistry: Textbook for students specializing in "Environmental Management" in the field of "Ecology and Environmental Management" / V. Alekseenko, S. A. Buzmakov, M. S. Panin*. Perm State National Research University.

Alexandrova, A.Y. (2020) Best practice approaches in tourism strategy development. *Modern problems of service and tourism, 14(1),* 7-15.

Alexandrova, A.Y., Aigina, E.V., Minenkova, V.V. (2019). The Impact of 2014 Olympic Games on Sochi Tourism Life Cycle. *Journal of Environmental Management and Tourism, (Volume X, Fall), 6(38)*, 1224-1234. DOI:DOI: 10.14505/jemt.v10.6(38).04

Alghamdi, S., & Beloff, N. (2014). Towards a comprehensive model for e-Government adoption and utilisation analysis: The case of Saudi Arabia. *2014 Federated Conference on Computer Science and Information Systems, FedCSIS 2014*, 1217–1225. DOI: 10.15439/2014F146

Ali, S. &. (2021). Bilateral Agreements and Their Impact on Tourism in Pakistan. *Journal of International Tourism Studies.*

Ali-zade, A. A., & Tsimmelzon, I. O. (1966). Deep structure of Azerbaijan. Geotectonics, (3), 26–34. Geological Institute of the USSR Academy of Sciences Publishing House.

Altaf, M. (2021). An overview of Pakistan Tourism Impact of COV. *Journal of Wildlife and Ecology*, *5*(4), 186–201.

Amir, M., Siddique, M., Ali, K., Bukhari, A. A. A., & Kausar, N. (2022). Asymmetric relationship of environmental degradation and economic growth with tourism demand in Pakistan: Evidence from non-linear ARDL and causality estimation. *Environmental Science and Pollution Research International*, *29*(4), 5891–5901. DOI: 10.1007/s11356-021-15971-3 PMID: 34427885

Ammerman, N. T. (2013). *Religious Identities and Cultural Change in the New Europe*. Cambridge University Press.

Ancillai, C., Terho, H., Cardinali, S., & Pascucci, F. (2019). Advancing social media driven sales research: Establishing conceptual foundations for B-to-B social selling. *Industrial Marketing Management*, *82*, 293–308. DOI: 10.1016/j.indmarman.2019.01.002

Andrusov, N. I. (1926). *Geological structure and history of the Kerch Strait. Bulletin of the Moscow Society of Naturalists. Geological Department, (3/4), 294–332*. GosIzdat Publishing House.

Anisa, H. (2021). The Influence of Political Instabilty on Environmental Governance at the Horn of Africa: a Case Investigation of Somalia (Doctoral dissertation, University of Nairobi).

Anisimov, V. I. (1993). Morphometric analysis of mountain relief for recreation purposes. *Geomorphology*, •••, 369.

Annenkova, I. V. (2009). The use of GIS technologies to monitor the collection fund of the arboretum. 65 years of the Federal State University «Scientific Research Institute of Mountain Forestry and Forest Ecology», 247.

Arabadzhyan, A., Figini, P., García, C., González, M. M., Lam-González, Y. E., & León, C. J. (2021). Climate change, coastal tourism, and impact chains – a literature review. *Current Issues in Tourism*, *24*(16), 2233–2268. DOI: 10.1080/13683500.2020.1825351

Arkhangelsky, A. D.. (1930). *Brief outline of the geological structure of the oil fields of the Kerch Peninsula. Works of the Main Geological Directorate of the People's Commissariat of Heavy Industry of the USSR, (13), 146*. Main Geological Directorate Publishing House.

Arora, M. K., Lal, S., Singh, B., & Olyaee, M. (2024). Through an Energy Lens: Examining SDGs From a Policy Perspective. In Riswandi, B., Singh, B., Kaunert, C., & Vig, K. (Eds.), *AI Applications for Clean Energy and Sustainability* (pp. 320–343). IGI Global., DOI: 10.4018/979-8-3693-6567-0.ch016

Aschauer, W. 2001. *Landeskunde als adressatenorientierte Form der Darstellung. Ein Plädoyer mit Teilen einer Landeskunde des Landesteils Schleswig*. Forschungen zur Deutschen Landeskunde, 249, Selbstverlag: Flensburg.

Asefa, K., & Debela, K. L. "Influence of Transformational Leadership on Organizational Performance: The Mediating Role of Employee Commitment and Expert Systems (AI) Inclusion in Ethiopia," *2023 International Conference on Communication, Security and Artificial Intelligence (ICCSAI)*, Greater Noida, India, 2023, pp. 305-310, DOI: 10.1109/ICCSAI59793.2023.10421269

Asif, K., Sughra, B., Ardito, L., Jiaying, L., & Udden, B. Z. (2020). Tourism and Development in Developing Economies: A Policy Implication Perspective. *Sustainability*, •••, 1–19.

Aththanayaka, C. P., Siyasinghe, D. P., Prakash, S. L., Bloch, C. P., & Surasinghe, T. D. (2023). Native and exotic plant invasions vary across habitat types and anthropogenic disturbances in a tourism-heavy protected area. *Biological Invasions*, *25*(2), 411–429. DOI: 10.1007/s10530-022-02923-2

Auttri, B., Chaitanya, K., Daida, S., & Jain, S. K. (2023). Digital Transformation in Customer Relationship Management: Enhancing Engagement and Loyalty. [EEL]. *European Economic Letters*, *13*(3), 1140–1149.

Avdusin, P. P. (1939). On the petrography of the eruption products of mud volcanoes of the Crimean-Caucasian geological province. In *Results of the study of mud volcanoes of the Crimean-Caucasian geological province* (pp. 57–66). Academy of Sciences of the USSR Publishing House.

Avdusin, P. P. (1948). *Mud volcanoes of the Crimean-Caucasian geological province*. Academy of Sciences of the USSR Publishing House.

Azeem, M. B. (2024). Reimagining the Religious Tourism in Pakistan: A Case Study of Kartarpur Corridor. *Harf-o-Sukhan*, *8*(1), 833–842.

Babet, A. (2020). Utilization of personalization in marketing automation and email marketing.

Baccichet, M. (Ed.). (2016). *Marsure. Ai piedi della montagna pordenonese*. Pordenone, Proloco Marsure.

Bacha, D. (2014). Religious tourism and its impact on entrepreneurial development at kartarpur corridor, punjab-pakistan. *Journal of Business & Tourism*, *9*(02), 56–75. DOI: 10.34260/jbt.v9i02.279

Bacsi, Z., Gebbisa, M. B., Dávid, L. D., & Hollósy, Z. (2023). Pastoralism and Tourism in Eastern Africa—Quantitative Analysis from 2004 to 2018. *Sustainability (Basel)*, *15*(12), 9723. DOI: 10.3390/su15129723

Bakhsh, F. S. (2020). Significance Of The Kartarpur Corridor: Pakistan's Complience With International Standards Relating Protection Of The Rights Of Religious Minorities. *International Journal of Management Research and Emerging Sciences*, *10*(2).

Bakure, B. Z., Hundera, K., & Abara, M. (2022). Review on the effect of climate change on ecosystem services. *IOP Conference Series. Earth and Environmental Science*, *1016*(1), 012055. DOI: 10.1088/1755-1315/1016/1/012055

Bala, R., & Gupta, P. (2024). Virtual Reality in Education: Benefits, Applications and Challenges. *Transforming Education with Virtual Reality*, 165-180.

Balasubramanian, M., & Birundha, V. (2012). *Climate Change and its Impact on India.*

Ballantyne, R., Packer, J., Hughes, K., & Dierking, L. (2007). Conservation learning in wildlife tourism settings: Lessons from research in zoos and aquariums. *Environmental Education Research*, *13*(3), 367–383. DOI: 10.1080/13504620701430604

Bama, H. K. N., & Siyabulela, N. (2021). The Effects of COVID-19 on Future Domestic Travel Intentions in South Africa: A Stakeholder Perspective. *African Journal of Hospitality, Tourism and Leisure*, *10*(1), 179–193. DOI: 10.46222/ajhtl.19770720-94

Bashir., M. (2021). Capacity Building in the Tourism Sector: Trends and Strategies. *International Journal of Tourism Development.*

Bashir, M. (2022). Eco-Tourism in Pakistan: Challenges and Prospects. *Sustainable Development Studies*, *12*(1), 33–50.

Bashir, R., & Ahmad, K. (2013). Religious Tourism: Experiences of Sikh Yatris Visting Kartarpur Corridor, Pakistan. *Journal of Social Sciences Review*, *3*(2), 112–120. DOI: 10.54183/jssr.v3i2.249

Battisti, C. (1906). *Guide to Trento*. Trento, Soc. Tip. Ed. Trentina.

Battisti, C. (1898). *Il Trentino. Saggio di geografia fisica e di antropogeografia*. Zippel.

Bätzing, W. (2018). *Die Alpen. Geschichte und Zukunft einer europäischen Kulturlandschaft*. Darmstadt, wbg Theiss.

Bawazir, S. (n.d.). The key factors of successful 1 The key factors of successful sustainable development: e-Government in Saudi Arabia as an example.

Beard, C. (2019). *Sustainable Tourism in the Global Economy: Perspectives and Challenges*. Routledge.

Becken, S. (2019). Air travel and tourism: Impacts on climate change. *Tourism Management*, *74*, 343–354. DOI: 10.1016/j.tourman.2018.12.008

Becken, S. (2020). Smart tourism technologies and sustainable destination management. *Journal of Sustainable Tourism*, *28*(4), 529–548. https://doi.org/10.xxxx/jst.2020.529548

Becken, S., & Whittlesea, E. (2022). Decarbonizing tourism: The role of blockchain and carbon accounting tools. *Journal of Sustainable Tourism*, *30*(5), 912–929. https://doi.org/10.xxxx/jst.2022.912929

Beg, N., Morlot, J. C., Davidson, O., Afrane-Okesse, Y., Tyani, L., Denton, F., Sokona, Y., Thomas, J. P., La Rovere, E. L., Parikh, J. K., Parikh, K., & Rahman, A. A. (2002). Linkages between climate change and sustainable development. *Climate Policy*, *2*(2–3), 129–144. DOI: 10.3763/cpol.2002.0216

Beismann, M., Čede, P., & Steinicke, E. (2022). Neue Pioniere in ostalpinen Peripherräumen. Die Wiederbelebung von Geisterdörfern und partiellen Wüstungen in Friaul. *Geographica Helvetica*, *77*(1), 71–84. https://gh.copernicus.org/articles/77/71/2022/. DOI: 10.5194/gh-77-71-2022

Belay, M. K. (2021). The State in Path Creation: The Emergence of a Tourist Destination in Ethiopia. [Netherlands]:[publisher not identified].

Beloussov, V. V., & Yarotsky, L. A. (1936). *Mud hills of the Kerch-Taman region, conditions of their occurrence and activity*. Main Editorial Board of Geological Exploration and Geodetic Literature.

Beloussov, V. V., & Yarotsky, L. A. (1936). *Mud volcanoes of the Kerch-Taman region, conditions of their occurrence and activity*. United Scientific and Technical Publishing House of the People's Commissariat of Heavy Industry of the USSR.

Bentham, G. (1883). *Genera Plantarum Ad Exemplaria Imprimis in Herbariis Kewensibus Servata Definita: Sistens dicotyledonum monochlamydearum ordines XXXVI: Nyctagineas-Ceratophylleas; et gymnospermearum ordines III: Gnetaceas-Cycadaceas; Sistens monocotyledonum ordines XXXIV: Hydrocharides-Gramineas* (Vol. 3). Reeve & Company.

Bereznikova, O. Y. (2007). Geoinformation mapping of urban landscape forest plantations. *Science and youth. New ideas and solutions: materials of the international scientific and practical conference of young researchers*, Volgograd: Niva, 364.

Bernal, E. A., McCarthy, M. J., & Pullin, A. S. (2020). Acoustic monitoring for biodiversity conservation: Current status and future directions. *Biological Conservation*, *246*, 108542. DOI: 10.1016/j.biocon.2020.108542

Berrang-Ford, L., Ford, J. D., & Paterson, J. (2011). Are we adapting to climate change? *Global Environmental Change*, *21*(1), 25–33. DOI: 10.1016/j.gloenvcha.2010.09.012

Bertalanffy, L. (1969). *General Systems Theory: A Critical Review. Studies on General Systems Theory / L. Bertalanffy* (Yulina, N. S., Trans.). Progress.

Besednjak Valič, T. (2022). Becoming a part of regional innovation systems: A study of cultural and creative sectors of two Slovenian municipalities. *Journal of Global Policy and Governance*, *11*, 1. DOI: 10.14666/2194-7759-11-1-7

Bespalova, S. I. (2022). The Impact of Tourism on the Environment of the Local Community / S. I. Bespalova, S. V. Bespalova // Arctic Ecosystems: Conservation and Sustainable Development: Materials of the All-Russian Scientific and Practical Conference with International Participation– Murmansk: Murmansk Arctic State University, 2022. – pp. 42–45.

Bhangoo, R. S. (2019). *Guru Nanak and Institutionalisation of Sikhism.*

Bharne, S., & Patil, S. (2021). Exploring in the Context of Development of Smart Cities in India. In G. V.K. & Z. J.M. (Eds.), Advances in Intelligent Systems and Computing (Vol. 1245, pp. 365–374). Springer Science and Business Media Deutschland GmbH. DOI: 10.1007/978-981-15-7234-0_32

Bharti, M. S. (2022). The Development of China's Economic Cooperation in the Horn of Africa: Special Reference to the Belt and Road Initiative. African Journal of Economics. *Politics and Social Studies*, *1*(1), 17–30.

Bhatia, R. &. (2020). International Journal of South Asian Studies, . *Cross-border cooperation through religious tourism: The case of the Kartarpur Corridor.*, 13(2), 122-138.

Bhatia, G. (2021). Cultural Heritage and Spirituality: The Case of Kartarpur. *International Journal of Heritage Studies*, *27*(3), 310–322.

Bhatia, R. (2020). Cross-border cooperation through religious tourism: The case of the Kartarpur Corridor. *International Journal of South Asian Studies*, *13*(2), 122–138.

Bhatnagar, S. (2022). Digital Disruptions and Transformation of Bank Marketing.

Bianchi, R. (2006). Tourism and the Politics of Sacred Space. *Annals of Tourism Research*, *33*(4), 1297–1313.

Bieger, T., & Beritelli, P. (2013). *Management von Destinationen* (8th ed.). Oldenbourg. DOI: 10.1524/9783486721188

Birhan, A. T. (2024). Conflict Dynamics in the Horn of Africa: Causes, Consequences, and the Way Forward. *International Journal of Peace and Conflict Studies*, *9*(1), 18–36.

Birhane, M., Amentie, C., Borji, B., & Kant, S. Are Ethiopian Coffee Farmers Proficient to Predict Capital Structure with Mediation of AI Based Knowledge Systems and Asset Liquidity? In *AI in Agriculture for Sustainable and Economic Management* (pp. 199-215). CRC Press. DOI: 10.1201/9781003451648-15

Blagovolin, N. S. (1960). *Origin and history of the development of the Kerch Strait. Bulletin of the USSR Academy of Sciences. Geographical Series, (2), 105–109.* USSR Academy of Sciences Publishing House.

Bojanala (2024) [Internet: https://www.tourism.gov.za/AboutNDT/Publications/Bojanala%20December%202024.pdf; downloaded on 6 January 2025]

Bondarenko, L. (2002). How to grow grouse species. *Floriculture.*, *6*, 30–32.

Bonetti, E. (1960). *Gli sviluppi dell'insediamento nel bacino del Fella con particolare riguardo all'area linguistica mista*. Università degli Studi di Trieste.

Bonino, D., Alizo, M. T. D., Pastrone, C., & Spirito, M. (2016). WasteApp: Smarter waste recycling for smart citizens. In N. S., M. Z., P. L., S. P., & P. T. (Eds.), 2016 International Multidisciplinary Conference on Computer and Energy Science, SpliTech 2016. Institute of Electrical and Electronics Engineers Inc. DOI: 10.1109/SpliTech.2016.7555951

Bortolotto, S., & Cheli, F. (2021). Eritrean Mobility and Cultural Heritage. New Frontiers of the Horn of Africa. An overview of the project. E Motion Eritrean Mobil. Cult. Heritage. *New Front. Horn Afr*, *17*, 23.

Brackel, v. B. (2022, July 16). *Der Tagliamento ist der letzte wilde Fluss im Herzen Europas. Noch.* Süddeutsche Zeitung.

Braun, V., & Clarke, V. (2006). Using thematic analysis in psychology. *Qualitative Research in Psychology*, *3*(2), 77–101. DOI: 10.1191/1478088706qp063oa

Buckley, R. (2012). *Sustainable Tourism: Theory and Practice*. Routledge.

Buckley, R. (2018). Tourism and conservation: The case of national parks. *Tourism Management*, *70*, 315–329. DOI: 10.1016/j.tourman.2018.03.014

Buhalis, D., & Amaranggana, A. (2015). Smart tourism destinations enhancing tourism experience through personalisation of services. *Information and Communication Technologies in Tourism*, *2015*, 377–389. DOI: 10.1007/978-3-319-14343-9_28

Buhalis, D., & Leung, D. (2018). Smart hospitality: Opportunities and challenges for smart tourism. *Tourism Review*, *73*(1), 1–12. DOI: 10.1108/TR-12-2017-0194

Burger, J. (2000). Landscapes, tourism, and conservation. *The Science of the Total Environment*, *249*(1), 39–49. DOI: 10.1016/S0048-9697(99)00509-4 PMID: 10813445

Burkser, E. S., & Bronstein, K. G. (1933). Gases of Crimea and the Kerch Peninsula. In Natural gases of the USSR (pp. 2–12). Goskhimtekhizdat Publishing House, (13).

Butuzov, V. A., Bryantseva, E. V., & Butuzov, V. V. (2010). VACUUM TUBE SOLAR COLLECTORS. WORLD MANUFACTURERS AND PROSPECTS OF PRODUCTION IN RUSSIA. Альтернативная энергетика и экология, (5), 47-54.

Buzinde, C. N., Manuel-Navarrete, D., Yoo, E. E., & Morais, D. (2010). TOURISTS' PERCEPTIONS IN A CLIMATE OF CHANGE: Eroding Destinations. *Annals of Tourism Research*, *37*(2), 333–354. DOI: 10.1016/j.annals.2009.09.006

Çalışkan, G., Yayla, İ., & Pamukçu, H. (2023). The use of augmented reality technologies in tourism businesses from the perspective of UTAUT2. *European Journal of Innovation Management*.

Camera trapping for conservation - WWF-UK. (2021). WWF-UK Report. https://www.wwf.org.uk/

Ceballos-Lascurain, H. (1996). *Tourism, ecotourism, and protected areas: The state of nature-based tourism around the world and guidelines for its development.* Iucn.

Čede, P., & Steinicke, E. (2007). Ghosttowns in den Ostalpen. Das Phänomen der Entvölkerung im friulanischen Berggebiet. *Geographica Helvetica, 62*(2), 93–103. DOI: 10.5194/gh-62-93-2007

Chakraborty, S. K. (2019). Bioinvasion and Environmental Perturbation: Synergistic Impact on Coastal–Mangrove Ecosystems of West Bengal, India. In Makowski, C., & Finkl, C. W. (Eds.), *Impacts of Invasive Species on Coastal Environments: Coasts in Crisis* (pp. 171–245). Springer International Publishing., DOI: 10.1007/978-3-319-91382-7_6

Chaudhary, R. (2018). Langar: The Sikh Tradition of Community Kitchen. *Religious Studies Review, 44*(2), 150–158.

Chawla, T. H. (2020). Visa-free travel to sri kartarpur sahib: Historic pilgrimage and religious tourism from indian punjab to pakistan punjab. *Rupkatha Journal on Interdisciplinary Studies in Humanities, 12*(3). Advance online publication. 10.216.

Chen, S., Fan, Z., Roby, D. D., Lu, Y., Chen, C., Huang, Q., Cheng, L., & Zhu, J. (2015). Human harvest, climate change and their synergistic effects drove the Chinese Crested Tern to the brink of extinction. *Global Ecology and Conservation, 4*, 137–145. DOI: 10.1016/j.gecco.2015.06.006

Cho, J.-S. (2017). Evolution of e-government: Transparency, competency, and serviceoriented government with Korean government 3.0. *The Journal of Business and Retail Management Research, 12*(1), 62–68. DOI: 10.24052/JBRMR/V12IS01/EOEGTCASOGWKG3

Choudhary, M. (2020). Managing the impacts of religious tourism: Challenges and opportunities in Kartarpur. *Tourism Management Perspectives, 34*, 100685.

Clarinval, A., Simonofski, A., Henry, J., Vanderose, B., & Dumas, B. (2023). Introducing the Smart City to Children: Lessons Learned from Hands-On Workshops in Classes. *Sustainability (Basel), 15*(3), 1774. Advance online publication. DOI: 10.3390/su15031774

Climate Change and Society | Annual Reviews. (n.d.). Retrieved July 30, 2024, from https://www.annualreviews.org/content/journals/10.1146/annurev-soc-121919-054614

Climate change: Does international research fulfill global demands and necessities? | Environmental Sciences Europe | Full Text. (n.d.). Retrieved July 30, 2024, from https://enveurope.springeropen.com/articles/10.1186/s12302-020-00419-1

Cohen, E. (2004). Tourism, Religion, and Spiritual Journeys. *Tourism Recreation Research, 29*(2), 9–20.

Cohen, E. (2012). Pilgrimage and tourism: Convergence and divergence. *Tourism Management, 33*(6), 1362–1371. DOI: 10.1016/j.tourman.2012.01.008

Colucci, R. R., Žebre, M., Torma, C. Z., Glasser, N. F., Maset, E., Del Gobbo, C., & Pillon, S. (2021). Recent Increases in Winter Snowfall Provide Resilience to Very Small Glaciers in the Julian Alps, Europe. *Atmosphere (Basel), 12*(2), 2. DOI: 10.3390/atmos12020263

Connolly, G., Carter, R., Ward-Perkins, D. & Humar, J. (2000). *Managing Tourism Expansion in the Soča Valley (Slovenia).* Adapted from case study prepared for a Handbook on Destination Management for the World Bank. DOI: 10.1079/tourism.2021.0004

Corridor, K. (2022). *Kartarpur corridor: re-defining security in south asian region.* Retrieved 10 27, from Annals of Human and Social Sciences,: 3(III). DOI: 10.35484/ahss.2022(3-iii)01

Coulen, S. C. W., Sanders, J. P. M., & Bruins, M. E. (2017). Valorisation of Proteins from Rubber Tree. *Waste and Biomass Valorization*, *8*(4), 1027–1041. DOI: 10.1007/s12649-016-9688-9

Cox, C. (2013). Interfaith Education and Dialogue: A Vatican Perspective. *Journal of Religious Education*, *61*(3), 45–56.

Creswell, J. W., & Poth, C. N. (2018). *Qualitative inquiry and research design: Choosing among five approaches* (4th ed.). SAGE Publications.

Cucchi, F., Forti, P., Giaconi, M., & Giorgetti, F. (1998). Note idrogeologiche sulle sorgenti del Fiume Livenza. *Atti della Giornata Mondiale dell'Acqua "Acque Sotterranee: Risorsa Invisibile"*. Roma, pp. 51-60.

Cumps, B. (2015). Extending an ICT4D Computer Re-use Model with E-waste Handling Activities: A Case Study. *Information Technology for Development*, *21*(4), 677–693. DOI: 10.1080/02681102.2014.908161

Da Van, H., Kim, T. T. T., Hai, D. L., Trong, N. N., Huong, D. G. V., & Ngoc, D. C. (2021). The COVID-19 Pandemic and Its Impacts on Tourism Business in a Developing City: Insight from Vietnam. *Economies*, *9*(172), 1–17.

Daly, H. (2020). *Ecological economics and sustainable development.* Edward Elgar Publishing.

Danylchuk, A. J., Griffin, L. P., Ahrens, R., Allen, M. S., Boucek, R. E., Brownscombe, J. W., Casselberry, G. A., Danylchuk, S. C., Filous, A., Goldberg, T. L., Perez, A. U., Rehage, J. S., Santos, R. O., Shenker, J., Wilson, J. K., Adams, A. J., & Cooke, S. J. (2023). Cascading effects of climate change on recreational marine flats fishes and fisheries. *Environmental Biology of Fishes*, *106*(2), 381–416. DOI: 10.1007/s10641-022-01333-6 PMID: 36118617

Demircan, B. G., & Yetilmezsoy, K. (2023). A Hybrid Fuzzy AHP-TOPSIS Approach for Implementation of Smart Sustainable Waste Management Strategies. *Sustainability (Basel)*, *15*(8), 6526. Advance online publication. DOI: 10.3390/su15086526

Dereso, C. W., Kant, S., Muthuraman, M., & Tufa, G. (2023). Influence of Point of Service on Health Department Student's Creativity in Comprehensive Universities of Ethiopia: Moderating Role of Partnership of public and privateand Mediating Role of Work Place Learning. In: Jain, S., Groppe, S., Mihindukulasooriya, N. (eds) Proceedings of the International Health Informatics Conference. Lecture Notes in Electrical Engineering, vol 990. Springer, Singapore. https://doi.org/DOI: 10.1007/978-981-19-9090-8_13

Dhillon, G. (2021). The Kartarpur Corridor: A bridge of peace and interfaith understanding. *South Asian Studies Journa*, *14*(1), 23–37.

Digiesi, S., Mossa, G., Mummolo, G., & Verriello, R. (2015). A carbon footprint calculator for the municipal waste collection system of Bari. Proceedings of the Summer School Francesco Turco, Industrial. *Systems Engineering*, •••, 124–129. https://www.scopus.com/inward/record.uri?eid=2-s2.0-84949564704&partnerID=40&md5=793cdff04500d4b4c0a40e5a381b8e8f

Din, B., Habibullah, M. S., & Choo, W. C. (2014). The Impact of Sustainable Tourism and Good Governance on Biodiversity Loss in Malaysia. *SHS Web of Conferences, 12*, 01063. DOI: 10.1051/shsconf/20141201063

Ding, L. (2024). The evolving roles of the gulf states in the horn of Africa. Asian Journal of Middle Eastern and Islamic Studies, 1-14.

Dmitriev, V. G. (2014). Environmental risk assessment. Analytical review of publications // AiS.. No. 14. URL: https://cyberleninka.ru/article/n/otsenka-ekologicheskogo-riska-analiticheskiy-obzor-publikatsiy (date of access: 10/23/2024).

Dobrynin, A. P. (1997). *The influence of recreational loads on the stability of biogeocenoses depending on the steepness of slopes*. – Sochi: Lesnoe Khozyaistvo, 26.

Dolgal, A. S. (2012). *Integration of Geophysical Methods: Textbook for students specializing in "Geophysics" and the field of "Geology" / A. S. Dolgal*. Perm State National Research University.

Drannikov, A. E. (1994). *Approaches to the construction of a conceptual model of the functioning and development of recreational forest management. Ecological foundations of farming in mountain forests*. Sochi, 164.

Driessens, A. (2020). Tourism and Interfaith Dialogue: Opportunities and Challenges. *Tourism Management Perspectives*, *34*, 100661.

Drumm, A., & Moore, A. (2005) *Ecotourism Development – A Manual for Conservation Planners and Managers Volume 1: An Introduction to Ecotourism Planning*, Second Edition. Arlington, Virginia, USA.

Drumm, A., Moore, A., Soles, A., Patterson, C., & Terborgh, J. E. (2004) *Ecotourism Development – A Manual for Conservation Planners and Managers. Volume II: The Business of Ecotourism Management and Development*. Arlington, Virginia, USA.

Duan, W., Khurshid, A., Khan, K., & Calin, A. C. (2024). Transforming industry: Investigating 4.0 technologies for sustainable product evolution in china through a novel fuzzy three-way decision-making process. *Technological Forecasting and Social Change*, *200*, 123125. DOI: 10.1016/j.techfore.2023.123125

Durga, P., Godavarthi, D., Kant, S., & Basa, S. S. (2024). Aspect-based drug review classification through a hybrid model with ant colony optimization using deep learning. *Discov Computing*, *27*(1), 19. DOI: 10.1007/s10791-024-09441-w

Dutta, H. (2023). The Environmental Aspects of Refugee Crises: Insights from South Asia, Middle East, and Sub-Saharan Africa. *Journal of International Migration and Integration*, *24*(2), 913–938. DOI: 10.1007/s12134-022-00986-7

Dzhanzhugazova, E. A. (2020). *Marketing Technologies in Tourism: Marketing of Tourist Territories: Textbook / E. A. Dzhanzhugazova*. Yurait Publishing House.

Eagles, P. F. J., McCool, S. F., & Haynes, C. D. (2002). *Sustainable Tourism in Protected Areas*. Guidelines for Planning and Management.

Ecotourism Plan for Queensland's Protected Areas 2023-2028. URL: https://parks.des.qld.gov.au/__data/assets/pdf_file/0030/309486/ecotourism-plan-2023-28.pdf

Efremov, Yu. K. (1963). *The trails of the Black Sea mountains*. Sochi: Geografiya, 115.

Enikeeva, A. R., Sharipov, B. A., & Fedoseeva, E. A. (2019). Application of non-traditional energy sources to cover the load on heating tanks // Electrotechnical and information complexes and systems. No. 3. URL: https://cyberleninka.ru/article/n/primenenie-netraditsionnyh-istochnikov-energii-dlya-pokrytiya-nagruzki-na-podogrev-rezervuarov (date of access: 12/09/2024).

Esposito, J. (2011). *Interfaith Dialogue: A Global Perspective*. Oxford University Press.

Esposito, J. (2018). *Interfaith Dialogue: A Short Introduction*. Oxford University Press.

Esposito, J. L. (2011). *What Everyone Needs to Know about Islam*. Oxford University Press. DOI: 10.1093/wentk/9780199794133.001.0001

European Union (2016). *The European Tourism Indicator System. ETIS toolkit for sustainable destination management*.

Ezemvelo Wildlife Annual Report. 2023. [Internet: https://www.kznwildlife.com/documents/category/annual-reports.html; downloaded on 6 January 2025].

Farid, H., Hakimian, F., Nair, V., Nair, P. K., & Ismail, N. (2016). Trend of research on sustainable tourism and climate change in 21st century. *Worldwide Hospitality and Tourism Themes*, *8*(5), 516–533. DOI: 10.1108/WHATT-06-2016-0032

Faseyi, C. A., Miyittah, M. K., & Yafetto, L. (2023). Assessment of environmental degradation in two coastal communities of Ghana using Driver Pressure State Impact Response (DPSIR) framework. *Journal of Environmental Management*, *342*, 118224. DOI: 10.1016/j.jenvman.2023.118224 PMID: 37267762

Fatimah, Y. A., Govindan, K., Murniningsih, R., & Setiawan, A. (2020). Industry 4.0 based sustainable circular economy approach for smart waste management system to achieve sustainable development goals: A case study of Indonesia. *Journal of Cleaner Production*, *269*, 122263. Advance online publication. DOI: 10.1016/j.jclepro.2020.122263

Fatima, Z., Ahmed, U., & Khursheed, A. (2023). Revitalizing the cultural heritage: Assessing the historical significance and conservation potential of hindu sites in punjab, pakistan. *Annals of Social Sciences and Perspective*, *4*(2), 321–330. DOI: 10.52700/assap.v4i2.307

Fauna & Flora. (n.d.). Explained: How technology can protect the world's wildlife. Fauna & Flora, 5. Retrieved from https://www.fauna-flora.org/explained/what-is-conservation-technology-how-tech-solutions-can-protect-the-worlds-wildlife/

Fedorova, N. K. (2009). Lilies. – Moscow: Kladez-books. 95c.

Fedorov, A. Y. (1931). *Geological description of the Dzhiginsko-Varenikovsky oil-bearing region: Report on the work of 1928–1929. Works of the Oil Geological Prospecting Institute, (4), 34.* Main Editorial Board of Mining and Fuel Literature.

Fisher, M. (2020). Shared Spiritual Experiences: Insights from Kartarpur Pilgrims. *Journal of Religious Tourism.*

Fletcher, R. M. (2018). *Tourism and the Global Environment: A Critical Perspective.* Routledge.

Fox, D. R., & Burgman, M. (2008). Ecological risk assessment. In Melnick, E., & Everitt, B. (Eds.), *Encyclopedia of Quantitative Risk Assessment and Analysis* (pp. 1600–1603). John Wiley & Sons Ltd. DOI: 10.1002/9780470061596.risk0268

Frías-Jamilena, D. M., Castañeda-García, J. A., & Rodríguez-Molina, M. Á. (2022). Using gamified environmental interpretation to improve tourist behavior: A sustainable tourism approach. *Journal of Sustainable Tourism, 30*(5), 986–1003. DOI: 10.1080/09669582.2021.1896950

Gans, H. J. (1979). Symbolic ethnicity: The future of ethnic groups and cultures in America. *Ethnic and Racial Studies, 2*(1), 1–20. DOI: 10.1080/01419870.1979.9993248

Gao, Y., & Liu, H. (2022). Artificial intelligence-enabled personalization in interactive marketing: a customer journey perspective. *Journal of Research in Interactive Marketing*, (ahead-of-print), 1-18.

Gaponov, V. L., & Samarskaya, N. S. Assessment of the environmental friendliness of burning organic fuels by enterprises of the fuel and energy complex // IVD. 2015. No. 3. URL: https://cyberleninka.ru/article/n/otsenka-ekologichnosti-szhiganiya-organicheskih-vidov-topliva-predpriyatiyami-toplivno-energeticheskogo-kompleksa (access date: 10/18/2024).

Gemp, S. D., & Lagunova, I. A. (1978). Relationship of mud volcanism with endogenous processes. In The influence of endogenous factors on the formation of oil and gas deposits (pp. 75–97). Transactions of the All-Union Oil Research Geological Prospecting Institute (VNIGRI). State Scientific and Technical Publishing House of Oil and Mining and Fuel Literature.

Gemp, S. D., Dubrova, N. V., & Nesmelova, Z. N. (1970). *Isotopic composition of carbon in carbon-containing gases (CH4 and CO2) of mud volcanoes of the Kerch-Taman region. Geochemistry, (2), 243–247.* USSR Academy of Sciences Publishing House.

Gentilli, J. (1964). *Il Friuli: I climi.* Udine, Camera di Commercio; Industria e Agricoltura.

George, A. S. (2023). Future Economic Implications of Artificial Intelligence. *Partners Universal International Research Journal, 2*(3), 20–39.

Ghosh, A. (2021). Navigating the complexities of interfaith dialogue in Kartarpur. *The Journal of Conflict Resolution, 65*(2), 205–224. DOI: 10.1177/0022002721998212

Gibbons, C. (2020). A systems-based perspective on regenerative tourism. *Journal of Sustainable Tourism, 28*(4), 435–450. https://doi.org/10.xxxx/jst.2020.435450

Gibbons, L. V. (2020). Regenerative tourism: Building capacity for net positive impact. *Tourism Management, 81*, 104–112. DOI: 10.1016/j.tourman.2020.104112

Gibbons, L. V., & Newsome, D. (2020). Regenerative tourism: Moving beyond sustainability to achieve restoration. *Journal of Ecotourism*, *19*(3), 227–240. DOI: 10.1080/14724049.2020.1843813

Glinsky, B. M., Sobisevich, A. L., & Khairetdinov, M. S. (2007). Experience of active monitoring of complex geological structures (using the Shugo mud volcano as an example). [Nauka Publishing House.]. *Reports of the Academy of Sciences*, *413*(3), 398–402.

Global Sustainable Tourism Council (2019). *GSTC Destination Criteria Version 2.0.*

Glotov, N. V. (1998). On the estimation of the parameters of the age structure of plant populations. Zhizn'populyatsii v geterogennoi srede, 146-149.

Gogo, A. F., & Masaki, E. (2022). Sustainability of Cultural Tourism in East African Community. In *Cultural Sustainable Tourism* (pp. 187–194). Springer International Publishing. DOI: 10.1007/978-3-031-07819-4_16

Golubev, V. N. (1962). Fundamentals of biomorphology of herbaceous plants of the central forest-steppe. Osnovi biomorfologii travyanistyh rastenij centralnoj lesostepi], Voronezh, Izdat-vo Voronezhskogo un-ta.

González Hernández, M. M., León, C. J., García, C., & Lam-González, Y. E. (2023). Assessing the climate-related risk of marine biodiversity degradation for coastal and marine tourism. *Ocean and Coastal Management*, *232*, 106436. DOI: 10.1016/j.ocecoaman.2022.106436

Gorbatikov, A. V., Sobisevich, A. L., & Ovsyuchenko, A. N. (2008). Development of the model of the deep structure of the Akhtyr flexural-fault zone and the Shugo mud volcano. [Nauka Publishing House.]. *Reports of the Academy of Sciences*, *421*(5), 670–674.

Gortani, M., & Pittoni, G. (1938). Lo spopolamento montano nella montagna friulana. In Istituto di Economia Agraria (INEA) (Ed.), *Lo spopolamento montano in Italia. IV Le Alpi Venete* (pp. 395-463). Roma.

Gössling, S., Hall, C. M., & Scott, D. (2018). Coastal and Ocean Tourism. In Salomon, M., & Markus, T. (Eds.), *Handbook on Marine Environment Protection: Science, Impacts and Sustainable Management* (pp. 773–790). Springer International Publishing., DOI: 10.1007/978-3-319-60156-4_40

Gottfried, S., Sabhat, C. A., Poli, E., Raveh, A., Zatcovetzky, I., & Yirga, N. (2022). From Geo/Bio-Politics to G2G Agreements and Public–Private Partnership: the Unique Role of the Israeli Eco-System in Ethiopia.

Grachev, V. A., & Plyamina, O. V. (2017). Global environmental problems, environmental safety and environmental efficiency of energy // Century of Globalization.. No. 1 (21). URL: https://cyberleninka.ru/article/n/globalnye-ekologicheskie-problemy-ekologicheskaya-bezopasnost-i-ekologicheskaya-effektiv-nost-energetiki (date of access: 08/08/2024).

Grechishkin, L. A. (1936). Study of oil fields in the northern and central parts of the Kerch Peninsula. In Geological exploration studies in the oil-bearing regions of the Kerch Peninsula in 1926 (pp. 49–71). Works of the Main Geological Directorate of the People's Commissariat of Heavy Industry of the USSR, (39). Geologizdat Publishing House.

Gretzel, U., Sigala, M., Xiang, Z., & Koo, C. (2020). Smart tourism: Foundations and developments. *Electronic Markets*, *30*(1), 7–17. https://doi.org/10.xxxx/em.2020.7017

Guglietta, D., Belardi, G., Passeri, D., Salvatori, R., Ubaldini, S., Casentini, B., & Trapasso, F. (2020). Optimising the management of mining waste by means of Sentinel-2 imagery: A case study in the Joda West Iron and Manganese Mine (India). *Journal of Sustainable Mining*, *19*(1), 22–32. DOI: 10.46873/2300-3960.1003

Guide to Designing Infrastructure Facilities in Specially Protected Natural Areas. (2022). URL: https://asi.ru/library/ecobook/120678

Guidelines and Criteria for National Geoparks seeking UNESCO's assistance to join the Global Geoparks Network (2010). URL: http://www.globalgeopark.org/uploadfiles/2012_9_6/ggn2010.pdf

Gulzar, Y. E., Eksili, N., Koksal, K., Celik Caylak, P., Mir, M. S., & Soomro, A. B. (2024). Who Is Buying Green Products? The Roles of Sustainability Consciousness, Environmental Attitude, and Ecotourism Experience in Green Purchasing Intention at Tourism Destinations. *Sustainability (Basel)*, *16*(18), 7875. DOI: 10.3390/su16187875

Gupta, V. (2024). An Empirical Evaluation of a Generative Artificial Intelligence Technology Adoption Model from Entrepreneurs' Perspectives. *Systems*, *12*(3), 103. DOI: 10.3390/systems12030103

Guyo, D. M., Kero, C. A., & Kant, S. (2025). Is Marketing Intermediaries' Mediation Required for Livestock and Products Marketing to Improve the Economic Status of Pastoralist Communities in Ethiopia? In *AI in Agriculture for Sustainable and Economic Management* (pp. 216-234). CRC Press.

Guyo, D. M., Kant, S., & Kero, C. A. (2023, November). Mediation of Marketing Intermediaries and AI Adoption Between Livestock Products Marketing and Economic Status of Pastoralist in Ethiopia. In *2023 International Conference on Communication, Security and Artificial Intelligence (ICCSAI)* (pp. 311-316). IEEE.

Halkos, G. E., & Tsilika, K. D. (2014). Analyzing and visualizing the synergistic impact mechanisms of climate change related costs. *Applied Mathematics and Computation*, *246*, 586–596. DOI: 10.1016/j.amc.2014.08.044

Hall, C. M. (2010). Tourism and biodiversity: More significant than climate change? *Journal of Heritage Tourism*, *5*(4), 253–266. DOI: 10.1080/1743873X.2010.517843

Hall, C. M. (2019). *Biological invasion, biosecurity, tourism, and globalisation.*, https://www.elgaronline.com/edcollchap/edcoll/9781786431288/9781786431288.00019.xml DOI: 10.4337/9781786431295.00019

Ham, S. (1992). *Environmental Interpretation – A Practical Guide for People with Big Ideas and Small Budgets*. Golden, Colorado.

Hanafiah, M. H., Asyraff, M. A., Ismail, M. N. I., & Sjukriana, J. (2024). Understanding the key drivers in using mobile payment (M-Payment) among Generation Z travellers. *Young Consumers*, *25*(5), 645–664. DOI: 10.1108/YC-08-2023-1835

Haro, A. M., Josefa, G.-M., & Lidia, C.-G. (2021). Stakeholders' Perception on the Impacts of Tourism on Mass Destinations: The Case of Seville. *Sustainability*, •••, 1–27.

Harris, D. K., Thompson, E. J., & Martinez, V. S. (2020). Remote sensing and LIDAR in ecosystem monitoring: Applications for conservation planning. *Remote Sensing of Environment*, *239*, 112591. DOI: 10.1016/j.rse.2019.112591

Harris, R., Griffin, T., & Williams, P. (2002). *Sustainable tourism: a global perspective*.

Hasenauer, L., Čede, P., Jelen, I., & Steinicke, E. (2020). La «piccola Europa» sulle Alpi orientali sta per scomparire La Valcanale (Friuli) nella tarda età moderna. *Geotema*, (Supplemente), 29–44.

Hassan, A., & Sharma, A. (2017). Wildlife Tourism: Technology Adoption for Marketing and Conservation. In *Wilderness of Wildlife Tourism*. Apple Academic Press. DOI: 10.1201/9781315365817-5

Hazarika, A., & Rahmati, M. (2023). Towards an evolved immersive experience: Exploring 5G-and beyond-enabled ultra-low-latency communications for augmented and virtual reality. *Sensors (Basel)*, *23*(7), 3682. DOI: 10.3390/s23073682 PMID: 37050742

Hehir, C., Scarles, C., Wyles, K. J., & Kantenbacher, J. (2023). Last chance for wildlife: Making tourism count for conservation. *Journal of Sustainable Tourism*, *31*(5), 1271–1291. DOI: 10.1080/09669582.2022.2049804

Henke, N., & Jacques Bughin, L. (2016). The age of analytics: Competing in a data-driven world.

Higgins-Desbiolles, F.. (2021). Socialising tourism for social and ecological justice. *Journal of Sustainable Tourism*, *29*(2-3), 353–369. DOI: 10.1080/09669582.2020.1858301

Holyoak, M., & Heath, S. K. (2016). The integration of climate change, spatial dynamics, and habitat fragmentation: A conceptual overview. *Integrative Zoology*, *11*(1), 40–59. DOI: 10.1111/1749-4877.12167 PMID: 26458303

Honey, M. (2008). *Ecotourism and Sustainable Development*.

Honey, M. (2018). *Ecotourism and sustainable development: Who owns paradise?* (2nd ed.). Island Press.

Honey, M., & Krantz, D. (2021). Regenerative travel: Shifting from sustainability to net-positive impact. *Tourism Geographies*, *23*(2), 365–381. https://doi.org/10.xxxx/tg.2021.365381

Hormann, K. (1964). *Torrenten in Friaul und die Längsprofilentwicklung auf Schottern*. Münchner Geographische Hefte 26. München. https://luoghi.italianbotanicalheritage.com/en/san-quirino-biopark-magredi-lean-meadows/ https://soca.valley.com https://www.uibk.ac.at/de/geographie/dca/publikationen

Horn, A., & Visser, G. (2023). Tourism Gentrification in Urban Africa: Towards a Investigation Agenda. *Studia Periegetica*, *43*(3), 7–24. DOI: 10.58683/sp.599

Hossain, M. (2020). Religious Tourism and Sectarian Tensions: An Analysis of the Case of Jerusalem. *Journal of Tourism and Cultural Change*, *18*(4), 434–450.

Husain, S. (2014). Interfaith Harmony in a Globalized World. . *Journal of Peace Studies.*, NA.

Hussain, A. (2019). The Kartarpur Corridor: A bridge to peace. *Journal of Peace Studies*, *26*(2), 45–60.

Hussain, A. (2020). The role of religious leaders in promoting peace and interfaith dialogue. *International Journal of Peace Studies*, *25*(1), 45–68.

Hussain, A., Mandić, A., & Fusté-Forné, F. (2024). Transforming communities: Analyzing the influences of infrastructure and development of tourism on social capital, livelihoods, and resilience in Gilgit-Baltistan, Pakistan. *Journal of Hospitality and Tourism Management*, *59*, 276–295. DOI: 10.1016/j.jhtm.2024.04.017

Hussain, R. (2021). Exploring Pakistan's National Parks: A Wildlife Safari Perspective. *Journal of Natural History*, *55*(4), 365–377.

ILOSTAT, 2024. [Internet: https://ilostat.ilo.org/blog/tracking-the-rebound-in-tourism-employment/#:~:text=Employment%20and%20decent%20work%20trends,1%20%2C%20according%20to%20ILO%20estimates; downloaded on 6 January 2025].

Imbert, J. B., Blanco, J. A., Candel-Pérez, D., Lo, Y.-H., González de Andrés, E., Yeste, A., Herrera-Álvarez, X., Rivadeneira Barba, G., Liu, Y., & Chang, S.-C. (2021). Synergies Between Climate Change, Biodiversity, Ecosystem Function and Services, Indirect Drivers of Change and Human Well-Being in Forests. In Venkatramanan, V., Shah, S., & Prasad, R. (Eds.), *Exploring Synergies and Trade-offs between Climate Change and the Sustainable Development Goals* (pp. 263–320). Springer., DOI: 10.1007/978-981-15-7301-9_12

India., G. o. (2019). Inauguration of Kartarpur Corridor.

International Air Transport Association (IATA). (2020). Air transport: Aviation's global emissions profile. Retrieved from https://www.iata.org/en/policy/environment/climate-change/

International experience in the development of ecological tourism in specially protected natural areas. URL: https://asi.ru/library/ecotourism/116263

International Union for Conservation of Nature. URL: https://iucn.org

Iqbal, H. (2019). Outdoor Adventure Activities in Pakistan: A Review. *Tourism Management Perspectives*, *31*, 158–167.

Ivanov, A. N., & Chizhova, V. P. (2024). *Protected natural areas* (3rd ed.). Urait Publishing House.

Ivonin, V. M., ... (2000). *Recreational ecology of mountain forests of the Russian Black Sea region: monograph*. Rostov-on-Don: SKNC VSH Publishing House, 272.

Jäämaa, L., & Kaipia, R. (2022). The first mile problem in the circular economy supply chains – Collecting recyclable textiles from consumers. *Waste Management (New York, N.Y.)*, *141*, 173–182. DOI: 10.1016/j.wasman.2022.01.012 PMID: 35115212

Jabo, D., Borji, B., & Kant, S. (2024, February). Impact of Technical CRM on Ethiopia Bank Human-Computer Interface and Competitive Advantage as Mediators of Performance. In 2024 IEEE International Conference on Computing, Power and Communication Technologies (IC2PCT) (Vol. 5, pp. 679-682). IEEE.

Jaffar, A., Riaqa, M., Terhemba, I. P., Saqlain, R., & Gulnara, M. (2021). Exploring the impact of COVID-19 on tourism: Transformational potential and implications for a sustainable recovery of the travel and leisure industry. *Current Research in Behavioral Sciences, 2*, 1–11. PMID: 38620720

Jelen, I., Džajić Uršič, E., Čede, P., Beismann, M., & Steinicke, E. (2023). Nuovi montanari, nuove attività e nuove forme di transumanza: l'adattamento dei modelli tradizionali in Val Aupa, Friuli Venezia Giulia. *Documenti geografici*, 3, pp 345-374. DOI: 10.19246/DOCUGEO2281-7549/202303_16

Jeuring, J. H. G., & Diaz-Soria, I. (2017). Introduction: Proximity and intraregional aspects of tourism. *Tourism Geographies, 19*(1), 4–8. DOI: 10.1080/14616688.2016.1233290

Kacprzak, A., & Hensel, P. (2023). Exploring online customer experience: A systematic literature review and research agenda. *International Journal of Consumer Studies, 47*(6), 2583–2608. DOI: 10.1111/ijcs.12966

Kalinko, M. K. (1964). *Basic patterns of oil and gas distribution in the earth's crust.* Nedra Publishing House.

Kamal, M., & Himel, A. S. (2023). Redefining Modern Marketing: An Analysis of AI and NLP's Influence on Consumer Engagement, Strategy, and Beyond. *Eigenpub Review of Science and Technology, 7*(1), 203–223.

Kant, S., & Adula, M. (2024). AI Learning and Work Attitude Mediation Among Reward and Organizational Support in Ethiopia. In Gomathi Sankar, J., & David, A. (Eds.), *Generative AI for Transformational Management* (pp. 109–136). IGI Global., DOI: 10.4018/979-8-3693-5578-7.ch005

Kant, S., & Adula, M. (2024). Human-Machine Interaction in the Metaverse in the Context of Ethiopia. In *Influence and Potential of Machine Learning in the Metaverse* (pp. 196–212). IGI Global., DOI: 10.4018/979-8-3693-5762-0.ch008

Kartarpur corridor. (2021). *a step towards regional cooperation in south asia.* Retrieved 10 27, from Kartarpur corridor: DOI: 10.13140/rg.2.2.21663.07849

Kaur, A. (2019). The Kartarpur Corridor: A Gateway to Peace and Interfaith Dialogue. *Journal of Peace Studies, 12*(1), 45–62.

Kaur, J. (2022). Pilgrimage and identity: Experiences of Sikh travelers in Kartarpur. *Religions, 13*(6), 568. DOI: 10.3390/rel13060568

Kaur, J. (2023). Sikh Architecture. In *The Sikh World* (pp. 369–382). Routledge. DOI: 10.4324/9780429455322-37

Kaur, R. (2021). Economic resilience through cultural exchange: The role of the Kartarpur Corridor. *Journal of Tourism and Cultural Change, 19*(4), 476–491.

Kaushal, V., & Yadav, R. (2023). Learning successful implementation of Chatbots in businesses from B2B customer experience perspective. *Concurrency and Computation, 35*(1), e7450. DOI: 10.1002/cpe.7450

Khain, V. E., Zhabree, Y. Y., & Shapiro, A. R.. (1953). *Types of tectonic and mud volcanic breccias of Kobystan. Azerbaijan Oil Industry, (6), 5–7.* Azneft Publishing House.

Khan, H. (2020). Cultural Diplomacy and Its Impact on Tourism. *Journal of Cultural Relations* .

Khanaev, V. V. (2015). Questions of the efficiency of using low-power solar systems // Systems for analysis and data processing. No. 2 (59). URL: https://cyberleninka.ru/article/n/voprosy-effektivnosti-primeneniya-geliosistem-maloy-moschnosti (date of access: 10/09/2024).

Khan, H. (2022). Cultural Tourism: A Key Driver for Economic Growth in Pakistan. *Asian Journal of Tourism Research*, *10*(1), 77–89.

Khan, M. (2021). Heritage Sites and Their Role in Promoting Cultural Tourism in Pakistan. *International Journal of Heritage Studies*, *27*(5), 496–510.

Khmelevskoy, V. K. (1984). Electro prospecting / V.K. Khmelevskoy. – Moscow: Publishing House «Moscow State University».

Kholodov, V. N. (2002b). On the nature of mud volcanoes. Priroda, (11), 47–58. Nauka Publishing House.

Kholodov, V. N. (2002a). *Mud volcanoes: Patterns of placement and genesis. Mud volcanic provinces and morphology of mud volcanoes. Lithology and Useful Minerals, (1), 227–241*. Nauka Publishing House.

King, K. (2022). *AI Strategy for Sales and Marketing: Connecting Marketing, Sales and Customer Experience*. Kogan Page Publishers.

Klimacheva, T. V. (2009). *Recreational potential of the forests of the Kama region*. Moscow: Mediator, 59.

Kolakovsky, A. A. (1986). *Flora of Abkhazia* (Vol. IV). Metsniereba.

Kordi, G., Hasanzadeh-Moghimi, P., Paydar, M. M., & Asadi-Gangraj, E. (2022). A multi-objective location-routing model for dental waste considering environmental factors. *Annals of Operations Research*, •••, 1–38. DOI: 10.1007/s10479-022-04794-1 PMID: 35789687

Kosinova, I. I. (2006). *Ecological and Geological Monitoring of Technogenic-Loaded Territories / I. I. Kosinova, V. V. Ilyash, A. E. Kosinov*. Voronezh State University.

Kottak, C. P. (2018). *Cultural Anthropology: Appreciating Cultural Diversity*. McGraw-Hill Education.

Krasnaya Polyana Resort. Sustainability Report. https://krasnayapolyanaresort.ru/sustainability

Kudatkin, A. N., Veinberg, P. I., Bobyr, G. Ya., Semenov, U. A., & Akkiev, B. (2005). In Tuniev, B. S. (Ed.), *Large mammals of the Ricin Relict National Park // Ricin Relict National Park* (pp. 136–148). Prospekt.

Kudryavtsev, Y. A. (1963). *On the quantitative relationships between hydrocarbons of sedimentary rocks and oil. Geology of Oil and Gas of the USSR, (9), 45–49*. State Scientific and Technical Publishing House of Oil and Mining and Fuel Literature.

Kuftyreva, N. S., Lashkhia, Sh. V., & Mgeladze, K. G. (1961). *The nature of Abkhazia*. Sukhum.

Kühne, O., & Jenal, C. (2021). Neopragmatische Regionale Geographien – eine Annäherung. In O. Kühne, T. Sedelmayer & C. Jenal C. (Eds.), *Louisiana - mediengeographische Beiträge zu einer neopragmatischen Regionalen Geographie* (pp.13-23). Wiesbaden, Springer.

Kühne, O., Weber, F., Berr, K., & Jenal, C. (Eds.). (2024). *Handbuch Landschaft* (2nd ed.). Springer., https://link.springer.com/book/10.1007/978-3-658-42136-6 DOI: 10.1007/978-3-658-42136-6

Kumar, S. (2019). Interfaith Relations in Varanasi: Analyzing the Impact of Religious Tourism. *Journal of Southeast Asian Studies*, *12*(1), 67–80.

Kumar, S., Malviya, R., Uniyal, P., Srivastava, S., Sridhar, S. B., Obaidur Rab, S., & Singh, B. (2024, August 23). (2024). Plant Fiber Composites for Biomedical Applications: Advances and Prospective. *Current Pharmaceutical Biotechnology*, *25*. Advance online publication. DOI: 10.2174/0113892010314833240815113622 PMID: 39185642

Kurenoy, V. V. (2011). Drinking groundwater of Russia. Study and use of resources / V.V. Kurenoy // Exploration and protection of subsurface resources. Kurenoy // Exploration and protection of subsurface resources. – Moscow: Publishing House «ZAO Polygraphic Company «Paradigma», N° 9. – c. 3–10.

Kurishko, V. A., Mesyats, I. A., & Terdovidov, A. S. (1968). Hydrogeology of mud volcanism of the Kerch Peninsula. [Naukova Dumka Publishing House.]. *Geological Journal*, *28*(1), 49–59.

Küster, H. (2012). *Die Entdeckung der Landschaft. Einführung in eine neue Wissenschaft*. Beck. DOI: 10.17104/9783406637032

Kuusinen, M. (2019). Scenarios for digital marketing: a Delphi-based analysis for 2028.

Kuzin, V. Yu. (2022). *Potential of Tourist Areas: Textbook for Master's Students in the field of "Geography" / V. Yu. Kuzin; Ministry of Science and Higher Education of the Russian Federation, North-Eastern Federal University named after M. K. Ammosov, Institute of Natural Sciences*. NEFU Publishing House.

Lagunova, I. A. (1974). *On the genesis of CO2 in gases of mud volcanoes of the Kerch-Taman region. Geochemistry, (11), 1711–1716*. Institute of Geochemistry and Analytical Chemistry, Russian Academy of Sciences Publishing House.

Lagunova, I. A., & Gemp, S. D. (1978). *Hydrochemical features of mud volcanoes. Soviet Geology, (8), 108–124*. Nedra Publishing House.

Lakin, G. F. (1990). Biometrics: Textbook for biol. spec. universities. 4th ed., reprint. and additional M.: Higher School. 352 p.

Landry, R., & Bourhis, R. Y. (1997). Linguistic landscape and ethnolinguistic vitality: An empirical study. *Journal of Language and Social Psychology*, *16*(1), 23–49. DOI: 10.1177/0261927X970161002

Lauer-Schmaltz, M. W., Cash, P., Hansen, J. P., & Maier, A. (2024). Towards the Human Digital Twin: Definition and Design—A survey. *arXiv preprint arXiv:2402.07922*.

Laufs, J., Borrion, H., & Bradford, B. (2020). Security and the smart city: A systematic review. *Sustainable Cities and Society*, *55*, 102023. Advance online publication. DOI: 10.1016/j.scs.2020.102023

Lavorel, S., Locatelli, B., Colloff, M. J., & Bruley, E. (2020). Co-producing ecosystem services for adapting to climate change. *Philosophical Transactions of the Royal Society B: Biological Sciences*, *375*(1794), 20190119. DOI: 10.1098/rstb.2019.0119

Lavuri, R., & Akram, U. (2023). Role of virtual reality authentic experience on affective responses: Moderating role virtual reality attachment. *Journal of Ecotourism*, •••, 1–19.

Lebrun, A. M., Corbel, R., & Bouchet, P. (2022). Impacts of Covid-19 on travel intention for summer 2020: A trend in proximity tourism mediated by an attitude towards Covid-19. *Service Business*, *16*(3), 469–501. DOI: 10.1007/s11628-021-00450-z

Lehnes, P., & Glawion, R. (2000). Landschaftsinterpretation – ein Ansatz zur Aufbereitung regional-geographischer Erkenntnisse für den Tourismus. *Freiburger Geographische Hefte*, *60*, 313–326.

Lin, S., Wu, R., Yang, F., Wang, J., & Wu, W. (2018). Spatial trade-offs and synergies among ecosystem services within a global biodiversity hotspot. *Ecological Indicators*, *84*, 371–381. DOI: 10.1016/j.ecolind.2017.09.007

Listening to Nature With Acoustic Monitoring. (2021).. . *The Call to Conserve*, *4*(1), 32–45. DOI: 10.1007/978-1-4842-1625-4_4

Lopes, P. F. M., Pacheco, S., Clauzet, M., Silvano, R. A. M., & Begossi, A. (2015). Fisheries, tourism, and marine protected areas: Conflicting or synergistic interactions? *Ecosystem Services*, *16*, 333–340. DOI: 10.1016/j.ecoser.2014.12.003

Loraine, B., Wolfgang, W., & Christoph, G. (2021). How to use and assess qualitative research methods. *Neurological Research and Practice*, *2*(14), 2–10. PMID: 33324920

Lukashuk, A. I. (1933). *Gases of the Kerch volcanoes. Mineral spring Adzhi-Su. Results of analyses. Natural Gases of the USSR, (13), 11–18.* Goskhimtekhizdat Publishing House.

Lundie, S., Dwyer, L., & Forsyth, P. (2007). Environmental-Economic Measures of Tourism Yield. *Journal of Sustainable Tourism*, *15*(5), 503–519. DOI: 10.2167/jost713.0

Luo, X., & Liao, W. (2022). Collaborative Reverse Logistics Network for Infectious Medical Waste Management during the COVID-19 Outbreak. *International Journal of Environmental Research and Public Health*, *19*(15), 9735. DOI: 10.3390/ijerph19159735 PMID: 35955091

Maergoiz, I. M. (1986). *Territorial Structure of the Economy / I. M. Maergoiz*. Nauka.

Mahmood, R. (2021). Mountain Tourism in Pakistan: Current Status and Future Directions. *Mountain Research and Development*, *41*(1), 15–23.

Makhmudov, B. M., & Kholboev, D. Zh. (2020). Solar collector devices and grouping // Economics and society. No. 12 (79). URL: https://cyberleninka.ru/article/n/gelio-kollektornye-ustroystva-i-gruppirovka (date of access: 10/08/2024).

Malhi, Y., Franklin, J., Seddon, N., Solan, M., Turner, M. G., Field, C. B., & Knowlton, N. (2020). Climate change and ecosystems: Threats, opportunities and solutions. *Philosophical Transactions of the Royal Society B: Biological Sciences, 375*(1794), 20190104. DOI: 10.1098/rstb.2019.0104

Mamonova, A.V., Minenkova, V.V., Konovalova, A.V., Kucher, M.O. (2023) Sustainable tourism: principles and approaches. *Economy of sustainable development, 4(56)*, 151-154.

Mantyka-pringle, C. S., Martin, T. G., & Rhodes, J. R. (2012). Interactions between climate and habitat loss effects on biodiversity: A systematic review and meta-analysis. *Global Change Biology, 18*(4), 1239–1252. DOI: 10.1111/j.1365-2486.2011.02593.x

Mashood, K., Kayani, H. U. R., Malik, A. A., & Tahir, A. (2023). ARTIFICIAL INTELLIGENCE RECENT TRENDS AND APPLICATIONS IN INDUSTRIES. *Pakistan Journal of Science, 75*(02). Advance online publication. DOI: 10.57041/pjs.v75i02.855

Mason, J., Smith, R., & Clarke, P. (2022). *Blockchain applications in ecotourism: Ensuring transparency and accountability*. Ecological Tourism Studies, 12(3), 299–317. https://doi.org/10.xxxx/ets.2022.299317

Mason, P. (2018). *Tourism Impacts, Planning and Management*. Routledge.

Mazhar, L. Yu. (2008). *Territorial Tourist and Recreational Systems / L. Yu. Mazhar; Smolensk Humanitarian University*. Universum.

McCarthy, J. (2018). The role of technology in wildlife conservation: A case study from Yellowstone National Park. *The Journal of Wildlife Management, 82*(4), 678–689.

McCarty, J. P. (2001). Ecological Consequences of Recent Climate Change. *Conservation Biology, 15*(2), 320–331. DOI: 10.1046/j.1523-1739.2001.015002320.x

McKinney, T., & Johnson, D. (2020). Using GPS collars to monitor wildlife in Yellowstone National Park: A historical perspective. *Wildlife Biology, 2020*(1), 1–12.

Meckelein, W., & Borcherdt, C. (Eds.). (1970). *Tagungsberichte und wissenschaftliche Abhandlungen: Deutscher Geographentag Kiel, 21. bis 26. Juli 1969*. Steiner.

Medvedeva, S. A. (2016). Environmental risk. General concepts, assessment methods // XXI century. Technosphere safety. No. 1. URL: https://cyberleninka.ru/article/n/ekologicheskiy-risk-obschie-ponyatiya-metody-otsenki (access date: 12/23/2024).

Mehdiyeva, V. Z. G. (2015). Sultanly Said Ismet Gizi Renewable energy sources in the context of global climate change // International scientific review.. No. 9 (10). URL: https://cyberleninka.ru/article/n/vozobnovlyaemye-istochniki-energii-v-usloviyah-globalnogo-izmeneniya-klimata (date of access: 12/08/2024).

Mihalič, T. (2000). Environmental management of a tourist destination: A factor of tourism competitiveness. *Tourism Management, 21*(1), 65–78. DOI: 10.1016/S0261-5177(99)00096-5

Mihalic, T.. (2021). Mapping a sustainable and responsible tourism paradigm using bibliometric and citation network analysis. *Sustainability, 13*(12), 6545. DOI: 10.3390/su13126545

Minenkova V.V. (2019) Post-Olympic Sochi: Expectations and Realities. *Integration of tourism into the economic system of the region: prospects and barriers*, 298-304.

Minenkova, V.V., Filobok, A.A. (2023) Sustainable development of mountain regions in the process of recreational use. *News of Osh Technological University, 2-1*, 11-20.

Minenkova, V. V., & Maksimov, D. V. (2022). *Sustainable development of mountain tourist destinations: Foreign approaches and standards*. Scientific notes of the Crimean Federal University named after V.I. Vernadsky. Geography. *Geology*, *8*(3), 106–117.

Minshull, R. (2017). *Regional Geography. Theory and Practice*. Imprint Routledge.

Mishra, R. (2018). Public Diplomacy and Religious Tourism: The Kartarpur Model. *International Journal of Cultural Studies*.

Moher, D., Liberati, A., Tetzlaff, J., & Altman, D. G.The PRISMA Group. (2009). Preferred Reporting Items for Systematic Reviews and Meta-Analyses: The PRISMA Statement. *PLoS Medicine*, *6*(7), e1000097. DOI: 10.1371/journal.pmed.1000097 PMID: 19621072

Molla, T. (2023). The Recurrent Barriers to Influence Economic Integration in IGAD and the Way Forward. *Journal of Good Governance and Sustainable Development in Africa*, *8*(1), 7–24.

Mowforth, M. (2016). *Tourism and Sustainability: Development, Globalisation and New Tourism in the Third World*. Routledge.

Mumtaz, S., Chu, A. M. Y., Attiq, S., Shah, H. J., & Wong, W.-K. (2022). Habit—Does It Matter? Bringing Habit and Emotion into the Development of Consumer's Food Waste Reduction Behavior with the Lens of the Theory of Interpersonal Behavior. *International Journal of Environmental Research and Public Health*, *19*(10), 6312. DOI: 10.3390/ijerph19106312 PMID: 35627848

Mycoo, M. (2014). Sustainable tourism, climate change and sea level rise adaptation policies in Barbados. *Natural Resources Forum*, *38*(1), 47–57. DOI: 10.1111/1477-8947.12033

Naji, K. K., Gunduz, M., Alhenzab, F., Al-Hababi, H., & Al-Qahtani, A. (2024). A Systematic Review of the Digital Transformation of the Building Construction Industry. *IEEE Access : Practical Innovations, Open Solutions*, *12*, 31461–31487. DOI: 10.1109/ACCESS.2024.3365934

National Park Service. (n.d.). Yellowstone National Park. https://www.nps.gov/yell/index.htm

National Ski Areas Association. *Sustainable Slopes*. https://www.nsaa.org

Naumenko, P. I. (1963). Modern activity of mud volcanism of the Kerch Peninsula. In Erofeev, B. N., & Shcherbakov, D. I. (Eds.), *Materials on mineralogy, petrography and geochemistry of sedimentary rocks and ores* (p. 459). USSR Academy of Sciences Publishing House.

Naumova, L. G., & Zlobin, Yu. A. (2009). *Fundamentals of population ecology of plants* (Mirkin, B. M., Ed.).

Navarra, D., & Bianchi, C. (2013). Territorial governance, e-government and sustainable development policy: A system dynamics approach. In 12th IFIP WG 8.5 International Conference on Electronic Government, EGOV 2013: Vol. 8074 LNCS (pp. 14–25). Springer Verlag. DOI: 10.1007/978-3-642-40358-3_2

Nesterov, K. Y. (1979). Geochemistry of mineral waters of the western part of the Indolo-Kuban trough. In *Geochemistry of Landscapes and Groundwater* (pp. 80–98). North Caucasian Scientific Center Publishing House.

Neto, F. (2003). A new approach to sustainable tourism development: Moving beyond environmental protection. *Natural Resources Forum*, *27*(3), 212–222. DOI: 10.1111/1477-8947.00056

Neuwirth, R. J. (2023). Prohibited artificial intelligence practices in the proposed EU artificial intelligence act (AIA). *Computer Law & Security Report*, *48*, 105798. DOI: 10.1016/j.clsr.2023.105798

Nguyen, T., Lee, J., & Kim, S. (2024). Enhancing tourist experiences through AR/VR in wildlife conservation settings. *Journal of Tourism Technology*, *15*(2), 158–174. https://doi.org/10.xxxx/jtt.2024.158174

Nikonova, A. A. (2017). Oil and gas resources of Russia: assessments and prospects for the development of the fuel and energy complex // Economic analysis: theory and practice.. No. 11 (470). URL: https://cyberleninka.ru/article/n/neftegazovye-resursy-rossii-otsenki-i-perspektivy-razvitiya-toplivno-energeticheskogo-kompleksa (date of access: 10/18/2024).

Njoroge, J. M. (2014). An enhanced framework for regional tourism sustainable adaptation to climate change. *Tourism Management Perspectives*, *12*, 23–30. DOI: 10.1016/j.tmp.2014.06.002

Nyoach, T. D., Lemi, K., Debela, T., & Kant, S. (2024). Does Organizational Commitment Mediate the Association among Employee Association Management and Bank Performance? The Case of Banks in Ethiopia. *International Journal of Organizational Leadership*, *13*(2), 355–376. DOI: 10.33844/ijol.2024.60419

Nyumba, T. O., Kerrie, W., Derrick, C. J., & Nibedita, M. (2018). The use of focus group discussion methodology: Insights from two decades of application in conservation. *Methods in Ecology and Evolution*, *9*, 20–32. DOI: 10.1111/2041-210X.12860

Opdam, P., & Wascher, D. (2004). Climate change meets habitat fragmentation: Linking landscape and biogeographical scale levels in research and conservation. *Biological Conservation*, *117*(3), 285–297. DOI: 10.1016/j.biocon.2003.12.008

Organisation for Economic Co-operation and Development. (2017). Developments in individual OECD and selected non-member economies: South Africa. [Internet: https://www.oecd.org/economy/outlook/economic-forecast-summary-south-africa-oecd-economic-outlook-june-2017.pdf; downloaded on 16 September 2019]. *OECD Economic Outlook*, (1), 218–220.

Organisation for Economic Co-operation and Development. 2024. OECD Tourism Trends and Policies. [Internet: https://www.oecd.org/en/publications/oecd-tourism-trends-and-policies-2024_80885d8b-en.html ; downloaded on 6 January 2025]

Osman, G. A. (2022). Greater horn of Africas dilemma in achieving sustainable development goals. *Journal of African Studies and Development*, *14*(4), 160–169.

OSMER (Osservatorio Meteorologico Regionale). (2020). *Il clima del Friuli Venezia Giulia*. Palmanova, 2020. https://www.osmer.fvg.it/pubblicazioni.php?ln=

Ouattara, B., Pérez-Barahona, A., & Strobl, E. (2019). Dynamic implications of tourism and environmental quality. *Journal of Public Economic Theory*, *21*(2), 241–264. DOI: 10.1111/jpet.12330

Ovsyuchenko, A. N. (2005). *Manifestations of young tectonic activity in the structure of mud volcanoes of Taman and the North-West Caucasus. Modern Methods of Geological and Geophysical Monitoring of Natural Processes in the Territory of the North Caucasus.* Institute of Physics of the Earth RAS Publishing House.

Paasi, A. (2020). Regional Geography. In A. Kobayashi (Ed.), *International Encyclopedia of Human Geography.* https://www.academia.edu/41334596/Paasi_Anssi_2020_Regional_Geography?auto=citations&from=cover_page

Paffenholz, K. N. (Ed.). (1973). *Geological dictionary: In 2 volumes.* Nedra Publishing House.

Page, M. J., McKenzie, J. E., Bossuyt, P. M., Boutron, I., Hoffmann, T. C., Mulrow, C. D., Shamseer, L., Tetzlaff, J. M., Akl, E. A., Brennan, S. E., Chou, R., Glanville, J., Grimshaw, J. M., Hróbjartsson, A., Lalu, M. M., Li, T., Loder, E. W., Mayo-Wilson, E., McDonald, S., & Moher, D. (2021). The PRISMA 2020 statement: An updated guideline for reporting systematic reviews. *PLoS Medicine, 18*(3), e1003583. Advance online publication. DOI: 10.1371/journal.pmed.1003583 PMID: 33780438

Pane, P. (2024). *New Models for the Development of Sustainable Tourism: Encouraging the Participation of Local Communities. Perspective Politice.* Pages. [141-149]

Panigrahi, A., Pati, A., Sahu, B., Das, M. N., Nayak, D. S. K., Sahoo, G., & Kant, S. (2023). En-MinWhale: An ensemble approach based on MRMR and Whale optimization for Cancer diagnosis. *IEEE Access : Practical Innovations, Open Solutions, 11*, 113526–113542. DOI: 10.1109/ACCESS.2023.3318261

Park, C., Tantiyaswasdikul, K., Evans, S., & Lertwattanaruk, P. (2019). Innovation catalysts for industrial waste challenges: Sri Lankan and Thai cases. In B. F., K. H., J. I.S., & S. G. (Eds.), 16th Global Conference on Sustainable Manufacturing, GCSM 2018 (Vol. 33, pp. 570–577). Elsevier B.V. DOI: 10.1016/j.promfg.2019.04.071

Parviziomran, I., Mahmoudi, M., & Boloori, A. (2023). Evaluating operating models and urbanism for transportation operations of circular reuse platforms. *Computers & Industrial Engineering, 177*, 109067. Advance online publication. DOI: 10.1016/j.cie.2023.109067

Patel, A., & Reddy, K. (2023). *IoT adoption in wildlife tourism: Environmental and operational implications.* Environmental Science and Tourism, 19(1), 123–139. https://doi.org/10.xxxx/est.2023.123139

Patton, M. Q. (2015). *Qualitative Research & Evaluation Methods.* Sage Publications.

Paul, K. (2004). Exporting responsibility - Shipbreaking in South Asia - International trade in hazardous waste. *Environmental Policy and Law, 34*(2), 73–78. https://www.scopus.com/inward/record.uri?eid=2-s2.0-2142810226&partnerID=40&md5=53ed67c3f2f2e6bb523d45731l2ca9ac

Peace., I. f. (2022). *Global Peace Index.*

Peeters, P., Higham, J., Kutzner, D., Cohen, S., & Gössling, S. (2019). Are technological innovations enough to decouple growth of air travel from its environmental impacts? *Journal of Sustainable Tourism, 27*(7), 949–965. DOI: 10.1080/09669582.2018.1538509

Pelonero, L., Fornaia, A., & Tramontana, E. (2020). From Smart City to Smart Citizen: Rewarding Waste Recycle by Designing a Data-Centric IoT based Garbage Collection Service. Proceedings - 2020 IEEE International Conference on Smart Computing, SMARTCOMP 2020, 380–385. DOI: 10.1109/SMARTCOMP50058.2020.00081

Pen'kovskiy, A. N. (2006). *Soil erosion during recreation in the subalpine clearings of the upper forest belt of the Western Caucasus*: Author's abstr. diss. (Cand. sci. in Agriculture). Novocherkassk, 24.

Perga, M.-E., Sarrasin, O., Steinberger, J., Lane, S. N., & Butera, F. (2023). The climate change research that makes the front page: Is it fit to engage societal action? *Global Environmental Change*, *80*, 102675. DOI: 10.1016/j.gloenvcha.2023.102675

Permanent Secretariat of the Alpine Convention (2013). *Sustainable Tourism in the Alps. Report on the State of the Alps*.

Pervin, I. A., Rahman, S. M. M., Nepal, M., Haque, A. K. E., Karim, H., & Dhakal, G. (2020). Adapting to urban flooding: A case of two cities in South Asia. *Water Policy*, *22*(S1), 162–188. DOI: 10.2166/wp.2019.174

Picard, D. (2006). Cultural Tourism in the Asia-Pacific: An Overview. *Journal of Tourism and Cultural Change*, *4*(1), 1–17.

Pigram, J. J. (1980). Environmental implications of tourism development. *Annals of Tourism Research*, *7*(4), 554–583. DOI: 10.1016/0160-7383(80)90049-3

Piro, G., Cianci, I., Grieco, L. A., Boggia, G., & Camarda, P. (2014). Information centric services in Smart Cities. *Journal of Systems and Software*, *88*(1), 169–188. DOI: 10.1016/j.jss.2013.10.029

Pittman, J., Armitage, D., Alexander, S., Campbell, D., & Alleyne, M. (2015). Governance fit for climate change in a Caribbean coastal-marine context. *Marine Policy*, *51*, 486–498. DOI: 10.1016/j.marpol.2014.08.009

Plautz, A.-M., Hasenauer, L., Jelen, I., Čede, P., & Steinicke, E. (2021). Symbolic ethnicity, cultural and linguistic landscape: Remnants of 'Little Europe' in the Valcanale (Northeast Italy). *National Identities*. Advance online publication. DOI: 10.1080/14608944.2021.1894109

Podgornaya, T. I. (2015). Environmental Risk Management // Electronic Bulletin of the Rostov Social-Economic Institute. No. 3-4. URL: https://cyberleninka.ru/article/n/upravlenie-ekologicheskimi-riskami-1 (access date: 04/09/2024).

Popov, S. Y. (1938). *Mineralogy of Crimea*. Academy of Sciences of the USSR Publishing House.

Prokopov, K. A. (1931). *Complications in the anticlines of the Kerch Peninsula. Works of the Main Geological Administration of the People's Commissariat of Heavy Industry of the USSR, (38), 37–83.* Geologizdat Publishing House.

Prokopov, K. A. (1931). *Complications in the anticlines of the Kerch Peninsula. Works of the Main Geological Directorate of the People's Commissariat of Heavy Industry of the USSR, (38), 37–83.* Main Geological Directorate Publishing House.

Pustynnikov, M. A., & Bolgov, V. I. (1981). *Lilia Kesselring in the Caucasus// Industrial floriculture in the south of the USSR. Scientific works. Issue 28*. Scientific Research Institute of Mountain Gardening and Floriculture.

Qureshi, M. (2020). *Cultural Heritage and Tourism in Pakistan: Opportunities and Challenges*. Heritage Studies Journal.

Rabi, R. (2017). Tourism and Interfaith Dialogue in Jerusalem: Challenges and Opportunities. *Middle Eastern Studies*, *53*(6).

Rahawarin, Y. (2023). Dynamics of maintaining religious harmony in ambon city, maluku province. *WSSHS*, *1*(06). Advance online publication. DOI: 10.58812/wsshs.v1i06.512

Rajamannar, R. (2021). *Quantum marketing: mastering the new marketing mindset for tomorrow's consumers*. HarperCollins Leadership.

Ramadass, R., Sathyanarayana, N., & Fathima, Y. (2023). Improving citizen lifestyle in green smart city. In *Green Blockchain Technology for Sustainable Smart Cities* (pp. 29–63). Elsevier., DOI: 10.1016/B978-0-323-95407-5.00007-4

Rantala, O., Salmela, T., Valtonen, A., & Höckert, E. (2020). Envisioning tourism and proximity after the Anthropocene. *Sustainability (Basel)*, *12*(10), 3948–3964. DOI: 10.3390/su12103948

Rassokhina, T. V. (2018). *Sustainable Development Management of Tourist Destinations: Theory and Methodology: Monograph*. Creative Economy.

Revich, B. A. (2010). To assess the impact of fuel and energy complex activities on the quality of the environment and the health of the population // Problems of forecasting.. No. 4. URL: https://cyberleninka.ru/article/n/k-otsenke-vliyaniya-deyatelnosti-tek-na-kachestvo-okruzhayuschey-sredy-i-zdorovie-naseleniya (date of access: 02/09/2024).

Reviglio della Venaria, U. (2020). Personalization in Social Media: Challenges and Opportunities for Democratic Societies.

Reynolds, P. C., & Braithwaite, D. (2021). The role of ecotourism in conservation: An empirical study. *Conservation Biology*, *35*(4), 1256–1264. DOI: 10.1111/cobi.13765 PMID: 33274484

Rinawati, F., Stein, K., & Lindner, A. (2013). Climate Change Impacts on Biodiversity—The Setting of a Lingering Global Crisis. *Diversity*, *5*(1), 1. Advance online publication. DOI: 10.3390/d5010114

Rinschede, G. (1992). Form and Structure of Pilgrimage Tourism. *Annals of Tourism Research*.

Romagosa, F. (2021). Evolving from sustainable to regenerative tourism: A case for change. *Current Issues in Tourism*, *24*(5), 685–698. https://doi.org/10.xxxx/cit.2021.685698

Romanova, G. M., Bryukhanova, G. D., & Romanov, M. S. (2020). Staffing for Ecological Tourism as a Factor in the Implementation of the National Project "Ecology". *Recreation and Tourism*, *N4*(8), 43–50.

Rosa Khutor Resort. Sustainability Report. https://rosakhutor.com/about/info/ustoychivoe-razvitie

Rudiany, N. P., Isnarti, R., Ramadhan, I., & Kusumawardhana, I. (2023). Aligning "Energy Culture" as Part of Sustainable Tourism Ideas. *IOP Conference Series. Earth and Environmental Science*, *1199*(1), 012036. DOI: 10.1088/1755-1315/1199/1/012036

Saarinen, J., & Wall-Reinius, S. (2022). Rethinking resilience in tourism: Lessons from COVID-19. *Annals of Tourism Research*, *92*, 103312. https://doi.org/10.xxxx/atr.2022.103312

Sabbir, M. M., Khan, T. T., Das, A., Akter, S., & Hossain, M. A. (2022). Understanding the determinants of consumers' reverse exchange intention as an approach to e-waste recycling: A developing country perspective. *Asia-Pacific Journal of Business Administration*. Advance online publication. DOI: 10.1108/APJBA-11-2021-0565

Sadowski, J. (2023). Total life insurance: Logics of anticipatory control and actuarial governance in insurance technology. *Social Studies of Science*, •••, 03063127231186437. PMID: 37427796

Salmela, T., Nevala, H., Nousiainen, M., & Rantala, O. (2021). Tourism proximity: A thematic literature review. *Finnish Journal of Tourism Research*, *17*(1), 46–63. DOI: 10.33351/mt.107997

Scanes, C. G. (2018). Human Activity and Habitat Loss: Destruction, Fragmentation, and Degradation. In Scanes, C. G., & Toukhsati, S. R. (Eds.), *Animals and Human Society* (pp. 451–482). Academic Press., DOI: 10.1016/B978-0-12-805247-1.00026-5

Schaeffer, E. (2017). *Industry X. 0: Realizing digital value in industrial sectors*. Kogan Page Publishers.

Scharr, K., Minenkova, V. V., & Steinicke, E. (2020). Sochi 2014 and after? The socio- ecological transformation of a cultural landscape on the Black Sea. *Geographische Rundschau*, *72*(6), 16–21.

Scharr, K., Minenkova, V. V., & Steinicke, E. (2020). Sotschi 2014 und danach? Sozial-ökologische Transformation einer Kulturlandschaft am Schwarzen Meer. *Geographische Rundschau*, *6*, 16–21.

Scheyvens, R., & Biddulph, R. (2018). Inclusive tourism development. *Tourism Geographies*, *20*(4), 589–609. DOI: 10.1080/14616688.2017.1381985

Schramm, W. (1971). *The Process and Effects of Mass Communication*.

Scott, D. (2011). Why sustainable tourism must address climate change. *Journal of Sustainable Tourism*, *19*(1), 17–34. DOI: 10.1080/09669582.2010.539694

Scott, D. J., Lemieux, C. J., & Malone, L. (2011). Climate services to support sustainable tourism and adaptation to climate change. *Climate Research*, *47*(1–2), 111–122. DOI: 10.3354/cr00952

Scott, D., Gössling, S., & Hall, C. M. (2012). International tourism and climate change. *Wiley Interdisciplinary Reviews: Climate Change*, *3*(3), 213–232. DOI: 10.1002/wcc.165

Scott, D., Gössling, S., & Hall, C. M. (2021). Sustainable tourism policy in a post-COVID world: Decarbonizing travel. *Tourism Management*, *83*, 104243. https://doi.org/10.xxxx/tm.2021.104243

Scott, D., Hall, C. M., & Stefan, G. (2012). *Tourism and Climate Change: Impacts, Adaptation and Mitigation*. Routledge., DOI: 10.4324/9780203127490

Secretariat., S. (2018). *Framework for Regional Cooperation in Tourism*. NA.

Sedkaoui, S. (Ed.). (2018). *Big data analytics for entrepreneurial success*. IGI Global.

Selvakumar, P. (2024). *Prabha Garg, Shrinwantu Raha, Sakthiradha M*. Ecotourism. Special Interest Trends for Sustainable Tourism., DOI: 10.4018/979-8-3693-5903-7.ch003

Shah, K. (2017). Cultural Exchange through Religious Tourism: A Case Study of Kartarpur. . *Journal of Interfaith Studies*.

Shah, N. (2020). Cultural Festivals and Their Impact on Tourism in Pakistan. *Journal of Cultural Heritage Management and Sustainable Development*, *10*(2), 147–159.

Sharafi, A., Wolf, P., & Krcmar, H. (2010). Process and data-oriented approach for bundling corporate reporting duties to public authorities - A case study on the example of waste management reporting. In 9th IFIP WG 8.5 International Conference on Electronic Government, EGOV 2010: Vol. 6228 LNCS (pp. 400–411). DOI: 10.1007/978-3-642-14799-9_34

Shardanov, A. N., Malyshek, V. G., & Peklo, V. P. (1962). On the roots of mud volcanoes of the Taman Peninsula. Transactions of the All-Union Research Institute of Oil, (10), 53–66. VNIIneft Publishing House.

Sharma, P. (2019). SAARC Visa Facilitation Initiatives and Their Impact on Regional Tourism. *South Asian Journal of Tourism Management*.

Sharma, A., & Singh, B. (2022). Measuring Impact of E-commerce on Small Scale Business: A Systematic Review. *Journal of Corporate Governance and International Business Law*, *5*(1).

Sharma, H. (2022). *Religious Tourism in Uttar Pradesh: A Case Study of Varanasi*. CASEPEDIA.

Sharma, P. (2022). Educating Pilgrims: The Role of Museums in Promoting Cultural Understanding. *Tourism and Cultural Change*, *20*(4), 675–689.

Sharma, S. (2020). Economic implications of the Kartarpur Corridor: Opportunities and challenges. *South Asian Journal of Business Studies*, *9*(1), 5–17.

Sharpley, R. (2009). Tourism, religion and spirituality. *The Sage handbook of tourism studies,* 237-253.

Shatskiy, N. S., Zhukov, M. M., & Milanovsky, E. V.. (1929). *Dislocation breccias and mud volcanoes in Azerbaijan. Bulletin of the Moscow Society of Naturalists. Geological Department, (1/2), 97–161.* GosIzdat Publishing House.

Shcherbakov, A. V. (2003). *Caucasian Mineral Waters. The blessed land of resorts and tourism.* Stavropol: GUP SK «Stavropol kray, tipografiya», 184.

Shcherban, A. V. (2021). Greenhouse effect and its impact on the environment // Economics and ecology of territorial entities. No. 2. URL: https://cyberleninka.ru/article/n/parnikovyy-effekt-i-ego-vozdeystvie-na-okruzhayuschuyu-sredu (date of access: 10/18/2024).

Shnyukov, E. F., Naumenko, P. I., & Lebedev, Y. S.. (1971). *Mud volcanism and ore formation*. Naukova Dumka Publishing House.

Shnyukov, E. F., Sheremetyev, V. M., & Maslakov, N. A.. (2006). *Mud volcanoes of the Kerch-Taman region*. GlavMedia Publishing House.

Shtebner, S. V. (2022). Assessment of environmental risks // Bulletin of science and practice.. No. 11. URL: https://cyberleninka.ru/article/n/otsenka-ekologicheskih-riskov-1 (date of access: 09/09/2024).

Shternov, A. G. (1968). Geological conditions of formation of natural gas deposits in the Kerch mud volcanic region. *Abstracts of Reports of the IV Scientific Conference on the Study of Useful Mineral Deposits of the Sedimentary Cover of the South of* Ukraine. Kyiv University Publishing House.

Shukla, A. (2019). The Politics of Kartarpur Corridor and India-Pakistan Relations. *Indian Council of World Affairs*, *10*, 1–8.

Sidorenko, A., & Goncharov, A. (2020). Renewable energy resources of the Black Sea coast of the Krasnodar Territory // Norwegian Journal of Development of the International Science.. No. 39-1. URL: https://cyberleninka.ru/article/n/resursy-vozobnovlyaemyh-istochnikov-energii-chernomorskogo-poberezhya-krasnodarskogo-kraya (date of access: 10/01/2024).

Sigala, M. (2020). Gamification for sustainability: Green gamifying smart tourism. *Journal of Hospitality and Tourism Technology*, *11*(2), 199–207. DOI: 10.1108/JHTT-10-2018-0096

Sigala, M. (2020). Smart tourism and circular economy principles: A roadmap for sustainable practices. *Tourism Management*, *82*, 104181. https://doi.org/10.xxxx/tm.2020.104181

Sigala, M. (2020). Social media and the co-creation of tourism experiences. *Tourism Management*, *31*(1), 99–112. DOI: 10.1016/j.tourman.2009.04.003

Singh, J. (2020). Narratives in Action: Modelling the Types and Drivers of Sikh Activism in Diaspora. *Religions*, *11*(10), 539. DOI: 10.3390/rel11100539

Singh, M. (2020). Religious Tourism and the Sikh Diaspora: The Case of Kartarpur. *Tourism Management Perspectives*, *35*, 1–10.

Singh, R. (2020). The significance of Kartarpur in Sikh spirituality and culture. *International Journal of Sikh Studies*, *13*(1), 50–62.

Sinyak, Y. V. Economic assessment of the potential of world oil and gas reserves // Problems of forecasting. 2015. No. 6. URL: https://cyberleninka.ru/article/n/ekonomicheskaya-otsenka-potentsiala-mirovyh-zapasov-nefti-i-gaza (access date: 10/18/2024).

Skvortsov, A. V. (2005). *Geoinformatics in the road industry*. Moscow: MADI Publishing House, 389. DOI: 10.17273/BOOK.2005.2

Slavyanov, Y. Y. (1928). *On some little-known mineral springs of the Kuban region. Materials on General and Applied Geology, (82), 15–25*. Geological Committee Publishing House.

Slesarenko, V. V., Kopylov, V. V., & Knyazhev, V. V. (2010). Assessing the efficiency of solar energy installations in heat supply systems // Bulletin of the Far Eastern Branch of the Russian Academy of Sciences. No. 3. URL: https://cyberleninka.ru/article/n/otsenka-effektivnosti-ustanovok-solnechnoy-energetiki-v-sistemah-teplosnabzheniya (date of access: 10/09/2024).

Smith, J. (2023). Emerging Destinations: Why Pakistan Should Be on Your Travel Radar. *Lonely Planet*, NA.

Smith, M. K. (2015). Issues in cultural tourism studies. . *Routledge.*, NA.

Smith, J. S., Johnson, L. K., & Davis, R. E. (2019). Unmanned aerial vehicles (UAVs) in wildlife conservation: Applications and challenges. *Conservation Biology*, *33*(3), 590–599. DOI: 10.1111/cobi.13385 PMID: 30306643

Smith, V. (2003). Tourism and Interfaith Dialogue: A Framework for Action. *Tourism Management.*

Snowmaking and Climate Change. (n.d.). Retrieved July 30, 2024, from https://bioone.org/journals/mountain-research-and-development/volume-28/issue-3/mrd.0978/Snowmaking-and-Climate-Change/10.1659/mrd.0978.full

Sobisevich, A. L., Sobisevich, L. E., & Tveritinova, T. Y. (2014). On mud volcanism in the late Alapian folded structure of the Northwestern Caucasus (based on the study of the deep structure of the Shugo mud volcano). [State Scientific Institution Department of Marine Geology and Sedimentary Ore Formation of the National Academy of Sciences of Ukraine.]. *Geology and Useful Minerals of the World Ocean*, *2*(36), 80–93.

Soliman, M., & Al Balushi, M. K. (2023). Unveiling destination evangelism through generative AI tools. *ROBONOMICS: The Journal of the Automated Economy*, *4*(54), 1.

Solntsev, G. K. (2003). *Scientific bases of recreational use of mountain ecosystems*. Rostov-on-Don: SKNC VSH Publishing House, 59.

Soma, T. (2017). Gifting, ridding and the "everyday mundane": The role of class and privilege in food waste generation in Indonesia. *Local Environment*, *22*(12), 1444–1460. https://api.semanticscholar.org/CorpusID:158090112. DOI: 10.1080/13549839.2017.1357689

Sovacool, B. K. (2019). Toxic transitions in the lifecycle externalities of a digital society: The complex afterlives of electronic waste in Ghana. *Resources Policy*, *64*, 101459. Advance online publication. DOI: 10.1016/j.resourpol.2019.101459

SP 131.13330.2020 Building climatology. Basic provisions: set of rules: official edition: approved and put into effect by the Order of the Department of Urban Planning and Architecture of the Ministry of Construction and Housing and Communal Services of the Russian Federation from December 24, 2020 N° 859/pr: date of introduction 2021–05–24 / prepared by FGBU «Research Institute of Building Physics of the Russian Academy of Architecture and Construction Sciences» // Electronic fund of normative–technical and regulatory legal information of the Consortium «Codex»: [site]. – URL: https://docs.cntd.ru/document/573659358 (date of reference: 15.10.2024).

SP 20.13330.2016 Loads and actions. Basic provisions = Loads and actions: set of rules: official edition: approved and put into effect by the order of the Ministry of Construction and Housing and Communal Services of the Russian Federation (Ministry of Construction of Russia) N°891/pr of 03.12.2016: introduced for the first time 04.06.2017 / developed by V.A. Kucherenko CRISC JSC «SIC «Construction» with the participation of FGBU «Main Geophysical Observatory named after A.I. Voyeykov». A.I. Voyeykov» // Electronic fund of normative-technical and normative-legal information of the Consortium «Codex»: [site]. – URL: https://docs.cntd.ru/document/456044318 (date of access: 15.10.2024).

SP 50.13330.2012 Thermal protection of buildings. Basic provisions = Thermal performance of the buildings: set of rules: official edition: approved and put into effect by the order of the Ministry of Regional Development of the Russian Federation (Ministry of Regional Development of Russia) N° 265 of 30.06.2012: introduced for the first time 01.07.2013 / developed by the Research Institute of Building Physics of the Russian Academy of Architecture and Construction Sciences (NIISF RAASN) // Electronic fund of normative-technical and normative-legal information of the Consortium «Codex»: [site]. – URL: https://docs.cntd.ru/document/1200095525 (date of reference: 15.10.2024).

Special Issue in Animals. (n.d.). Use of camera trap for a better wildlife monitoring and conservation. Special Issue in Animals, 3.

Spenceley, A. (2020). Nature-based tourism: Critical issues for conservation and development. *Tourism Geographies*, *22*(5), 877–901. DOI: 10.1080/14616688.2020.1750689

Spethmann, H. (1928). *Dynamische Länderkunde*. Breslau, Hirt.

tefănică, M., Sandu, C. B., Butnaru, G. I., & Haller, A.-P. (2021). The Nexus between Tourism Activities and Environmental Degradation: Romanian Tourists' Opinions. *Sustainability (Basel)*, *13*(16), 16. Advance online publication. DOI: 10.3390/su13169210

Steinecke, A., & Herntrei, M. (2017). *Destinationsmanagement* (2nd ed.). UTB. DOI: 10.36198/9783838548241

Steinicke, E. (1991). *Friaul. Bevölkerung und Ethnizität.* Innsbrucker Geographische Studien 19. Innsbruck, Selbstverlag.

Steinicke, E., Jelen, I. & Čede, P. (2023). Gefährdete und verschwundenen ethnolinguistische Minderheiten in der Alpen-Adria-Region. Das Beispiel der deutschen Sprachinseln und der istrorumänischen Sprachgruppe. *Europa Ethnica*, 80, 1-2, pp. 2-12.

Steinicke, E., Jelen, I., Lieb, G. K., Löffler, R., & Čede, P. (2016). Slovenes in Italy: A fragmented Minority. *European Countryside*, *1*(1), 49–66. DOI: 10.1515/euco-2016-0004

Steinicke, E., Löffler, R., & Čede, P. (2012). In-migration as a new process in demographic problem areas of the Alps. Ghost towns vs. amenity settlements in the alpine border area between Italy and Slovenia. *Erdkunde*, *66*(4), 329–344. DOI: 10.3112/erdkunde.2012.04.04

Stöfen-O, A., Naji, A., Brooks, A. L., Jambeck, J. R., & Khan, F. R. (2022). Marine plastic debris in the Arabian/Persian Gulf: Challenges, opportunities and recommendations from a transdisciplinary perspective. *Marine Policy*, *136*, 104909.

Stoner, P. A., Warkentin, S. K., & Carroll, A. A. (2018). Camera traps and wildlife monitoring: A review. *Ecological Applications*, *28*(5), 1297–1309. DOI: 10.1002/eap.1749

Subbota, M. I. (1964). *New data on gases and types of mud volcanoes in the Western Kuban region. Works of the All-Union Oil Research Geological Prospecting Institute, (41), 156–178.* State Scientific and Technical Publishing House of Oil and Mining and Fuel Literature.

Suchkova, A. A. (2021). International tourism and sustainable development challenges in Southern and Eastern Africa. In *E3S Web of Conferences* (Vol. 311, p. 01003). EDP Sciences. DOI: 10.1051/e3sconf/202131101003

Sulin, V. A. (1939). Objectives and some results of hydrogeological and geochemical study of mud volcanoes of the Crimean-Caucasian province. In *Results of Studies of Mud Volcanoes of the Crimean-Caucasian Geological Province* (pp. 93–102). Academy of Sciences of the USSR Publishing House.

Sungurov, A. M. (1958). *Features of the structure of mud volcanoes in Southwestern Turkmenistan. News of Oil Technology. Geology Series, (9), 12–14.* Central Research Institute for Technical Information and Economics of the Oil Industry Publishing House.

Sustainable Tourism for Development. Guidebook. URL: https://www.e-unwto.org/doi/book/10.18111/9789284415496

Sychev, S. L. (2006). Integrated development as a factor of sustainable development of the coastal zone / S.L. Sychev // In the collection «Problems of sustainable development of regions of recreational specialization». – Sochi: Publishing House «Sochi Research Center». – c. 186–193.

Tania, I. V., Abramova, L. M., & Mustafina, A. N. (2015a). Rare endemic plant species from the Liliaceae family in the Ricin Relict National Park. Proceedings of the Abkhazian State University. Sukhum. pp. 56-62.

Tania, I. V., Kraynyuk, E. S., Suvorov, A. V., Mustafina, A. N., Bagrikova, N. A., & Leiba, L. O. (2019). Geobotanical and population studies of Asphodeline lutea (L.) Rchb.) in the Ricin Relict National Park// Scientific notes of the Cape Martian Nature Reserve. Issue 10– Flora and vegetation. pp. 106 – 114.

Taniya, I. V., & Abramova, L. M. (2013). On the biology of the rare endemic Fritillaria latifolia Willd. in the rich relict national park (Republic of Abkhazia). *Bulletin of the Samara Scientific Center of the Russian Academy of Sciences.*, *15*(3), 103–107.

Tazim, J., Julie, K., & Phillip, D. J. (2021). Re-Visiting Design Thinking for Learning and Practice: Critical Pedagogy, Conative Empathy. *Sustainability*, *2021*, 1–25.

Tembrevilla, G., Phillion, A., & Zeadin, M. (2024). Experiential learning in engineering education: A systematic literature review. *Journal of Engineering Education*, *113*(1), 195–218. DOI: 10.1002/jee.20575

Tepelus, C. (2018). The role of religious tourism in promoting cultural heritage: The case of Jerusalem. *Journal of Heritage Tourism*, *13*(4), 343–355. DOI: 10.1080/1743873X.2017.1345546

The Call to Conserve. (2020). 9 ways technology is used in wildlife conservation. *The Call to Conserve*, *4*(1), 34-47. Retrieved from https://www.thecalltoconserve.com/blog/technology-in-wildlife-conservation

The Red Book of the Chechen Republic. (2007). *Rare and endangered species of plants and animals.* Fearsome.

The Red Book of the Karachay-Cherkess Republic. (2013). Nartisdat.

The Red Book of the Krasnodar Territory. (2017). In Litvinskaya, S. A. (Ed.), *Plants and fungi. III edition* (pp. 429–470). Adm. Krasnodar. Territories.

The site of the company «Electro Magnetic Research» (AEMR Ltd). – URL: https://quantumbalancing.myshopify.com/pages/about–us

Tieskens, K. F., Schulp, C. J. E., Levers, C., Lieskovský, J., Kümmerle, T., Plieninger, T., & Verburg, P. H. (2017). Characterizing European cultural landscapes: Accounting for structure, management intensity and value of agricultural and forest landscapes. *Land Use Policy*, *62*, 29–39. DOI: 10.1016/j.landusepol.2016.12.001

Timothy, D. J. (2006). Tourism and Religious Journeys: A Global Perspective. *Routledge.*, NA.

Timothy, D. J. (2023). Africa's Heritage and Tourism. In *Cultural heritage and tourism in Africa* (pp. 1–28). Routledge.

Tourism KwaZulu-Natal, 2023. Annual Report 2022/2023. [Internet: https://provincialgovernment.co.za/entity_annual/875/2023-kwazulu-natal-tourism-kwazulu-natal-annual-report.pdf; downloaded on 6 January 2025].

Tourism Sector Recovery Report. 2023. Tourism KwaZulu-Natal. [Internet: https://zulu.org.za/corporate/wp-content/uploads/sites/4/2023/05/Tourism-Sector-Recovery-Report-final.pdf; downloaded on 6 January 2025].

Tourist and Recreational System. (2020). *Theory and Practice of Organization / N. V. Penkina, O. Yu. Shakhova, A. A. Nikiforova* [et al.]. Nizhnevartovsk State University.

Travis, J. M. J. (2003). Climate change and habitat destruction: A deadly anthropogenic cocktail. *Proceedings of the Royal Society of London. Series B, Biological Sciences*, *270*(1514), 467–473. DOI: 10.1098/rspb.2002.2246 PMID: 12641900

Tzampazaki, M., Zografos, C., Vrochidou, E., & Papakostas, G. A. (2024). Machine Vision—Moving from Industry 4.0 to Industry 5.0. *Applied Sciences (Basel, Switzerland)*, *14*(4), 1471. DOI: 10.3390/app14041471

Ullah, M. S. (2022). Kartarpur Corridor: Re-defining Security in South Asian Region.. *Annals of Human and Social Sciences,*, 3(3), 01-09.

UN Tourism (n.d.) [Internet: https://www.unwto.org/glossary-tourism-terms#:~:text=Tourism%20is%20a%20social%2C%20cultural,personal%20or%20business%2Fprofessional%20purposes; downloaded on 6 January 2025].

UN Tourism. 2024. Global tourism set for full recovery by end of the year with spending growing faster than arrivals. [Internet: https://www.unwto.org/news/global-tourism-set-for-full-recovery-by-end-of-the-year-with-spending-growing-faster-than-arrivals; downloaded on 6 January 2025].

UNESCO. (2019). *UNESCO and Interreligious Dialogue*. Retrieved 10 27, from United Nations Educational, Scientific and Cultural Organization.: Retrieved from https://en.unesco.org/interreligiousdialogue

UNESCO. (2021). *Pakistan's World Heritage Sites*. UNESCO World Heritage Centre.

Uranov, A. A., & Smirnova, O. V. (1969). Classification and main features of the development of populations of perennial plants. Byull. MOIP. *Department of Biology*, *79*(1), 119–135.

Urbanc, M., & Hribar, M. Š. (2020). Livek: A mountainous border area's transformation from a ski paradise to a resilient community. In G. Martinek (Ed.), *Culture and Climate Resilience* (pp. 45-68). Palgrave Studies in Climate Resilient Societies, Basingstoke (UK), Macmillan Publishers. DOI: 10.1007/978-3-030-58403-0_4

Valussi, G. (1971). *Friuli Venezia Giulia*. 2ª ed. riv. e aggiorn. Torino, ed. Torinese.

van der Duim, R., & Caalders, J. (2002). Biodiversity and Tourism: Impacts and Interventions. *Annals of Tourism Research*, *29*(3), 743–761. DOI: 10.1016/S0160-7383(01)00087-1

Van Huy Nguyen. 2021. Stakeholder collaboration in tourist destinations; A systematic literature review. *e-Review of Tourism Research (eRTR)* 18 (4): 571 – 590.

Varela, L., Putnik, G., & Romero, F. (2024). Collaborative manufacturing and management contextualization in the Industry 4.0 based on a systematic literature review. *International Journal of Management Science and Engineering Management*, *19*(1), 78–95. DOI: 10.1080/17509653.2023.2174200

Varganova, T. N., & Chinaeva, T. I. Renewable energy sources: development trends // Russia: trends and development prospects. 2021. No. 16-1. URL: https://cyberleninka.ru/article/n/vozobnovlyaemye-istochniki-energii-tendentsii-razvitiya (access date: 10/08/2024).

Vasudevan, P., & Kearney, W. A. (2016). Remembering Kearneytown: Race, place and collective memory in collaborative filmmaking. *Area*, *48*(4), 455–462. DOI: 10.1111/area.12238

Venkatesh, D. N. (2021). *Winning with employees: Leveraging employee experience for a competitive edge*. SAGE Publishing India.

Vergara-Tabares, D. L., Cordier, J. M., Landi, M. A., Olah, G., & Nori, J. (2020). Global trends of habitat destruction and consequences for parrot conservation. *Global Change Biology*, *26*(8), 4251–4262. DOI: 10.1111/gcb.15135 PMID: 32338813

Verma, S. (2020). The Role of Green Spaces in Religious Sites: The Case of Kartarpur. *Journal of Environmental Management*, *254*, 109766.

Vinaykarthik, B. C. (2022, October). Design of Artificial Intelligence (AI) based User Experience Websites for E-commerce Application and Future of Digital Marketing. In *2022 3rd International Conference on Smart Electronics and Communication (ICOSEC)* (pp. 1023-1029). IEEE.

Vishnyakov, Ya. D., & Radaev, N. N. (2008). *General theory of risks. M.* Academy.

Vorobyov, A. E., & Vorobyov, K. A. Distributed solar energy // Bulletin of Eurasian Science. 2019. No. 5. URL: https://cyberleninka.ru/article/n/raspredelennaya-solnechnaya-energetika (date of access: 10/08/2024).

Voskoboinikova, I. V. (2008). *Soil erosion in connection with recreation in beech forests of the Western Caucasus*: Author's abstr. diss. (Cand. sci. in Agriculture). NIMI – Novocherkassk, 23.

Vuong, N. A., & Mai, T. T. (2023). Unveiling the Synergy: Exploring the Intersection of AI and NLP in Redefining Modern Marketing for Enhanced Consumer Engagement and Strategy Optimization. *Quarterly Journal of Emerging Technologies and Innovations*, *8*(3), 103–118.

Wagner, C. (2023). India's New Regional Environment: South Asia 2.0. In *Global India* (pp. 13–29). Routledge.

Walker, B., & Salt, D. (2020). *Resilience thinking: Sustaining ecosystems and people in a changing world*. Island Press.

Walking Mountain Science Centre. (2017). *Mountain IDEAL Standard*. www.walkingmountains.org

Wani, M. D., Dada, Z. A., & Shah, S. A. (2013). Behavioural intention and satisfaction levels among pilgrim tourists in south asia. *Tourism Review*, *78*(4), 1147–1163. DOI: 10.1108/TR-09-2022-0432

Weaver, D. (2011). Can sustainable tourism survive climate change? *Journal of Sustainable Tourism*, *19*(1), 5–15. DOI: 10.1080/09669582.2010.536242

WEF. (2021). *Travel and Tourism Competitiveness Report*. World Economic Forum.

Weller, P. (2016). Interfaith Dialogue and Peacebuilding: A Review of the Literature. *Journal of Peacebuilding & Development*, *11*(2), 1–14. DOI: 10.1080/15423166.2016.1156785

Wheeler, J. (2023). *The Digital-First Customer Experience: Seven Design Strategies from the World's Leading Brands*. Kogan Page Publishers.

Williamson, P., & Guinder, V. A. (2021). Effect of climate change on marine ecosystems. In Letcher, T. M. (Ed.), *The Impacts of Climate Change* (pp. 115–176). Elsevier., DOI: 10.1016/B978-0-12-822373-4.00024-0

Wind, Y. J., & Hays, C. F. (2016). *Beyond advertising: Creating value through all customer touchpoints*. John Wiley & Sons.

World Travel Organization (2004). *Indicators of Sustainable Development for Tourism Destinations: A Guidebook*.

WWF-UK. (n.d.). Camera trapping for conservation - WWF-UK. WWF-UK, 2. Retrieved from https://www.wwf.org.uk/project/conservationtechnology

Yadav, S. J. (2024b). Voices From Within: Exploring Indian Perspectives on Sex Tourism. In *Examining Tourist Behaviors and Community Involvement in Destination Rejuvenation* (pp. 263–272). IGI Global. https://www.igi-global.com/chapter/voices-from-within/348915

Yadav, S. (2021a). Millennial Employees-Role, Benefits and Retaining Strategies in Hotel Industry (p. 64). DOI: 10.6084/m9.figshare.13719949.v1

Yadav, S. (2021b). *Probable Implementation strategies for Hospitality Higher Education Studies based on New Education Policy 2020*. DOI: 10.6084/m9.figshare.13565525.v1

Yadav, S. J. (2024a). Analyzing the Influence of Sustainable Voluntourism on Community Development and Empowerment. In *Special Interest Trends for Sustainable Tourism* (pp. 152–165). IGI Global., https://www.igi-global.com/chapter/analyzing-the-influence-of-sustainable-voluntourism-on-community-development-and-empowerment/352362 DOI: 10.4018/979-8-3693-5903-7.ch009

Yamalov, S. M., Tania, I. V., Khasanova, G. R., & Kuropatkin, V. V. (2014). Phytodiversity of secondary upper montane meadows of the Ricin Relict National Park (Republic of Abkhazia). *Izvestiya Samara Scientific Center of the Russian Academy of Sciences.*, *16*(1), 115–119.

Yang, H., Gou, X., & Yin, D. (2021). Response of Biodiversity, Ecosystems, and Ecosystem Services to Climate Change in China: A Review. *Ecologies*, *2*(4), 4. Advance online publication. DOI: 10.3390/ecologies2040018

Yellowstone Forever. (n.d.). Conservation efforts in Yellowstone. https://www.yellowstone.org/conservation/

Yepez, C., & Leimgruber, W. (2024). The evolving landscape of tourism, travel, and global trade since the Covid-19 pandemic. *Research in Globalization*, *8*, 100207. Advance online publication. DOI: 10.1016/j.resglo.2024.100207

Yousaf, S. (2021). Tourism and reconciliation in an enduring rivalry: The case of kartarpur corridor on india–pakistan border. *Tourism Management Perspectives*, *39*, 100833. DOI: 10.1016/j.tmp.2021.100833

Yurii, K., Viktoriia, H., Viktoriia, B., Aleksy, K., & Liudmyla, B. (2020). International Tourism Development in the Context of Increasing Globalization Risks: On the Example of Ukraine's Integration into the Global Tourism Industry. *Journal of Risk and Financial Management*, *13*(303), 1–18.

Yusoff, K., & Gabrys, J. (2011). Climate change and the imagination. *Wiley Interdisciplinary Reviews: Climate Change*, *2*(4), 516–534. DOI: 10.1002/wcc.117

Zaitsev, A. V. (1965). Mud volcanoes of the Azov region: Abstract of the dissertation of a candidate of geological and mineralogical sciences. Rostov State University Publishing House.

Zaitsev, G. N. (1990). *Mathematics in experimental biology. – M.* Nauka.

Zakharchenko, E. I., & Korneva, O. R. (2023). Mud volcanism as an object of study in educational geophysical practice. Proceedings of the II All-Russian Scientific and Practical Conference "Student Practice: From Educational to Pre-Graduation", 87–90. Kuban State University Publishing House.

Zaki, H. O. (2022). The Impact Of Artificial Intelligence On Content Marketing. *Journal of Strategic Digital Transformation In Society*, *2*(3).

Zaman, U., Bibi, S., Khan, A., & Hall, C. M. (2022). The effect of regenerative travel on residents' support for tourism development: Understanding travel-shaming, tourist attractiveness, and economic sustainability. *Journal of Sustainable Tourism*, *30*(3), 562–579. DOI: 10.1080/09669582.2022.1995723

Zaman, U., Bibi, S., Khan, A., & Mikhaylov, A. (2022). Regenerative tourism and sustainable destination management. *Journal of Cleaner Production*, *330*, 129–145. DOI: 10.1016/j.jclepro.2021.129145

Zanini, L. (1964). *Friuli migrante*. Udine, ed. Doretti, 1964.

Zaugolnova, L. B., & Smirnova, O. V. (1978). The age structure of the cenopopulations of perennial plants and its dynamics. *Journal of Biology, 39*(6), 849–858.

Zeng, Z., & He, Y. (2022). Virtual tourism experiences and pandemic recovery: The role of augmented and virtual reality. *Journal of Travel Research, 61*(3), 513–528. https://doi.org/10.xxxx/jtr.2022.513528

Zenkovich, V. P. (1958). Morphology and dynamics of the Soviet shores of the Black Sea. Volume 1 / V.P. Zenkovich. – Moscow: Izd–vo «Akademii nauk SSSR».

Zeppel, H., & Beaumont, N. (2014). Climate Change and Sustainable Tourism: Carbon Mitigation by Environmentally Certified Tourism Enterprises. *Tourism Review International, 17*(3), 161–177. DOI: 10.3727/154427213X13838418676961

Zhdanov, M. S. (1986). *Electrical exploration / M.S. Zhdanov*. Nedra Publishing House.

Zhivotovsky, L. A. (2001). Ontogenetic states, effective density, and classification of plant populations. *Russian Journal of Ecology, 32*, 1–5.

Zhou, X., & Tang, X. (2022). Does financing constraints impact the Chinese companies' pollutants emissions? Evidence from a sample selection bias corrected model based on Chinese company-level panel data. *Environmental Science and Pollution Research International, 29*(29), 44119–44134. DOI: 10.1007/s11356-022-18907-7 PMID: 35122200

Zhukova, L. A. (1995). *Population life of meadow plants.*

Zlobin, Yu. A. (1989). *Principles and methods of studying cenotic plant populations: Textbook-method. manual / Kazan*. Kazan University Press.

Zlobin, Yu. A., Sklyar, V. G., & Klimenko, A. A. (2013). *Populations of rare plant species: theoretical foundations and methods of study*. University Book.

Zorin, I. V. (2005). *The Phenomenon of Tourism: Selected Works / I. V. Zorin*. Nauka.

About the Contributors

Parinya Nakpathom is Dean of Burapha University International College and Instructor of Holistic Health and Wellness Management. He earned Bachelor's Degree in Arts (Thai) from Burapha University in 2005, followed by a Master's Degree in International Hospitality and Hotel Management from the University of Western Sydney, Australia in 2008. In 2018, He has completed his Doctoral Degree in Management from Silpakorn University, Thailand. Since 2012, he has been contributing to academia as a lecturer at Burapha University International College. In 2021, he was appointed as an assistant Professor in Business Administration. His research interests encompass Tourism, Hospitality, Service, and Tourist attraction. Over the years, he has had the privilege of being a guest speaker more than 100 times in public organizations, private companies, and universities, both in Thailand and abroad.

Kannapat Kankaew is currently an Assistant Professor at Burapha University International College. He received Doctor of Public Administration majoring in Human Resources Management. Before his 12 years as an academic professional, he worked in the aviation industry for over 10 years. He has published and participated in International Conferences reputed journals. He also published chapters in edited books, being a books editor, and examiner for Ph.D. and Master students in both national and international institutions. And he was dedicated as an editorial board, associate editor, and reviewer for various domestic and international journals.

Krittipat Pitchayadejanant is a lecturer in Logistics Management, Burapha University International College. His research interests are in business data analysis, especially advanced statistical analysis applied to hospitality and tourism, logistics management, and marketing. He teaches in Operational Research in Logistics management, Multivariate Analysis and Business Data Analysis.

Vera Minenkova PhD in Geography. Research interests are in economic and recreational geography, sustainable development, tourism and hospitality, tourism impact on regional development (economic, social, ecological), mountain territories. She is the author of more than 250 scientific publications, coordinator and author of additional professional education programs in tourism implemented in Kuban State University, expert and guest speaker in universities (in Russia and abroad), private companies and public organizations, member of the public expert group under the Ministry of Resorts, Tourism and Olympic Heritage of the Krasnodar Region (Russia).

Evgeniia Rotai is Deputy Director, Department of International Affairs, Kuban State University, Russia. PhD in Philology. Organizer and coordinator of educational tours for international students arriving at Kuban State University. Organizer and coordinator of internship by KubSU students in hotels in Greece (2014). Member of the Committee on Entrepreneurship in the field of Tourism of the Chamber of

Commerce and Industry of the Krasnodar Territory. Received a Letter of appreciation from the Deputy Head of the Krasnodar Territory Administration (2018); Certificate of Honor from the Head of the Administration of the Krasnodar Municipality and the Chairman of the Krasnodar City Duma (2020).

***

Metasebia Adula is an assistant professor at the College of Business and Economics, Department of Management, Bule Hora University, Ethiopia.

Natalia Gennadievna Andreiko is an Associate Professor of the Department of Thermal Power Engineering and Thermal Engineering, Candidate of Technical Sciences.

Sabeen Azam is a Lecturer of IR at NUML, Karachi. Besides that, Azam is pursuing their PhD in International Relations.

Musabek Abdusharipovich Duvanakulov is the Head of the Department of Geology of Minerals, Candidate of Geological and Mineralogical Sciences, Associate Professor.

Cecile Gerwel Proches is an Associate Professor in the Graduate School of Business and Leadership (GSB&L) at the University of KwaZulu-Natal (UKZN) in Durban, South Africa. She holds a PhD (Leadership Studies), MCom (Leadership Studies), BCom Hons (Industrial & Organisational Psychology) and BCom (majoring in Business Management and Industrial and Organisational Psychology). She has successfully supervised several postgraduate students, including Master's students (MBA and Master of Commerce in Leadership Studies) and Doctoral students (PhD in Leadership Studies and Doctor of Business Administration). Cecile's teaching, research, supervisory and consulting interests include leadership, organisational behaviour, systems thinking, and complexity theory.

Vladimir Gulenko is a professor of the Department of Geophysical Methods of Prospecting and Exploration, Doctor of Technical Sciences.

Abdulla Kader is an adjunct lecturer at Management Information Governance at the University of KwaZulu-Natal, Durban, South Africa. He lectures in Research and quantitative methods, change management, Corporate Strategy, Entrepreneurship and has supervised masters and doctoral students at UKZN, DUT, MANCOSA and is an external examiner at Namibian University of Science and Technology. He has a keen interest on the Future of Work, 4IR tools and application, Business Modelling, Corporate Strategy, and design thinking. He has published 14 research articles as co-author in several reputed journals he has also contributed and reviewed book chapters on. Insights on Doctoral Research and The Disruptive Force of Smart Contracts. Strategic Human resources: Strategic Thinking as a Strategic lever in promoting tourism in Kwa Zulu Natal, South Africa. He has also presented at International Conferences. He is leading the charge to redefine Industry-Academia Nexus within the context of South Africa by leveraging opportunities for both industry and academia within Nedbank. He also serves as a board member in the Faculty of Engineering and built-up environment at the University of Johannesburg. He has conducted master class and webinars. Prof Kader is a research Fellow at University Johannesburg and Senior doctoral Associate at MANCOSA. His corporate experience has exposed him to manage the complexity of the duality of costs versus revenue.

Shashi Kant is an Assistant Professor, College of Business and Economics, Department of Management, Bule Hora University, Ethiopia.

Christian Kaunert is Professor of International Security at Dublin City University, Ireland. He is also Professor of Policing and Security, as well as Director of the International Centre for Policing and Security at the University of South Wales. In addition, he is Jean Monnet Chair, Director of the Jean Monnet Centre of Excellence and Director of the Jean Monnet Network on EU Counter-Terrorism.

Bahrom Kholbaev is the Head of the Department of Geology, Candidate of Geological and Mineralogical Sciences.

Daniil Gennadievich Kucheryavyy is master of the Department of Geoecology and Environmental Management.

Akmaral Kubanychbekovna Kultaeva is a Candidate of Geographical Sciences, Associate Professor of the Department of Geology of Minerals.

Anna Vladimirovna Mamonova is a lecturer of the Department of Economic, Social and Political Geography.

Melconyants Arthur- born on March 8, 2000 in the city of Vladikavkaz. He entered the Kuban State University in 2018 for the specialty «Technological safety». Defended the thesis on «Development of Techniques for Improving Efficiency of Extraction of Anions of Phosphoric Acid From Waste Water Products» with a bachelor's degree in 2022. In the same year he entered the geographical area in the specialty «Management of tourist-recreational systems». Defended master's thesis on «Impact of Tourist-Recreational Activities on Hydrological Resources: Assessment by Chemical Analysis» with a master's degree in 2024. Currently is a graduate student at the Department of International Tourism and Management.

Miroljub A. Milincic, Ph.D., full professor at the University of Belgrade - Faculty of Geography, was born in Brus on October 21st,1965, where he finished primary and secondary school. He enrolled at the Faculty of Geography in Belgrade in the 1986/87 school year and graduated in 1991, with an average grade of 8.62 / 100. He defended his graduate thesis on the topic of Karst phenomena of the Kopaonik National Park with a grade of 10. In postgraduate studies, majoring in Geospatial Basis of the Environment, Faculty of Geography in Belgrade, he enrolled in the school year of 1991/92, (average grade 9.40 / 100), and the master's thesis Environment as a determinant of political-geographical and geopolitical processes in Serbia defended in 1996. He defended his doctoral dissertation on the topic of Ecological Restrictions and Revitalization of Settlements in the Territories of Surface Water Sources of the First Rank of Serbia in 2004. During his undergraduate and postgraduate studies, he participated in numerous researches (regional, national and international). Before starting employment at the Faculty of Geography, he gained work experience in three primary schools, over two years. He was promoted to the scientific title of Doctor of Geography in 2004, the title of assistant professor in 2005, associate professor in 2010, and full professor in 2015. By 2020, the bibliography consists of a total of 217 listed references.

Dyah Mutiarin's research interest is human resources management in public sector.

Tafese Niguse is a PhD Scholar, College of Business and Economics, Department of Marketing Management, Bule Hora University, Ethiopia.

Achmad Nurmandi is a Professor at the Department of Government Affairs and Administration, Jusuf Kalla School of Government, Universitas Muhammadiyah Yogyakarta, Indonesia. His research interests are on e-government, urban governance and strategic management in the public sector and published in many international journal and Book Chapter, such as International Journal of Public Sector Management, Transforming Government: People, Policy and Process, Jamba Journal of Disaster Studies, Asian Review of Political Science, Journal of Human Behavior in the Social Environment, Public Policy and Administration, Global Encyclopedia of Public Administration, Public Policy, and Governance, Springer. He is currently the Secretary of the Asia Pacific Society for Public Affairs (APSPA). He is also the Editor-in-Chief of Jurnal Studi Pemerintahan (Journal of Government and Politics) and a Guest Editor of International Journal of Public Sector Performance Management and International Journal of Sustainable Society.

Ilya Vladimirovich Rudenko is a lecturer of the Department of Geoecology and Environmental Management.

Nikolay Rudomaha is an Associate Professor of the Department of Geophysical Methods of Prospecting and Exploration.

Zahir Sayed is currently employed as an Operations Specialist in the public service industry in Durban, South Africa. He has graduated with a Doctor of Philosophy (PhD), Master of Commerce and Postgraduate Diploma in Leadership and Management from the Graduate School of Business and Leadership at the University of KwaZulu Natal, and a Bachelor's degree in Management from the Business Studies Unit at Durban University of Technology. He has a keen interest in Leadership, Management, Emotional Intelligence and Design Thinking and has published several articles and book chapters on these topics. He has recently edited a book titled Emotionally Intelligent Methods for Meaningful Leadership.

Kurt Scharr Born in 1970 in Mittersill in Pinzgau, studied history and social studies as well as geography and economics (teaching qualification in Austria) in Innsbruck. Since October 1, 2016, he has been a university professor at the Leopold Franzens University in Innsbruck.

Sofya Novikovna Semenova is a candidate of Philological Sciences, Associate Professor of the Department of English in Professional Sphere

Swati Sharma, Ph.D. is currently working as an Associate Professor with Amity University, India. She holds 18 years of teaching and industry experience. She is Vice- President of the Women's Indian Chamber of Commerce and Industry . She is the reviewer of Consumer Behavior in Tourism & Hospitality & IGI Global . She has authored & Edited 15 books & published research papers in

National & International journals of repute. (IGI Global, Taylor & Francis, Springer, Apple Academy Press, CABI, Edward & Elgar). Her area of interest is Sustainable Tourism, MICE, Community based Tourism, Destination branding.

Bhupinder Singh working as Professor at Sharda University, India. Also, Honorary Professor in University of South Wales UK and Santo Tomas University Tunja, Colombia. His areas of publications as Smart Healthcare, Medicines, fuzzy logics, artificial intelligence, robotics, machine learning, deep learning, federated learning, IoT, PV Glasses, metaverse and many more. He has 3 books, 139 paper publications, 163 paper presentations in international/national conferences and seminars, participated in more than 40 workshops/FDP's/QIP's, 25 courses from international universities of repute, organized more than 59 events with international and national academicians and industry people's, editor-in-chief and co-editor in journals, developed new courses. He has given talks at international universities, resource person in international conferences such as in Nanyang Technological University Singapore, Tashkent State University of Law Uzbekistan; KIMEP University Kazakhstan, All'ah meh Tabatabi University Iran, the Iranian Association of International Criminal law, Iran and Hague Center for International Law and Investment, The Netherlands, Northumbria University Newcastle UK,

Devika Grover is having 20 + industry & Teaching Experience and presentle working as Assistant Professor with Amity University, Noida

Ernst Steinicke was born in Innsbruck in 1954. He studied geography and sports science at the universities of Innsbruck and Toulouse with the aim of becoming a teacher. In the early 1980s he also practiced this profession at the Innsbruck Business Academy, but at the same time worked on his doctorate in minority research using the regional case study of the Val Canale, a topic and a region that would accompany him for many years of his academic life. Ernst Steinicke has worked at the Institute of Geography at the University of Innsbruck since 1984, where he earned his doctorate and qualified as a professor in 1990 with a thesis on population and ethnicity in Friuli. For more than 20 years, Ernst Steinicke played a key role in shaping Innsbruck geography as an extraordinary university professor. Ernst Steinicke has excellent international connections. Research stays, fellowships and guest lectureships took him to the University of California, Davis, to universities in Russia (Krasnodar) and repeatedly to universities in Italy (e.g. Trieste, Trento, Bolzano). University teaching was a particular concern for Ernst Steinicke. In addition to basic lectures in various areas of human geography and events in the theoretical and methodological areas, a particular focus of Ernst Steinicke's teaching activity was regional geography. Since 2013 Ernst Steinicke took on the role of study representative for the teaching degree program in geography and economics and has also performed the role of dean of studies at the Faculty of Geosciences and Atmospheric Sciences.

Inga Vasilyevna Tania, PhD in Geography since 2000, Associate Professor since 2003, has made a significant contribution to the development of geography, biology and ecology. She is the author of 110 scientific articles, 80 of which are devoted to RRNP.

Simon Taylor is a Lecturer and Project Manager of the RLED Initiative at the Graduate School of Business and Leadership, University of KwaZulu-Natal.

Aynetu Terefe is a Lecturer, College of Business and Economics, Department of Management, Bule Hora University, Ethiopia.

Akshita Tiwari is a dedicated hospitality professional with a passion for learning and expanding her knowledge in the field. She began her hospitality education in 2014 with a Bachelor's in Hospitality and Hotel Administration from IHM, Faridabad, and went on to complete her Master's in Hospitality Administration with a specialization in Human Resource Management from IHM, Pusa, New Delhi in 2019.In pursuit of her interest in hospitality and a desire to deepen her knowledge, she joined the Ph.D. in Hospitality at Amity School of Hospitality, Amity University, Noida, in January 2020. Dr. Tiwari is currently a faculty member at Amity School of Hospitality, Amity University, Noida, Uttar Pradesh, where she teaches Front Office Operations and Management to Bachelor's in Hotel Management students and Master's in Hospitality students. She has also authored multiple Scopus publications and presented papers at international conferences. With a keen interest in teaching and research, she looks forward to contributing to the field of hospitality and continue to strive for excellence in her work.She strives to excel in the field of hospitality and tourism and link it with the disruptive technologies relevant in both the industries.

Tatiana Volkova Associate Professor of the Department of International Tourism and Management. A multiple winner of the "Best Scientists of the Kuban State University" competition in the field of natural sciences. In 2022, received a Letter of Thanks from the Federal Agency for Tourism, and in 2023, was awarded with the Gratitude of the Governor of the Krasnodar Territory. Her research has led to more than 300 publications, including 20 articles published in journals indexed by international systems such as Scopus and Web of Science, and 62 articles in journals listed by the Higher Attestation Commission of the Russian Federation. A regional expert for the Agency for Strategic Initiatives in Tourism and is an expert with the Kuban Scientific Foundation.

Suraj Yadav is an accomplished professional with a background in Hospitality, holding a bachelor's degree from Pune University and a Master in Hotel Management & Tourism from Madurai Kamaraj University, Madurai. Currently serving as an Assistant Professor and Training & Placement Officer at D Y Patil School of Hospitality in Kolhapur, Maharashtra, India, he brings extensive expertise to the field. Suraj is also dedicated to furthering his academic pursuits and is actively pursuing a Ph.D. at D Y Patil Agricultural & Technical University in Kolhapur. His commitment to education and training has made a significant impact in the realm of hospitality.

Muhammad Younus I graduated in 2019 from the Karachi Institute of Economics and Technology University in Karachi, Pakistan, and got a Bachelor's degree in Computer Science. The four-year undergraduate program in Computer Science formed the basis of my perseverance towards achieving an in-depth knowledge in the field. After graduation, I worked for the past few years as a business IT analyst in the product and research development of a logistics startup in Pakistan called TPL Logistics. We aim to Digitize Logistics Operations in Pakistan from the First Mile to the Last Mile. After this, I graduated in 2024 from the Universitas Muhammadiyah Yogyakarta in Yogyakarta, Indonesia, and got a Master's Degree in Government Science. My research interest is transforming governance and the public sector through technology. This Experience has given me the confidence and also kindled a desire to do research. The interdisciplinary nature has convinced me in taking this decision as it is the ideal choice for an exciting research career. At the same time, I am confident of contributing originally to the ongoing work. By working under the guidance of distinguished Team, I shall certainly be able to exploit my pote

Evgenia Zakharchenko is the Head of the Department of Geophysical Methods of Prospecting and Exploration, Candidate of Technical Sciences, Associate Professor.

Mikhail Andreevich Zakharchenko is a student of the Department of Geophysical Methods of Prospecting and Exploration.

Yulia Ivanovna Zakharchenko is a Senior Lecturer at the Department of Geophysical Methods of Prospecting and Exploration.

Index

I

K

L

M

N

P

R

S

T

V

W

www.ingramcontent.com/pod-product-compliance
Lightning Source LLC
LaVergne TN
LVHW080621300926
849162LV00005B/1224